Current Progress in Signal Processing

Current Progress in Signal Processing

Edited by **George Pilato**

Willford Press

New York

Published by Willford Press,
118-35 Queens Blvd., Suite 400,
Forest Hills, NY 11375, USA
www.willfordpress.com

Current Progress in Signal Processing
Edited by George Pilato

International Standard Book Number: 978-1-68285-234-7 (Hardback)

Printed in China.

Contents

Preface

From audio processing and video processing to seismology, one can find many applications of signal processing. As the discipline of signal processing is progressing at a rapid pace, this innovative and comprehensive book integrates the well-developed theory and practical applications of this discipline. Comprising researches by an eminent panel of scholars, this book discusses topics of utmost importance, like signal processing theory, algorithms, audio signal processing, signal processing tools and techniques, signal processing for sensors, etc. It will serve as a valuable source of reference for graduate and post graduate students. This book is a complete source of knowledge on the present status of this important field.

The information contained in this book is the result of intensive hard work done by researchers in this field. All due efforts have been made to make this book serve as a complete guiding source for students and researchers. The topics in this book have been comprehensively explained to help readers understand the growing trends in the field.

I would like to thank the entire group of writers who made sincere efforts in this book and my family who supported me in my efforts of working on this book. I take this opportunity to thank all those who have been a guiding force throughout my life.

Editor

1

EEG seizure detection and prediction algorithms: a survey

Turkey N Alotaiby[1], Saleh A Alshebeili[2], Tariq Alshawi[3], Ishtiaq Ahmad[4] and Fathi E Abd El-Samie[5,6*]

Abstract

Epilepsy patients experience challenges in daily life due to precautions they have to take in order to cope with this condition. When a seizure occurs, it might cause injuries or endanger the life of the patients or others, especially when they are using heavy machinery, e.g., deriving cars. Studies of epilepsy often rely on electroencephalogram (EEG) signals in order to analyze the behavior of the brain during seizures. Locating the seizure period in EEG recordings manually is difficult and time consuming; one often needs to skim through tens or even hundreds of hours of EEG recordings. Therefore, automatic detection of such an activity is of great importance. Another potential usage of EEG signal analysis is in the prediction of epileptic activities before they occur, as this will enable the patients (and caregivers) to take appropriate precautions. In this paper, we first present an overview of seizure detection and prediction problem and provide insights on the challenges in this area. Second, we cover some of the state-of-the-art seizure detection and prediction algorithms and provide comparison between these algorithms. Finally, we conclude with future research directions and open problems in this topic.

1 Review

1.1 Introduction

Epilepsy, which is classified as a neurological disorder that affects the brain, impacts about 2% of the world population leading to a reduction in their productivity and imposing restrictions on their daily life [1]. Diagnosis of epilepsy is done by analyzing electroencephalogram (EEG) signals, as well as patient behavior. EEG signals have two types: scalp EEG and intracranial EEG (iEEG). Scalp EEG signals are usually collected with electrodes placed on the scalp using some sort of conductive gel after treating the scalp area with light abrasion in order to decrease the impedance resulting from dead skin cells. Commonly, 19 recording electrodes in addition to a ground and system reference are placed on the scalp area according to specifications by the International 10–20 system. However, fewer electrodes are used when the EEG signals are recorded for neonates [2]. Each of these electrodes collects an EEG signal, which is centrally recorded for post-processing. On the other hand, in iEEG, electrodes are placed directly on the exposed surface of the brain during a surgery to record electrical activity from the cerebral cortex. For the seizure detection task in the case of epilepsy patients, it is required to analyze these EEG signals towards a decision of the existence, or absence, of an epileptic seizure. If a seizure exists, further analysis could be made for more understanding of seizure behavior. Most of the research work in the field of seizure detection depends on scalp EEG signals, which are acquired with non-invasive techniques. So, we will adopt the term EEG to refer to scalp EEG throughout the paper.

The seizure detection process can be made on a single- or multi-channel basis [3,4]. Single-channel seizure detection requires selecting the channel containing the strongest EEG signal collected from the closest point to the seizure spot. This selection process depends mainly on activity measures evaluated for the different channels instantaneously such as the local variance. A better treatment to the seizure detection issue depends on incorporating the information from all EEG signals available into the seizure detection process through data fusion, or multi-channel processing techniques [5].

Several studies have been conducted on EEG seizure detection [3,4]. To perform seizure detection on time-domain waveforms, there is a need to create models for the EEG signals, from which features are extracted,

* Correspondence: fathi_sayed@yahoo.com
[5]KACST-TIC in Radio Frequency and Photonics for the e-Society (RFTONICS), King Saud University, Riyadh 11362, Saudi Arabia
[6]Faculty of Electronic Engineering, Menoufia University, Menouf, 32952, Egypt
Full list of author information is available at the end of the article

because treating the signal as discrete-time sequence is not a robust approach for extracting characterizing features. Some algorithms create models for normal and abnormal EEG signals of the patients and use these models in the training process. New signals are tested against these models leading to the decision of seizure or non-seizure activities. Most of the time-domain seizure detection algorithms are patient specific; that is, they are applied to the patient for which training data are extracted. This is intuitively meaningful as each patient has a different nature for his/her EEG signals.

Efforts have also been made to develop seizure prediction algorithms [6,7]. Note that with seizure prediction, it is possible to alert ambulatory patients or caregivers before seizure occurs and causes injury. In addition, seizure prediction could initiate time-sensitive clinical procedures necessary for characterizing epileptic syndromes. It could, also, help in initiating delivery of therapy early enough to attenuate seizure attack and reduce its duration or may even terminate it completely. The endeavor towards fast and simple seizure prediction algorithms has led to the development of a variety of algorithms which will be covered in this paper.

Signal processing has shown a breakthrough with the evolution of several discrete transforms and signal decomposition techniques, because all of these transforms and decompositions have their unique nature. For example, the discrete Fourier transform (DFT) and discrete wavelet transform (DWT) have found popularity in seizure detection and prediction applications. Similarly, the singular value decomposition (SVD) and empirical mode decomposition (EMD) have also found a role in these applications. Even the principal component analysis (PCA) and the independent component analysis (ICA) have found a pre-processing noise removal role in these applications.

Several attempts have been presented in the literature to classify EEG seizure detection and prediction methods. For example, classification into linear and non-linear methods considers variance-based, correlation-based, and simple power spectrum-based methods as linear methods and all other methods as non-linear methods [8-10]. We did not adopt this classification as most of the reviewed methods in this paper are non-linear techniques. Another classification for seizure detection and prediction methods depends on the type of EEG signals used: whether scalp EEG or iEEG. We mentioned in this paper the type of database used for testing each method and the type of the signals in the database.

Tzallas et al. presented a classification of EEG seizure detection methods into pattern recognition methods, morphological analysis methods, parametric methods, decomposition methods, clustering methods, and data mining methods [11]. In this paper, we review some of the recently developed seizure detection and prediction algorithms along with a comparison study between them adopting another basis for classification of seizure detection and prediction methods depending on the transform domain of operation. It seems that seizure detection and prediction are two different trends, but we look at them from a signal processing perspective with different transform domains. We investigate in our description and classification in this paper the most important seizure detection and prediction algorithms operating in each transform domain. Figure 1 shows a classification of the seizure detection and prediction methods based on the processing domain.

Therefore, we organize the paper as follows. We first begin by the most common methods for seizure detection and prediction: time-domain methods, which are addressed in Section 1.2. Section 1.3 is devoted to frequency-domain seizure detection methods. Wavelet-domain seizure detection and prediction methods are covered in Section 1.4. Sections 1.5 and 1.6 are devoted for seizure detection methods that are based on the EMD and the SVD, respectively. Section 1.7 presents the role of ICA and PCA in seizure detection. A comparison study illustrating seizure detection and prediction methods is given in Section 2.

1.2 Time-domain methods

In general, local activities of EEG waveforms vary from patient to patient; therefore, seizure detection and prediction algorithms are preferred to be patient specific.

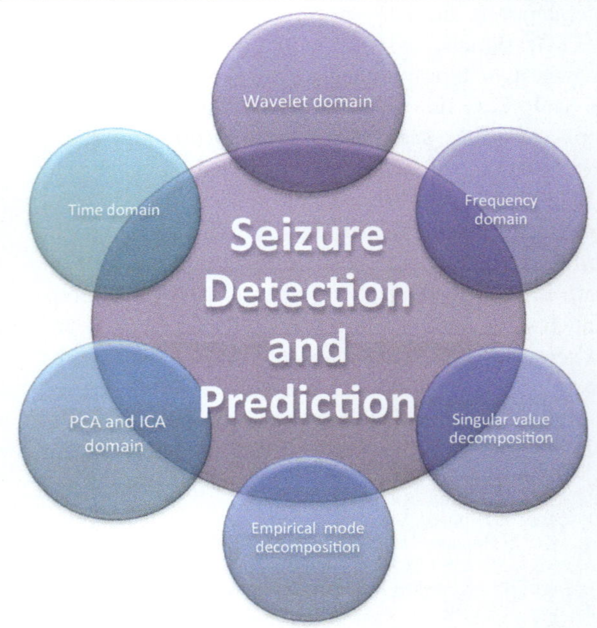

Figure 1 Classification of seizure detection and prediction methods.

Figure 2 shows an example of an EEG signal including a seizure period. It is clear that there is a difference between seizure and non-seizure intervals. As we are able to differentiate between these intervals visually, time-domain detection and prediction methods attempt to differentiate between them automatically, and evaluate the performance using different metrics such as the sensitivity, specificity, accuracy, and false-positive value. These metrics are defined as follows [12]:

$$\text{Sensitivity} = \frac{\text{TP}}{\text{TP} + \text{FN}} \times 100 \quad (1)$$

$$\text{Specificity} = \frac{\text{TN}}{\text{TN} + \text{FP}} \times 100 \quad (2)$$

$$\text{Accuracy} = \frac{\text{TP} + \text{TN}}{\text{TN} + \text{FP} + \text{TP} + \text{FN}} \times 100 \quad (3)$$

$$\text{False positive value} = \frac{\text{TP}}{\text{TP} + \text{FP}} \times 100, \quad (4)$$

where true positive (TP) is the number of seizure epochs determined by both algorithm and experienced physicians, false positive (FN) is the number of seizure epochs which are missed by the algorithm but have been determined by experienced physicians, true negative (TN) is the number of non-seizure epochs recognized by both algorithm and experienced physicians, and false positive (FP) is the number of non-seizure epochs recognized as seizure by the algorithm.

1.2.1 Seizure detection methods

To detect EEG seizures in time domain, there is a need to analyze discrete time sequences of EEG epochs. This analysis can be accomplished through histograms of the epochs. Runarsson and Sigurdsson presented a simple time-domain seizure detection method that is based on tracing consecutive peaks and minima in the signal segment at hand and estimating the histograms for two variables: the amplitude difference and time separation between peak values as well as minima [13]. The features used for classification of an epoch as a seizure or non-seizure are the estimated values of the histogram bins. The authors used a support vector machine (SVM) classifier for this task and achieved an average sensitivity of about 90% on self-recorded data.

Another approach to deal with the EEG seizure detection method in time domain is to compute the signal energy during seizure and non-seizure periods. A better treatment to the energy estimation approach is to estimate the energies of the signal sub-bands not the signal as a whole in order to build a more discriminative feature vector. Yoo et al. adopted this approach and presented an eight-channel EEG acquisition system-on-chip (SoC) that can detect and record patient-specific epileptic seizures [14]. They used a bank of seven band-pass filters covering the frequency range from 2 to 26 Hz on eight channels of the processed EEG signals. Eight highly dynamic analog channels, classification processor, and a 64-KB SRAM have been integrated in the SoC. With this approach, a long-term seizure monitoring and storage

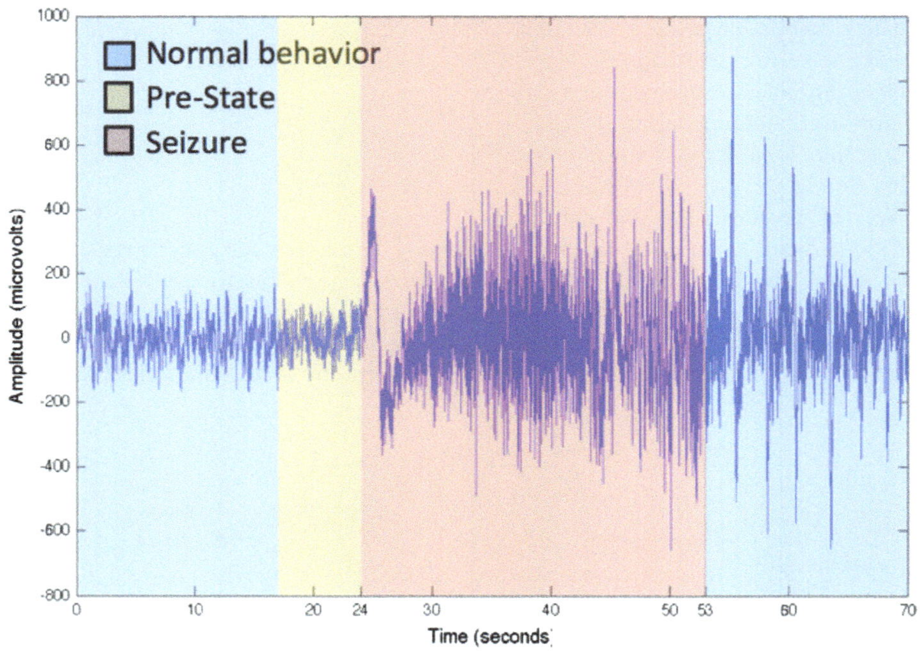

Figure 2 An EEG signal containing a seizure.

device was built. The authors used an SVM as a classifier with a gain and bandwidth (GBW) controller to perform real-time gain and bandwidth adaptation to analog front end (AFE) in order to keep a high accuracy. This classifier is well suited for the hardware implementation. The SoC was tested on CHB-MIT scalp EEG database [15], and it was verified in the presence of a rapid eye blink giving an accuracy of 84.4% with 2.03 μJ/classification energy.

Another approach to deal with time-domain seizure detection is to exploit some discriminating statistics between seizure and non-seizure epochs. Dalton et al. developed a body senor network (BSN) that can monitor and detect epileptic seizures based on statistics extracted from time-domain signals [16]. These statistics include the mean, variance, zero-crossing rate, entropy, and autocorrelation with template signals. For auto-correlation estimation, they adopted a dynamic time warping (DTW) approach for best alignment between the signal segment to be tested and the template signal. The authors presented a network to be embedded in wearable kinematic sensors and an N810 Internet tablet. Data were recorded from the subjects using kinematic sensors such as triaccelerometer, gyroscope, and magnetic sensor for physical activity monitoring. This algorithm was, then, commercially distributed and a BSN was developed on a Mercury platform. The sensitivity of the proposed algorithm for a dataset of 21 seizures was found to be 91% with a specificity of 84% and battery lifetime of 10.5 h.

1.2.2 Seizure prediction methods

The research work on the issue of time-domain seizure prediction is richer than time-domain seizure detection due to the importance of the seizure prediction problem. We can think of the seizure prediction problem as a detection problem of the pre-ictal state on seizure records. This requires a considerable long inter-ictal state for good prediction results. Similar statistics to those used in seizure detection like the zero-crossing rate can be used for seizure prediction. Zandi et al. used the zero-crossing rate of EEG signal segments to develop a patient-specific seizure prediction method [17,18]. A moving window analysis is used in this method. The histograms of the different window intervals are estimated, and selected histogram bins are used for classification into pre-ictal and inter-ictal states based on comparison with reference histograms. A variational Bayesian Gaussian mixture model has been used for classification. In this method, a combined index for the decisions taken on selected bins is computed and compared with a pre-defined patient-specific threshold to raise an alarm for coming seizures as shown in Figure 3. This method has been tested on 561 h of scalp EEG containing 86 seizures for 20 patients. It achieved a sensitivity of 88.34%, a false prediction rate of 0.155 h^{-1}, and an average prediction time of 22.5 min.

Aarabi and He [19] developed a time-domain rule-based patient-specific seizure prediction method which consists of three stages: pre-processing, feature extraction, and rule-based decision making. In the pre-processing stage, the iEEG data is filtered using a 0.5- to 100-Hz band-pass filter in addition to a 50-Hz notch filter. Then, the filtered signal is segmented into non-overlapping 10-s segments. Five univariate features (correlation entropy, correlation dimension, Lempel-Ziv complexity, noise level, and largest Lyapunov exponent) and one bivariate feature (non-linear independence) were extracted from each segment in the second stage.

Based on the theory of chaos, the correlation dimension (denoted by ν) represents a dimensionality measure of the space having a set of random points; in our case, EEG signals. For an m-dimensional space containing a set of N points, we have:

$$\vec{x}\,(i) = [x_1(i), x_2(i), \;\ldots\ldots., x_m(i)], \quad i = 1, 2, \;\ldots., N \tag{5}$$

The correlation integral $C(\varepsilon)$ can be estimated as [20]:

$$C(\varepsilon) = \lim_{N\to\infty} \frac{g}{N^2}, \tag{6}$$

where g represents the total number of pairs of signals or points having a distance less than ε. As the number of points increases and tends to infinity and the distance tends to be shorter or close to zero, the correlation integral, in turn, for small values of ε becomes:

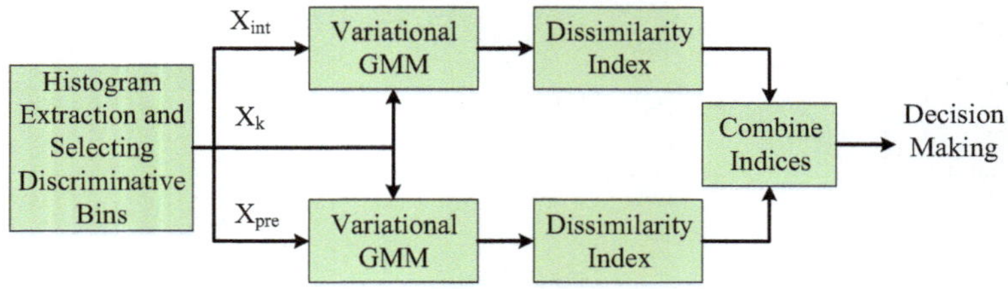

Figure 3 Zandi et al. seizure prediction method.

$$C(\varepsilon) \approx \varepsilon^{\nu} \tag{7}$$

If a large number of evenly distributed points exists, a log-log graph of the correlation integral versus ε can be used to estimate ν. For objects with higher dimensions, several ways exist for points to be close to each other, and hence, the number of pairs which are close to each other jumps rapidly for higher dimensions [20].

Correlation entropy is a Kolmogorov entropy variant, which is similar to the mutual information between two sequences of data. Large mutual information between an available data segment and stored segments with specific patterns is an indication that the segment at hand belongs to a dataset with similar characteristics to the stored pattern [21]. The Lempel-Ziv complexity is a measure of randomness of data sequences [22]. It counts the number of data patterns with certain characteristics in data segments. For example, if we find enough short patterns with specific mean, variance, or higher-order statistics in an EEG segment, we can classify this segment as a seizure segment.

The Lyapunov exponent of a dynamical system determines the separation rate of very closely related trajectories. Hence, two signal vectors in the phase space with an initial separation of $\delta\mathbf{Z}_0$ will eventually diverge at a rate given by:

$$|\delta\mathbf{Z}(t)| \approx e^{\lambda t} |\delta\mathbf{Z}_0|, \tag{8}$$

where λ is the Lyapunov exponent. This can be achieved if the divergence can be dealt with within the linearized approximation.

The separation rate differs based on the initial separation vector orientation. The maximal Lyapunov exponent can be estimated as [23]:

$$\lambda = \lim_{t\to\infty} \lim_{\delta\mathbf{Z}_0\to 0} \frac{1}{t} \ln \frac{|\delta\mathbf{Z}(t)|}{|\delta\mathbf{Z}_0|} \tag{9}$$

The limit $\delta\mathbf{Z}_0 \to 0$ ensures the validity of the linear approximation at any time.

Considering a given segment and feature, if N channels exhibit behaviors like those expected in the feature or segment spatiotemporal profile describing the pre-ictal state of the patient, and if N was greater than a threshold N_{ch}, then this segment is temporarily categorized as a seizure precursor, and hence a flag, labeled as *I*, is raised for the segment as shown in Figure 4. A value is given to this flag by an averaging process for the channels' absolute normalized feature values. The channels selected are those that show similar behavior to the one expected in the spatiotemporal profile of the feature characterizing the pre-ictal state of the patient. The values and location of all flag *Is* are hence stored and given as input into the feature integrator, which integrates decisions for every segment to accurately locate the seizure precursors. For any signal segment, if M flag *Is* with values that are higher than a certain significance threshold T_{c1} are raised and if M is larger than a certain threshold N_{f}, then another flag which is labeled as *II* is raised. This flag is determined as the average of the values of the flag *Is*. Flag *IIs* ensure a higher likelihood of accurate seizure prediction. Flag *IIs* are used as input to the feature integrator *II* to obtain a higher-level decision. This feature integrator integrates flag *Is* and flag *IIs* as shown in Figure 4. If both flag *I* and flag *II* are raised, then a flag *III* representing a definitive seizure precursor is also raised.

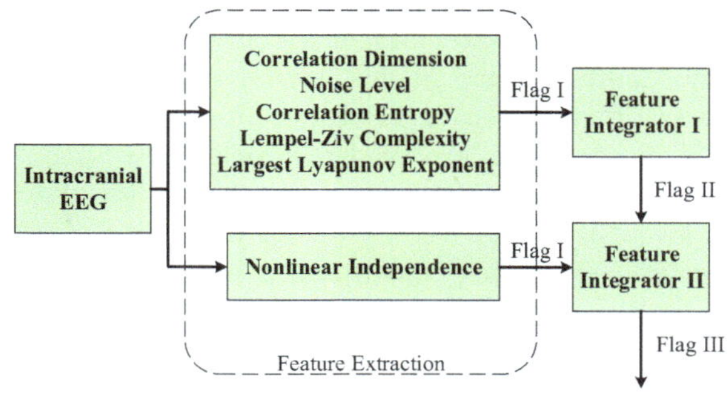

Figure 4 Method of Aarabi and He for seizure prediction.

This method has been evaluated using iEEG data from two patients (frontal, temporal lobe origin) from Freiburg Seizure Prediction EEG (FSPEEG) database with a 256-Hz sampling rate and a total of 58 h, and 10 seizures with 50-min pre-ictal at least [24]. The results demonstrated average sensitivities of 90% and 96.5% for patient one and patient two, respectively. The average false prediction rates were 0.06/h and 0.055/h for both patients for prediction horizons of 30 and 60 min.

Schelter et al. presented a new method to minimize the false alarms adopting circadian concepts [25]. A circadian rhythm is defined as any biological operation that reveals an endogenous, entrainable oscillation for 24 h. The authors used the output of the mean phase coherent algorithm, which measures the interaction between pairs of EEG signals, as a seizure predictor, which causes an alarm to be raised if it exceeds a certain threshold. It was assumed that the seizures occur while the patient is sleeping. The false alarms display a circadian dependency with most of the seizure prediction algorithms. The seizure predectability is increased during night due to the large number of seizures. Accordingly, threshold adaptation can be used in day and night to enhance predictability. The authors evaluated this method utilizing iEEG data from eight patients and a total of 1400 h, which include 172 seizures and reported a good prediction performance for 40% of the patients.

Wang et al. proposed an adaptive learning system that interactively learns from the patient and improves its seizure predictability over time [26]. It is based on reinforcement learning and online monitoring, in addition to adaptive control theory. In this system, a sliding window size of 10 min is used to read continuous multichannel EEG data with a 50% overlap at each move. Then, k-nearest neighbor (KNN) method is adopted for the classification of the windowed epochs to normal or pre-seizure states based on pre-constructed baselines for both states using pre-specified baseline for normal and pre-seizure states. Finally, according to the prediction feedbacks, the two baselines are updated. This method was evaluated using iEEG data for five patients having temporal lobe epilepsy. The EEG data consisted of 26 channels with a duration range from 3 to 13 h. This method achieved an accuracy of 70% compared to 50% for the Poisson random predictor with a mean interval of λ minutes.

Researchers have proved that symptoms like sleep problems or headaches are observable from the analysis of the iEEG. These symptoms can be utilized as a major tool for seizure prediction. Bedeeuzzaman et al. have presented a seizure prediction algorithm with a statistical feature set consisting of mean absolute deviation (MAD) and inter-quartile range (IQR) to predict epileptic seizures [27]. A linear classifier has been used to find the seizure prediction time in pre-ictal iEEGs. A sensitivity of 100% with zero false-positive rate (FPR) in 12 patients and low values of FPR for the rest were achieved using Freiburg iEEG dataset. Average prediction time varied between 51 and 96 min.

The envelope of the EEG signal can be exploited to distinguish between different activities. Li et al. presented a time-domain method for seizure prediction that is based on spike rate estimation [28]. Morphological operations and averaging filters are applied to transform each signal segment to a train of spikes in a way similar to the process of envelope detection. Based on the spike rate, ictal, inter-ictal, and pre-ictal states can be identified through comparison with a certain threshold. This method was applied on 21 patients from Freiburg database, and it achieved a sensitivity of 75.8% and an average false-alarm prediction rate of 0.09/h.

Due to the inherent intra- and inter-patient seizure heterogeneities and EEG non-stationarity, it is difficult to measure the modulations of neuronal network interactions using seizure precursors. To solve this problem, Stamoulis et al. studied the pre-ictal neuro-modulations corresponding to temporal and/or frontal lobe seizures. They used information theoretic features like entropy and mutual information extracted from two frequency ranges (the range that is less than or equal to 100 Hz and the range that is greater than 100 Hz) of awake scalp EEGs [29]. They succeeded to measure pre-ictal modulations at frequencies greater than 100 Hz with high specificity.

Another approach to process EEG signals in the time domain in order to detect or predict seizure is to create models from the EEG signal segments corresponding to different activities. One of such models is the autoregressive (AR) model, which can be thought of as a data reduction model that transforms the EEG signal segment into few coefficients. Chisci et al. studied the implantation of monitoring and control units on drug-resistant epilepsy patients with AR modeling [30]. They adopted AR modeling with a least-squares parameter estimator for EEG feature extraction in addition to a binary SVM classifier to distinguish between pre-ictal, ictal, and inter-ictal states as shown in Figure 5. This algorithm is computationally simple enabling real-time implementation. Simulation results on the Freiburg database have shown 100% sensitivity with low false-alarm rate. This is attributed to the regularization strategy of the SVM classifier with Kalman post-processing.

The realization of implantable seizure prediction devices that can be used for alerting the patient and taking an action is a very challenging task. Cellular non-linear networks (CNNs), which represent a paradigm for high-speed computations, can be used for this task. Tetzlaff and Senger presented four different CNN-based approaches for epileptic seizure prediction towards an implantable seizure warning device working on any type of simple time-domain features [31]. This method can be used with any of the abovementioned features. The CNNs have been used in these approaches because they consist of locally coupled dynamical systems that can simulate the non-linear phenomena encountered in physical communication.

1.3 Frequency-domain methods

Frequency-domain techniques have been used for EEG seizure detection. Both of the Fourier transform magnitude and phase can be exploited for this purpose. Rana et al. presented a frequency-domain epileptic seizure

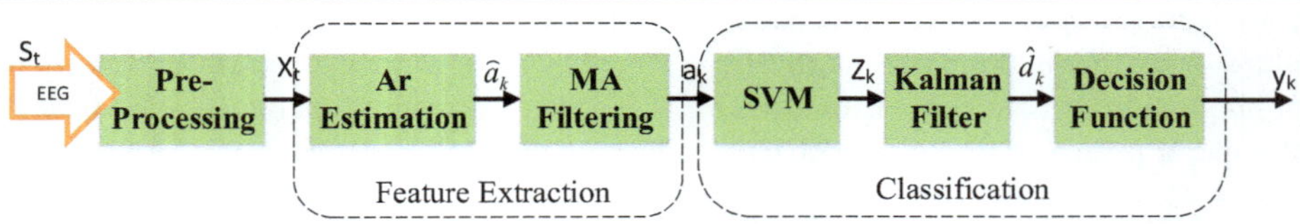

Figure 5 Chisci et al. seizure prediction method.

detection approach depending on the phase-slope index (PSI) of multi-channel EEG signals [32]. If we consider signals $z_i\,[n]$ and $z_j\,[n]$, their cross spectrum is given by:

$$S_{ij}(f) = E\left[Z_i(f)Z_j^*(f)\right], \tag{10}$$

where $Z_i\,(f)$ and $Z_i\,(f)$ are the Fourier transforms of $z_i\,[n]$ and $z_j\,[n]$. Hence, the complex coherence is given by:

$$C_{ij}(f) = \frac{S_{ij}(f)}{\sqrt{S_{ii}(f)S_{jj}(f)}} \tag{11}$$

An unnormalized PSI metric can be defined using complex coherence as follows:

$$\tilde{\Psi}_{ij} = \mathrm{Im}\left(\sum_{f\in F} C_{ij}^*(f)C_{ij}(f+\delta f)\right), \tag{12}$$

where δf is the frequency resolution and F is the frequency band of interest. We can deduce that $\tilde{\Psi}_{ij}$ measures a weighted sum of the slopes of the phase between $z_i\,[n]$ and $z_j\,[n]$ over the selected band F [32]. Normalization with the standard deviation is used to determine whether causal influence from $z_i\,[n]$ to $z_j\,[n]$ is of significant extent or not.

The PSI computes the measure of interaction between two channels. The authors used the PSI metric to distinguish between seizure and normal activities. The detection performance has been evaluated over five patients having different types of epilepsy with 47 seizures in 258 h of recorded data. The simulation results showed that this algorithm succeeded in the detection of all seizures for four out of five patients, and it achieved a lower false detection rate than two per hour. The results also showed that the channels with strong activity can be determined for each patient.

Khamis et al. used frequency-moment signatures for building a patient-specific seizure detection method [33]. Firstly, experienced electroencephalographs have marked the collected scalp EEG data with seizure events. After that, a filtering process has been performed on the windowed EEG data from electrode differences T6-P4 for the right hemisphere and T5-P3 for the left hemisphere. Power spectral densities of the signals on both hemispheres have been computed and a background removal technique has been used. Moments of these spectra have been used as features for signal classification as seizure or non-seizure. Results reported a sensitivity of 91% and false-alarm rate of 0.02 false positives per hour.

EEG signals are in general non-linear and non-stationary. So, there is a difficulty to characterize different activities of EEG signals with certain mathematical models. To tackle this problem, Acharya et al. presented a modified method for the detection of normal, pre-ictal, and ictal conditions from recorded EEG signals [34]. This method is based on four entropy features for classification: phase entropy 1 (S_1), phase entropy 2 (S_2), approximate entropy (ApEn), and sample entropy (SampEn). The phase entropies are estimated from the higher-order spectra of EEG signal epochs as discriminating features for ictal, pre-ictal, and inter-ictal activities. The approximate and sample entropies are logarithmic metrics that determine the closeness and matching between the incoming EEG signal pattern and the recorded templates. These features are extracted from EEG signals and fed to seven different classifiers for comparison: SVM, fuzzy Sugeno classifier (FSC), probabilistic neural network (PNN), KNN, naive Bayes classifier (NBC), decision tree (DT), and Gaussian mixture model (GMM). The obtained results with this method showed the superiority of the Fuzzy classifier, which achieved an accuracy of 98.1%.

1.4 Wavelet-domain methods

Wavelets have been widely used in the field of EEG signal analysis, especially for seizure detection and prediction. The wavelet transform in itself can be regarded as some sort of sub-band decomposition, but with down-sampling. The wavelet transform can be implemented on analog as well as digital signals. We are more interested in the DWT. The DWT can be implemented with low-pass (LP) and high-pass (HP) filtering in addition to a decimation process, and it must be invertible as shown in Figure 6a [35]. The DWT can be implemented with a single level or multi levels as shown in Figure 6b, c. For the multi-level wavelet decomposition, further decompositions up to the required level are performed on the low-pass branch only. Another implementation of wavelet analysis is the wavelet packet transform, which performs further decomposition on the low-pass and high-pass branches.

The wavelet decomposition and reconstruction filters H_0, H_1, G_0, and G_1 must satisfy the perfect reconstruction (PR) condition. These filters can be obtained by solving the following equation [35]:

$$\begin{aligned} Y(z) &= \frac{1}{2}\{X_0(z)+X_0(-z)\}G_0(z)+\frac{1}{2}\{X_1(z)+X_1(-z)\}G_1(z) \\ &= \frac{1}{2}X(z)\{H_0(z)G_0(z)+H_1(z)G_1(z)\} \\ &\quad +\frac{1}{2}X(-z)\{H_0(-z)G_0(z)+H_1(-z)G_1(z)\} \end{aligned} \tag{13}$$

The main idea of using the wavelet analysis for EEG seizure detection and prediction is extracting discriminating features from appropriate sub-bands to be used for further classification.

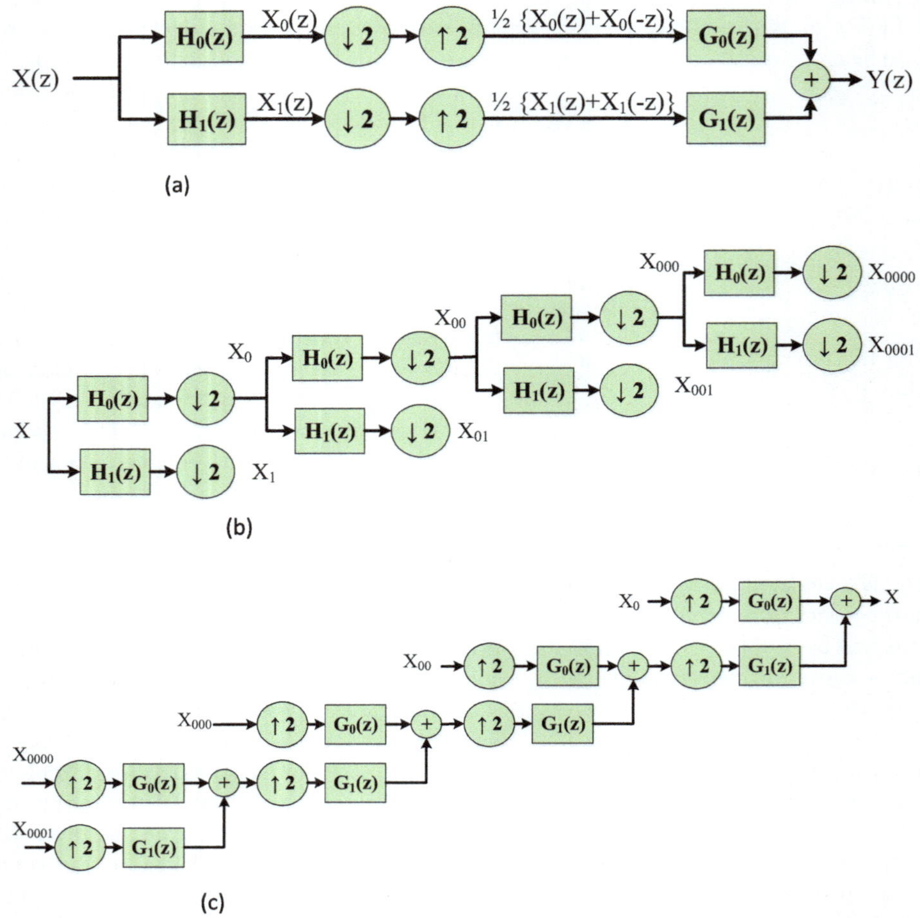

Figure 6 Wavelet decomposition and reconstruction. The down arrow is down-sampling by 2, and the up arrow is up-sampling by 2. **(a)** The two-band decomposition-reconstruction filter bank. **(b)** Multi-level wavelet decomposition tree. **(c)** Multi-level wavelet reconstruction tree.

1.4.1 Wavelet-domain seizure detection

The main challenge in wavelet-based EEG seizure detection is the determination of the appropriate wavelet decomposition level and the selection of the features from certain sub-bands for discrimination between seizure and non-seizure periods. Zhou et al. presented a wavelet-based seizure detection method that depends on lacunarity and fluctuation index as features with Bayesian linear discriminant analysis (BLDA) [12]. The lacunarity is a measure of heterogeneity used in fractal analysis. In this method, the authors first decompose the EEG epochs into five wavelet sub-bands. Three sub-bands with scales 3, 4, and 5 are chosen for further processing. For feature extraction, lacunarity and fluctuation index are computed within the frequency bands. If $p(m,l)$ denotes the probability of points, whose amplitude is equal to m, l represents the length of EEG epochs, A is the minimum amplitude value, and B is the maximum amplitude value, then the equation below holds [12]:

$$\sum_{m=A}^{B} p(m,l) = 1 \tag{14}$$

After calculating,

$$M_1(l) = \sum_{m=A}^{B} mp(m,l) \tag{15}$$

$$M_2(l) = \sum_{m=A}^{B} m^2 p(m,l) \tag{16}$$

The lacunarity can be obtained as:

$$la(l) = \frac{M_2(l) - [M_1(l)]^2}{[M_1(l)]^2} \tag{17}$$

After feature extraction, BLDA is used to classify the seizures. The BLDA tries to minimize the risk associated with the classification decision. It can deal with high dimensionality and noisy datasets, assuming a linear relation between targets t and feature vectors $\mathbf{x}$ from one side and additive white Gaussian noise n from the other side in Bayesian regression. Post-processing, such as smoothing, and multi-channel decision fusion are applied to enhance the detection accuracy of the BLDA. The performance of this method was investigated on Freiburg EEG database, and it achieved a 96.25% sensitivity with 0.13/h false detection rate.

Another five-level wavelet decomposition method for seizure detection was developed by Liu et al. [36]. This method works on multi-channel iEEG signals. Three wavelet sub-bands are selected for further processing. The extracted features from these sub-bands are the relative amplitude, relative energy, coefficient of variation, and fluctuation index from the selected frequency bands. The coefficient of variation is the ratio between the standard deviation of a decomposed sub-band and the square of its mean. The fluctuation index is a measure of the intensity of a decomposed sub-band. An SVM classifier is used in this approach, and some sort of post-processing is implemented to enhance the detection performance with smoothing, multi-channel decision fusion, and collar processing. The collar processing is technique aiming to maintain the data continuity during processing. The performance of this method has been tested on 509 h for 21 epilepsy patients. Experimental results have shown that this method achieved a 94.46% sensitivity, a 95.26% specificity, and a 0.58/h false detection rate on long-term iEEG.

The five-level wavelet decomposition was also adopted by Panda et al. with an SVM classifier for seizure detection from background EEGs [37]. This classifier was tested on a healthy subject with open eyes, a healthy subject with closed eyes, and an epilepsy patient. The extracted features for signal classification are energy, standard deviation, and entropy. The simulation results revealed an accuracy of 91.2% in seizure activity detection. Khan et al. proposed a similar approach for seizure detection, but with relative energy and a normalized coefficient of variation (NCOV) as features [38]. It works on wavelet coefficients acquired in the frequency range of 0 to 32 Hz as follows:

1) The COV is estimated as:

$$\mathrm{COV} = \frac{\sigma^2}{\mu^2} \tag{18}$$

where σ^2 is the epoch variance and μ is the epoch mean.

2) The RCOV is estimated as:

$$\mathrm{RCOV} = \frac{COV_e(n)}{COV(l)} \tag{19}$$

where $\mathrm{COV}_e(n)$ is the epoch coefficient of variation and $\mathrm{COV}(l)$ is the background coefficient of variation.

3) The NCOV is evaluated as:

$$\mathrm{NCOV} = \frac{\sigma^2}{\mu_a}, \tag{20}$$

where μ_a represents the mean of the absolute values of the wavelet coefficients. The obtained results with the NCOV are better than those obtained using the COV only.

Wang et al. used Neyman-Pearson rules and an SVM classifier for seizure detection as shown in Figure 7 [39]. This method depends on the wavelet coefficients in addition to the ApEn in the wavelet domain as extracted features, and the detection is performed using Neyman-Pearson rules with an SVM. The approximate entropy is an entropy metric that takes into consideration the ordering of the points of the discrete time sequence at hand, and hence, it is a good measure for the regularity of the data sequence. Simulation results have shown a detection accuracy of 98% and a false detection rate of 6%.

Zainuddin et al. investigated the use of wavelet neural networks (WNNs) based on wavelet basis functions for seizure detection [40]. Firstly, the wavelet transform of EEG signals is estimated, and maximum, minimum, and standard deviation of the absolute values of the wavelet coefficients in each sub-band are extracted as features. These features are then fed to trained WNNs. The Gaussian, Mexican Hat, and Morlet wavelet activation functions have been investigated for classification. A cross-validation approach have been adopted in the

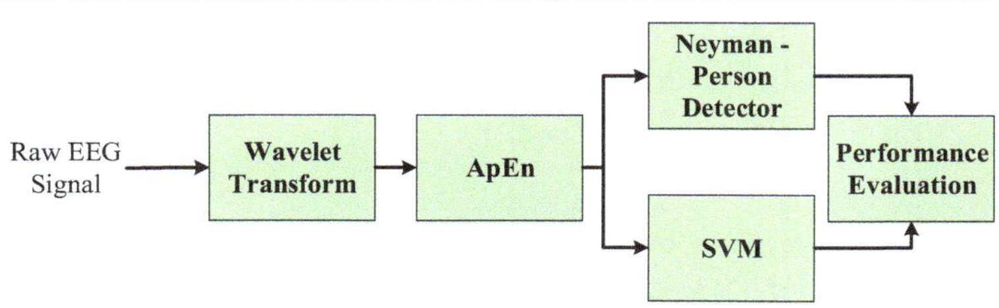

Figure 7 Wang et al. method for seizure detection.

simulation experiments. Simulation results revealed that the best performance was obtained with WNNs employing a Morlet wavelet activation function with order 4 Daubechies wavelet for feature extraction. The authors have carried simulation experiments on the University of Bonn database for scalp EEG signals [41]. The simulation results reported sensitivity and accuracy up to 98% with such combination.

Niknazar et al. presented a wavelet-based method for epileptic seizure detection that adopts recurrence quantification analysis (RQA) on EEG recordings and their delta, theta, alpha, beta, and gamma sub-bands extracted through a four-level Daubechies wavelet transform [42]. The RQA is well-suited for non-linear data analysis. It quantifies the number and duration of recurrences of the EEG signals based on phase space trajectories. The phase space is built on estimating a time delay and an embedding dimension, which are the features corresponding to each EEG signal state. The authors adopted an error-correcting output coding (ECOG) classifier for discriminating between three states: healthy, inter-ictal, and ictal. This method achieved an accuracy of 98.67%.

Daou and Labeau presented a wavelet-based approach for EEG signal compression and seizure detection, simultaneously [43]. The compression is performed with the set partitioning in hierarchical trees (SPIHT) algorithm. The SPIHT codes the LP and HP components of the signals with binary sequences. A dictionary is built for these binary sequences of normal and seizure activities. The seizure detection process is performed for each incoming segment by correlation estimation with the binary codes of normal sequences in the dictionary. If low correlation is detected, correlation estimation is performed with binary sequences of the HP components of seizure segments to ensure seizure activity. This method achieved an accuracy of about 90%.

Statistical analysis revealed that the spectra of normal EEG signals follow an inverse attenuation law over the bands of interest, which means that there are self-similar fluctuations in EEG signals over a multiplicity of scales. Based on this idea, Mehta et al. presented a scale-invariant seizure detection method [44]. In this method, the onset is detected, when the scale invariance is lost, which means that the slope of the regression line on a logarithmic plot for the wavelet scales 6 to 2 decreases.

Shoaib et al. presented a low-energy scalable processor for directly analyzing EEG signals acquired through compressive sensing [45]. This work adopts wavelet energy features. It studies the effect of compressive sensing on seizure detection performance showing a reduction of only 4% in detection sensitivity and an increase of only 0.15/h in false-alarm rate and 1 s in latency compared to baseline performance. The main advantage of this method is energy saving archived through compressive sensing.

Zandi et al. presented a wavelet packet real-time seizure detection algorithm working on scalp EEG signals [46]. They developed a patient-specific metric to differentiate between seizure and non-seizure states in the 1- to 30-Hz frequency range based on wavelet coefficients of seizure and non-seizure references. A combined seizure index (CSI) is derived from all EEG signal channels depending on the rhythmicity and relative energy of signal epochs for classification as seizure or non-seizure. A majority voting rule is used on the decisions from all channels to generate the CSI and raise an alarm when a seizure is detected. This algorithm was tested on 14 patients having 75.8 h with 63 seizures, and it achieved a sensitivity of 90.5%, a median detection delay of 7 s, and a false-alarm rate of 0.51 h^{-1}.

1.4.2 Wavelet-domain seizure prediction

The same concepts of wavelet signal analysis used for seizure detection can be extended to seizure prediction, but with the target this time as the detection of the pre-ictal state. In general, EEG signals containing seizures are build up of constantly changing bursting levels. This signal nature enables discrimination between different signal activities from wavelet sub-bands. The residual sub-band wavelet entropy (RSWE) can be directly used to estimate the entropy of bursts from the sub-bands as proposed by Paul et al. [47]. The wavelet decomposition equation for an EEG signal using a sliding window of index m is given by:

$$s(t) = \sum_{\tau=-\infty}^{\infty} a_L^m(\tau)\phi(2^{-L}t-\tau) + \sum_{l=1}^{L}\sum_{\tau=-\infty}^{\infty} C_l^m(\tau)\psi(2^{-l}t-\tau), \tag{21}$$

where $C_1^m(\tau), C_2^m(\tau), \ldots, C_L^m(\tau)$ are the wavelet coefficients. The sequence $\{a_L^m(\tau)\}$ is the coarser-resolution signal for a high-level decomposition. The authors experimented lower and higher numbers of levels and found that the five levels are the optimum choice.

The relative wavelet energy (RWE) of the wavelet coefficients is used to derive a sub-band wavelet entropy (SWE) feature. For a sliding window with index m, the field potential (FP) is given by [48]:

$$E_l^m(\tau) = \left|C_l^m(\tau)\right|^2 \tag{22}$$

The wavelet coefficients total energy is given by:

$$E_{rmtotal}^m = \sum_l \sum_\tau E_l^m(\tau) \tag{23}$$

The RWE can be expressed with normalization as:

$$p_l^m(\tau) = \frac{E_l^m(\tau)}{E_{rmtotal}^m} \tag{24}$$

The SWE at scale l can be defined with a probabilistic approach as:

$$H(C_l^m) = \sum_{\tau} p_l^m(\tau) \log p_l^m(\tau) \tag{25}$$

The lth level RSWE of the nth frame is given by:

$$J(C_l^{m,n}) = H(C_l^{m,n}) - \frac{1}{N_o} \sum_{n=1}^{N_o} H(C_l^{m,n}), \tag{26}$$

where $C_l^{m,n}$ represents the wavelet coefficients of the nth frame in the mth window.

For the estimation of the RSWE, a sliding window of 1-min length has been used assuming that the FP is stationary during this window. It was also assumed that the background FP fluctuations stay statistically unchanged for frames of 5-s length during the main sliding window. On the other hand, the FP fluctuations have another component that is statistically variant over the 5-s frames. Hence, an averaging process is performed on the background entropy components estimated in the main window. Consequently, resulting residual entropy corresponds only to the bursting components. A key observation on this approach is that there is an increment in the mean cortical RSWE during the transition from the pre-ictal to inter-ictal period. This increment can be used with an efficient slope change detector and used for early seizure prediction.

Hung et al. developed a very large scale integration (VLSI) setup of wavelet-based seizure prediction algorithm using the correlation dimension (D_c) and its correlation coefficient [48]. Their system comprises arithmetic functional and control units. The arithmetic functional units are the DWT, correlation dimension, correlation coefficient, and seizure prediction. The DWT of the pre-processed signal is estimated to decompose it into four sub-bands (0 to 63 Hz, 64 to 128 Hz, 0 to 1 Hz, and 32 to 64 Hz). The higher-frequency sub-bands are then represented in the phase space. The correlation dimension and correlation coefficient are estimated in the phase space as seizure prediction features. The authors evaluated their method utilizing iEEG data from 11 patients of the Freiburg database with 256-Hz sampling rate. Their method achieved an average of 87% sensitivity, 0.24/h false prediction rate, and in average a 27-min warning time ahead the ictal.

Chiang et al. developed an online wavelet-domain retraining method to improve the seizure prediction by enlarging the training dataset gradually [49]. Their method is based on the method of Mirowski et al. [50] that uses non-linear interdependence, cross-correlation, difference of Lyapunov exponents, and phase locking. Post-processing is used in this method to reduce the false-alarm rate if two consecutive patterns are classified as pre-ictal. The authors evaluated their method using three datasets: Freiburg database, CHB-MIT database (eight patients), and National Taiwan University Hospital database for scalp EEG (one patient) [51]. This method achieved sensitivities of 74.2% and 52.2% on intracranial and scalp databases, respectively. It also achieved an improvement in the sensitivity of off-line training on both databases by 29.0% and 17.4%, respectively.

Rojas et al. presented a seizure prediction method depending on brain excitability recognized with couplings between low-frequency phases (delta: 0.5 to 33 Hz, theta: 3 to 8 Hz) and high-frequency amplitudes (low gamma: 40 to 70 Hz, high gamma 70 to 140 Hz) of brain waves [52]. They evaluated this method on 20 patients from EPILEPSIAE scalp EEG database [53]. The EEG data was recorded using either depth or subdural stereotactic electrodes at a sampling rate 1,024 Hz with a total of 267 seizures and more than 3,400 h of inter-ictal activities. They found that in 50% of the cases, their predictor performed better than the random predictor, which is based on Poisson process, with an average sensitivity 98.9%, false rate of 1.84/h, and pre-ictal window length of 10 min.

Gadhoumi et al. presented an iEEG-based seizure prediction method based on measuring the similarity with a reference state as shown in Figure 8 [54]. In the validation of this method, 1,565 h of continuous iEEG data for 17 patients having mesial temporal lobe epilepsy were used. These recordings have 175 seizures. Distance, inclusion, and persistence features are extracted in this method from the continuous wavelet transform of EEG recordings to discriminate between pre-ictal and inter-ictal states. The distance is a Euclidean distance between the center energy and the rest of energy points computed from the wavelet coefficients. Inclusion is the percentage of energy points with entropy confined to a pre-specified distance. Inclusion represents the time interval in which the maximum number of successive 2-s points has energy and entropy profiles confined to the pre-specified distance. An SVM classifier has been used in the signal discrimination and a comparison has been performed with a random statistical predictor. Simulation results have shown a sensitivity of 85% and a false-alarm rate of 0.35/h. The authors came to a conclusion that state similarity measures can be used for seizure prediction above chance in 41% of the patients.

Gadhoumi et al. also presented a method to discriminate between pre-ictal and inter-ictal states using features extracted from the high-frequency sub-bands of iEEG signals [55]. The selected features are the wavelet energy and entropy extracted from different signal epochs. The

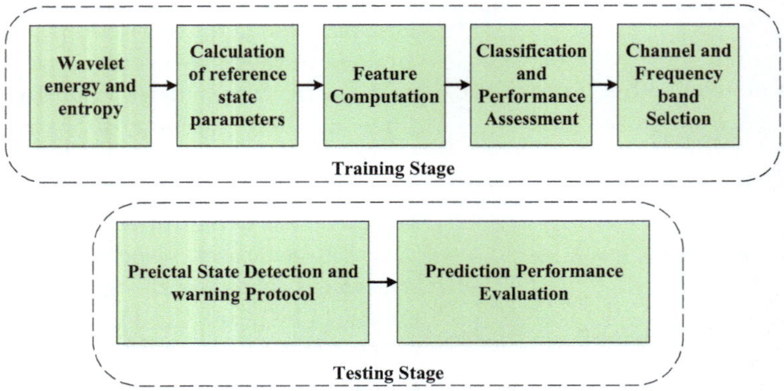

Figure 8 Gadhoumi et al. method showing training and testing stages.

estimated energy and entropy are compared to reference energy and entropy for pre-defined signals with pre-state and seizure activities, and based on distance metrics, the decisions are taken from different channels and fused. Sensitivity and false-alarm rate evaluation revealed a sensitivity of about 80% and a false positive rate ranging from 0.09 to 0.7 for different patients.

Wang et al. presented a wavelet-based online adaptive seizure prediction system [56]. This system adopts Lyapunov exponent, correlation dimension, Hurst exponent, and entropy features extracted from the wavelet transform of EEG recordings. It adopts also a KNN classifier. The adaptation is performed in this system with a reinforcement learning algorithm. A 150-min prediction horizon used on ten patients and the system achieved the best prediction results of 73% sensitivity and 67% specificity.

Soleimani et al. presented a simple and fast adaptive online method for the detection of pre-ictal patterns depending on multiple features as illustrated in Figure 9 [57]. These features are time domain and wavelet domain. The time-domain features are the curve length, the average energy, non-linear energy, six-order power at time n, kurtosis, skewness, and variance. The wavelet-domain feature include mean of absolute values, average power, standard deviation, absolute mean of sub-bands

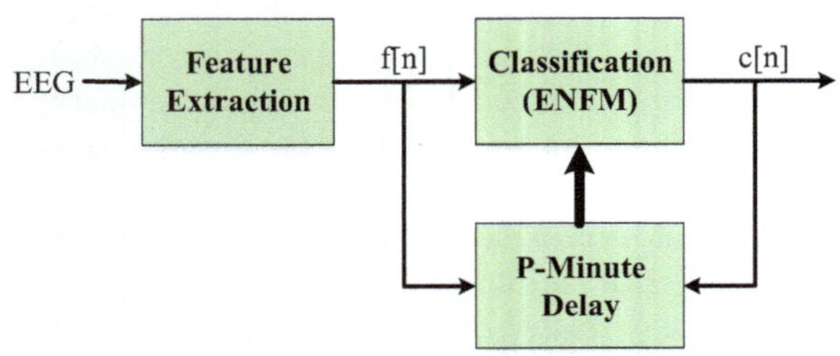

Figure 9 Soleimani et al. seizure prediction method.

of a fifth level Daubechies wavelet transform. In this method, a neuro-fuzzy model is used for combined features learning in an adaptive manner. An adaptive tuning process is used in the classifier operation to build a personal seizure predictor. Freiburg database of intracranial recordings for 21 patients has been used with this classifier. Simulation results revealed that on-line adaptive seizure prediction achieves better results than off-line non-adaptive seizure prediction. The percentage of prediction was 99.52%, and the FPR was 0.1417/h.

It is possible to use some time-domain features with wavelet-domain features to enhance the detectability of seizures. Costa et al. have selected 14 features for efficient seizure prediction from EEG recordings [58]. These features include energy estimated from time-domain signals, energy variation, energy level, non-linear statistics, and sub-band energies extracted from the wavelet sub-bands. They used a neural network classifier for signal classification into four states: pre-ictal, ictal, inter-ictal, and post-ictal. They achieved an average specificity of 99%, an average sensitivity of 83%, and an average accuracy of 96% on patient records from Freiburg database. Moghim and Corne compared the Costa et al. results with multi-class SVM and evolved neural network classifiers [59]. They carried out this comparative study on one patient (patient 2: 135 min, 3 seizures, 30 to 40 min before and 10 min after each seizure) from Freiburg database, and they reported that 8- to 10-min detection before the onset can be achieved with reasonable specificity and sensitivity.

1.5 Empirical mode decomposition

The EMD is a signal decomposition, which transforms a signal into a group of intrinsic mode functions (IMFs). For EEG seizure detection, these IMFs show different behavior with normal and abnormal activities in the signals. Features can also be extracted from the IMFs and tested for seizure detection and prediction. The steps to extract these IMFs from an original signal $x(t)$ are summarized below [60]:

1. Determine the signal envelope by maxima and minima point estimation and interpolate between them.
2. The local mean, $m(t)$ is estimated using the envelope as follows:

$$m(t) = \frac{e_{\min}(t) + e_{\max}(t)}{2} \tag{27}$$

   where $e_{\min}(t)$ and $e_{\max}(t)$ are the envelope minimum and maximum values respectively.
3. The mean is subtracted from the original signal to get $h(t) = x(t) - m(t)$.
4. If $h(t)$ does not satisfy the IMF criteria of a difference of at most 1 between the number of extrema (maxima and minima) and number of zero crossings, we go to step 1 with $h(t)$ as a new input.
5. The obtained $h(t)$ is stored as an IMF in case it satisfies the criteria of an IMF, $c_i(t) = h(t)$. Hence, this IMF is removed from the original signal by subtraction, $r(t) = x(t) - c_i(t)$, where i is the ith IMF.
6. Similar steps are repeated beginning from step 1.

The Hilbert transform (HT) is performed on each IMF to get $\tilde{c}_i(t)$ provided that $z(t) = c_i(t) + j\tilde{c}_i(t)$ is analytic. Both amplitude and phase of $z(t)$ are given by:

$$A(t) = \sqrt{c_i^2(t) + \tilde{c}_i^2(t)} \tag{28}$$

$$\phi(t) = \arctan\left[\frac{\tilde{c}_i^2(t)}{c_i^2(t)}\right] \tag{29}$$

The instantaneous frequency is estimated as:

$$\omega = \frac{d\phi(t)}{dt} \tag{30}$$

1.5.1 EMD seizure detection

Eftekhar et al. used the EMD approach for seizure detection [60]. They adopted features such as the frequency rise at the seizure onset with the EMD in a patient-specific manner. Their simulation results have shown that the Hilbert transform can be used to decompose EEG signals into components, from which features can be extracted for seizure onset detection.

Tafreshi et al. evaluated the performance of the EMD in discriminating epileptic seizure data from normal data using means of the absolute of the IMFs as features [61]. They compared this approach for feature extraction with wavelet features using both multi-layer perceptron (MLP) and self-organizing map (SOM) neural networks. Results have shown that the EMD approach is better in performance than the wavelet approach with success rates up to 95%, and they revealed that four empirical modes are enough to obtain good results. In addition, the results have shown that a window length of 1,500 samples is appropriate to obtain good recognition rates. The MLP networks are superior in performance to the SOM networks.

Orosco et al. presented a seizure detection approach based on the energies of IMFs as discriminating features between seizure and non-seizure activities as shown in Figure 10 [62]. In this approach, the IMF energies are compared with certain thresholds for decision making. It was tested on nine patient records from Freiburg database with invasive nature. The obtained sensitivity and specificity were 56.41% and 75.86%, respectively, using 39 seizure records.

Guarnizo and Delgado presented a modified EMD approach, in which mutual information is used for feature selection in the EMD domain [63]. These features include the average or instantaneous frequency and amplitude for all EMD components. Higher-order statistics such as the skewness and kurtosis in addition to Shannon's entropy have been selected as features extracted from the energy estimated with the Teager energy operator (TEO) over all EMD components [64-66]. This approach adopts a linear Bayes classifier and it achieved an accuracy of 98%.

Alam and Bhuiyan presented a seizure detection method depending on extracting kurtosis, skewness, largest Lyapunov exponent, variance, approximate entropy, and correlation dimension from the EMD of EEG signals with artificial neural network classifiers [67,68]. This method achieved a 100% sensitivity in seizure detection and has shown a superiority as compared to time-frequency techniques and band-limited techniques in the computational complexity.

Bajaj and Pachori presented an EMD-based seizure detection method to detect focal temporal lobe epilepsy [69]. In this method, they used Hilbert transformation of IMFs which were obtained by an EMD process. Epileptic seizures are then detected based on the instantaneous area estimated from the trace of analytic IMFs of EEG signals. The performance of this epileptic detection method was evaluated on Freiburg database and achieved a sensitivity of 90%, a specificity of 89.31%, and an error detection rate of 24.25%.

The local mean of EMD can be used as a statistical feature for seizure detection. This local mean is obtained by averaging the lower and upper envelopes obtained through interpolation between the local maxima and minima. If these local maxima and minima are not defined as in the case of multivariate signals, another alternative which is the multivariate EMD (MEMD) can be used instead. The MEMD depends on signal projections

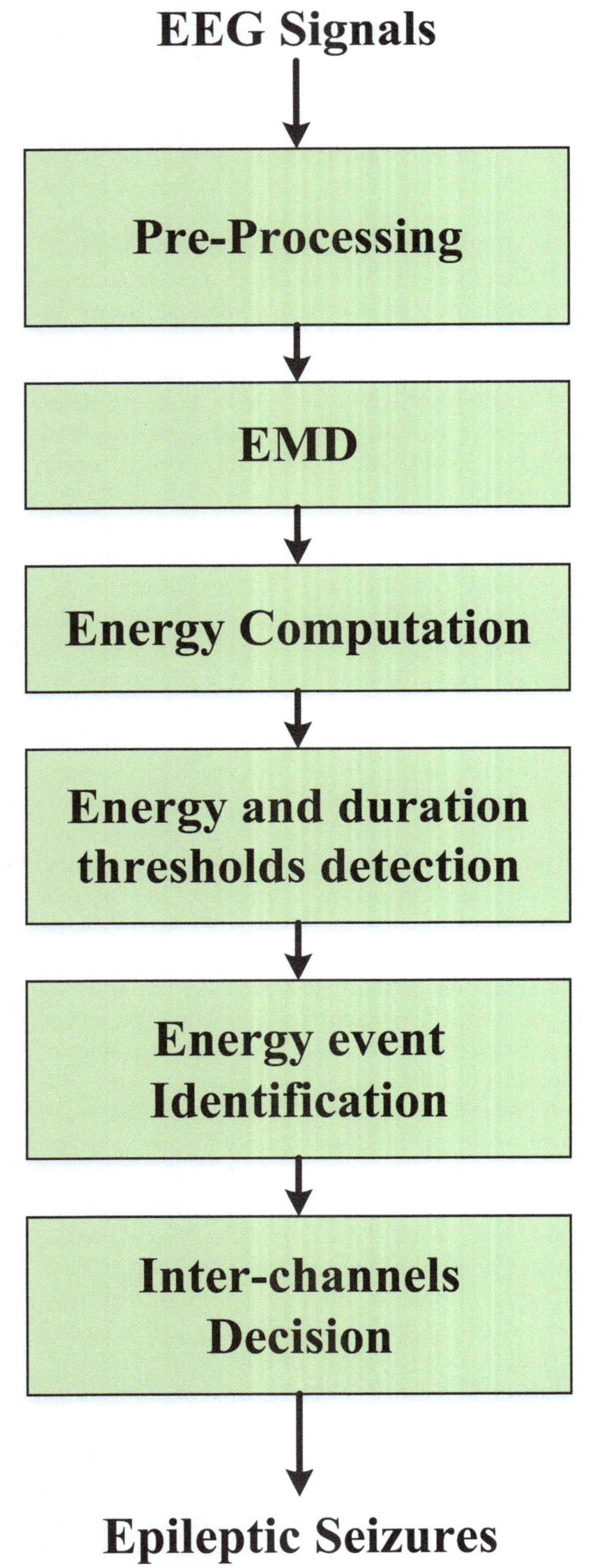

Figure 10 Orosco et al. method for epileptic seizure detection with EMD.

along n-dimensional spaces to generate n-dimensional envelopes, which are then averaged to get the local mean. The projection direction vectors are based on spherical and polar coordinates. To sample the n-sphere, quasi-Monte Carlo low-discrepancy sequences are exploited for sampling enhancement. This style of sampling is non-uniform but not random. Its objective is to get the most useful samples over a certain space. The other steps of the algorithm to estimate the MEMD are similar to the standard EMD algorithm. The use of the MEMD enables the multi-channel EEG signal decomposition into narrow frequency bands that can be analyzed separately for better detectability of seizures. In addition, the mean frequency of the signal segments can be used as a discriminative feature for seizure detection. This mean frequency can be calculated by applying the Hilbert transform on the resulting IMFs. Rehman et al. investigated the use of the MEMD with the mean frequency as possible features for EEG seizure detection [70], but this proposal needs further studies.

1.5.2 EMD seizure prediction

Qi et al. suggested the use of EMD for seizure prediction with a new feature, called IMF-VoE. This feature combines three IMFs obtained from the EMD of an EEG signal in addition to the range variance between upper and lower envelopes (VoE) in the EMD domain [71]. The authors tested this method using 80.4 h of EEG data containing ten seizures for four patients. This method achieved a sensitivity of 100% and a false detection rate of 0.16/h. The estimated average time delays were 19.4, 13.2, and 10.7 s for false detection rates of 0.16, 0.27, and 0.41/h, respectively, using different thresholds.

1.6 SVD methods

The SVD operating on a matrix **A** decomposes it into matrices **U**, **S**, and **V** [72,73]:

$$\mathbf{A} = \mathbf{USV}^{\mathrm{T}} \tag{31}$$

The matrices **U** and **V** are orthogonal such that $\mathbf{U}^{\mathrm{T}}\mathbf{U} = \mathbf{I}$, and $\mathbf{V}^{\mathrm{T}}\mathbf{V} = \mathbf{I}$. $\mathbf{S} = \mathrm{diag}(\sigma_1,\ldots,\sigma_P)$, where $\sigma_1 \geq \sigma_2 \geq \ldots\sigma_P \geq 0$ is the matrix of SVs of **A**. The matrix **U** contains the left singular vectors of **A**, while the matrix **V** contains the right singular vectors of **A**. An important characteristic of the SVD that can be used in EEG seizure detection is that the SVs are not largely affected by little disturbances in the matrix **A**. This property can be used for the detection of seizure activities taking into consideration the slight variations between seizures [72,73].

Vanrumste et al. developed an SVD-based algorithm for EEG seizure detection [74]. In this algorithm, the EEG is segmented into overlapping epochs, which are formulated as 2-D arrays to allow SVD. Most of the signal energy, which differs from seizure to non-seizure activities, is concentrated in the first SVs. After that, an EEG dipole source analysis adopting a single-dipole model is implemented on the EEG segments. This analysis assumes that each spike results from a dipole in the human brain and performs a localization process for the position of this source dipole [75]. This yields dipole parameters in addition to a relative residual energy (RRE). A seizure alarm on an epoch is raised if the SVD indicates a dominant source and the RRE has a low value. The algorithm has been investigated on synthetic EEG signals obtained from two sources, synchronous and asynchronous. In the synchronous case, the main metric is the RRE, and in the asynchronous case, the SVD and RRE are of major concern. The same algorithm has been investigated with real EEG signals having two spikes and an eye-blink artifact. The algorithm has shown that the RRE was always low for all three events.

Shahid et al. presented an algorithm based on SVD for the detection of seizures [76]. In this algorithm, the SVD is applied sequentially on a sliding window of the EEG signal of 1-s width and then r singular values are obtained and used to indicate sudden changes in the signals. EEG recordings of four pediatric patients with 20 seizures have been used to validate this algorithm, and the results indicated the sensitivity of the SVs to the changes in the EEG signals due to epileptic seizure.

1.7 ICA and PCA methods

The SVD methods presented in above section adopt a strategy of transforming the 1-D data to a 2-D array, but the PCA and ICA techniques in this section adopt a covariance approach to obtain the PCs. ICA is signal decomposition that transforms a multivariate signal into additive, non-Gaussian, and statistically independent components [77]. The main principle that the ICA depends on is the maximization of the statistical independence between the estimated components. This can be guaranteed through either mutual information minimization or non-Gaussianity maximization. For mutual information minimization, measures like Kullback–Leibler divergence and maximum entropy are exploited. On the other hand, for non-Gaussianity maximization, central limit theorem, higher-order statistics such as kurtosis, and negentropy are used [77]. The ICA depends mainly on some pre-processing techniques such as mean removal, whitening, and dimensionality reduction. Both whitening and dimensionality reduction can be achieved with PCA or SVD. Whitening ensures an equal treatment to all dimensions before the algorithm is run.

1.7.1 Seizure detection

Both ICA and PCA have been used as signal separation algorithms to enhance the detectability of seizures from EEG signals by separating artifacts from these signals [78]. Xie and Krishnan proposed a multi-scale PCA method combining both DWT and PCA for de-noising and EEG signal decomposition [79]. They developed an empirical classification (EC) method using spatial and temporal features. This method achieved high classification accuracy.

1.7.2 Seizure prediction

Miri and Nasrabadi proposed a seizure prediction technique based on chaos theory and non-linear dynamics to create a return map depending on zero crossings [80]. This method consists of nine steps: raw EEG data recording from six channels, baseline noise reduction through filtering, signal segmentation to non-overlapping 5-min sections, PCA decomposition of the signal, zero-crossing rate estimation for the first PCA component, resampling of the second component based on the estimated zero-crossing rate, creation of a series of samples to construct a Poincare map, feature extraction, and finally estimation of the seizure risk. They evaluated this method with iEEG data from six patients of Freiburg database with 256-Hz sampling rate, 50-min pre-ictal, and 24 h of non-seizure data. Their method achieved a 20 ± 5-min prediction time, an 86.6% sensitivity, and a 0.067/h false-alarm rate.

Williamson et al. presented a seizure prediction method that is based on estimating eigenspectral features from space-delay correlation and covariance matrices of EEG signal blocks with multiple delay scales [81]. An SVM classifier has been used in this method. An averaging process over 15-min windows is used as a post-processing step to get the final prediction score. This method was tested on the recordings of 19 patients from the Freiburg database having at least three seizures. It succeeded in predicting 71 out of 83 seizures. The false prediction rate was 13.8/h.

2 Conclusions

Table 1 shows a comparison between the different seizure detection and prediction methods. It is clear from this table that most seizure detection and prediction methods adopt time- or wavelet-domain features. For wavelet-domain method, several papers have adopted five-level decomposition for robust feature extraction. Few methods have used hybrid features like time and wavelet-domain features and achieved better performance than using features from one domain only. It is clear also that the EMD is a promising trend for seizure detection and prediction that needs further investigation. Also, we can say that seizure detection and prediction methods based on classifiers are better in performance than methods that do not implement classifiers. Multi-

Table 1 Summary of seizure detection and prediction methods

Method	Domain and algorithm classification	Single or multi-channel	Database	Frame length	Features	Classifier	Performance metrics	Hardware implementation
Runarsson and Sigurdsson [9]	Time, detection	Single channel	Self-recorded data	Variable length frames	Histogram bin amplitudes for amplitude difference and separation time between peaks and minima	SVM	Sensitivity: 90%	No
Yoo et al. [14]	Time, detection	Multi-channel	MIT database (scalp EEG)	2 s	Energies of sub-bands	SVM	Accuracy: 84.4%, classification energy 2.03 μJ	Yes
Dalton et al. [16]	Time, detection	Single channel	Dataset of 21 seizures	12 to 25 s	Signature of seizure	Template matching	Sensitivity: 91 %, specificity: 84 %, battery lifetime: 10.5 h	Yes
Zandi et al. [17,18]	Time, prediction	Multi-channel	561 h of scalp EEG with 86 seizures for 20 patients	15 s	Combined index	Non-linear SVM with Gaussian classifier	Sensitivity: 88.34%, false prediction rate: 0.155 h^{-1}, average prediction time: 22.5 min	No
Aarabi and He [19]	Time, prediction	Single channel	316 h of iEEG data containing 49 seizures for 11 patients from Freiburg database	10 s	Correlation dimension, correlation entropy, noise level, Lempel-Ziv complexity, largest Lyapunov exponent	No classifier	Average sensitivity: 79.9%, 90.2% with average false prediction rate: 0.17 and 0.11/h	No
Schelter et al. [25]	Time, prediction	Single channel	Eight patients with focal epilepsy	30 s	Probabilistic threshold	No classifier	Sensitivity: 60%	No
Wang et al. [26]	Time, prediction	Single channel	Data for five patients	10 min	Area under the ROC curve	KNN	Average accuracy: 70%	No
Bedeeuzzaman et al. [27]	Time, prediction	Single channel	21 patients from Freiburg database (scalp and iEEG)	1 min	MAD, IQR	Binary linear classifier	Sensitivity: 100%, FPR: 0 (for 12 patients), average prediction time: 51 to 96 min	No
Li et al. [28]	Time, prediction	Multi channel	21 patients from Freiburg database (scalp and iEEG)	5 s	Spike rate	No classifier	Accuracy: 75.8%, false prediction rate: 0.09	No
Chisci et al. [30]	Time, prediction	Multi channel	21 patients from Freiburg database (scalp and iEEG)	2 s	AR coefficients	SVM	Sensitivity: 100%, average prediction time: 60 min	No
Rana et al. [32]	Frequency, detection	Multi-channel	EEG and ECoG time series, BW: 0.5 to 100 Hz	10 to 60 s	Phase slope index	No classifier	Detection accuracy: 100%	No
Khamis et al. [33]	Frequency, detection	Single channel	12 patients with six data records (R1 to R6)	32 s	Range of frequencies, range of frequencies used for background subtraction, number of points, hemisphere parameters, moment parameters, artifact removal parameters, *p*-norm power, cutoff point	Powell's direction set method	Sensitivity: 91%, false positives per hour: 0.02	No
Acharya et al. [34]	Frequency, detection	Single channel	Self-recorded data	23.6 s	Phase entropy 1 (S_1), phase entropy 2 (S_2), approximate entropy (ApEn), and sample entropy (SampEn)	SVM, FSC, PNN, KNN, NBC, DT, and GMM	Average accuracy: 98.1%	No

Table 1 Summary of seizure detection and prediction methods *(Continued)*

Zhou et al. [12]	Wavelet, detection	Multi-channel	Freiburg database of 21 patients; six channels: channels 1, 2, and 3 near the epileptic focus, and channels 4, 5, and 6 in remote locations	4 s without overlapping	Lacunarity and fluctuation index on wavelet scales	Bayesian linear discriminant analysis (BLDA) classifier	Sensitivity: 96.25%, false detection rate: 0.13/h, mean delay time: 13.8 s	No
Liu et al. [36]	Wavelet, detection	Multi-channel	Dataset of 509 h from 21 epileptic patients	4 s	Relative energy, relative amplitude, coefficient of variation, fluctuation index	SVM	Sensitivity: 94.46%, specificity: 95.26%, false detection rate: 0.58/h	No
Panda et al. [37]	Wavelet, detection	Single channel	500 epochs of EEG data from five different brain activities (100 signals per epoch)	0.5 s	Energy, entropy, standard deviation	SVM	Accuracy: 91.2%	No
Khan et al. [38]	Wavelet, detection	Single channel	Self-recorded data for five patients	25 s	Energy and normalized coefficient of variation	Simple LDA classifier	Sensitivity: 83%, specificity: 100%, accuracy: 92%, overall precision: 87%	No
Wang et al. [56]	Wavelet, detection	Single channel	Changhai Hospital database (scalp EEG)	70 to 200 ms	Approximate entropy	Neyman-Pearson criteria and SVM	Detection accuracy: 98%, false detection rate: 6%	No
Zainuddin et al. [40]	Wavelet, detection	Single channel	University of Bonn database (scalp EEG)	23.6 s	Maximum, minimum, standard deviation of absolute wavelet coefficients	WNN	Sensitivity: 98%, accuracy: 98%	No
Niknazar et al. [42]	Wavelet, detection	Single channel	100 single-channel segments from Bonn University Database (scalp EEG)	23.6 s	Time delay, embedding dimension	ECOG	Accuracy: 98.67%	No
Daou and Labeau [43]	Wavelet, detection	Multi-channel	Nine patients from MIT database, and nine patients from Montreal Neurological Institute database (scalp EEG)	1 s	SPIHT codes	No classifier	Accuracy: 90%	No
Mehta et al. [44]	Wavelet, detection	Single channel	Self-recorded	10 s	Slope of the regression line on a logarithmic plot for the wavelet scales 6 to 2	No classifier	Slope decreases with onset	No
Shoaib et al. [45]	Wavelet, detection	Multi-channel	MIT database (scalp EEG)	2 s	Wavelet energy	SVM	Sensitivity: 91% to 96%, latency: 4.7 to 5.3 s, false-alarm rate: 0.17 to 0.3/h	Yes
Zandi et al. [46]	Wavelet, detection	Multi-channel	EEG recordings from 14 patients approximately 75.8 h with 63 seizures	10 to 40 s	Combined seizure index	No classifier	Sensitivity: 90.5%, false detection rate: 0.51 h^{-1}, median detection delay: 7 s	No
Paul et al. [47]	Wavelet, prediction	Single channel	Data for rats	5 s	RSWE	No classifier	Detection: 90%	No
Hung et al. [48]	Wavelet, prediction	Multi-channel	11 patients from Freiburg database (scalp and iEEG)	1 h	Chaos features, wavelet features, correlation dimension	No classifier	Accuracy: 87%, false prediction rate: 0.24/h, average warning time: 27 min ahead the ictal	Yes

Table 1 Summary of seizure detection and prediction methods *(Continued)*

Chiang et al. [49]	Wavelet, prediction	Single channel	Freiburg database, CHB-MIT database (eight patients), National Taiwan University Hospital database (one patient)	60 s	Wavelet coherence	SVM	74.2% and 52.2% sensitivity on intracranial and scalp databases, sensitivity improvement by 29.0% and 17.4% for both databases	No
Rojas et al. [52]	Wavelet, prediction	Single channel	20 patients (267 seizures, 3,400 h)	Up to 22 h	Cross-frequency coupling	No classifier	Sensitivity: 100%	No
Gadhoumi et al. [54]	Wavelet, prediction	Multi-channel	1,565 h iEEG, 17 patients with mesial temporal lobe epilepsy with 175 seizures	1 min with 75% overlap	Distance, inclusion, persistence	SVM	Sensitivity: >85%, warning rate: < 0.35/h, proportion of time under warning: <30%	No
Gadhoumi et al. [55]	Wavelet, prediction	Multi-channel	Self-recorded dataset for six patients	2 s, disjoint	Wavelet energy and entropy	Binary classifier	Average sensitivity: >80%, FP rate: from 0.09 to 0.7 FP/h	No
Wang et al. [56]	Wavelet, prediction	Multi-channel	Ten patients (self-recorded data)	10 min with 50% overlap	Lyapunov exponent, correlation dimension, Hurst exponent, entropy	KNN	Sensitivity: 73%, specificity: 67%	No
Soleimani et al. [57]	Wavelet and time, prediction	Single channel	21 patients Freiburg database (scalp and iEEG)	10 s	Time- and wavelet-domain features	Neuro-fuzzy	Prediction: 99.52%, FPR: 0.1417/h	No
Moghim and Corne [59]	Wavelet and time, prediction	Single channel	One patient from Freiburg database (135 min, three seizures, 10 min following each seizure) (scalp and iEEG)	Up to 5 min	Time-domain energies, wavelet-domain energies, correlation dimension, max Lyapunov exponent	MC-SVM and EANN	Sensitivity and specificity with different values according to simulation assumptions	No
Tafreshi et al. [61]	EMD, detection	Single channel	Five patients of Freiburg database (scalp and iEEG)	4 to 6 s	Mean of the absolute of each IMF, wavelet features	Neural network	Accuracy: 95%	No
Orosco et al. [62]	EMD, detection	Single channel	90 EEG segments acquired for nine patients	1 h	Energies of the IMFs	No classifier	Sensitivity: 56%, specificity: 75%, positive predictive value: 61%, negative predictive value: 72%	No
Guarnizo and Delgado [63]	EMD, detection	Single channel	Five groups with 100 single-channel registers sampled at 173.61 Hz	Whole signal	Instantaneous frequency, amplitude of each EMD component, skewness, kurtosis, Shannon's entropy	Linear Bayes classifier	Accuracy: 98%	No
Alam and Bhuiyan [67,68]	EMD, detection	Single channel	Bonn University database (scalp EEG)	23.6 s	Skewness, kurtosis, variance, largest Lyapunov exponent, correlation dimension, approximate entropy from intrinsic modes	ANN	Accuracy: 100%	No
Bajaj and Pachori [69]	EMD, detection	Single channel	Freiburg database for 21 patients (scalp and iEEG)	15 s	Modified central tendency measure	No classifier	Sensitivity: 90%, specificity: 89.31%, error detection rate: 24.25%	No
Qi et al. [71]	EMD, prediction	Single channel	Pre-surgical epilepsy monitoring database with 80.4 h of EEG data with ten seizures of four patients	Whole signal	IMF-VoE	No classifier	Sensitivity: 100%, false detection rate: 0.16/h, average time delay: 10.7 to 19.4 s are 19.4 s	No
Vanrumste et al. [74]	SVD, detection	Single channel	Simulated EEG signals	10 s	Dipole parameters, relative residual energy (RRE)	No classifier	Visual inspection of parameters	No

Table 1 Summary of seizure detection and prediction methods *(Continued)*

Shahid et al. [76]	SVD, detection	Single channel	Recordings of four pediatric patients with 20 seizures	1 s	Largest *r*-singular values, Euclidean distance from the *r*-singular values of a baseline window located 1 h before the seizure	No classifier	Visual inspection of results	No
Xie and Krishnan [79]	PCA in wavelet domain, detection	Single channel	Data for eight patients	4,097 s	Adjusted random index (ARI)	Empirical classifier	ARI up to 1	No
Miri and Nasrabadi [80]	PCA, prediction	Multi-channel	Six patients of Freiburg database (scalp and iEEG)	5 min	Zero-crossing rate, a statistical index	No classifier	Prediction time: 20 ± 5 min, sensitivity 86.6%, false-alarm rate: 0.067%	No
Williamson et al. [81]	PCA, FFT, prediction	Multi-channel	Freiburg database (scalp and iEEG)	15 s	Eigen spectral features	SVM	Sensitivity: 85.5%, false prediction rate: 0.033/h	No

channel techniques are also more superior in performance than single-channel techniques.

Competing interests
The authors declare that they have no competing interests.

Author details
[1]KACST, Riyadh 11442, Kingdom of Saudi Arabia. [2]Electrical Engineering Department, KACST-TIC in Radio Frequency and Photonics for the e-Society (RFTONICS), King Saud University, Riyadh 11362, Saudi Arabia. [3]Department of Electrical Engineering, King Saud University, Riyadh 11362, Saudi Arabia. [4]Institute for Telecommunications Research, University of South Australia, South Australia 5095, Australia. [5]KACST-TIC in Radio Frequency and Photonics for the e-Society (RFTONICS), King Saud University, Riyadh 11362, Saudi Arabia. [6]Faculty of Electronic Engineering, Menoufia University, Menouf, 32952, Egypt.

References
1. National Institute of Neurological Disorders and Stroke. http://www.ninds.nih.gov/. Accessed 15 Sept 2014
2. Trans Cranial Technologies Ltd, *10/20 System Positioning Manual* (Trans Cranial Technologies Ltd, Kowloon, 2012)
3. UR Acharya, SV Sree, G Swapna, RJ Martis, JS Suri, Automated EEG analysis of epilepsy: a review. Knowl.-Based Syst. **45**, 147–165 (2013)
4. DK Binder, SR Haut, Toward new paradigms of seizure detection. Epilepsy Behav. **26**(3), 247–252 (2013)
5. B Hunyadi, M Signoretto, WV Paesschen, JAK Suykens, SV Huffel, MD Vos, Incorporating structural information from the multichannel EEG improves patient-specific seizure detection. Clin. Neurophysiol. **123**, 2352–2361 (2012)
6. J Rasekhi, MRK Mollaei, M Bandarabadi, CA Teixeira, A Dourado, Preprocessing effects of 22 linear univariate features on the performance of seizure prediction methods. J. Neurosci. Methods **217**, 9–16 (2013)
7. PR Carney, S Myers, JD Geyer, Seizure prediction: methods. Epilepsy Behav. **22**, S94–S101 (2011)
8. PE McSharry, T He, LA Smith, L Tarassenko, Linear and non-linear methods for automatic seizure detection in scalp electro-encephalogram recordings. Med. Biol. Eng. Comput. **40**, 447–461 (2002)
9. PE McSharry, LA Smith, L Tarassenko, Comparison of predictability of epileptic seizures by a linear and a nonlinear method. IEEE Trans. Biomed. Eng. **50**(5), 628–633 (2003)
10. V Sakkalis, G Giannakakis, C Farmaki, A Mousas, M Pediaditis, P Vorgia, M Tsiknakis, Absence seizure epilepsy detection using linear and nonlinear EEG analysis methods, in *Proceedings of the 35th Annual International Conference of the IEEE EMBS* (Osaka, Japan), pp. 6333–6336. 3–7 July 2013
11. AT Tzallas, MG Tsipouras, DG Tsalikakis, EC Karvounis, L Astrakas, S Konitsiotis, M Tzaphlidou, Automated epileptic seizure detection methods: a review study, in *Epilepsy - Histological, Electroencephalographic and Psychological Aspects*, ed. by D Stevanovic (InTech Europe, Rijeka, 2012), p. 276
12. W Zhou, Y Liu, Q Yuan, X Li, Epileptic seizure detection using lacunarity and Bayesian linear discriminant analysis in intracranial EEG. IEEE Trans. Biomed. Eng. **60**(12), 3375–3381 (2013)
13. TP Runarsson, S Sigurdsson, On-line detection of patient specific neonatal seizures using support vector machines and half-wave attribute histograms, in *The International Conference on Computational Intelligence for Modelling, Control and Automation, and International Conference on Intelligent Agents, Web Technologies and Internet Commerce (CIMCA-IAWTIC)* (Vienna), pp. 673–677. 28–30 Nov 2005
14. J Yoo, L Yan, D El-Damak, MA Bin Altaf, AH Shoeb, AP Chandrakasan, An 8 channel scalable EEG acquisition SoC with patient-specific seizure classification and recording processor. IEEE J. Solid State Circuits **48**(1), 214–228 (2013)
15. AL Goldberger, LAN Amaral, L Glass, JM Hausdorff, P Ivanov, RG Mark, JE Mietus, GB Moody, C-K Peng, HE HE Stanley, PhysioBank, PhysioToolkit, and PhysioNet: components of a new research resource for complex physiologic signals. Circulation **101**(23), e215–e220 (2000). http://physionet.org/pn6/chbmit/
16. A Dalton, S Patel, AR Chowdhury, M Welsh, T Pang, S Schachter, G Olaighin, P Bonato, Development of a body sensor network to detect motor patterns of epileptic seizures. IEEE Trans. Biomed. Eng. **59**(11), 3204–3211 (2012)
17. AS Zandi, R Tafreshi, M Javidan, GA Dumont, Predicting temporal lobe epileptic seizures based on zero-crossing interval analysis in scalp EEG, in *Proceedings of the 32nd Annual International Conference of the IEEE EMBS* (Buenos Aires), pp. 5537–5540. 31 Aug–4 Sept 2010
18. AS Zandi, R Tafreshi, M Javidan, GA Dumont, Predicting epileptic seizures in scalp EEG based on a variational Bayesian Gaussian mixture model of zero-crossing intervals. IEEE Trans. Biomed. Eng. **60**(5), 1401–1413 (2013)
19. A Aarabi, B He, A rule-based seizure prediction method for focal neocortical epilepsy. Clin. Neurophysiol. **123**(6), 1111–1122 (2012)
20. JC Sprott, G Rowlands, Improved correlation dimension calculation. Int J. Bifurcation. Chaos. **11**(7), 1865–1880 (2001)
21. SJ Gu, CP Sun, HQ Lin, Universal role of correlation entropy in critical phenomena. J. Phys. A Math. Theor. **41**(2), 1–10 (2008)
22. A Doganaksoy, F Gologlu, On the Lempel-Ziv complexity of sequences. Lect. Notes Comput. Sci **4086**, 180–189 (2006)
23. JB Dingwell, *Lyapunov Exponents* (Wiley Encyclopedia of Biomedical Engineering, New York, 2006)
24. *University of Freidberg, Seizure Prediction Project Freidburg* (University of Freiburg, 2014). http://epilepsy.uni-freiburg.de/freiburg-seizure-prediction-project/eeg-database. Accessed 15 Sept 2014
25. B Schelter, HF Drentrup, M Ihle, AS Bonhage, J Timmer, Seizure prediction in epilepsy: from circadian concepts via probabilistic forecasting to statistical evaluation, in *Proceedings of the IEEE International Conference of IEEE EMBS* (Boston, MA, USA), pp. 1624–1627. 30 Aug 30–3 Sept 2011
26. S Wang, WA Chaovalitwongse, S Wong, A novel reinforcement learning framework for online adaptive seizure prediction, in *Proceedings of the IEEE International Conference on Bioinformatics and Biomedicine* (Hong Kong), pp. 499–504. 18–21 Dec 2010
27. M Bedeeuzzaman, T Fathima, YU Khan, O Farooq, Seizure prediction using statistical dispersion measures of intracranial EEG. Biomed. Signal. Process. Control. **10**, 338–341 (2014)
28. S Li, W Zhou, Q Yuan, Y Liu, Seizure prediction using spike rate of intracranial EEG. IEEE Trans. Neural Syst. Rehabil. Eng. **21**(6), 880–886 (2013)
29. C Stamoulis, LJ Gruber, DL Schomer, BS Chang, High-frequency neuronal network modulations encoded in scalp EEG precede the onset of focal seizures. Epilepsy Behav. **23**(4), 471–480 (2012)
30. L Chisci, A Mavino, G Perferi, M Sciandrone, C Anile, G Colicchio, F Fuggetta, Real-time epileptic seizure prediction using AR models and support vector machines. IEEE Trans. Biomed. Eng. **57**(5), 1124–1132 (2010)
31. R Tetzlaff, V Senger, The seizure prediction problem in epilepsy: cellular nonlinear networks. IEEE Circuits Syst. Mag. **12**(4), 8–20 (2012)
32. P Rana, J Lipor, H Lee, WV Drongelen, MH Kohrman, BV Veen, Seizure detection using the phase-slope index and multichannel ECoG. IEEE Trans. Biomed. Eng. **59**(4), 1125–1134 (2012)
33. H Khamis, A Mohamed, S Simpson, Frequency–moment signatures: a method for automated seizure detection from scalp EEG. Clin. Neurophysiol. **124**(12), 2317–2327 (2013)
34. UR Acharya, F Molinari, SV Sree, S Chattopadhyay, KH Ng, JS Suri, Automated diagnosis of epileptic EEG using entropies. Biomed. Signal. Process. Control. **7**(4), 401–408 (2012)
35. A Prochazka, J Uhlir, PJW Rayner, NJ Kingsbury, *Signal Analysis and Prediction* (Birkhauser Inc., Switzerland, 1998)
36. Y Liu, W Zhou, Q Yuan, S Chen, Automatic seizure detection using wavelet transform and SVM in long-term intracranial EEG. IEEE Trans. Neural Syst. Rehabil. Eng. **20**(6), 749–755 (2012)
37. R Panda, PS Khobragade, PD Jambhule, SN Jengthe, PR Pal, TK Gandhi, Classification of EEG signal using wavelet transform and support vector machine for epileptic seizure diction, in *Proceedings of International Conference on Systems in Medicine and Biology* (Kharagpur), pp. 405–408. 16–18 Dec 2010
38. YU Khan, N Rafiuddin, O Farooq, Automated seizure detection in scalp EEG using multiple wavelet scales, in *Proceedings of the IEEE International Conference on Signal Processing, Computing and Control (ISPCC)* (Waknaghat Solan), pp. 1–5. 15–17 March 2012
39. C Wang, C Zhang, J Zou, J Zhang, Performance evaluation for epileptic electroencephalogram (EEG) detection by using Neyman–Pearson criteria and a support vector machine. Physica. A. **391**(4), 1602–1609 (2012)
40. Z Zainuddin, LK Huong, O Pauline, On the use of wavelet neural networks in the task of epileptic seizure detection from electroencephalography signals. Proc Comput. Sci. **11**(2012), 149–159 (2012)
41. RG Andrzejak, EEG time series download page, in (University of Bonn, 2003). http://epileptologie-bonn.de/cms/upload/workgroup/lehnertz/eegdata.html. Accessed

42. M Niknazar, SR Mousavi, BV Vahdat, M Sayyah, M Niknazar, SR Mousavi, BV Vahdat, M Sayyah, A new framework based on recurrence quantification analysis for epileptic seizure detection. IEEE J. Biomed Health Inform. **17**(3), 572–578 (2013)
43. H Daou, F Labeau, Dynamic dictionary for combined eeg compression and seizure detection. IEEE J. Biomed Health Inform. **18**(1), 247–256 (2014)
44. S Mehta, B Onaral, R Koser, Detection of seizure onset using wavelet analysis, in *Proceedings of the IEEE Conference of Engineering in Medicine and Biology Society*, vol. 2 (Baltimore, MD, USA), pp. 1220–1221. 03–06 Nov 1994
45. M Shoaib, KH Lee, NK Jha, N Verma, A 0.6–107 μW energy-scalable processor for directly analyzing compressively-sensed EEG. IEEE Trans. Circuits Syst I. Reg. Papers **61**(4), 1105–1118 (2014)
46. AS Zandi, M Javidan, GA Dumont, R Tafreshi, Automated real-time epileptic seizure detection in scalp EEG recordings using an algorithm based on wavelet packet transform. IEEE Trans. Biomed. Eng. **57**(7), 1639–1651 (2010)
47. JS Paul, CB Patel, H Al-Nashash, N Zhang, WC Ziai, MA Mirski, DL Sherman, Prediction of PTZ-induced seizures using wavelet-based residual entropy of cortical and subcortical field potentials. IEEE Trans. Biomed. Eng. **50**(5), 640–648 (2003)
48. SH Hung, CF Chao, SK Wang, BS Lin, CT Lin, VLSI implementation for Epileptic Seizure Prediction System based on Wavelet and Chaos Theory, in *Proceedings of the IEEE TENCON* (Fukuoka), pp. 364–368. 21–24 Nov 2010
49. CY Chiang, NF Chang, TC Chen, HH Chen, LG Chen, Seizure prediction based on classification of EEG synchronization patterns with on-line retraining and post-processing scheme, in *Proceedings of the International Conference of the IEEE EMBS* (Boston, MA, USA), pp. 7564–7569. 30 Aug–3 Sept 2011
50. PW Mirowski, Y LeCun, D Madhavan, R Kuzniecky, Comparing SVM and convolutional networks for epileptic seizure prediction from intracranial EEG, in *IEEE Workshop on Machine Learning for Signal Processing* (Cancun), pp. 244–249. 16–19 Oct 2008
51. National Taiwan University Medical Library, 2013. http://ntuml.mc.ntu.edu.tw/Index.action. Accessed 15 Sept 2014
52. CA Rojas, M Valderrama, A Witon, V Navarro, MLV Quyen, Probing cortical excitability using cross-frequency coupling in intracranial EEG recordings: a new method for seizure prediction, in *Proceedings of the International Conference of the IEEE EMBS* (Boston, MA, USA), pp. 1632–1635. 30 Aug–3 Sept 2011
53. Epilepsiae, in, 2014. http://www.epilepsiae.eu/project_outputs/european_database_on_epilepsy. Accessed 15 Sept 2014
54. K Gadhoumi, JM Lina, J Gotman, Seizure prediction in patients with mesial temporal lobe epilepsy using EEG measures of state similarity. Clin. Neurophysiol. **124**(9), 1745–1754 (2013)
55. K Gadhoumi, JM Lina, J Gotman, Discriminating preictal and interictal states in patients with temporal lobe epilepsy using wavelet analysis of intracerebral EEG. Clin. Neurophysiol. **123**(10), 1906–1916 (2012)
56. S Wang, WA Chaovalitwongse, S Wong, Online seizure prediction using an adaptive learning approach. IEEE Trans. Knowl. Data Eng. **25**(12), 2854–2866 (2013)
57. HB Soleimani, C Lucas, BN Araabi, L Schwabe, Adaptive prediction of epileptic seizures from intracranial recordings. Biomed Signal. Process. Control. **7**(5), 456–464 (2012)
58. RP Costa, P Oliveira, G Rodrigues, B Direito, A Dourado, Epileptic seizure classification using neural networks with 14 features. Lect Notes Comput Sci **5178**, 281–288 (2008)
59. N Moghim, D Corne, Evaluating bio-inspired approaches for advance prediction of epileptic seizures, in *Proceedings of the Word Congress on Nature and Biologically Inspired Computing (NaBIC)* (Salamanca), pp. 540–545. 19–21 Oct 2011
60. A Eftekhar, F Vohra, C Toumazou, EM Drakakis, K Parker, Hilbert-Huang transform: preliminary studies in epilepsy and cardiac arrhythmias, in *Proceedings of the IEEE Biomedical Circuits and Systems Conference, BioCAS* (Baltimore, MD, USA), pp. 373–376. 20–22 Nov 2008
61. AK Tafreshi, AM Nasrabadi, AH Omidvarnia, Epileptic seizure detection using empirical mode decomposition, in *Proceedings of the IEEE International Symposium on Signal Processing and Information Technology, ISSPIT* (Sarajevo), pp. 238–242. 16–19 Dec 2008
62. L Orosco, E Laciar, AG Correa, A Torres, JP Graffigna, An epileptic seizures detection algorithm based on the empirical mode decomposition of EEG, in *Proceedings of the International Conference of the IEEE EMBS* (Minneapolis, MN, USA), pp. 2651–2654. 3–6 Sept 2009
63. C Guarnizo, E Delgado, EEG single-channel seizure recognition using empirical mode decomposition and normalized mutual information, in *Proceedings of the IEEE International Conference on Signal Processing (ICSP)* (Beijing), pp. 1–4. 24–28 Oct 2010
64. P Maragos, J Kaiser, T Quatieri, On separating amplitude from frequency modulations using energy operators, in *Proceedings of IEEE International Conference on Acoustics, Speech, and Signal Processing, ICASSP-92*, vol. 2 (San Francisco, CA, USA), pp. 1–4. 23–26 Mar 1992
65. HM Teager, Some observations on oral air flow during phonation. IEEE Trans Acoust. Speech. Signal. Process. **28**(5), 599–601 (1980)
66. HM Teager, SM Teager, Evidence for nonlinear sound production mechanisms in the vocal tract, in *Speech Production and Speech Modelling*, ed. by WJ Hardcastle, A Marchal (Springer, Amsterdam, 1990), pp. 241–261
67. SMS Alam, MIH Bhuiyan, Detection of epileptic seizures using chaotic and statistical features in the EMD domain, in *Proceedings of the India Conference (INDICON)* (Hyderabad), pp. 1–4. 16–18 Dec 2011
68. SMS Alam, MIH Bhuiyan, Detection of seizure and epilepsy using higher order statistics in the EMD domain. IEEE J. Biomed Health Inform. **17**(2), 312–318 (2013)
69. V Bajaj, RB Pachori, Epileptic seizure detection based on the instantaneous area of analytic intrinsic mode functions of EEG signals. Biomed. Eng. Lett. **3**(1), 17–21 (2013)
70. NU Rehman, Y Xia, DP Mandic, Application of multivariate empirical mode decomposition for seizure detection in EEG signals, in *Proceedings of the International Conference of the IEEE EMBS* (Buenos Aires), pp. 1650–1653. 31 Aug–4 Sept 2010
71. Y Qi, Y Wang, X Zheng, J Zhang, J Zhu, J Guo, Efficient epileptic seizure detection by a combined IMF-VoE feature, in *Proceedings of the International Conference of the IEEE EMBS* (San Diego, CA, USA), pp. 5170–5173. 28 Aug–1 Sept 2012
72. JM Shieh, DC Lou, MC Chang, A semi-blind watermarking scheme based on singular value decomposition. Comp. Stand. Inter. **28**(4), 428–440 (2006)
73. FE Abd El-Samie, *Information Security for Automatic Speaker Identification*, 1st edn. (Springer, Berlin, 2011)
74. B Vanrumste, RD Jones, PJ Bones, Detection of focal epileptiform activity in the EEG: an SVD and dipole model approach, in *Proceedings of the Second IEEE Joint EMBS/BMES Conference*, vol. 3 (Houston, TX, USA), pp. 2031–2032. 23–26 Oct 2002
75. GV Hoey, B Vanrumste, R Van de Walle, P Boon, I Lemahieu, MB D'Hav, K Vonck, Detection and localization of epileptic brain activity using an artificial neural network for dipole source analysis, in *Proceedings of EUSIPCO* (Tampere, Finland, 2000), pp. 4–8
76. A Shahid, N Kamel, AS Malik, MA Jatoi, Epileptic Seizure Detection using the singular values of EEG signals, in *Proceedings of the IEEE Conference on Complex Medical Engineering* (Beijing), pp. 652–655. 25–28 May 2013
77. P Comon, Independent component analysis: a new concept. Signal Process. **36**(3), 287–314 (1994)
78. YH Shih, TJ Chen, CH Yang, H Chiueh, Hardware-efficient EVD processor architecture in FastICA for epileptic seizure detection, in *Proceedings of the Signal & Information Processing Association Annual Summit and Conference (APSIPA ASC)* (Hollywood, CA, USA), pp. 1–4. 3–6 Dec 2012
79. S Xie, S Krishnan, Signal decomposition by multi-scale PCA and its applications to long-term EEG signal classification, in *Proceedings of the IEEE International Conference on Complex Medical Engineering* (Harbin Heilongjiang), pp. 532–537. 22–25 May 2011
80. MR Miri, AM Nasrabadi, A new seizure prediction method based on return map, in *Proceedings of the Iranian Conference on BioMedical Engineering* (Tehran), pp. 244–248. 14–16 Dec 2011
81. JR Williamson, DW Bliss, DW Browne, JT Narayanan, Seizure prediction using EEG spatiotemporal correlation structure. Epilepsy Behav. **25**(2), 230–238 (2012)

2

Soft context clustering for F0 modeling in HMM-based speech synthesis

Soheil Khorram[1*], Hossein Sameti[1] and Simon King[2]

Abstract

This paper proposes the use of a new binary decision tree, which we call a *soft decision tree*, to improve generalization performance compared to the conventional 'hard' decision tree method that is used to cluster context-dependent model parameters in statistical parametric speech synthesis. We apply the method to improve the modeling of fundamental frequency, which is an important factor in synthesizing natural-sounding high-quality speech. Conventionally, hard decision tree-clustered *hidden Markov models (HMMs)* are used, in which each model parameter is assigned to a single leaf node. However, this 'divide-and-conquer' approach leads to data sparsity, with the consequence that it suffers from poor generalization, meaning that it is unable to accurately predict parameters for models of unseen contexts: the hard decision tree is a weak function approximator. To alleviate this, we propose the soft decision tree, which is a binary decision tree with soft decisions at the internal nodes. In this soft clustering method, internal nodes select both their children with certain membership degrees; therefore, each node can be viewed as a fuzzy set with a context-dependent membership function. The soft decision tree improves model generalization and provides a superior function approximator because it is able to assign each context to several overlapped leaves. In order to use such a soft decision tree to predict the parameters of the HMM output probability distribution, we derive the smoothest (maximum entropy) distribution which captures all partial first-order moments and a global second-order moment of the training samples. Employing such a soft decision tree architecture with maximum entropy distributions, a novel speech synthesis system is trained using *maximum likelihood (ML)* parameter re-estimation and synthesis is achieved via maximum output probability parameter generation. In addition, a soft decision tree construction algorithm optimizing a log-likelihood measure is developed. Both subjective and objective evaluations were conducted and indicate a considerable improvement over the conventional method.

Keywords: Context clustering; Decision tree-based clustering; F0 modeling; Hidden Markov model; HMM; HMM-based speech synthesis; Maximum entropy model; Soft decision tree; Soft context clustering; Statistical parametric speech synthesis

1 Introduction

Demand for natural and high-quality speech-based human-computer interaction is increasing due to applications including speech-based virtual assistants for mobile devices. Speech synthesis plays a significant role, not only in transmitting factual information, but also as the outward 'face' of the application: the naturalness of the synthesis affects overall user satisfaction. Speech synthesis from text is usually achieved via an intermediate linguistic specification [1], which can be thought of as a collection of contextual factors - such as phonetic and prosodic properties of the current, preceding, and following segment - which have been derived from the text. Here, we are concerned only with the conversion of this linguistic specification to a speech waveform. In order to perform this conversion, several methods have been proposed [2], of which *statistical parametric speech synthesis (SPSS)* [3,4] has been dominant, at least in research terms, for the last decade or more.

Figure 1 portrays the overall architecture of a typical SPSS system, which comprises two distinct phases [3,4], namely, *training* and *synthesis*. The training phase starts with the extraction of acoustic features and the linguistic specification (i.e., contextual factors) for all training

* Correspondence: khorram@ce.sharif.edu
[1]Department of Computer Engineering, Sharif University of Technology, Azadi Avenue, Tehran 14588, Iran
Full list of author information is available at the end of the article

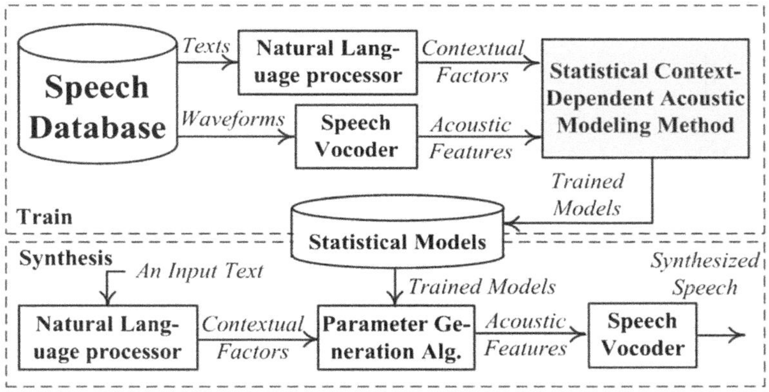

Figure 1 Overall block diagram of a typical SPSS.

utterances. Waveforms are converted to a compact set of acoustic features using a vocoder (such as *MELP* [5], *STRAIGHT* [6], *DSM* [7,8], or *HNM* [9]), and simultaneously, all texts are expanded into contextual factors using a natural language pre-processing front end [10]. Thereafter, the training phase proceeds to the context-dependent statistical modeling step in which the dependencies between extracted acoustic features and contextual factors are modeled through context-dependent statistical models [11]. It is important to note that, because of the extreme sparsity in the contextual feature space, the decision tree (DT) used to cluster the model parameters is the critical component in the statistical modeling.

In the synthesis phase, contextual factors are obtained for the input text, and the decision tree is used to obtain the corresponding trained model parameters, using which a *parameter generation (PG)* algorithm [11-14] generates acoustic feature trajectories. These are then converted to a speech waveform using the vocoder.

In contrast to concatenative synthesis [15], which stores speech waveforms, the parametric representation in SPSS has several potential advantages, including flexibility in changing voice characteristics [3], speaker and style adaptation [16-19], easier multilingual support [20-22], superior coverage of acoustic space [3], reduced memory footprint [3], and better robustness to low-quality speech recordings [23].

Though compressing a human voice into a compact statistical model offers the abovementioned advantages over concatenation of waveforms, there remains one major shortcoming: lower quality synthetic speech. This is caused by one or more of the blocks shown in Figure 1, e.g., inadequate acoustic coverage of training utterances, noisy speech database [24,25], errors in natural language processing, inadequate contextual factors [26], inaccurate statistical modeling [3], the PG algorithm [11-14], or vocoding distortion [5-9]. Here, we propose improvements to the statistical modeling (the shaded block in Figure 1) and specifically for the F0 stream.

Hidden Markov model (HMM)-based speech synthesis [27-33] models not only the spectrum, but also the excitation and duration in a unified framework of *context-dependent* [34,35] *multi-space probability distribution* [36] *hidden semi-Markov models (HSMMs)* [37]. More precisely, an independent binary-branching hard decision tree is constructed for each stream of acoustic features (spectrum, aperiodic energy, and fundamental frequency). In the case of F0, a multi-space probability distribution [36] is associated with each leaf of the decision tree. Contextual space (which is very large and very sparse due the great number of contextual factors typically employed) is divided by the decision trees into multiple hard (i.e., non-overlapping) clusters; each cluster is a group of context-dependent HMM states that share the same output probability distribution.

Hard decision tree-based context clustering, which is the standard approach to F0 modeling, has *poor generalization* [38]. In other words, this structure cannot accurately predict the parameters of models of unseen contexts, given the very limited subset of contexts observed in the training data. In order to predict acoustic variations with high generalization capability, the model has to be able to express a large number of robust distributions (i.e., a large number of distributions, but such that each one can be trained from a sufficient number of training samples). In hard decision trees, increasing the number of distributions by growing the depth of the tree reduces the number of training samples assigned to each leaf, and thus, the robustness of the distributions is weakened. This problem stems from the fact that the hard decision tree structure assigns each model parameter to exactly one cluster (corresponding to a small part of the large contextual space): each training sample contributes to the estimation of only one set of model parameters (one mean vector and one covariance matrix). Our hypothesis is that by enabling each training sample to influence multiple sets of model parameters (and thus cover a larger portion of contextual space), generalization to unseen contexts would be improved.

1.1 Related work

Several attempts have already been made to alleviate the limitations of F0 modeling in standard decision tree-clustered H(S)MMs. One of these is the use of *deep neural networks (DNNs)* [38,39] which are able to approximate complex acoustic feature-to-linguistic context dependencies by employing many hidden layers - contrast this with decision trees that cannot efficiently represent something as simple as XOR or multiplexing [38], i.e., they must use an excessive number of leaves to capture such relationships and thus over-fragment the already sparse training data [38]. DNNs are also able to represent non-binary contextual features, whereas decision trees generally only use binary splits. Other deep learning approaches such as *restricted Boltzmann machines (RBMs)* and *deep belief networks (DBNs)* have also been demonstrated to be effective generative models when applied to speech synthesis [40,41].

Speech synthesis based on *Gaussian process regression (GPR)* [42] is another new technique that has recently been introduced to alleviate basic limitations of HMM-based speech synthesis. The goal of GPR is to remove the incorrect stationarity assumption of state output distribution in HMM-based speech synthesis. GPR uses frame-level contextual factors - such as position of the frame within the phone, and articulatory features - to estimate frame-level acoustic trajectories [42]. GPR can directly express complex context dependencies without needing decision tree structures and is able to use all contextual factors of all types simultaneously; therefore, it has the potential to provide better generalization.

In [43], a new system is proposed that replaces the usual *maximum likelihood (ML)* point estimate of the model parameters with a *variational Bayesian* method. Their system outperforms the usual approach when the amount of training data is limited, thus demonstrating superior generalization.

F0 modeling with *additive* structures has also been used to express the relationship between contextual factors and the F0 trajectory [44-54]. *Contextual additive* modeling [45-48] assumes model parameters to be a sum of multiple independent components, each having different context dependencies; therefore, different decision trees have to be trained for them. The contextual additive model is able to exploit contextual factors more efficiently, because mean vectors and covariance matrices of the predicted distributions are the sum of mean vectors and covariance matrices of the additive components [45]: each training sample contributes to more than one model parameter. Takaki et al. [46-48] used an additive structure for spectral modeling and reported that it has a high computational cost. To alleviate this, they proposed *covariance parameter tying* and a simplified likelihood calculation algorithm using the *matrix inversion lemma*. Though the contextual additive model [45-48] was originally proposed for spectral modeling, it could be used for F0 trajectories.

Zen et al. [49] also developed an additive F0 modeling structure with multiple components for mean vectors and a single component for variance values. Accordingly, multiple decision trees were trained for the mean vectors, and just one decision tree was built for the variance values. In their system, different sets of contextual factors were used for different additive components and all trees were built concurrently. Similarly, [50] proposed another additive structure with multiple decision trees, but a *minimum generation error (MGE)* measure was used as the selection criterion instead of the more common *maximum likelihood (ML)* measure. In [51], an additive model with three different layers, including intonational phrase, word level, and pitch accent, was designed. All three components were trained concurrently using a *regularized least square error* measure. Qian et al. [52] proposed to use a new gradient-based tree-boosting approach with a view to training multiple additive regression trees. Their decision trees were built in successive stages to minimize the squared error.

Some studies [53,54] have also highlighted another important problem of the common decision tree-based F0 modeling: its *deficiency in capturing the effect of contextual features that are poorly represented in the training database*. These features (i.e., questions used in the decision tree splits) have little influence on the likelihood criterion and hence will not be selected by the usual decision tree construction algorithms [53,54]. One obvious technique to solve this problem is to build the decision tree using a two-stage algorithm [53]. In the first stage, all splits are made only with these under-represented contextual factors. This stage captures the influence of such factors, even though they are rare. In the second stage, the well-represented factors are employed. This procedure is not efficient, since the first stage reduces the amount of the training data available for modeling the dependency between well-represented contextual factors and F0 [54]. Context adaptive training with factorized decision trees [54] is another method designed to exploit rare features more effectively. There, cluster adaptive training [55] is employed such that an average model is built and then this general model is adapted using a set of transforms. In fact, well-represented contextual factors contribute to generate the average model, and rare contextual questions are taken into account for the transforms. Due to the use of cluster adaptive training, this structure also is able to improve context generalization.

1.2 Scope of the paper

Numerous binary and non-binary contextual factors are generally taken into account in modeling F0. Conventional HMM-based speech synthesis converts all non-binary contextual factors to multiple binary questions (i.e., potential decision tree splits). As mentioned earlier, this structure may suffer from inadequate context generalization. To alleviate this deficiency, we propose the direct use of non-binary contextual factors in a soft decision tree framework [56,57]. The proposed soft decision tree structure is an innovative binary decision tree with soft questions at each non-terminal node. Both children are selected with a specific membership degree. In contrast to a hard decision tree that partitions contextual factor space into hard contextual regions, the proposed soft decision tree is able to provide soft - i.e., overlapping - clusters. In this structure, each context will be assigned to several terminal leaves with certain membership functions, and consequently, each training sample affects multiple model parameters, and generalization should be improved.

The rest of the paper is organized as follows: Section 2 presents the classical hard decision tree approach to F0 modeling in statistical parametric speech synthesis. The proposed soft context-clustered HMM structure and details of the associated speech synthesis system that employs such trees are provided in Section 3. Section 4 reports the experiments and results, and Section 5 concludes the paper.

2 F0 modeling using hard decision trees

This section describes the predominant framework for F0 modeling in HMM-based speech synthesis, which is the same framework used for the spectral envelope, aperiodic energy, and duration. This section also sets out the notation, algorithms, and structures required for subsequent sections.

2.1 F0 modeling in the HMM framework

Typically, F0 along with its delta and delta-delta derivatives form three streams[a] of a *context-dependent* [34,35] *multi-space probability distribution (MSD)* [36] *left-to-right without skip transitions HSMM* [58,37] (which for obvious reasons, we shorten to simply 'HMM' in this paper). This model structure generates acoustic trajectories of a unit (e.g., phoneme) by emitting observations from hidden states. The output distribution of the state is a context-dependent multi-space Gaussian distribution [36], and these are clustered into groups of related contexts using a decision tree in order to reduce the number of free parameters and allow the modeling of unseen contexts. For notational simplicity, we limit our discussion here to an HMM with just one stream. Generalizing this to the multi-stream case is straightforward.

Figure 2 shows the equivalent dynamic Bayesian network (DBN) for such an HMM [59]. In this figure, q_t, o_t, and g_t respectively represent the state index, the acoustic feature vector, and the space index [36] in time t.

When using MSD output distributions with two spaces - for defined and undefined values - the space index is an observed value equal to the voicing label. The figure also introduces c_j, d_j, and t_j which are the contextual factors, the duration, and the last frame index of the jth state (clearly, $d_j = t_j - t_{j-1}$). Note that state boundaries are latent variables and have to be trained in an unsupervised manner using the expectation maximization (EM) [60] algorithm.

According to this figure, the HMM is simply specified through three sets of fundamental distributions: i) state duration probability distribution ($p_j(d_j|c_j)$), ii) voicing (space) probability distribution ($w_j(g_t|c_j)$), and iii) output probability distribution given voicing labels ($b_j(o_t|g_t,c_j)$). Using these fundamental distributions and considering the graphical model represented by Figure 2, the likelihood of a given utterance with observations (o, g, c) can be factorized as

$$p(o,g|c;\lambda) = \sum_{t_1,t_2,\ldots,t_J} \prod_{j=1}^{J} p_j(d_j|c_j) \prod_{j=1}^{J} w_j(g_t|c_j) b_j(o_t|g_t,c_j), \tag{1}$$

where J and λ denote the total number of states and the model parameters, respectively.

Now, assume g_t takes two values: '1' for voiced frames and '0' for unvoiced regions; also, let b_j and p_j be expressed through Gaussian distributions. Therefore, the above utterance likelihood can be rewritten as

$$p(o,g|c;\lambda) = \sum_{(t_1,t_2,\ldots,t_J)} \prod_{j=1}^{J} \mathcal{N}\left(d_j;\bar{m}_j,\bar{\sigma}_j^2\right) \prod_{t=t_{j-1}}^{t_j} \left[g_t \bar{w}_j \mathcal{N}\left(o_j;\bar{\mu}_j,\bar{\Sigma}_j\right) + (1-g_t)(1-\bar{w}_j)\right], \tag{2}$$

where $\mathcal{N}(.;\mu,\Sigma)$ represents a Gaussian distribution with mean vector μ, and covariance matrix Σ. In this equation, duration and output distributions are parameterized by duration mean $(\bar{m}_j)$, duration variance $\left(\bar{\sigma}_j^2\right)$, voicing probability $(\bar{w}_j)$, output mean vector $\left(\bar{\mu}_j\right)$, and observation covariance matrix $(\bar{\Sigma}_j)$.

As previously mentioned, a canonical decision tree structure is used to express the fundamental distributions. Assume $I_l^d(c_j)$ and $I_l^o(c_j)$ are defined as binary indicator functions of decision trees trained for duration and output distributions where l and c_j are, respectively, the leaf index and the contextual factors extracted for the jth state. In other words, $I_l^d(c_j)$ and $I_l^o(c_j)$ determine whether the jth state is assigned to the lth leaf of the duration and observation decision trees or not. Using these decision tree indicator functions, the HMM parameters can be expressed by

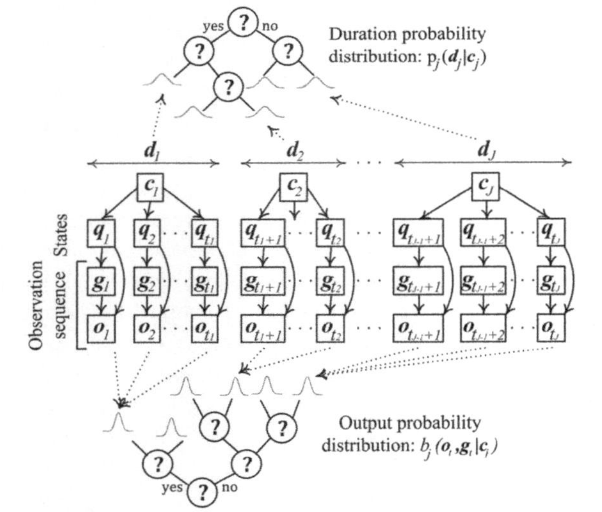

Figure 2 Graphical model of the common HMM. Rectangular nodes and solid arrows demonstrate random variables and their dependencies. Dotted arrows show the tying structure through decision trees.

$$\begin{aligned}&\bar{m}_j = \sum\nolimits_l I_l^d(c_j) m_l,\ \ \bar{\sigma}_j^2 = \sum\nolimits_l I_l^d(c_j)\sigma_l^2,\\ &\bar{w}_j = \sum\nolimits_l I_l^o(c_j) w_l,\ \ \bar{\mu}_j = \sum\nolimits_l I_l^o(c_j)\mu_l,\ \ \bar{\Sigma}_j = \sum\nolimits_l I_l^o(c_j)\Sigma_l,\end{aligned} \tag{3}$$

where m_l and σ_l^2 are duration mean and variance values lying in the lth leaf of the duration decision tree. Similarly, w_l, μ_l, and Σ_l represent parameters of the voicing and output probability distributions that are trained for the lth leaf of the output decision tree.

2.2 HMM parameter re-estimation

The ML criterion is commonly used to estimate model parameters of HMM. However, state boundaries are latent, and therefore, the EM algorithm has to be adopted. Given N i.i.d. utterances, $\{(o^n,\ g^n)\}_{n=1}^N$, along with their corresponding contextual factors, $\{c^n\}_{n=1}^N$, the EM algorithm leads to the following re-estimation formulas:

$$\begin{aligned}
\hat{m}_l &= \frac{\sum_{n=1}^N \sum_{j=1}^{J^n} I_l^d\left(c_j^n\right) \sum_{t_j,t_{j-1}} \chi_j^n(t_j, t_{j-1})\left[t_j - t_{j-1}\right]}{\sum_{n=1}^N \sum_{j=1}^{J^n} I_l^d\left(c_j^n\right) \sum_{t_j,t_{j-1}} \chi_j^n(t_j, t_{j-1})},\\
\hat{\sigma}_l^2 &= \frac{\sum_{n=1}^N \sum_{j=1}^{J^n} I_l^d\left(c_j^n\right) \sum_{t_j,t_{j-1}} \chi_j^n(t_j, t_{j-1})\left[t_j - t_{j-1} - \hat{m}_l\right]^2}{\sum_{n=1}^N \sum_{j=1}^{J^n} I_l^d\left(c_j^n\right) \sum_{t_j,t_{j-1}} \chi_j^n(t_j, t_{j-1})},\\
\hat{\mu}_l &= \frac{\sum_{n=1}^N \sum_{j=1}^{J^n} I_l^o\left(c_j^n\right) \sum_t \gamma_j^n(t) g_t^n\left[o_t^n\right]}{\sum_{n=1}^N \sum_{j=1}^{J^n} I_l^o\left(c_j^n\right) \sum_t \gamma_j^n(t) g_t^n},\\
\hat{\Sigma}_l &= \frac{\sum_{n=1}^N \sum_{j=1}^{J^n} I_l^o\left(c_j^n\right) \sum_t \gamma_j^n(t) g_t^n\left[(o_t^n - \hat{\mu}_l)(o_t^n - \hat{\mu}_l)^{\mathrm{T}}\right]}{\sum_{n=1}^N \sum_{j=1}^{J^n} I_l^o\left(c_j^n\right) \sum_t \gamma_j^n(t) g_t^n},\\
\hat{w}_l &= \frac{\sum_{n=1}^N \sum_{j=1}^{J^n} I_l^o\left(c_j^n\right) \sum_t \gamma_j^n(t) g_t^n}{\sum_{n=1}^N \sum_{j=1}^{J^n} I_l^o\left(c_j^n\right) \sum_t \gamma_j^n(t)},
\end{aligned} \tag{4}$$

where $\hat{m}_l$, $\hat{\sigma}_l^2$, $\hat{\mu}_l$, $\hat{\Sigma}_l$, and $\hat{w}_l$ are new values of m_l, σ_l^2, μ_l, Σ, and w_l during EM algorithm. Also, $\chi_j(t_j, t_{j-1})$ is the probability of occupying the jth state from time t_{j-1} to t_j, and $\gamma_j(t)$ denotes the posterior probability of being in state j at time t. These probabilities are calculated through the well-known *forward-backward* algorithms. It should be noted that the publically available *HMM-based speech synthesis system (HTS)* [61] has been implemented based on the algorithms expressed in [62]. These algorithms were originally proposed by Ferguson [63] and were refined by Levinson [64]. A more efficient version of the forward-backward algorithm has recently been proposed by Yu et al. [65].

2.3 Decision tree-based state clustering

In order to capture the context dependencies inherent in the acoustic features, canonical decision trees are typically incorporated in the HMM framework. Decision trees are constructed iteratively through a greedy and top-down procedure which maximizes the log-likelihood criterion [34,35]. The procedure starts with a single root node representing all contexts. In each iteration, an optimum pair of terminal node and question is selected so that splitting the terminal node by the selected question results in the largest log-likelihood increase. The splitting procedure is continued until a termination criterion (such as *minimum description length (MDL)* [66]) is satisfied. The overall log-likelihood increase $\delta\mathcal{L}$, achieved by splitting a parent node l_1 into two children l_2 and l_3, is simply obtained by the following equation [34]:

$$\begin{aligned}\delta\mathcal{L} &= \frac{1}{2}\log\left(\left|\hat{\Sigma}_{l_1}\right|\right)\sum\nolimits_{n=1}^N \sum\nolimits_{j=1}^{J^n} I_{l1}^o\left(c_j^n\right)\sum\nolimits_t \gamma_j^n(t) -\\ &\sum\nolimits_{l\in\{l_2,l_3\}} \frac{1}{2}\log\left(\left|\hat{\Sigma}_l\right|\right)\sum\nolimits_{n=1}^N \sum\nolimits_{j=1}^{J^n} I_l^o\left(c_j^n\right)\sum\nolimits_t \gamma_j^n(t),\end{aligned} \tag{5}$$

where superscript n is an index defined for the number of training utterances. It should be noted that in order to obtain the above likelihood increase expression, the following simplifying assumptions have to be made [34]: 1 - The values of occupation probabilities are assumed to be fixed during the clustering procedure [34]. 2 - The overall likelihood measure is supposed to be approximated by a simple average of the log likelihoods weighted by the posterior probabilities [34]. These assumptions make the calculation of $\delta\mathcal{L}$ possible for all pairs of terminal nodes and questions.

3 Soft context-clustered HMM

Generally, decision tree is the term for a hierarchical structure consisting of internal nodes and terminal leaves. In a canonical hard binary decision tree, used for acoustic modeling, each terminal node carries a distribution that captures the statistical characteristics of a context cluster. Also, for a given context c, each internal node m applies a binary test $f_m(c)$ and chooses one of its children based on the result of the test. Let $I_m(c)$ be the indicator function defined for the node m. Also, assume that $I_{m_L}(c)$ and $I_{m_R}(c)$ represent the indicator functions lying on its left and right children; $I_{m_L}(c)$ and $I_{m_R}(c)$ are obtained by

$$I_{m_L}(c) \stackrel{\text{def}}{=} \begin{cases} I_m(c) & \text{if } f_m(c) = \text{true} \\ 0 & \text{if } f_m(c) = \text{false} \end{cases},$$
$$I_{m_R}(c) \stackrel{\text{def}}{=} \begin{cases} 0 & \text{if } f_m(c) = \text{true} \\ I_m(c) & \text{if } f_m(c) = \text{false} \end{cases}. \tag{6}$$

Accordingly, to determine the distribution of a given contextual factor, we need to start from the root node and recursively apply the test at each internal node and select one of the two branches depending on the outcome. This process is repeated iteratively until a leaf node is hit at which point the distribution of the leaf is considered as the output probability distribution. Therefore, for each context, just one path from the root to a terminal node is always traversed, and each context is hereby assigned to one leaf and affects the distribution of that single leaf. In order to improve the performance of the canonical decision tree, this paper proposes the soft binary decision tree structure which is able to establish several fuzzy paths from the root to multiple leaves.

3.1 Soft context-clustered HMM structure

The soft decision tree applies soft decisions $\tilde{f}_m(c)$ in its internal nodes m and redirects all samples to both children, but with certain membership degrees computed by $\tilde{f}_m(c)$ and $1\text{-}\tilde{f}_m(c)$. In fact, each node of a soft decision tree represents a fuzzy subset of contextual factor space; therefore, each context belongs to several nodes with a membership degree. More precisely, when we are traversing the node m for the given context c, a soft question $\tilde{f}_m(c)$ represents the membership grade of the left child, and clearly, $1\text{-}\tilde{f}_m(c)$ computes the degree of selecting the right child.

In both hard and soft decision tree-based HMMs, initially, a set of contextual factors have to be defined and extracted for all training utterances. Thereafter, as opposed to the hard decision tree that requires hard questions $f_m(c)$, here, we have to design a great number of soft questions (soft tests) $\tilde{f}_m(c)$ for each contextual factor. These questions are finally assigned to the internal nodes of the decision tree and make fuzzy decisions to select among their children instead of the common crisp decisions.

As it is realized from the above discussion, all terminal leaves may be active for an arbitrary context; as a consequence, it is necessary to generalize the indicator function $I_m(c)$ expressed by Equation 3 to the membership function of assigning context c to the node m. This membership function is denoted by $\tilde{I}_m(c)$ and can be computed through the following recursion:

$$\text{Initialization}: \; \tilde{I}_{\text{root}}(c) = 1,$$
$$\text{Recursion}: \begin{Bmatrix} \tilde{I}_{m_L}(c) = \tilde{f}_m(c)\tilde{I}_m(c) \\ \tilde{I}_{m_R}(c) = \left(1\text{-}\tilde{f}_m(c)\right)\tilde{I}_m(c) \end{Bmatrix} \tag{7}$$

where m_L and m_R are the left and right children of node m. According to the above recursion, all the membership degrees can be calculated by traversing the tree in a preorder style. The traversing procedure starts with setting the membership degree of the root to 1. After observing a node m and determining its membership degree $\tilde{I}_m(j)$, its left m_L and right m_R children are observed. If the node is a left child, its membership degree is calculated through $\tilde{f}_m(c)\tilde{I}_m(c)$; otherwise, the procedure returns $\left(1\text{-}\tilde{f}_m(c)\right)\tilde{I}_m(c)$, in which m is the parent node.

In the training phase, soft decisions $\tilde{f}_m(c)$ are selected from a set of predefined contextual functions. These functions must hold the following limitation for all contextual factors:

$$\forall \; m, \; c, \; 0 \leq \tilde{f}_m(c) \leq 1. \tag{8}$$

The above constraint has to be taken into account during the procedure of defining soft questions. That is, we are not allowed to employ soft questions with a value greater than 1 or less than 0; thus, a normalization step is required for some questions before starting decision tree-based clustering.

3.2 Soft context-clustered HMM distribution

The proposed soft context-clustered HMM exploits the same structure and graphical model as the original hard decision tree-based HMM, and thus, the model likelihood expression given by Equation 1 is also valid for the proposed model. The only difference between the conventional and the proposed approaches lies in the method of capturing context dependencies inherent in the F0 trajectory. More specifically, the method of representing output distribution $b_j(\cdot)$ in Equation 1 is different. The goal of this section is to find this probability distribution for the soft decision tree structure described in the previous section. With a view to providing an efficient context generalization, this section derives the smoothest distribution that is able to accurately express the behavior of the F0 trajectory. To estimate the smoothest distribution, the *maximum entropy model (MEM)* [67,68], presented in the next subsection, is employed.

3.2.1 Maximum entropy-based distributions

Our task is to estimate the distribution of the observation vectors. The maximum entropy principle states that an efficient estimate is the one that maximizes entropy (uncertainty) subject to our knowledge about the observation vectors. This knowledge normally appears in the form of some constraints that make the distribution consistent with sufficient statistics of the observation vectors [67]. Let us now derive a simple maximum entropy model for the output distribution given voicing labels, $b_j(o_t|g_t, c_j)$. Suppose the training utterances consist of T i.i.d. voicing labels $\{g_t\}_{t=1}^{T}$ and D-dimensional output feature vectors $\{o_t\}_{t=1}^{T}$ that may be influenced by some contextual information $\{c_t\}_{t=1}^{T}$. Also, the contextual information is clustered through a soft decision tree structure with the total number of L leaves partitioning the contextual factor space through the membership functions $\{\tilde{I}_l(.)\}_{l=1}^{L}$. The maximum entropy principle first imposes a set of constraints on the distribution and then chooses a distribution as close as possible to a uniform distribution by optimizing the entropy criterion [67]. Indeed, this modeling scheme finds the least biased distribution among all distributions that satisfy our constraints. In other words,

$$b(o|g,c) \stackrel{\text{def}}{=} \arg\max_b \mathcal{H}\{b(o|g,c)\} \;\; S.T. \;\; \text{constraints}. \tag{9}$$

where $\mathcal{H}$ is the entropy measure which is defined by

$$\mathcal{H}\{b(o|g,c)\} \stackrel{\text{def}}{=} -\sum\nolimits_{t=1}^{T} \int_o b(o|g_t, c_t) \log\; b(o|g_t, c_t) do. \tag{10}$$

The constraints play a crucially important role in maximum entropy modeling. They ensure that the model captures the statistical characteristics of the training samples. In this paper, the following constraints are taken into account:

$$b(o|g,c) \stackrel{\text{def}}{=} \arg\max_b \mathcal{H}\{b(o|g,c)\} \;\; S.T. \left\{ \begin{array}{c} \forall\; c,g \;\; \int_o b(o|g,c) = 1 \\ E\{goo^T\} = \bar{E}\{goo^T\} \\ \forall 1 \le l \le L \;\; E\{\tilde{I}_l(c)go\} = \bar{E}\{\tilde{I}_l(c)go\} \end{array} \right\} \tag{11}$$

The first constraint ensures that the distributions sum to 1. Also, E and $\bar{E}$ indicate real and empirical mathematical expectations given by the following equations:

$$\begin{aligned} E\{\tilde{I}_l(c)go\} &= \sum\nolimits_{t=1}^{T} \tilde{I}_l(c_t) g_t \int_o o b(o|g_t, c_t) do, \\ \bar{E}\{\tilde{I}_l(c)go\} &= \sum\nolimits_{t=1}^{T} \tilde{I}_l(c_t) g_t o_t, \\ E\{goo^T\} &= \sum\nolimits_{t=1}^{T} g_t \int_o oo^T b(o|g_t, c_t) do, \\ \bar{E}\{goo^T\} &= \sum\nolimits_{t=1}^{T} g_t o_t o_t^T \end{aligned} \tag{12}$$

These constraints make the estimated distribution capture the partial first-order moments $E\{\tilde{I}_l(c)go\}$ and the global second-order moment $E\{goo^T\}$ of the training data in voiced frames (i.e., in frames where observation features o_t are defined and voicing label g_t is 1); therefore, the training phase of the maximum entropy model estimates the smoothest distribution that preserves the first- and second-order moments, expressed in Equation 9, of the training database. Moreover, the selected constraints lead to a simple expression for the output probability distributions that can be estimated efficiently.

In order to solve optimization problems with equality constraints, the Lagrange multipliers method can be applied. This method defines a new optimization function as follows:

$$\begin{aligned} & b(o|g,c) \stackrel{\text{def}}{=} \arg\max_b \mathcal{J}(b), \\ & \mathcal{J}(b) = \mathcal{H}\{b(o|g,c)\} + \lambda_{b0}\left[\int_o b(o|g,c)do - 1\right] + \\ & [E\{go^T \Lambda o\} - \bar{E}\{go^T \Lambda o\}] + \sum\nolimits_{l=1}^{L} \lambda_{bl}^T [E\{\tilde{I}_l(c)go\} \\ & \qquad - \bar{E}\{\tilde{I}_l(c)go\}] \end{aligned} \tag{13}$$

where $\mathcal{J}(b)$ represents the new optimization function; Also, λ_{b0}, λ_{b1}, and Λ are Lagrange multiplayers incorporated in the optimization function to remove the equality constraints.

Taking the derivatives of the optimization function $\mathcal{J}(b)$ with respect to the output probability distribution $b(o|g,c)$, and setting it to zero leads to the following equation:

$$\frac{\partial \mathcal{J}(b)}{\partial b} = \sum\nolimits_{t=1}^{T} \int_o [-\log b(o|g_t, c_t) - 1 + \lambda_{b0} + g_t o^T \Lambda o\;. + \sum\nolimits_{l=1}^{L} \lambda_{bl}^T \tilde{I}_l(c_t) g_t o] do = 0$$

An obvious solution satisfying the above equality is

$$\log\; b\;(o|g_t,\; c_t) = g_t o^T \Lambda o + \sum\nolimits_{l=1}^{L} \lambda_{bl}^T \tilde{I}_l(c_t) g_t o + \lambda_{b0} - 1.$$

Therefore, $b(o|g_t,c_t)$ is a simple Gaussian distribution that can be expressed by

$$b(o|g,c) = \mathcal{N}\left(o; \sum_{l=1}^{L} \tilde{I}_l(c)\mu_l, \Sigma\right) \tag{14}$$

where $\mathcal{N}$ indicates the Gaussian distribution; μ_l is a D-dimensional vector of mean parameters defined for the lth leaf; Also, Σ is a D-by-D covariance matrix that is used for all leaves.

In sum, each leaf of the soft decision tree carries a set of model parameters represented by μ_l that contributes to express the output probability distribution $b(o|g,c)$. The output probability $b(o|g,c)$ is simply approximated by a Gaussian distribution. This Gaussian distribution uses a unique context-independent covariance matrix Σ and a context-dependent mean vector. The mean component is obtained by linearly combining μ_l parameters (i.e., $\sum_{l=1}^{L} \tilde{I}_l(c)\mu_l$) and the weights of the linear combination are determined by the membership functions $\tilde{I}_l(c)$. In fact, the proposed maximum entropy-based output probability distribution is remarkably similar to the distribution expressed by the contextual additive structure that ties all covariance matrixes [46-48]. In the contextual additive method, similar to the proposed method, the output distribution has the form of Equation 14, but the contextual additive method exploits multiple hard decision trees [46] or a hard decision tree with overlapped leaves [47] instead of the proposed soft decision tree. In other words, in contextual additive structure, $\tilde{I}_l(c)$ indicates a leaf indicator function that may be 1 for multiple overlapped leaves, but in the proposed model, $\tilde{I}_l(c)$ is a real number, ranging from 0 to 1, that represents the membership degrees of a soft decision tree terminal node.

3.3 Parameter re-estimation

Having described the soft context-clustered HMM structure, it is now time to discuss its parameter re-estimation procedure. In the training phase, we are given a set of N i.i.d. training utterances containing acoustic features $\{o^n\}_{n=1}^{N}$, voicing labels $\{g^n\}_{n=1}^{N}$, and contextual factors $\{c^n\}_{n=1}^{N}$. The goal is to find the optimum set of model parameters $\hat{\lambda}$ which maximizes the log-likelihood measure:

$$\begin{aligned} \hat{\lambda} &\stackrel{\text{def}}{=} \operatorname{argmax}_{\lambda} \mathcal{L}(\lambda), \\ \mathcal{L}(\lambda) &\stackrel{\text{def}}{=} \sum_{n=1}^{N} \ln\, p(o^n, g^n | c^n; \lambda). \end{aligned} \tag{15}$$

This section assumes that the soft decision tree structure has been trained earlier and we just try to find the maximum log-likelihood estimate of its parameters λ, including $\{\mu_l\}_{l=1}^{L}$ and Σ. Training the optimum soft decision tree structure will be described in the next section. Similar to the classical HMM, the likelihood expression of Equation 1 leads to an extremely complex optimization problem with seemingly impossible direct solution. The main problem is that the distribution depends on the state boundaries which are latent. The EM technique offers an iterative algorithm which is able to overcome this problem. According to the EM technique, $\hat{\lambda}$ is obtained by iteratively maximizing an axillary function $\mathcal{Q}(\lambda;\lambda^r)$:

$$\begin{aligned} \lambda^{r+1} &= \operatorname{argmax}_{\lambda} \mathcal{Q}\ (\lambda;\ \lambda^r) \\ \mathcal{Q}(\lambda;\lambda^r) &= \sum_n \Big[\sum_{t_{j-1},t_j} \chi_j^n(t_j, t_{j-1}; \lambda^r) \log\, p_j(t_j - t_{j-1} | c_j^n) + \\ &\quad \sum_t \sum_j \gamma_j^n(t;\lambda^r) \Big\{ \log\, w_j(g_t^n | c_j^n) \\ &\quad + \log\, b_j(o_t^n | g_t^n, c_j^n) \Big\} \Big], \end{aligned} \tag{16}$$

where χ_j and γ_j are occupation probabilities defined in Section 2.2. Also, r is the index of the EM iterations, and n ranges over the utterance numbers. In order to estimate the optimum set of parameters, the partial derivatives of $\mathcal{Q}$ with respect to all model parameters λ have to be set to zero. These partial derivatives are calculated by considering the distribution introduced in Section 2.2 as follows:

$$\begin{aligned} \frac{\partial \mathcal{Q}(\lambda;\lambda^r)}{\partial \mu_\ell} &= \Sigma^{-1} \sum_n \sum_t \sum_j \gamma_j^n(t;\lambda^r) \\ &\quad \tilde{I}_l\left(c_j^n\right)\left(o_t^n - \sum_{l=1}^{L} \tilde{I}_l\left(c_j^n\right)\mu_l\right), \\ \frac{\partial \mathcal{Q}(\lambda;\lambda^r)}{\partial \Sigma} &= \frac{1}{2}\Sigma^{-1} \sum_n \sum_t \sum_j \gamma_j^n(t;\lambda^r) \\ &\quad \left\{ -1 + \left(o_t^n - \sum_{l=1}^{L} \tilde{I}_l\left(c_j^n\right)\mu_l\right)^T \right. \\ &\quad \left. \left(o_t^n - \sum_{l=1}^{L} \tilde{I}_l\left(c_j^n\right)\mu_l\right)\Sigma^{-1} \right\} \end{aligned} \tag{17}$$

By setting these equations to zero, the maximum likelihood estimate of model parameters is obtained. According to these equations, the optimum vectors for mean parameters $\{\hat{\mu}_l\}_{l=1}^{L}$ can be simply calculated through solving the following system of equations:

$$R\hat{\mu} = P, \tag{18}$$

where $\hat{\mu}$ is a L-by-D matrix containing all mean parameters as

$$\hat{\mu} = [\hat{\mu}_1,\ \hat{\mu}_2, \ldots \hat{\mu}_L]^T. \tag{19}$$

Also, R and P are L-by-L and L-by-D matrixes defined by

$$R = [r_{uv}]_{L\times L},$$
$$r_{uv} = \sum_n \sum_j \tilde{I}_u\left(c_j^n\right)\tilde{I}_v\left(c_j^n\right)\sum_t \gamma_j^n(t;\lambda^r), \quad (20)$$
$$P = [p_u]_{L\times L},\ p_u = \sum_n \sum_j \tilde{I}_u\left(c_j^n\right)\sum_t o_t^n \gamma_j^n(t;\lambda^r).$$

As it is realized from Equation 20, R represents the cross-correlation matrix of membership functions. This matrix is symmetric and positive definite; therefore, it is possible to solve the above system of equations efficiently using Cholesky decomposition.

Furthermore, by setting zero the partial derivatives of the auxiliary function $\mathcal{Q}(\lambda;\ \lambda^r)$ with respect to the globally tied covariance matrix Σ, the maximum likelihood estimate of Σ is calculated as follows:

$$\hat{\Sigma} = \frac{\sum_n \sum_t \sum_j \gamma_j^n(t;\lambda^r)\left(o_t^n - \sum_{l=1}^{L}\tilde{I}_l\left(c_j^n\right)\mu_l\right)^T\left(o_t^n - \sum_{l=1}^{L}\tilde{I}_l\left(c_j^n\right)\mu_l\right)}{\sum_n \sum_t \sum_j \gamma_j^n(t;\lambda^r)}. \quad (21)$$

The above equations introduce a straightforward procedure to train the parameters of the output probability distribution factorized by a soft decision tree.

The next section discusses the procedure of constructing the proposed soft decision tree. In order to conduct a soft decision tree clustering algorithm, it is required to calculate the log-likelihood measure for the optimum model parameters. This optimum log-likelihood measure is expressed by

$$\mathcal{L} \propto -\frac{1}{2}\ \log\left(|\hat{\Sigma}|\right)\sum_n \sum_t \sum_j \gamma_j^n(t) \quad (22)$$

where $|.|$ denotes the matrix determinant operator.

3.4 Soft context clustering algorithm

To automatically capture the dependencies between acoustic features and contextual factors, this section proposes a soft decision tree construction algorithm. Similar to the classical hard decision tree building algorithm, the soft decision tree is built iteratively through a greedy and top-down procedure which maximizes the log-likelihood measure.

The major advantage of the classical hard decision tree construction algorithm is that its terminal nodes can be split independently. In hard decision tree, terminal nodes represent non-overlapped regions of the contextual factor space; therefore, after splitting a leaf, all values obtained for other leaves are still valid, and it is not required to calculate them once again. This advantage causes the algorithm to be computationally tractable. However, in the soft decision tree construction procedure, the different terminal nodes may cover overlapped regions of the contextual space and splitting a leaf using a soft question affects the parameters of all other leaves. Consequently, as opposed to the conventional hard decision tree structure, here, after splitting a leaf, it is required to update all values obtained for all terminal nodes, and it needs tremendous amount of computations.

The procedure of the proposed soft decision tree construction algorithm is stated as follows:

Step 1. Create the root node embarrassing all samples of the training database.

Step 2. Split all terminal nodes using all questions and compute their optimum log-likelihood value. To compute the optimum log-likelihood value for each possible pair of leaf and question, the maximum likelihood estimate of mean parameters $\hat{\mu}$ has to be first obtained by Equation 18. Then, $\hat{\Sigma}$ is calculated through Equation 21, and Equation 22 is finally employed to find the optimum log-likelihood value.

Step 3. Select the best pair of terminal node and question that provides the maximum increase in log-likelihood measure. Thereafter, split the node using the question and estimate the maximum likelihood estimate of all model parameters.

Step 4. Stop the splitting procedure, if a predefined condition is satisfied (e.g., the increase in log-likelihood falls below a certain threshold).

Algorithm 1 summarizes the overall procedure of the proposed soft context clustering. As it is realized from the explained clustering algorithm, the proposed soft clustering procedure is dramatically similar to the classical clustering algorithm. Their main difference is in the number of evaluations that has to be performed during each iteration of the clustering procedure. In hard clustering, both newly generated leaves are just required to be evaluated, but in the soft clustering, all leaf nodes have to be evaluated. This fact increases the computational complexity of the soft clustering by an order of magnitude. Assume we intend to build a decision tree with L leaves. Also, we have defined Q questions. In this case, hard clustering requires $(2L-3)Q$ likelihood calculations to be performed, while soft clustering will be finished after $[L(L-1)/2]Q$ likelihood calculations.

It should be noted that the likelihood calculation in soft decision tree-based clustering is more complicated than the hard clustering; it is mainly due to the fact that calculating the inverse of the matrix R to solve the system of equations expressed by Equation 18 is computationally intractable. Takaki et al. [46] proposed a method to reduce the computational complexity of calculating this inverse problem. Their method exploits the matrix inversion lemma and can also be incorporated in the soft decision tree clustering procedure.

Algorithm 1 Proposed soft clustering algorithm

Inputs:

- c_j^n: Contextual factors of training utterances, $\forall 1 \leq n \leq N$ and $\forall 1 \leq j \leq J^n$, where n and j are utterance and state indexes.
- o_t^n : Observation features of training utterances, $\forall 1 \leq n \leq N$ and $\forall 1 \leq t \leq T^n$, where t denotes the frame index.
- $\gamma_j^n(t)$: State occupation probabilities, $\forall 1 \leq n \leq N$, $\forall 1 \leq j \leq J^n$, and $\forall 1 \leq t \leq T^n$.
- $\tilde{f}_m(c)$: predefined soft decisions, $\forall 1 \leq m \leq M$.

Outputs:

- L: Total number of decision tree terminal nodes.
- $\{\tilde{I}_l(c)\}_{l=1}^{L}$: Context-dependent membership functions of soft decision tree terminal nodes.

Soft context clustering algorithm:

Initialize $L=1$ and $f_1(c)=1$;

While the stopping criterion is not satisfied

1) For $l=1,2\ldots L$ (for each leaf)

For $m=1,2\ldots M$ (for each soft decision)

1') $\forall j,n \quad \tilde{I}_{L+1}(c_j^n)=(1-\tilde{f}_m(c_j^n))\tilde{I}_l(c_j^n)$;

2') $\forall j,n \quad \tilde{I}_l(c_j^n)=\tilde{f}_m(c_j^n)\tilde{I}_l(c_j^n)$;

3') Compute $R$and P according to Equation 20;

4') Compute $\hat{\mu}$ according to Equation 18;

5') Compute $\hat{\Sigma}$ according to Equation 21;

6') Compute $\mathcal{L}(l,m)$ according to Equation 22;

2) $\hat{l}\,,\hat{m}=\arg\max_{l,m}\mathcal{L}(l,m)$;

3) $\forall j,n \quad \tilde{I}_{L+1}(c_j^n)=\tilde{f}_{\hat{m}}(c_j^n)\tilde{I}_{\hat{l}}(c_j^n)$;

4) $\forall j,n \quad \tilde{I}_{\hat{l}}(c_j^n)=\tilde{f}_{\hat{m}}(c_j^n)\tilde{I}_{\hat{l}}(c_j^n)$;

3.5 Simple sinusoidal regression

In order to clarify the soft clustering advantages, a simple sinusoidal regression problem is solved using both soft and hard decision tree structures in this section. Assume we have just one continuous contextual factor named c ranging from 0 to 1, and our goal is to approximate the following sinusoidal function:

$$o(c) = \sin(3\pi c - \pi) \times \sin(\pi c) + 0.05 \times r(c) \tag{23}$$

where $o(c)$ represents the observation value for a given context c, and $r(c)$ is a normally distributed random noise with zero mean and unit variance. The 200 training samples shown in Figure 3a are independently drawn from Equation 23. Nineteen different contextual questions are defined to train the hard decision tree as follows:

$$\forall i \in \{1, 2, \ldots 19\} q_i(c) = \begin{cases} 1 c < i/20 \\ 0 c \geq i/20 \end{cases} \tag{24}$$

Therefore, each internal node of the hard decision tree structure has to select one of these hard

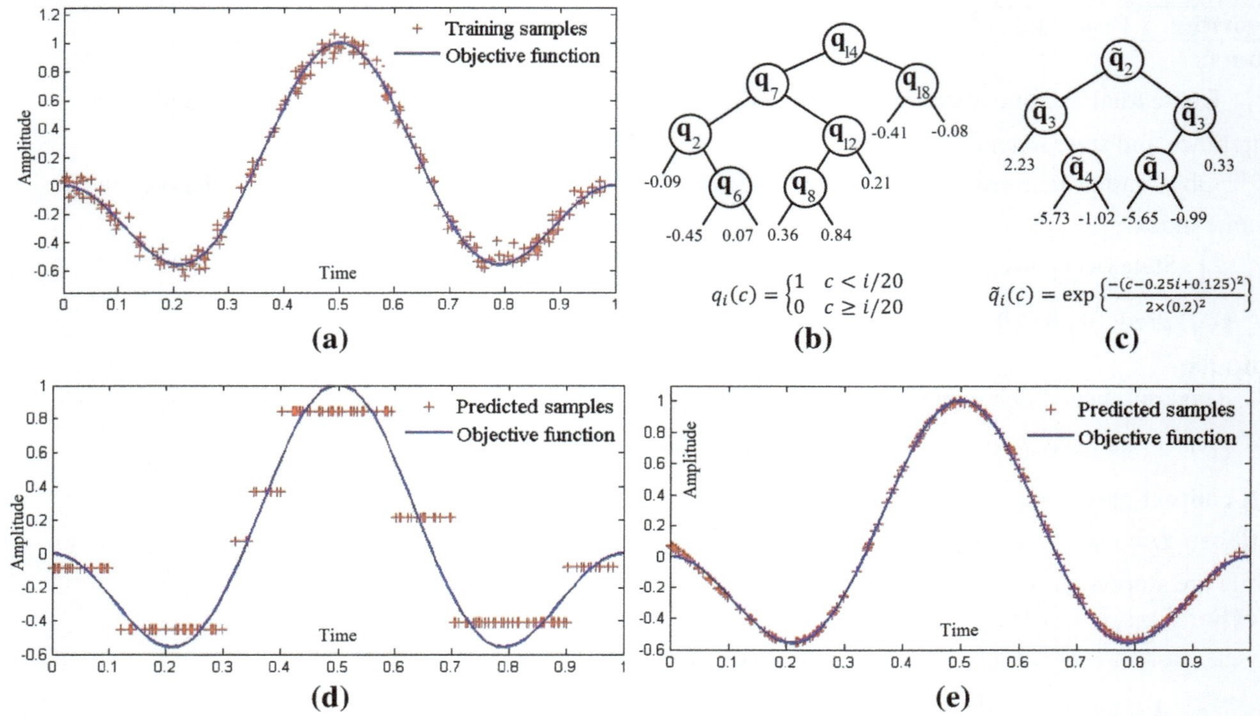

Figure 3 Samples and decision tree structures. **(a)** Data samples used to train the decision trees. The red dots and blue line are training samples and objective function, respectively. **(b)** Trained hard decision tree structure. **(c)** Trained soft decision tree structure. **(d)** Predicted samples using the hard decision tree. **(e)** Predicted samples using the soft decision tree.

questions denoted by $\{f_i(c)\}_{i=1}^{19}$. Additionally, the soft decision tree is trained by exploiting four distinct soft questions defined by

$$\forall i \in \{1,2,3,4\}\ \tilde{q}_i(c) = \exp\left\{\frac{-(c-0.25i+0.125)^2}{2\times(0.2)^2}\right\} \tag{25}$$

Figure 3b,c shows the hard and soft decision tree structures trained based on the maximum likelihood decision tree construction algorithms. As can be seen from the figures, the hard decision tree requires eight leaves to have an acceptable mean square error of 0.015, but the soft decision tree is able to accurately estimate the objective function with a small mean square error of 0.0002 using just six terminal nodes.

Figure 3d,e shows the approximated functions using hard and soft decision trees, respectively. As an obvious consequence of this simple experiment, the hard decision tree structures are not efficient to exploit the continuous attributes (contextual factors), and incorporating the soft decisions in their internal nodes significantly improves their predictive capabilities.

3.6 Defining soft questions

As it was mentioned earlier, in order to construct the soft decision tree structure, a set of basic contextual factors has to be extracted initially for all training and test datasets. Section 4.1.1 gives the details of the basic contextual factors employed in our experiments. These basic contextual factors have been denoted by c in this paper and can be grouped into two types of factors: categorical and numerical factors. 'Phoneme identity' is a sample of categorical factors, and 'Position of the current phoneme' is an example of the numerical factors. In fact, a numerical factor returns some ordered values, but a categorical factor provides some unordered symbols. For the categorical factors, we cannot define meaningful soft questions, and therefore, we have no choices but to exploit the conventional hard questions. However, for the numerical factors, it is possible to define a large number of soft questions. This subsection introduces the procedure of defining these soft questions in our experiments.

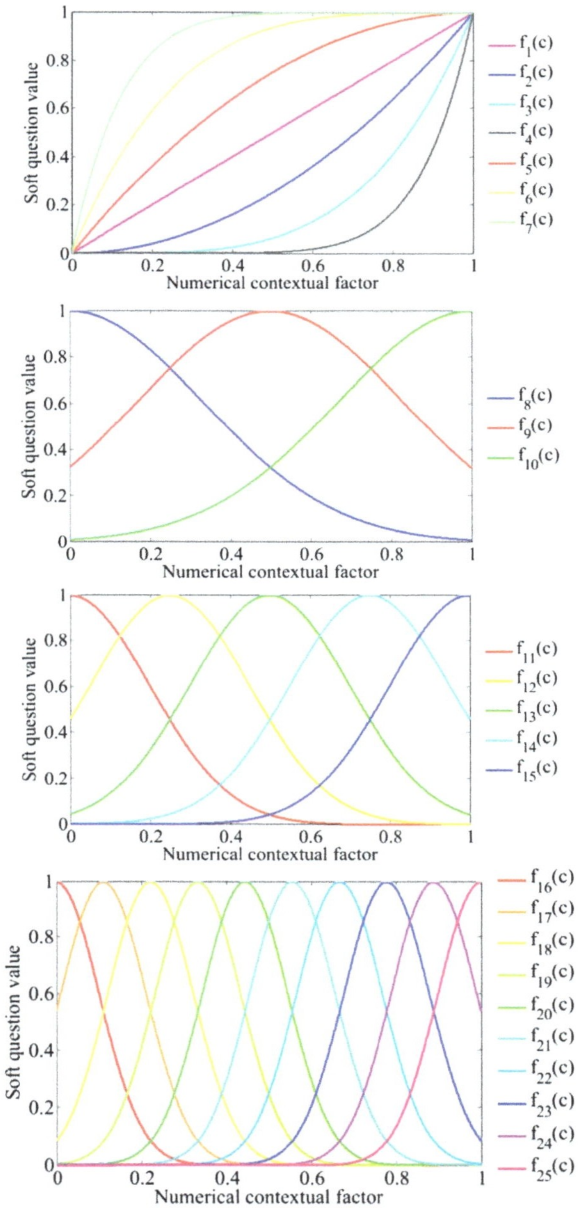

Figure 4 Soft questions defined for each numerical contextual factor in our experiments.

In this study, we first normalize all numerical contextual factors to range between 0 and 1, and then soft questions are obtained by applying a fixed set of candidate functions to the normalized contextual factors. Assume $\tilde{c}$ represents a normalized numerical contextual factor and $\tilde{f}_k(\tilde{c})$ is the kth soft question extracted for $\tilde{c}$. In this study, 25 soft questions have been defined for each numerical contextual factor. These soft questions are shown in Figure 4, and their mathematical expressions are given by Equation 26:

$$\begin{array}{lll}
\tilde{f}_1(\tilde{c}) = \tilde{c}, & \tilde{f}_2(\tilde{c}) = \tilde{c}^2, & \tilde{f}_3(\tilde{c}) = \tilde{c}^4, \\
\tilde{f}_4(\tilde{c}) = \tilde{c}^8, & \tilde{f}_5(\tilde{c}) = 1-(1-\tilde{c})^2, & \tilde{f}_6(\tilde{c}) = 1-(1-\tilde{c})^4 \\
\tilde{f}_7(\tilde{c}) = 1-(1-\tilde{c})^8 & \tilde{f}_8(\tilde{c}) = G_{0,\frac{1}{3}}(\tilde{c}), & \tilde{f}_9(\tilde{c}) = G_{0.5,\frac{1}{3}}(\tilde{c}), \\
\tilde{f}_{10}(\tilde{c}) = G_{1,\frac{1}{3}}(\tilde{c}), & \tilde{f}_{11}(\tilde{c}) = G_{0,\,0.2}(\tilde{c}), & \tilde{f}_{12}(\tilde{c}) = G_{\frac{1}{4},0.2}(\tilde{c}), \\
\tilde{f}_{13}(\tilde{c}) = G_{\frac{2}{4},0.2}(\tilde{c}), & \tilde{f}_{14}(\tilde{c}) = G_{\frac{3}{4},0.2}(\tilde{c}), & \tilde{f}_{15}(\tilde{c}) = G_{1,0.2}(\tilde{c}), \\
\tilde{f}_{16}(\tilde{c}) = G_{0,0.2}(\tilde{c}), & \tilde{f}_{17}(\tilde{c}) = G_{\frac{1}{9},0.2}(\tilde{c}), & \tilde{f}_{18}(\tilde{c}) = G_{\frac{2}{9},0.1}(\tilde{c}), \\
\tilde{f}_{19}(\tilde{c}) = G_{\frac{3}{9},0.1}(\tilde{c}), & \tilde{f}_{20}(\tilde{c}) = G_{\frac{4}{9},0.2}(\tilde{c}), & \tilde{f}_{21}(\tilde{c}) = G_{\frac{5}{9},0.1}(\tilde{c}), \\
\tilde{f}_{22}(\tilde{c}) = G_{\frac{6}{9},0.1}(\tilde{c}), & \tilde{f}_{23}(\tilde{c}) = G_{\frac{7}{9},0.1}(\tilde{c}), & \tilde{f}_{24}(\tilde{c}) = G_{\frac{8}{9},0.1}(\tilde{c}), \\
\tilde{f}_{25}(\tilde{c}) = G_{1,0.2}(\tilde{c}), & \text{where } G_{\mu,\sigma}(\tilde{c}) = & \exp\left(-\frac{1}{2}\left(\frac{\tilde{c}-\mu}{\sigma}\right)^2\right).
\end{array} \tag{26}$$

In conclusion, all contextual factors were divided into two groups, namely, categorical and numerical. According to the above procedure, a set of soft questions were extracted for numerical factors, and a number of hard questions were obtained for categorical contextual factors. Thereafter, all of the extracted hard and soft questions were grouped together and competed against each other during the soft clustering procedure.

4 Experiments

This section aims to compare the performance of fundamental frequency modeling approaches based on the conventional hard decision tree and the proposed soft clustering method.

4.1 Experimental conditions

Before presenting the experimental results, this section describes the experimental conditions, including database characteristics and employed contextual factors, in detail.

An English speech database called Nick [69] consisting of approximately 2,500 utterances from a British male speaker was used in our experiments. This database is collected in Edinburgh University for the purpose of speech synthesis research. Sentences range in length from 3 to 36 words with an average length of 7.3 words per sentence. Also, the sentences cover most frequent English words, bi-phoneme combinations, and syllables. Totally, 2,944 different words are covered in the sentences.

Speech waveforms were sampled at 48 kHz, windowed by a 25-ms Blackman window with 5-ms shift. The speech analysis and synthesis conditions expressed in CSTR/EMIME HTS 2010 [69] were used in this experiment. In this platform, Bark-cepstrum was extracted from smooth STRAIGHT trajectories [6], since it outperforms predominant Mel-cepstrum coefficients. Also, the widely used log-F0 and five aperiodicity sub-bands (0 to 1, 1 to 2, 2 to 4, 4 to 6, and 6 to 8 kHz) were replaced with pitch in Mel and auditory-scale motivated frequency bands for aperiodicity measure [69]. The analysis process generated 40 bark cepstrum coefficients, 1 Mel in pitch value, and 25 auditory-scale motivated aperiodicity frequency sub-bands for each frame of training signals. These parameters along with their delta and delta-delta derivatives formed five streams of our observation vectors.

For the baseline system, a five-state multi-stream left-to-right without skip path MSD-HSMM was trained. A conventional maximum likelihood-based decision tree construction algorithm was used to tie HMM states. In the conventional HMM-based speech synthesis framework, a unique tying structure (decision tree) is normally incorporated for both voicing probabilities and F0 output probabilities. As opposed to the conventional HMM-based synthesis system, the proposed method uses a soft decision tree structure for the output probability distribution and a hard decision tree for voicing probabilities; therefore, we cannot apply the same tying structure for both voicing and output probabilities in the proposed system. With a view to having a fair comparison, the baseline system was implemented with two different decision trees for F0 trajectories, one for the voicing labels and the other for the output probability distributions.

The same structure with just one different part was also implemented for the proposed synthesis system. In the proposed system, the soft decision tree structure is trained for F0 and its derivatives output probability distributions instead of the hard decision tree. All other decision trees, including the decision trees trained for state duration, Bark-cepstrum, aperiodicity, and voicing probability, are completely equal to the ones trained for the baseline system. Therefore, all parameters generated for them are equal to the parameters generated for the baseline system. It should be noted that both baseline and proposed synthesis systems employ the MDL criterion [66] to determine the size of all decision trees.

We considered four sets including 100, 200, 400, and 800 utterances for training, and 400 sentences that were not included in the training sets were used as a test data.

4.1.1 Employed contextual factors

Specific information about the contextual factors is presented in this subsection. Employed contextual factors can be categorized into five levels, including phonetic, syllable, word, phrase, and sentence levels. In each of these levels, all important features were considered.

➢ Phonetic-level factors
- Phoneme identity before the preceding phoneme, preceding, current, succeeding phoneme, and phoneme identity after the next phoneme
- Position of the current phoneme in the current syllable, word, phrase, and sentence

➢ Syllable-level factors
- Stress level of previous, current, and next syllable (three different stress levels are defined for this database)
- Position of the current syllable in the current word, phrase, and sentence
- Number of the phonemes of the previous, current, and next syllable
- Whether the previous, current, and next syllable is accented or not
- Number of the stressed syllables before and after the current syllable in the current phrase
- Number of syllables from the previous stressed syllable to the current syllable
- Number of syllables from the previous accented syllable to the current syllable

➢ Word-level factors
- Part of speech (POS) tag of the preceding, current, and succeeding word.
- Position of the current word in the current phrase and sentence (forward and backward)
- Number of syllables of the previous, current, and next word
- Number of content words before and after current word in the current phrase
- Number of words from previous and next content word

➢ Phrase-level factors
- Number of syllables and words of the preceding, current, and succeeding phrase
- Position of the current phrase in the sentence.
- Current phrase ToBI end tone.

➢ Sentence-level factors
- Number of phonemes, syllables, words, and phrases in the current utterance
- Type of the current sentence

4.2 Experimental results

Both objective and subjective tests are conducted to evaluate the proposed F0 modeling method. The results of these tests are given in the following subsections.

4.2.1 Objective evaluation

Figure 5 shows the learning curves obtained during building the hard and soft decision trees for 800 training utterances and 400 test sentences. Normalized log-likelihood measure, depicted in this figure, was computed through the following expression:

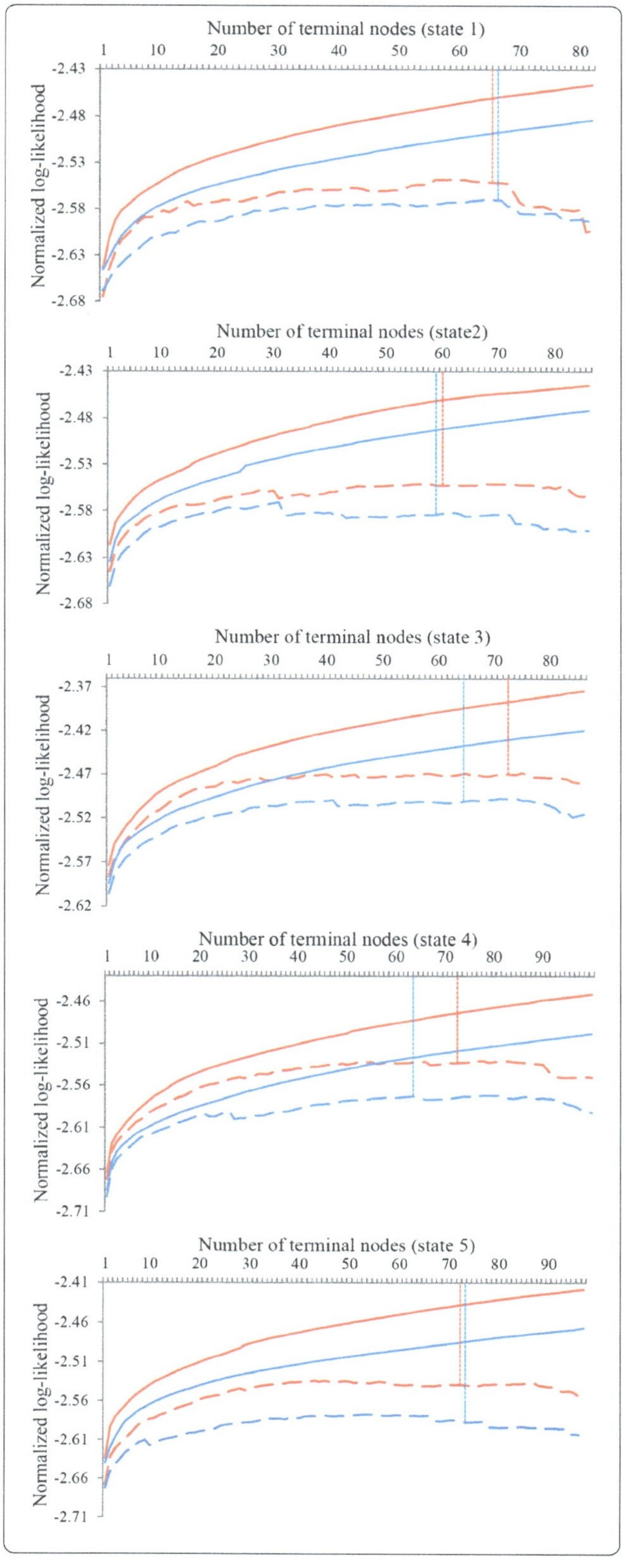

Figure 5 Normalized log-likelihood with respect to the number of leaves computed for each state of HMM. Blue and red curves are the learning curves of the hard and soft decision trees, respectively. In addition, solid curves illustrate the log-likelihood of the training set, and dashed curves are the log-likelihood computed for test data. MDL-based stop points are also shown through vertical dotted lines.

$$\mathcal{L} = \frac{1}{\sum_t \sum_l g_{tl}} \sum_t \sum_l g_{tl} \log\ b(o_{tl}|g_{tl}), \tag{27}$$

where the F0 and its derivatives are represented by o_{tl}, and their voicing labels are denoted by g_{tl}. t is the frame index and l represents the dynamic or static features ranging from 1 to 3. In this figure, the above measure is depicted for both test and train data. Red and blue curves are related to the proposed soft and the conventional hard decision tree structures, respectively. Solid curves are the normalized log-likelihood measure of the training sets, and the dashed curves represent the normalized log-likelihood measure computed for the test data. Also, the optimum number of terminal leaves calculated by the MDL principle is illustrated through vertical dotted lines. As it is realized from Figure 5, all red curves surpass their corresponding blue curves; therefore, the soft decision tree is able to provide superior log-likelihood measure with a smaller number of model parameters. All learning curves confirm the fact that the soft decision tree structure is able to provide better generalization in contrast to the canonical hard decision tree structure.

Another well-known objective measure, reported in this section, is the *root-mean-square error (RMSE)* between synthesized and natural log-F0 trajectories. In order to compute this measure, first, all test utterances were synthesized with natural voicing labels and natural durations (durations obtained through applying the Viterbi algorithm to natural acoustic trajectories). Thereafter, the RMSE measure is computed through the following expression:

$$\mathrm{RMSE} = \sqrt{\frac{1}{\sum_t g_t} \sum_t g_t \left(f_t^P - f_t^T\right)^2}, \tag{28}$$

where g_t, f_t^T, and f_t^P are voicing label, target log-F0 value, and predicted log-F0 value of the tth frame. This measure is computed for four training datasets including 100, 200, 400, and 800 training utterances. Figure 6 shows the calculated RMSE values in terms of cent. As it is realized from this figure, the log-F0 trajectories generated from the proposed approach is more similar to the natural log-F0 trajectories, and therefore, the proposed soft decision tree structure improves the performance of log-F0 modeling. However, by increasing the size of database,

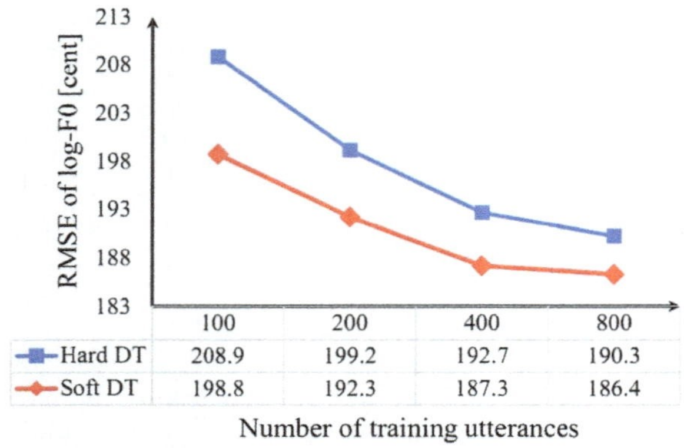

Figure 6 RMSE as an objective measure to compare log-F0 trajectories generated by hard and soft decision trees. Blue and red lines are the results of hard and soft decision trees, respectively.

the amount of this improvement is slightly reduced. Hence, it can be implied that the effect of applying soft clustering for small databases is relatively more than its effect on large databases.

4.2.2 Subjective evaluation

Two subjective tests have been selected in order to assess the effectiveness of the proposed system in comparison with the conventional synthesis system. The *comparative mean opinion score (CMOS)* test [7] with a 7-point scale, ranging from −3 to 3, and the paired comparison test [70] have been used to evaluate the subjective similarity of the synthesized and the natural utterances. Eighteen evaluators participated in our subjective evaluations, and each of them was asked to listen to 20 randomly chosen pairs of synthesized waveforms generated by two different synthesizers (i.e., the soft decision tree-based system and the conventional system).

In paired comparison tests, listeners are presented with a number of pairs of waveforms and they are asked to identify which one is more similar to its corresponding natural speech signal. If the two utterances sound equal, listeners are allowed to choose the equality option. The paired comparison test simply reports the percentage of comparisons that a certain synthesizer outperforms the other.

In CMOS tests, listeners not only select the better utterance, but also determine the difference level between two utterances. Four levels are normally defined for this purpose (namely, 0, 1, 2, and 3 which respectively have the meaning of about the same, slightly different, different, and much different). These difference levels are mainly useful in computing CMOS scores which have to be calculated in each comparison for each synthesizer separately. More precisely, a positive score equal to the difference level is computed for the winner of the comparison, and a negative score with equivalent absolute is assigned to the loser. Finally, the value of the CMOS score is obtained by taking an average over all scores.

The results of CMOS and paired comparison evaluations are respectively shown in Figures 7 and 8. Remarkably, the proposed soft context-clustered HMM is noticed to outperform the conventional hard decision tree structure for all training utterances. This result is completely in line with the conclusion of the objective assessments. For small datasets (i.e., 100 and 200 training utterances), more than 58% of the comparisons are in favor of the proposed method and the average CMOS score of the proposed system is more than one unit higher than the baseline system. These results show that the proposed soft decision tree structure is able to improve the F0 estimation accuracy of the baseline system significantly in small training datasets, and therefore, an important application of the proposed system is in low-resource languages when limited amount of data is available for training.

Another considerable conclusion that can be drawn from the results presented in this section is that by increasing the number of training utterances, the

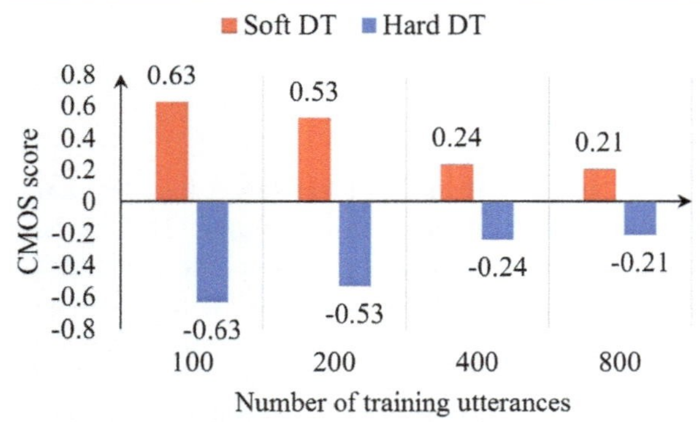

Figure 7 Subjective evaluation of soft context-clustered HMM through CMOS test. Blue and red bars are the CMOS scores of the conventional and the proposed systems.

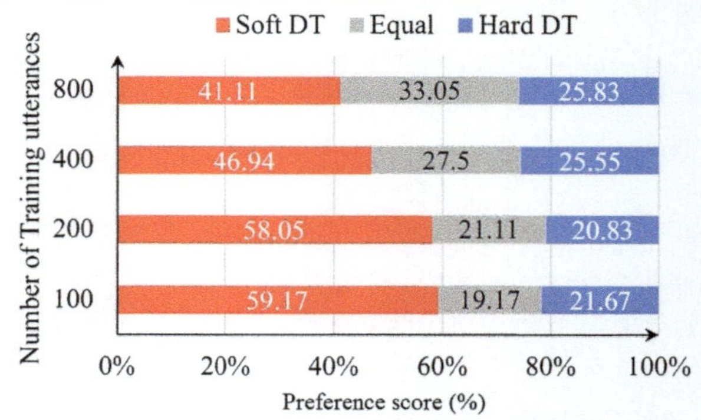

Figure 8 Paired comparison test as a subjective comparison between soft and hard decision tree structures. The scores of hard and soft decision trees are shown with blue and red colors.

improvement achieved through applying soft clustering is slightly reduced; thus, it is more efficient to employ the proposed structure in limited training datasets.

5 Conclusions

This paper addressed one of the most important shortcomings of hard decision tree-based context-dependent F0 modeling, namely, poor context generalization. In the hard decision tree structure, each acoustic feature vector is associated with modeling only one contextual cluster, and it is the main reason of poor generalization. In order to alleviate this problem, the capability of exploiting soft questions was added to the conventional decision tree architecture. The resulting structure, which is called soft decision tree, splits the contextual factor space into several soft clusters; therefore, each context is assigned to several leaves and it can provide superior generalization. In this paper, a maximum entropy model was used to drive the distribution expressed by the soft decision tree architecture. Relying on maximum entropy-based distribution, a speech synthesis system with all details was designed and implemented. Experimental results using both objective and subjective criteria showed that the proposed system outperforms the conventional hard decision tree-based system.

Endnote

[a]The unfortunate need for three separate streams only arises when using MSD output distributions to model F0: it is possible (at the onset or offset of voicing) for the dimensionality of the delta stream to be 0 in the same frame that the dimensionality of F0 is 1. That is, F0 exists, but its delta is undefined.

Competing interests

The authors declare that they have no competing interests.

Author details

[1]Department of Computer Engineering, Sharif University of Technology, Azadi Avenue, Tehran 14588, Iran. [2]The Centre for Speech Technology Research, University of Edinburgh, Informatics Forum, 10 Crichton Street, Edinburgh EH8 9AB, UK.

References

1. S King, An introduction to statistical parametric speech synthesis. Sadhana **36**(5), 837–852 (2011)
2. T Dutoi, *An Introduction to Text-to-Speech Synthesis, vol. 3* (Springer book (Kluwer Academic Publishers), The Netherlands, 1997)
3. H Zen, K Tokuda, AW Black, Statistical parametric speech synthesis. Speech Comm **51**(11), 1039–1064 (2009)
4. AW Black, H Zen, K Tokuda, *Statistical Parametric Speech Synthesis* (IEEE International Conference on Acoustics, Speech and Signal Processing (ICASSP), Honolulu, Hawaii, USA, 2007), vol 4, pp. IV-1229
5. T Yoshimura, K Tokuda, T Masuko, T Kobayashi, T Kitamura, *Mixed Excitation for HMM-Based Speech Synthesis* (European Conference on Speech Communication and Technology INTERSPEECH, Aalborg, Denmark, 2001). pp. 2263–2266
6. H Kawahara, I Masuda-Katsuse, A Cheveigné, Restructuring speech representations using a pitch-adaptive time–frequency smoothing and an instantaneous-frequency-based F0 extraction: possible role of a repetitive structure in sounds. Speech Comm **27**(3), 187–207 (1999)
7. T Drugman, T Dutoit, The deterministic plus stochastic model of the residual signal and its applications. IEEE Transactions on Audio, Speech and Language Processing **20**(3), 968–981 (2012)
8. T Drugman, G Wilfart, T Dutoit, *A Deterministic Plus Stochastic Model of the Residual Signal for Improved Parametric Speech Synthesis* (INTERSPEECH, Brighton, United Kingdom, 2009). pp. 1779–1782
9. Y Stylianou, Applying the harmonic plus noise model in concatenative speech synthesis. IEEE Transactions on Speech and Audio Processing **9**(1), 21–29 (2001)
10. MY Liberman, KW Church, Text analysis and word pronunciation in text-to-speech synthesis. Advances in Speech Signal Processing, 791–831 (1992)
11. K Tokuda, T Yoshimura, T Masuko, T Kobayashi, T Kitamura, *Speech Parameter Generation Algorithms for HMM-Based Speech Synthesis* (IEEE International Conference on Acoustics, Speech and Signal Processing (ICASSP), Istanbul, 2000), pp. 1315–1318
12. T Toda, K Tokuda, Speech parameter generation algorithm considering global variance for HMM-based speech synthesis. IEICE - Transactions on Information and Systems **E90-D**(5), 816–824 (2007)
13. S Takamichi, T Toda, Y Shiga, S Sakti, G Neubig, S Nakamura, *Parameter Generation Methods With Rich Context Models for High-Quality and Flexible Text-to-Speech Synthesis* (IEEE International Conference on Acoustics, Speech and Signal Processing (ICASSP), Vancouver, British Columbia, Canada, 2013)
14. M Shannon, W Byrne, *Fast, low-Artifact Speech Synthesis Considering Global Variance* (IEEE International Conference on Acoustics, Speech and Signal Processing (ICASSP), Vancouver, British Columbia, Canada, 2013), pp. 7869–7873
15. AJ Hunt, AW Black, *Unit Selection in a Concatenative Speech Synthesis System Using a Large Speech Database* IEEE International Conference on Acoustics, Speech, and Signal Processing (ICASSP), Atlanta, Georgia, USA, (373, 376, 1996)
16. J Yamagishi, T Kobayashi, Average-voice-based speech synthesis using HSMM-based speaker adaptation and adaptive training. IEICE - Transactions on Information and Systems **90**(2), 533–543 (2007)
17. J Yamagishi, T Nose, H Zen, ZH Ling, T Toda, K Tokuda, S King, S Renals, Robust speaker-adaptive HMM-based text-to-speech synthesis. IEEE Transactions on Audio, Speech, and Language Processing **17**(6), 1208–1230 (2009)
18. J Yamagishi, T Kobayashi, Y Nakanc, K Ogata, J Isogai, Analysis of speaker adaptation algorithms for HMM-based speech synthesis and a constrained SMAPLR adaptation algorithm. IEEE Transactions on Audio, Speech, and Language Processing **17**(1), 66–83 (2009)
19. H Zen, N Braunschweiler, S Buchhclz, MJ Gales, K Knill, S Krstulovic, J Latorre, Statistical parametric speech synthesis based on speaker and language factorization. IEEE Transactions on Audio, Speech, and Language Processing **20**(6), 1713–1724 (2012)
20. YJ Wu, Y Nankaku, K Tokuda, *State Mapping Based Method for Cross-Lingual Speaker Adaptation in HMM-Based Speech Synthesis* (INTERSPEECH, Brighton, United Kingdom, 2009), pp. 528–531
21. H Liang, J Dines, L Saheer, *A Comparison of Supervised and Unsupervised Cross-Lingual Speaker Adaptation Approaches for HMM-Based Speech Synthesis* (IEEE International Conference on Acoustics Speech and Signal Processing (ICASSP), Dallas, Texas, USA, 2010). pp. 4598–4601
22. M Gibson, T Hirsimaki, R Karhila, M Kurimo, W Byrne, *Unsupervised Cross-Lingual Speaker Adaptation for HMM-Based Speech Synthesis Using two-Pass Decision Tree Construction* (IEEE International Conference on Acoustics Speech and Signal Processing (ICASSP), Dallas, Texas, USA, 2010), pp. 4642–4645
23. J Yamagishi, Z Ling, S King, *Robustness of HMM-Based Speech Synthesis* (INTERSPEECH, Brisbane, Australia, 2008), pp. 581–584
24. R Karhila, U Remes, M Kurimo, *HMM-Based Speech Synthesis Adaptation Using Noisy Data: Analysis and Evaluation Methods* (IEEE International Conference on Acoustics, Speech and Signal Processing (ICASSP), Vancouver, British Columbia, Canada, 2013), pp. 6930–6934
25. K Yanagisawa, J Latorre, V Wan, MJ Gales, S King, *Noise Robustness in HMM-TTS Speaker Adaptation* (8th ISCA Speech Synthesis Workshop, Barcelona, Spain, 2013), pp. 119–124
26. M Cernak, P Motlicek, PN Garner, *On the (un) Importance of the Contextual Factors in HMM-Based Speech Synthesis and Coding* (IEEE International Conference on Acoustics, Speech and Signal Processing (ICASSP), Vancouver, British Columbia, Canada, 2013), pp. 8140–8143

27. T Yoshimura, K Tokuda, T Masuko, T Kobayashi, T Kitamura, *Simultaneous Modeling of Spectrum, Pitch and Duration in HMM-Based Speech Synthesis* (Proceedings of Eurospeech, Budapest, Hungary, 1999), pp. 2347–2350
28. T Yoshimura, K Tokuda, T Masuko, T Kobayashi, T Kitamura, *Duration Modeling in HMM-Based Speech Synthesis System* (Proceedings of ICSLP, Sydney. Australia, 1998), pp. 29–32
29. H Zen, T Nose, J Yamagishi, S Sako, T Masuko, A Black, T Keiichi, *The HMM-Based Speech Synthesis System (HTS) Version 2.0* (6th ISCA Workshop on Speech Synthesis (SSW), Bonn, Germany, 2007), pp. 294–299
30. K Tokuda, H Zen, AW Black, *An HMM-Based Speech Synthesis System Applied to English* (IEEE Workshop on Speech Synthesis, Scotland, 2002), pp. 227–230
31. H Zen, T Toda, M Nakamura, K Tokuda, Details of the Nitech HMM-based speech synthesis system for the Blizzard Challenge 2005. IEICE Trans Inf Syst **90**(1), 325–333 (2007)
32. H Zen, T Toda, K Tokuda, The Nitech-NAIST HMM-based speech synthesis system for the Blizzard Challenge 2006. IEICE Trans Inf Syst **91**(6), 1764–1773 (2008)
33. K Tokuda, Y Nankaku, T Toda, H Zen, J Yamagishi, K Oura, Speech synthesis based on hidden Markov models. Proc IEEE **101**(5), 1234–1252 (2013)
34. JJ Odell, *The use of Context in Large Vocabulary Speech Recognition* (PhD dissertation, Cambridge University, 1995)
35. SJ Young, JJ Odell, PC Woodland, *Tree-Based State Tying for High Accuracy Acoustic Modeling* (Proceedings of the Workshop on Human Language Technology, Association for Computational Linguistics, Stroudsburg, PA, USA, 1994), pp. 307–312
36. K Tokuda, T Masuko, N Miyazaki, T Kobayashi, Multi-space probability distribution HMM. IEICE Trans Inf Syst **85**(3), 455–464 (2002)
37. H Zen, T Keiichi, T Masuko, T Kobayashi, T Kitamura, A hidden semi-Markov model-based speech synthesis system. IEICE Transactions on Information and Systems, E series **D-90**(5), 825–834 (2007)
38. H Zen, A Senior, M Schuster, Statistical Parametric Speech Synthesis Using Deep Neural Networks, in *IEEE International Conference on Acoustics, Speech and Signal Processing, ICASSP 2013* (IEEE, 2013), pp. 7962–7966
39. H Lu, S King, O Watts, *Combining a Vector Space Representation of Linguistic Context With a Deep Neural Network for Text-to-Speech Synthesis* (8th ISCA Speech Synthesis Workshop, Barcelona, Spain, 2013), pp. 261–265
40. ZH Ling, L Deng, D Yu, *Modeling Spectral Envelopes Using Restricted Boltzmann Machines for Statistical Parametric Speech Synthesis* (IEEE Acoustics, Speech and Signal Processing (ICASSP), Vancouver, British Columbia, Canada, 2013), pp. 7825–7829
41. S Kang, X Qian, H Meng, *Multi-Distribution Deep Belief Network for Speech Synthesis* (IEEE International Conference on Acoustics, Speech and Signal Processing (ICASSP), Vancouver, British Columbia, Canada, 2013), pp. 8012–8016
42. T Koriyama, T Nose, T Kobayashi, Statistical parametric speech synthesis based on Gaussian process regression. IEEE Journal of Selected Topics in Signal Processing **99**, 1–11 (2013)
43. K Hashimoto, Y Nankaku, K Tokuda, *A Bayesian Approach to Hidden Semi Markov Model Based Speech Synthesis* (Proceedings of Interspeech, Brighton, United Kingdom, 2009), pp. 1751–1754
44. S Khorram, H Sameti, F Bahmaninezhad, S King, T Drugman, Context-Dependent Acoustic Modeling Based on Hidden Maximum Entropy Model for Statistical Parametric Speech Synthesis. EURASIP Journal on Audio, Speech, and Music Processing **12**, 1–21 (2014)
45. Y Nankaku, K Nakamura, H Zen, K Tokuda, *Acoustic Modeling With Contextual Additive Structure for HMM-Based Speech Recognition* (IEEE International Conference on Acoustics, Speech and Signal Processing (ICASSP), Las Vegas, Nevada, USA, 2008), pp. 4469–4472
46. S Takaki, Y Nankaku, K Tokuda, *Spectral Modeling With Contextual Additive Structure for HMM-Based Speech Synthesis* (Proceedings of 7th ISCA Speech Synthesis Workshop, Kyoto, Japan, 2010), pp. 100–105
47. S Takaki, Y Nankaku, K Tokuda, *Contextual Partial Additive Structure for HMM-Based Speech Synthesis* (IEEE International Conference on Acoustics, Speech and Signal Processing (ICASSP), Vancouver, British Columbia, Canada, 2013), pp. 7878–7882
48. S Takaki, Y Nankaku, K Tokuda, Contextual additive structure for HMM-based speech synthesis. IEEE J Selected Topics in Signal Processing **8**(2), 229–238 (2014)
49. H Zen, N Braunschweiler, *Context-Dependent Additive log F0 Model for HMM-Based Speech Synthesis* (INTERSPEECH, Brighton, United Kingdom, 2009), pp. 2091–2094
50. YJ Wu, F Soong, *Modeling pitch trajectory by hierarchical HMM with minimum generation error training* (IEEE International Conference on Acoustics, Speech and Signal Processing (ICASSP), Kyoto, Japan, 2012), pp. 4017–4020
51. S Sakai, Additive modeling of English F0 contour for speech synthesis. Proceedings of ICASSP **1**, 277–280 (2008)
52. Y Qian, H Liang, FK Soong, *Generating Natural F0 Trajectory With Additive Trees* (INTERSPEECH, Brisbane, Australia, 2008). pp. 2126–2129
53. K Yu, F Mairesse, S Young, *Word-Level Emphasis Modeling in HMM-Based Speech Synthesis* (IEEE International Conference on Acoustics, Speech and Signal Processing (ICASSP), Dallas, Texas, USA, 2010), pp. 4238–4241
54. K Yu, H Zen, F Mairesse, S Young, Context adaptive training with factorized decision trees for HMM-based statistical parametric speech synthesis. Speech Comm **53**(6), 914–923 (2011)
55. MJ Gales, Cluster adaptive training of hidden Markov models. IEEE Transactions on Speech and Audio Processing **8**(4), 417–428 (2000)
56. C Olaru, L Wehenkel, A complete fuzzy decision tree technique. Fuzzy Set Syst **138**(2), 221–254 (2003)
57. Y Yuan, MJ Shaw, Induction of fuzzy decision trees. Fuzzy Set Syst **69**(2), 125–139 (1995)
58. L Rabiner, BH Juang, An introduction to hidden Markov models. IEEE ASSP Mag **3**(1), 4–16 (1986)
59. K Yu, S Young, Continuous F0 modeling for HMM based statistical parametric speech synthesis. IEEE Transactions on Audio, Speech, and Language Processing **19**(5), 1071–1079 (2011)
60. TK Moon, The expectation-maximization algorithm. IEEE Signal Process Mag **13**(6), 47–60 (1996)
61. *HMM-based speech synthesis system (HTS)* http://hts.sp.nitech.ac.jp/
62. H Zen, *Implementing an HSMM-Based Speech Synthesis System Using an Efficient Forward-Backward Algorithm, Nagoya Institute of Technology, Technical Report TR-SP-0001*, 2007
63. JD Ferguson, *Variable Duration Models for Speech* (Proceedings of the Symposium on the Application Hidden Markov Models to Text and Speech, USA, 1980), pp. 143–179
64. SE Levinson, Continuously variable duration hidden Markov models for automatic speech recognition. Computer Speech and Language **1**(1), 29–45 (1986)
65. SZ Yu, H Kobayashi, An efficient forward-backward algorithm for an explicit-duration hidden Markov model. IEEE Signal Processing Letters **10**(1), 11–14 (2003)
66. K Shinoda, T Watanabe, *Speaker Adaptation With Autonomous Model Complexity Control by MDL Principle* (Proceedings of ICASSP, Atlanta, Georgia, USA, 1996), pp. 717–720
67. AL Berger, VJD Pietra, SAD Pietra, A maximum entropy approach to natural language processing. Computational Linguistics **22**(1), 39–71 (1996)
68. A Ratnaparkhi, *A Maximum Entropy Model for Part-of-Speech Tagging* (Proceedings of the Conference on Empirical Methods in Natural Language Processin, PA, USA, 1996), pp. 133–142
69. J Yamagishi, O Watts, *The CSTR/EMIME HTS System for Blizzard Challenge* (Proceedings of Blizzard Challenge 2010, Japan, 2010), pp. 1–6
70. J Yamagishi, *Average-Voice-Based Speech Synthesis* (PhD thesis, Tokyo Institute of Technology, 2006)

3

A collaborative adaptive Wiener filter for image restoration using a spatial-domain multi-patch correlation model

Khaled M Mohamed† and Russell C Hardie* †

Abstract

We present a new patch-based image restoration algorithm using an adaptive Wiener filter (AWF) with a novel spatial-domain multi-patch correlation model. The new filter structure is referred to as a collaborative adaptive Wiener filter (CAWF). The CAWF employs a finite size moving window. At each position, the current observation window represents the reference patch. We identify the most similar patches in the image within a given search window about the reference patch. A single-stage weighted sum of all of the pixels in the similar patches is used to estimate the center pixel in the reference patch. The weights are based on a new multi-patch correlation model that takes into account each pixel's spatial distance to the center of its corresponding patch, as well as the intensity vector distances among the similar patches. One key advantage of the CAWF approach, compared with many other patch-based algorithms, is that it can jointly handle blur and noise. Furthermore, it can also readily treat spatially varying signal and noise statistics. To the best of our knowledge, this is the first multi-patch algorithm to use a single spatial-domain weighted sum of all pixels within multiple similar patches to form its estimate and the first to use a spatial-domain multi-patch correlation model to determine the weights. The experimental results presented show that the proposed method delivers high performance in image restoration in a variety of scenarios.

Keywords: Image restoration; Wiener filter; Correlation model; Patch-based processing

1 Introduction

1.1 Image restoration

During image acquisition, images are subject to a variety of degradations. These invariably include blurring from diffraction and noise from a variety of sources. Restoring such degraded images is a fundamental problem in image processing that has been researched since the earliest days of digital images [1,2]. A wide variety of linear and non-linear methods have been proposed. Many methods have focused exclusively on noise reduction, and others seek to address multiple degradations jointly, such as blur and noise.

A widely used method for image restoration, relevant to the current paper, is the classic Wiener filter [3]. The standard Wiener filter is a linear space-invariant filter designed to minimize mean squared error (MSE) between the desired signal and estimate, assuming stationary random signals and noise. It is important to note that there are many disparate variations of Wiener filters. These include finite impulse response, infinite impulse response, transform-domain, and spatially adaptive methods. Within each of these categories, a wide variety of statistical models may be employed. Some statistical models are very simple, such as the popular constant noise-to-signal power spectral density model, and others are far more complex. In the case of the empirical Wiener filter [4], no explicit statistical model is used at all. Rather, a pilot or prototype estimate is used in lieu of a parametric statistical model. While all of these methods may go by the name of 'Wiener filter', they can be quite different in their character.

Recently, a form of adaptive Wiener filter (AWF) has been developed and successfully applied to super-resolution (SR) and other restoration applications by one of the current authors [5]. This AWF approach employs a spatially varying weighted sum to form an estimate of

*Correspondence: rhardie@udayton.edu
†Equal contributors
University of Dayton, Department of Electrical and Computer Engineering, 300 College Park, Dayton, 45469-0226 OH, USA

each pixel. The Wiener weights are determined based on a spatially varying spatial-domain parametric correlation model. This particular brand of AWF SR emerged from earlier work, including that in [6-8]. This kind of AWF is capable of jointly addressing blur, noise, and undersampling and is well suited to dealing with a non-stationary signal and noise. The approach is also very well suited to dealing with non-uniformly sampled imagery and missing or bad pixels. This AWF SR method has been shown to provide best-in-class performance for nonuniform interpolation-based SR [5,9-11] and has also been used successfully for demosaicing [12,13] and Nyquist sampled video restoration [14]. Under certain conditions, the method can also be very computationally efficient [5]. The key to this method lies in the particular correlation model used and how it is employed for spatially adaptive filtering.

A different approach to image restoration, also relevant to the current paper, is based on fusing multiple similar patches within the observed image. This patch-based approach is used primarily for noise reduction applications and exploits spatial redundancy that may express itself within an image, locally and/or non-locally. The method of non-local means (NLM), introduced in [15], may be the first method to directly fuse non-local patches from within the observed image based on vector distances for the purpose of image denoising. A notable early precursor to the NLM is the vector detection method [16,17]. In the vector detection approach, a codebook of representative patches from training data is used, rather than patches from the observed image itself [16,17]. A number of NLM variations have been proposed, including [18-26]. The basic NLM method forms an estimate of a reference pixel as a weighted sum of non-local pixels. The weights are based on the vector distances of the patch intensities between various non-local patches and the reference patch. In particular, the center samples of a non-local patches are weighted in proportion to the negative exponential of the corresponding patch distance. The NLM algorithm can be viewed as an extension of the bilateral filter [27-31], which forms an estimate by weighting neighboring pixels based on both spatial proximity and intensity similarity of individual pixels (rather than patches).

Improved performance for noise reduction is obtained with the block matching and 3D filtering (BM3D) approach proposed in [32-34]. The BM3D method also uses vector distances between patches, but the filtering is performed using a transform-domain shrinkage operation. By utilizing all of the samples within selected patches and aggregating the results, excellent noise reduction performance can be achieved with BM3D. Another related patch-based image denoising algorithm is the total least squares method presented in [35]. In this method, each ideal patch is modeled as a linear combination of similar patches from the observed image. Another type of patch-based Wiener denoising filter is proposed in [36], and a globalized approach to patch-based denoising is proposed in [37]. While such patch-based methods perform well in noise reduction, most are not capable of addressing blur and noise jointly. However, there are a few recent methods that do treat both blur and noise and incorporate multi-patch fusion. These include BM3D deblurring (BM3DDEB) [38] and iterative decoupled deblurring-BM3D (IDD-BM3D) [39]. The deblurring in these algorithms is not achieved by patch fusion alone. Rather, the patch fusion component of these algorithms serves as a type of signal model used for regularization. Note that multi-patch methods have also been developed and applied to SR [40-45]. However, the focus of this paper is on image restoration without undersampling/aliasing.

1.2 Proposed method and novel contribution

In this paper, we propose a novel multi-patch AWF algorithm for image restoration. In the spirit of [32], we refer to this new filter structure as a collaborative adaptive Wiener filter (CAWF). It can be viewed as an extension of the AWF in [5], with the powerful new feature of incorporating multiple patches for each pixel estimate. As with other patch-based algorithms, we employ a moving window. At each position, the current observation window represents the reference patch. Within a given search window about the reference, we identify the most similar patches to the reference patch. However, instead of simply weighting just the center pixels of these similar patches, as with NLM, or using transform-domain shrinkage like BM3D, we use a spatial-domain weighted sum of all of the pixels within all of the selected patches to estimate the one center pixel in the reference patch. The weights used are based on a novel spatially varying spatial-domain multi-patch correlation model. The correlation model takes into account each pixel's spatial distance to the center of its corresponding patch, as well as the intensity vector distances among the similar patches. The 'collaborative' nature of the CAWF springs from the fusion of multiple, potentially non-local, patches. One key advantage of the CAWF approach is that it can jointly handle blur and noise. Furthermore, the CAWF method is able to accommodate spatially varying signal and noise statistics.

Our approach is novel in that we use a single-pass spatial-domain weighted sum of all pixels within all of the similar patches to form the estimate each desired pixel. In the case of NLM, only the center pixel of each similar patch is given a weight [15].This is simple and effective for denoising, but deconvolution is not possible within the basic NLM framework, and all of the available information in the patches may not be exploited. While BM3D does fuse all of the pixels in the similar patches, the fusion in BM3D is based on a wavelet shrinkage operation and not a spatial-domain correlation model [32]. Because of the nature of wavelet shrinkage, the standard BM3D is also unable to perform deblurring [32]. On the other hand, the CAWF structure can jointly address blur and noise and does not employ separate transform-domain inverse filtering as in [38] or iterative processing like that in [39]. To the best of our knowledge, this is the first multi-patch algorithm to use a single-pass weighted sum of all pixels within multiple similar patches to jointly address blur and noise. It is also the first to use a spatial-domain multi-patch correlation model.

The remainder of this paper is organized as follows. The observation model is described in Section 2. The CAWF algorithm is presented in Section 3. This includes the basic algorithm description as well as the new spatial-domain multi-patch correlation model. Computational complexity and implementation are also discussed in Section 3. Experimental results for simulated and real data are presented and discussed in Section 4. Finally, conclusions are offered in Section 5.

2 Observation model

The proposed CAWF algorithm is based on the observation model shown in Figure 1. The model input is a desired 2D discrete grayscale image, denoted $d(n_1, n_2)$, where n_1, n_2 are the spatial pixel indices. Using lexicographical notation, we shall represent all of the pixels in this desired image with a single column vector $\mathbf{d} = [d_1, d_2, \ldots, d_N]^T$, where N is the total number of pixels in the image. Next in the observation model, we assume the desired image is convolved with a specified point spread function (PSF), yielding

$$f(n_1, n_2) = d(n_1, n_2) * h(n_1, n_2), \tag{1}$$

where $h(n_1, n_2)$ is the PSF, and $*$ is 2D spatial convolution. In matrix form, this can be expressed as

$$\mathbf{f} = \mathbf{Hd} = [f_1, f_2, \ldots, f_N]^T, \tag{2}$$

where $\mathbf{H}$ is an $N \times N$ matrix containing values of PSF, and the vector $\mathbf{f}$ is the image $f(n_1, n_2)$ in lexicographical form. The PSF can be designed to model different types of blurring, such as diffraction from optics, spatial detection

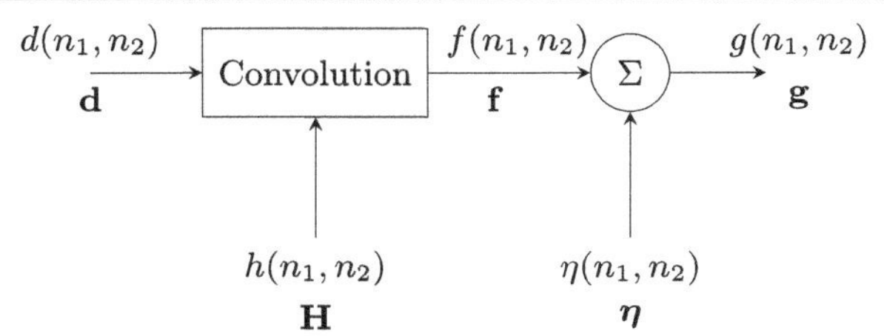

Figure 1 Observation model block diagram.

integration, atmospheric effects, and motion blurring. In the experimental results presented in this paper, we use a simple Gaussian PSF.

With regard to noise, our model assumes zero-mean additive Gaussian noise, with noise standard deviation of σ_η. Thus, the observed image is given by

$$g(n_1, n_2) = f(n_1, n_2) + \eta(n_1, n_2), \tag{3}$$

where $\eta(n_1, n_2)$ is a Gaussian noise array. In lexicographic form, this is given by

$$\mathbf{g} = \mathbf{f} + \boldsymbol{\eta}, \tag{4}$$

where $\mathbf{g} = [g_1, g_2, \ldots, g_N]^T$ and $\boldsymbol{\eta} = [\eta_1, \eta_2, \ldots, \eta_N]^T$ are the observed image and noise vectors, respectively. The random noise vector is Gaussian such that $\boldsymbol{\eta} \sim \mathcal{N}\left(\mathbf{0}, \sigma_\eta^2 \mathbf{I}\right)$.

3 Collaborative adaptive Wiener filter

3.1 CAWF overview

The CAWF employs a moving window approach with a moving reference patch and corresponding moving search window, each centered about pixel i, where $i = 1, 2, \ldots, N$. The reference patch spans $K_1 \times K_2 = K$ pixels symmetrically about pixel i. All of the pixels that lie within the span of this reference patch are placed into the reference patch vector defined as $\mathbf{g}_i = [g_{i,1}, g_{i,2}, \ldots, g_{i,K}]^T$. The search window is of size $L_1 \times L_2 = L$ pixels. Let the set $S_i = [S_i(1), S_i(2), \ldots S_i(L)]^T$ contain the indices of the pixels within the search window.

The next step in the CAWF algorithm is identifying the M patches from the search window that are most similar to the reference. This is done using a simple squared Euclidean distance. That is, we compute $\|\mathbf{g}_i - \mathbf{g}_j\|_2^2$, for $j \in S_i$. We select the M patches corresponding to the M smallest distances and designate these as 'similar patches'. All of the pixels from the similar patches shall be concatenated into a single $KM \times 1$ column vector, $\tilde{\mathbf{g}}_i = \left[\mathbf{g}_{s_{i,1}}^T, \mathbf{g}_{s_{i,2}}^T, \ldots, \mathbf{g}_{s_{i,M}}^T\right]^T$, where $\mathbf{s}_i = \left[s_{i,1}, s_{i,2}, \ldots s_{i,M}\right]^T$ contains the indices of the similar patches in order from smallest corresponding distance to largest. The minimum distance will always be zero and will correspond to the reference patch itself, such that $s_{i,1} = i$. This selection of similar patches is common to many patch-based algorithms, such as those in [32-34]. Examples of the similar patch selection is illustrated in Figure 2. The red square represents the reference patch, and the green squares represent selected similar patches. Note that there will generally be variability with regard to how similar the selected patches are to the reference and to each other. Some reference patches will have numerous low distance similar patches, and others will not. It is this variability that we wish to capture and account for with our multi-patch correlation model in Section 3.2.

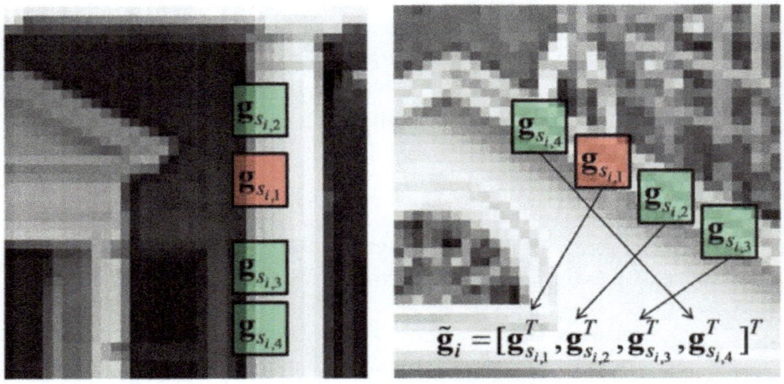

Figure 2 Selection of similar patches (green) from a given reference patch (red) within a search window.

With the collection of similar patches defined, we can now express the CAWF output as a weighted sum of the values in $\tilde{\mathbf{g}}_i$. In particular, the CAWF estimate for desired pixel i is given by

$$\hat{d}_i = \mathbf{w}_i^T \tilde{\mathbf{g}}_i, \tag{5}$$

where $\mathbf{w}_i = [w_1, w_2, \ldots w_{KM}]^T$ is a vector of weights. Note that this approach is a one-pass weighted-sum operation that incorporates all of the pixels in $\tilde{\mathbf{g}}_i$ for the estimate of d_i. To minimize the MSE, the Wiener filter weights [5] may be used such that

$$\mathbf{w}_i = \tilde{\mathbf{R}}_i^{-1} \tilde{\mathbf{p}}_i, \tag{6}$$

where $\tilde{\mathbf{R}}_i = E\left\{\tilde{\mathbf{g}}_i \tilde{\mathbf{g}}_i^T\right\}$ is a $KM \times KM$ autocorrelation matrix for the multi-patch observation vector $\tilde{\mathbf{g}}_i$, and $\tilde{\mathbf{p}}_i = E\left\{\tilde{\mathbf{g}}_i d_i\right\}$ is a $KM \times 1$ cross-correlation vector between the desired pixel d_i and $\tilde{\mathbf{g}}_i$. The statistics used to fill $\tilde{\mathbf{R}}_i$ and $\tilde{\mathbf{p}}_i$ are found using the new multi-patch correlation model described in Section 3.2.

3.2 Spatial-domain multi-patch correlation model

The multi-patch correlation model provides the values for $\tilde{\mathbf{R}}_i$ and $\tilde{\mathbf{p}}_i$, so that we may generate the weights in Equation 6. The model attempts to capture the spatial relationship among the pixels within a given patch, which is essential for deconvolution. Furthermore, it also seeks to incorporate knowledge of redundancy among the similar patches. Finally, the model captures the local desired signal variance, as well as the noise variance of each observed pixel.

To begin, first let $\tilde{\mathbf{f}}_i$ be the noise-free version of $\tilde{\mathbf{g}}_i$. In other words, we have $\tilde{\mathbf{g}}_i = \tilde{\mathbf{f}}_i + \tilde{\boldsymbol{\eta}}_i$, where $\tilde{\boldsymbol{\eta}}_i$ is the random noise vector associated with the samples within multi-patch observation vector i. We shall assume the noise is zero mean and uncorrelated with the signal. Furthermore, we will assume the noise samples in $\tilde{\boldsymbol{\eta}}_i$ are independent and identically distributed (i.i.d.) with a noise variance of σ_η^2. In this case, it is straightforward to show that

$$\tilde{\mathbf{R}}_i = E\left\{\tilde{\mathbf{g}}_i \tilde{\mathbf{g}}_i^T\right\} = E\left\{\tilde{\mathbf{f}}_i \tilde{\mathbf{f}}_i^T\right\} + \sigma_\eta^2 \mathbf{I} \tag{7}$$

and

$$\tilde{\mathbf{p}}_i = E\left\{\tilde{\mathbf{g}}_i d_i\right\} = E\left\{\tilde{\mathbf{f}}_i d_i\right\}. \tag{8}$$

Now, the problem reduces to modeling $E\left\{\tilde{\mathbf{f}}_i d_i\right\}$ and $E\left\{\tilde{\mathbf{f}}_i \tilde{\mathbf{f}}_i^T\right\}$.

In our new correlation model, the multi-patch statistics will be expressed in terms of statistics for a single patch and a distance matrix for all of the similar patches. In particular, we use the model

$$E\left\{\tilde{\mathbf{f}}_i \tilde{\mathbf{f}}_i^T\right\} = \hat{\sigma}_{d_i}^2 e^{-\mathbf{D}_i/(\alpha\sigma_\eta)} \otimes \mathbf{R}, \tag{9}$$

where $\otimes$ is a Kronecker product, and $\mathbf{R}$ is a $K \times K$ autocorrelation matrix of a single noise-free patch obtained from a variance-normalized desired image. We will say more about this shortly. The matrix

$$\mathbf{D}_i = \begin{bmatrix} D_{s_{i,1},s_{i,1}} & D_{s_{i,1},s_{i,2}} & \cdots & D_{s_{i,1},s_{i,M}} \\ D_{s_{i,2},s_{i,1}} & D_{s_{i,2},s_{i,2}} & \cdots & D_{s_{i,2},s_{i,M}} \\ \vdots & \vdots & \ddots & \vdots \\ D_{s_{i,M},s_{i,1}} & D_{s_{i,M},s_{i,2}} & \cdots & D_{s_{i,M},s_{i,M}} \end{bmatrix}. \tag{10}$$

is an $M \times M$ distance matrix among the M similar patches. In our notation, the exponential term in Equation 9 is a matrix whose elements are made of the exponential values of the corresponding distance matrix elements. The variable α is a tuning parameter that controls the correlation decay as a function of the distances between similar patches, and σ_η is noise standard deviation. The parameter $\hat{\sigma}_{d_i}^2$ is the estimated desired signal variance associated with the reference patch. By substituting Equation 9 into Equation 7, we get the autocorrelation matrix for $\tilde{\mathbf{g}}_i$ as

$$\tilde{\mathbf{R}}_i = \hat{\sigma}_{d_i}^2 e^{-\mathbf{D}_i/(\alpha\sigma_\eta)} \otimes \mathbf{R} + \sigma_\eta^2 \mathbf{I}. \tag{11}$$

In a similar manner, we model the cross-correlation vector as

$$\tilde{\mathbf{p}}_i = E\left\{\tilde{\mathbf{f}}_i\, d_i\right\} = \hat{\sigma}^2_{d_i} e^{-[\mathbf{D}_i]_1/(\alpha\sigma_\eta)} \otimes \mathbf{p}, \tag{12}$$

where $\mathbf{p}$ is the $K \times 1$ cross-correlation vector for a single normalized patch, and $[\mathbf{D}_i]_1$ is the first column of the distance matrix $\mathbf{D}_i$. The correlation models in Equations 11 and 12 capture the spatial correlations among pixels within each patch using $\mathbf{R}$ and $\mathbf{p}$. The patch similarities, captured in the distance matrix, are used to 'modulate' these correlations with the Kronecker product to provide the full multi-patch correlation model. In this manner, pixels belonging to patches with smaller inter-patch distances will be modeled with higher correlations among them. Potential changes in the underlying desired image variance are captured in the model with the term $\hat{\sigma}^2_{d_i}$. In addition, a spatially varying noise variance can easily be incorporated if appropriate.

The specific distance metric used to populate the distance matrix $\mathbf{D}_i$ here is a scaled and shifted l^2-norm. This type of metric has been used successfully to quantify similarity between image patches corrupted with additive Gaussian noise [15]. In particular, the distance between patches centered about pixels i and j is given by

$$D_{i,j} = \begin{cases} \frac{\|\mathbf{g}_i - \mathbf{g}_j\|_2}{\sigma_\eta\sqrt{2K}} - D_0 & \frac{\|\mathbf{g}_i - \mathbf{g}_j\|_2}{\sigma_\eta\sqrt{2K}} > D_0 \\ 0 & \text{otherwise} \end{cases} \tag{13}$$

where $\|\mathbf{g}_i - \mathbf{g}_j\|_2$ is the l^2-norm distance, K is total number of pixels in each patch, and σ_η is noise standard deviation. The scaling by $\sigma_\eta\sqrt{2K}$ normalizes the distance with respect to K and σ_η, and D_0 can be used as a tuning parameter in the correlation model to adjust for distance due to noise. To see how the scaling works, and understand the potential role of D_0, consider the distance with $D_0 = 0$ defined as

$$\bar{D}_{i,j} = \frac{\|\mathbf{g}_i - \mathbf{g}_j\|_2}{\sigma_\eta\sqrt{2K}}. \tag{14}$$

It can be shown that for identical patches with distance due only to i.i.d. Gaussian noise, the probability density function (pdf) for $\bar{D}_{i,j}$ is that of a scaled Chi random variable and is given by

$$f_{\bar{D}_{i,j}}(x) = \sqrt{K}\chi_K\left(x\sqrt{K}\right). \tag{15}$$

This pdf is plotted in Figure 3 for four values of K. Note that with our scaling, the pdf is not a function of σ_η and the mean is close to 1 for all K. Also, note that D_0 can be used to shift the pdf.

Let us now turn our attention to $\mathbf{R}$, $\mathbf{p}$, and $\hat{\sigma}^2_{d_i}$. Note that $\mathbf{R}$ and $\mathbf{p}$ correspond to a single patch derived from a desired image with zero mean and variance of one, denoted $\bar{d}(n_1, n_2)$. After PSF blurring, the resulting image is denoted $\bar{f}(n_1, n_2)$, following the model shown in Figure 1. Since these statistics are for only a single patch, they can be modeled in a fashion similar to that in [5]. To begin, consider the 2D wide sense stationary (WSS) autocorrelation function model for $\bar{d}(n_1, n_2)$ given by

$$r_{\bar{d}\bar{d}}(n_1, n_2) = \rho^{\sqrt{n_1^2 + n_2^2}}, \tag{16}$$

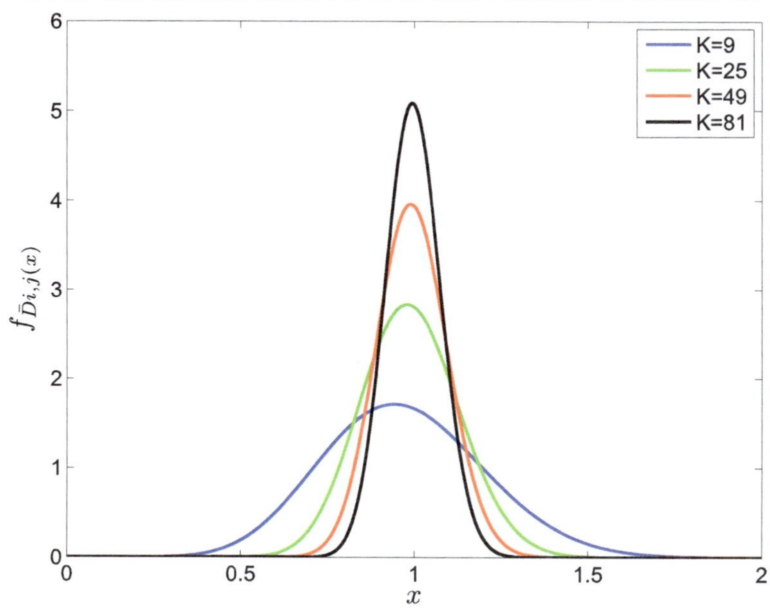

Figure 3 Probability density function for $\bar{D}_{i,j}$ for identical patches with Gaussian noise. The distance here is due exclusively to noise.

where ρ is the one pixel step correlation value. The cross correlation function between $\bar{d}(n_1, n_2)$ and $\bar{f}(n_1, n_2)$ can then be expressed as

$$r_{\bar{d}\bar{f}}(n_1, n_2) = r_{\bar{d}\bar{d}}(n_1, n_2) * h(n_1, n_2). \tag{17}$$

The auto-correlation function for $\bar{f}(n_1, n_2)$ can also be expressed in terms of the desired autocorrelation function as

$$r_{\bar{f}\bar{f}}(n_1, n_2) = r_{\bar{d}\bar{d}}(n_1, n_2) * h(n_1, n_2) * h(-n_1, -n_2). \tag{18}$$

Figure 4 shows an example of $r_{\bar{d}\bar{f}}(n_1, n_2)$ and $r_{\bar{f}\bar{f}}(n_1, n_2)$ for Gaussian blur PSF with standard deviation of 1.5 pixels, and $\rho = 0.7$. As expected, the correlation drops with distance, as controlled by ρ.

Now consider a $K_1 \times K_2 = K$ patch from $\bar{f}(n_1, n_2)$, denoted $\bar{\mathbf{f}} = [\bar{f}_1, \bar{f}_2, \ldots, \bar{f}_K]^T$, and the corresponding desired pixel value, $\bar{d}$. The $K \times K$ autocorrelation matrix for $\bar{\mathbf{f}}$ is $\mathbf{R} = E\left\{\bar{\mathbf{f}}\bar{\mathbf{f}}^T\right\}$, and the $K \times 1$ cross-correlation vector is $\mathbf{p} = E\left\{\bar{\mathbf{f}}\bar{\mathbf{d}}\right\}$. Expressing all of the terms in the autocorrelation matrix, we obtain

$$\mathbf{R} = E\left\{\bar{\mathbf{f}}\bar{\mathbf{f}}^T\right\} = \begin{bmatrix} E\left\{\bar{f}_1\bar{f}_1\right\} & E\left\{\bar{f}_1\bar{f}_2\right\} & \cdots & E\left\{\bar{f}_1\bar{f}_K\right\} \\ E\left\{\bar{f}_2\bar{f}_1\right\} & E\left\{\bar{f}_2\bar{f}_2\right\} & \cdots & E\left\{\bar{f}_2\bar{f}_K\right\} \\ \vdots & \vdots & \ddots & \vdots \\ E\left\{\bar{f}_K\bar{f}_1\right\} & E\left\{\bar{f}_K\bar{f}_2\right\} & \cdots & E\left\{\bar{f}_K\bar{f}_K\right\} \end{bmatrix}. \tag{19}$$

Assuming a 2D WSS model, as in [5], this matrix can be populated from Equation 18 as follows

$$\mathbf{R} = \begin{bmatrix} r_{\bar{f}\bar{f}}(0,0) & r_{\bar{f}\bar{f}}(\Delta(1,2)) & \cdots & r_{\bar{f}\bar{f}}(\Delta(1,K)) \\ r_{\bar{f}\bar{f}}(\Delta(2,1)) & r_{\bar{f}\bar{f}}(0,0) & \cdots & r_{\bar{f}\bar{f}}(\Delta(2,K)) \\ \vdots & \vdots & \ddots & \vdots \\ r_{\bar{f}\bar{f}}(\Delta(K,1)) & r_{\bar{f}\bar{f}}(\Delta(K,2)) & \cdots & r_{\bar{f}\bar{f}}(0,0) \end{bmatrix}, \tag{20}$$

where $\Delta(m, n) = [\Delta_x(m, n), \Delta_y(m, n)]$, and $\Delta_x(m, n)$ and $\Delta_y(m, n)$ are the x and y distances between $\bar{f}_m$ and $\bar{f}_n$ in $\bar{f}(n_1, n_2)$ in units of pixels. In a similar fashion, we can populate $\mathbf{p}$ using Equation 17 as follows

$$\mathbf{p} = E\{\bar{\mathbf{f}}\bar{\mathbf{d}}\} = \begin{bmatrix} E\{\bar{f}_1\bar{d}\} \\ E\{\bar{f}_2\bar{d}\} \\ \vdots \\ E\{\bar{f}_K\bar{d}\} \end{bmatrix} = \begin{bmatrix} r_{\bar{d}\bar{f}}\left(\Delta\left(1, \frac{K+1}{2}\right)\right) \\ r_{\bar{d}\bar{f}}\left(\Delta\left(2, \frac{K+1}{2}\right)\right) \\ \vdots \\ r_{\bar{d}\bar{f}}\left(\Delta\left(K, \frac{K+1}{2}\right)\right) \end{bmatrix}, \tag{21}$$

where $\bar{f}_{\frac{K+1}{2}}$ corresponds to the spatial position of $\bar{d}$.

The final term needed for the correlation model is $\hat{\sigma}^2_{d_i}$, which corresponds to the underlying desired signal variance associated with the reference patch. Since $\mathbf{R}$ and $\mathbf{p}$ are based on a desired signal with unit variance, we scale these by an estimate of the desired signal variance for each reference patch in the observed image to obtain the appropriate values. To obtain this estimate, we first compute the sample variance estimate of the pixels in $\tilde{\mathbf{g}}_i$ and we denoted this as $\hat{\sigma}_{g_i}$. We then subtract the noise variance to give an estimate of the noise-free observed signal variance as

$$\hat{\sigma}^2_{f_i} = \hat{\sigma}^2_{g_i} - \sigma^2_\eta. \tag{22}$$

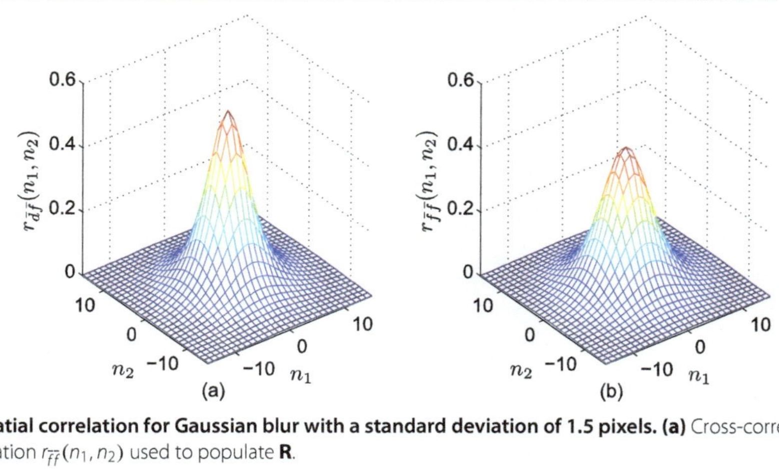

Figure 4 Single-patch spatial correlation for Gaussian blur with a standard deviation of 1.5 pixels. (a) Cross-correlation $r_{\bar{d}\bar{f}}(n_1, n_2)$ used to populate **p**. **(b)** Autocorrelation $r_{\bar{f}\bar{f}}(n_1, n_2)$ used to populate **R**.

Note that in practice, we do not allow the value in Equation 22 to go below a specified minimum value. Using Equation 17, it can be shown that the relationship between $\sigma^2_{d_i}$ and $\sigma^2_{f_i}$ is given by [5]

$$\sigma^2_{d_i} = \frac{1}{C(\rho)}\sigma^2_{f_i}, \quad (23)$$

where

$$C(\rho) = \sum_{-\infty}^{\infty}\sum_{-\infty}^{\infty} \rho^{\sqrt{n_1^2+n_2^2}}\,\tilde{h}(n_1, n_2), \quad (24)$$

and $\tilde{h}(n_1, n_2) = h(n_1, n_2) * h(-n_1, -n_2)$. Thus, our desired signal variance estimate is $\hat{\sigma}^2_{d_i} = \hat{\sigma}^2_{f_i}/C(\rho)$.

By substituting Equations11 and 12 into Equation 6, and dividing through by $\hat{\sigma}^2_{d_i}$, the CAWF weight vector can be computed as

$$\mathbf{w}_i = \tilde{\mathbf{R}}_i^{-1}\tilde{\mathbf{p}}_i = \left[e^{-\mathbf{D}_i/(\alpha\sigma_\eta)} \otimes \mathbf{R} + \frac{\sigma^2_\eta}{\hat{\sigma}^2_{d_i}}\mathbf{I}\right]^{-1} e^{-[\mathbf{D}_i]_1/(\alpha\sigma_\eta)} \otimes \mathbf{p}. \quad (25)$$

Note that the DC response of the CAWF filter is not guaranteed to be one (i.e., the weights may not sum to 1). To prevent artifacts when processing an image that is not zero mean, we normalize the weights to sum to one by dividing the weight vector by the sum of the weights for each i before computing the weighted sum in Equation 5. From Equation 25, it is clear that CAWF weights adapt spatially based on the local signal variance, the variance of the noise, and the distance matrix among the similar patches. There are two tuning parameters in the correlation model, ρ, which controls the correlation between samples within a given patch expressed in $\mathbf{R}$ and $\mathbf{p}$, and α which controls the correlation between patches. We have found that the algorithm is not highly sensitive to these tuning parameters, and good performance can be obtained for a wide range of images using a specified fixed value for these. Note that in addition to providing an estimate of the desired image, an estimate of the MSE itself can be readily generated based on the correlation model. This estimated MSE is given by [9]

$$\hat{J}_i = E\left\{\left(d_i - \hat{d}_i\right)^2\right\} = \hat{\sigma}^2_{d_i} - 2\mathbf{w}_i^T\tilde{\mathbf{p}}_i + \mathbf{w}_i^T\tilde{\mathbf{R}}_i\mathbf{w}_i. \quad (26)$$

A block diagram showing all of the key steps in the CAWF filter is shown in Figure 5.

To better understand the workings of the collaborative correlation model in determining the filter weights, consider Examples 1 and 2 shown in Figures 6 and 7, respectively. These examples are for the case of blur with a Gaussian PSF with a 1 pixel standard deviation and Gaussian noise of standard deviation 20. The patch size

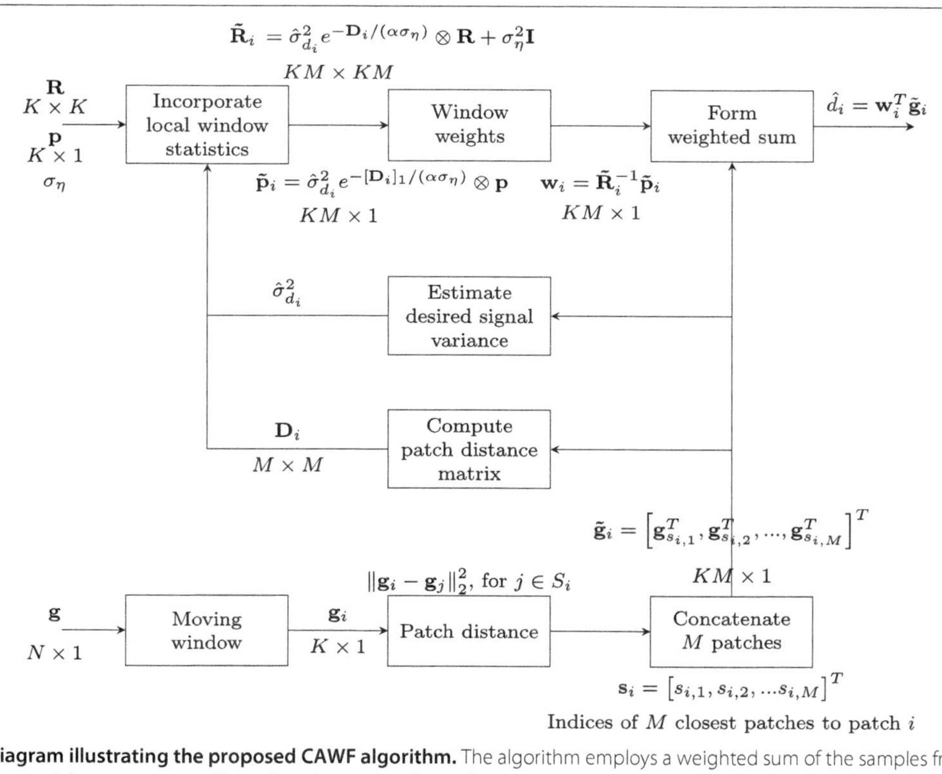

Figure 5 Block diagram illustrating the proposed CAWF algorithm. The algorithm employs a weighted sum of the samples from similar patches. The Wiener weights are computed based on the new multi-patch correlation model as shown.

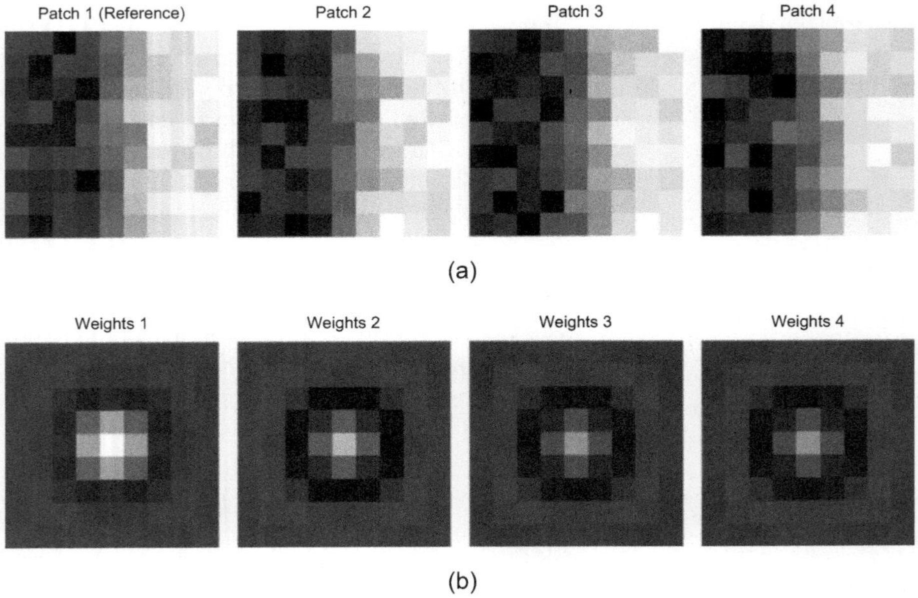

Figure 6 Example 1 showing a group of similar patches with small inter-patch distances. (a) Spatial domain similar patches; **(b)** CAWF weights corresponding to the four patches.

is $K = 9 \times 9 = 81$, and there are $M = 4$ patches. Example 1 in Figure 6 shows patches for an edge region where the multiple patches have a very similar underlying structure, and hence small inter-patch distances. In contrast, Example 2 in Figure 7 shows the case of dissimilar patches with large inter-patch distances. In Figures 6 and 7, (a) shows the spatial domain patches, and (b) shows the CAWF weights corresponding to these patches. Note that in Example 1, pixels in all of the patches get significant weight as shown in Figure 6b. In contrast, for the dissimilar patches in Example 2, only the reference patch pixels get significant weight, as shown in Figure 7b. Figure 8 provides additional insight by comparing the filter weights, summed over patches, for Examples 1 and

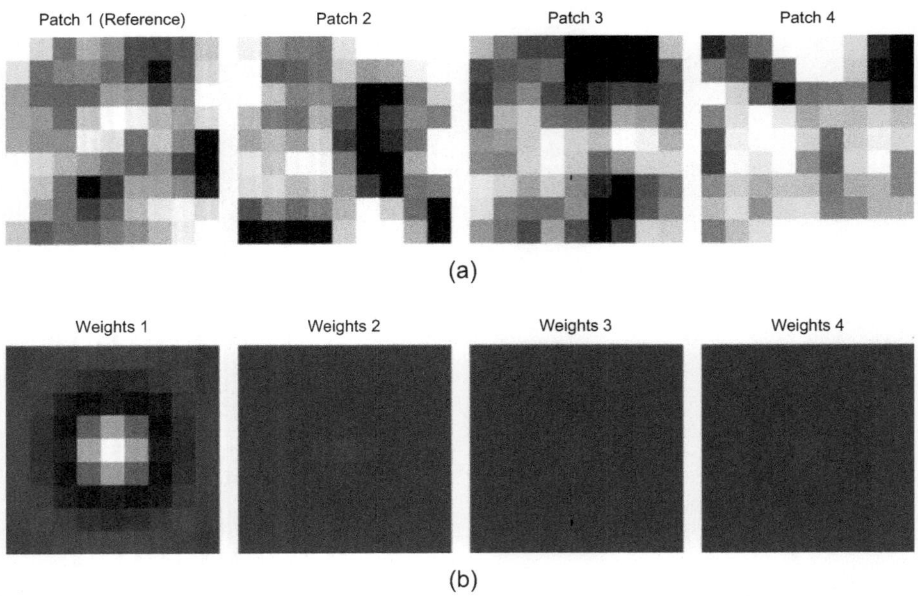

Figure 7 Example 2 showing a group of dissimilar patches. (a) Spatial domain patches; **(b)** CAWF weights corresponding to the four patches.

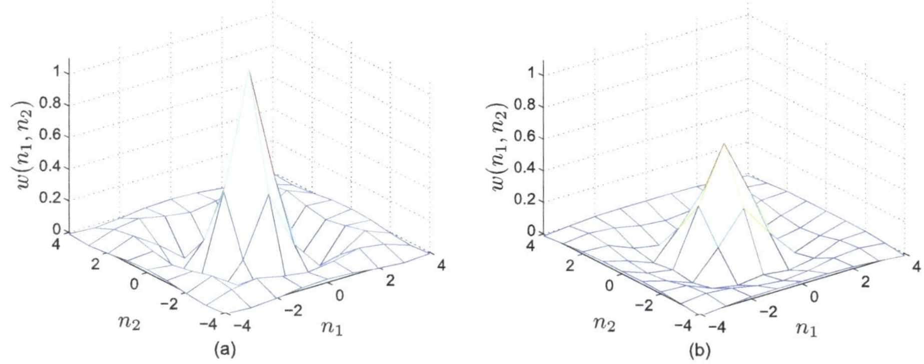

Figure 8 Mesh plots of the CAWF weights summed over patches. (a) Example 1 with similar patches and **(b)** Example 2 with dissimilar patches. Because of the high patch similarity, the weights for Example 1 represent a more aggressive deconvolution than in Example 2.

2 side by side. When many similar patches are available, more noise reduction is possible. In turn, this allows the deconvolution to be more aggressive. This can be seen in the more aggressive deconvolution weights in Figure 8a, compared with those in Figure 8b. On the other hand, if no low-distance patches can be found for the reference, the CAWF algorithm essentially reverts to a single-patch AWF filter, giving non-zero weights only to the reference patch. Because less noise reduction can be achieved this way, the filter automatically becomes less aggressive in its deconvolution, so as to not exaggerate noise.

These examples show how the distance matrix plays an interesting role in the joint deblurring/denoising aspect of the CAWF. This type of spatially adaptive deconvolution is unlike anything we have seen before in other multi-patch restoration algorithms. A similar process occurs with noise only. In that case, the CAWF balances the amount of spatial low-pass filtering employed. When good patch matches are found, the CAWF relies more on patch fusion for noise reduction. When no good matches are found, it resorts to more spatial smoothing. All of this is also balanced by the estimate of the local desired signal variance, $\hat{\sigma}^2_{d_i}$. Generally, more smoothing is done in low signal variance areas. One last point of interest with regard to these examples relates to the structure of the multi-patch autocorrelation matrix $\tilde{\mathbf{R}}_i$. These matrices are shown in Figure 9 for Examples 1 and 2. Note that based on Equation 11, these are both block matrices made up of a 4×4 grid of scaled 81×81 $\mathbf{R}$ submatrices. The $\mathbf{R}$ matrix itself has an apparent 9×9 substructure, due to the column-ordered lexicographical representation of each patch. Note that the off diagonal blocks of $\tilde{\mathbf{R}}_i$ for Example 2 in Figure 9b are essentially zero. In most cases, the inter-block correlations will lie between Examples 1 and 2.

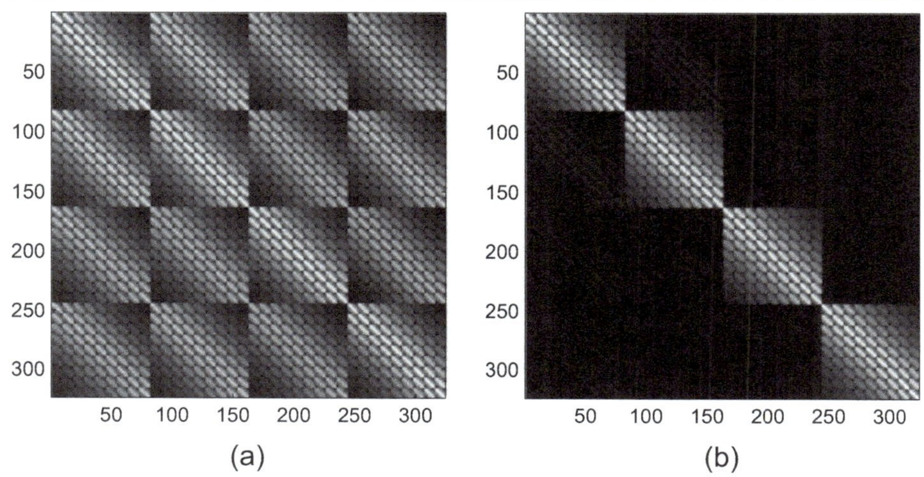

Figure 9 Multi-patch autocorrelation matrix $\tilde{\mathbf{R}}_i$. (a) Example 1 with similar patches and **(b)** Example 2 with dissimilar patches.

3.3 Multi-pixel estimation and aggregation

In the CAWF algorithm described in Section 3.1, one pixel is estimated for each reference patch. However, in a manner similar to that in [5], it is possible to estimate multiple desired pixels from each multi-patch observation vector $\tilde{\mathbf{g}}_i$. In fact, all of the desired pixels corresponding to $\tilde{\mathbf{g}}_i$ can be estimated. Let this full $KM \times 1$ vector of desired pixels be denoted $\tilde{\mathbf{d}}_i$. If all multi-patch observation vectors are used in this fashion, many estimates of each desired pixel are obtained. These can be aggregated by a simple average. In the case of noise only, we have observed that aggregation yields improved results. For joint deblurring and denoising with any significant amount of blur, the aggregation does not appear to provide any advantage. However, this multi-pixel estimation approach can be used to reduce the computational complexity, since not every multi-patch observation vector must be processed in order to form a complete image estimate.

To perform the multi-pixel estimation, the CAWF filter output is expressed as

$$\hat{\tilde{\mathbf{d}}}_i = \mathbf{W}_i^T \tilde{\mathbf{g}}_i, \tag{27}$$

where $\hat{\tilde{\mathbf{d}}}_i$ is the estimate of $\tilde{\mathbf{d}}_i$, and $\mathbf{W}_i$ is a $KM \times KM$ matrix of weights. The weight matrix is given by

$$\mathbf{W}_i = \tilde{\mathbf{R}}_i^{-1} \tilde{\mathbf{P}}_i, \tag{28}$$

where

$$\tilde{\mathbf{P}}_i = E\left\{\tilde{\mathbf{f}}_i \tilde{\mathbf{d}}_i^T\right\} = \hat{\sigma}_{d_i}^2 e^{-\mathbf{D}_i/(\alpha\sigma_\eta)} \otimes \mathbf{P}, \tag{29}$$

$\mathbf{P} = E\left\{\bar{\mathbf{f}}\bar{\mathbf{d}}^T\right\}$ is a $K \times K$ normalized cross-correlation matrix, and $\bar{\mathbf{d}}$ is the $K \times 1$ desired vector corresponding to $\bar{\mathbf{f}}$.

3.4 Computational complexity and implementation

Here, we briefly address the computational complexity of the CAWF filter by tracking the number of floating point operations (flops), where a flop is defined as one multiply plus add operation. The first action of the CAWF filter is finding similar patches. This requires computing L distances of K dimensional vectors (note that L is the search window size, and K is the patch size in pixels). The next step is computing the distance matrix based on Equation 13 for the M selected patches. This requires computing $M^2/2$ scaled and shifted distances for K dimensional vectors. The Kronecker products for $\tilde{\mathbf{R}}_i$ and $\tilde{\mathbf{p}}_i$ require $(KM)^2$ and KM multiplies, respectively. However, the main computational burden of the CAWF filter comes next with the computation of the weights in Equation 6. This can be done using Cholesky factorization, which requires $(KM)^3/3$ flops to perform LU decomposition for the $KM \times KM$ autocorrelation matrix $\tilde{\mathbf{R}}_i$. Computing the weights from the LU decomposition requires $2(KM)^2$ flops using forward and backward substitution. The final weighted sum operation is accomplished with KM flops. Since the dominant term in the computational complexity is the Cholesky factorization, we might conclude that the complexity of the CAWF filter is $O((KM)^3)$. Thus, the complexity of the CAWF algorithm goes up significantly with larger windows sizes, K, and more similar patches, M. However, an important thing to note is that the CAWF algorithm is completely parallel at the output pixel level. Unlike most variational image restoration methods, each output pixel can be computed independently and in parallel. Also, the CAWF approach requires only one pass over the data.

To put the CAWF computational complexity into context, consider that the AWF method employed here, with a spatially varying signal-to-noise ratio (SNR) estimate, may be viewed as a special case of the CAWF with $M = 1$. Thus, increasing M for CAWF causes a corresponding increase in complexity according to $O((KM)^3)$. The NLM method shares the same distance computations and comparisons and CAWF. However, in contrast to CAWF, NLM only requires L flops per output in the weighted sum, since it only weights the center sample of each patch in the search window. Although significantly simpler computationally, NLM does not fully exploit all of the information in the patches and it cannot perform deconvolution. Also, AWF is not able to exploit multi-patch information.

For pure denoising application, we have found that good results can be obtained with CAWF for $M = 10$, and $K = 3 \times 3 = 9$ for light noise and $K = 5 \times 5 = 25$ for moderate to heavy noise. In the case of joint deblurring and denoising, a larger window size is needed for adequate deconvolution. We have found that $K = 9 \times 9 = 81$ is a reasonable choice for light to moderate blurring. Our implementation uses MATLAB with no parallel acceleration or mex files, and processing is done on a PC with Intel®Xeon®Processor 3.7 GHz. CAWF processing time for a pure denoising application with a 512×512 image using $K = 9$ and $M = 10$ is 155 s. For context, the AWF processing takes 33 s, and NLM takes 3.2 s.

4 Experimental results

In this section, we demonstrate the efficacy of the proposed CAWF algorithm using images with a variety of simulated degradations and using real video frames. We also present a parameter sensitivity analysis. The filter parameters used for all of the experimental results are listed in Table 1. Note that for a given scenario, the same parameters are used for processing all of the test images.

4.1 Simulated data

In this section, we present quantitative results using simulated data. We consider two cases: noise only and blur with noise. For each case, we consider four specific scenarios

Table 1 CAWF parameters used in experimental results

Parameter name	Variable	Case 1: noise only		Case 2: blur and noise
		Selected value	Selected value	Selected value
		$\sigma_\eta < 20$	$\sigma_\eta \geq 20$	
Patch size	K	$3 \times 3 = 9$	$5 \times 5 = 25$	$9 \times 9 = 81$
Search window size	L	$17 \times 17 = 289$	$11 \times 11 = 121$	$9 \times 9 = 81$
Number of patches	M	10	10	8
Autocorrelation decay	ρ	0.65	0.70	0.65
Patch similarity decay	α	2.00	1.40	1.20
Distance offset	D_0	0.25	0.50	0.00
Aggregation	N/A	Averaging	Averaging	None

and use six test images. Also, for each case, we compare against state-of-the art methods for which MATLAB implementations are publicly available.

The test images are shown in Figure 10. These are 8-bit uncompressed images with a high level of detail. We use two quantitative performance metrics to evaluate the restorations. The first is the commonly used peak signal-to-noise ratio (PSNR), defined as

$$\mathrm{PSNR}\left(\mathbf{d}, \hat{\mathbf{d}}\right) = 10 \log_{10}\left(\frac{255^2}{\frac{1}{N}\sum_{i=1}^{N}\left(d_i - \hat{d}_i\right)^2}\right). \quad (30)$$

We also use the structural similarity (SSIM) index [46], which many argue is more consistent with subjective perception than PSNR. When reporting PSNR, we also include the improvement in PSNR (ISNR) for the reader's convenience. This is given by

$$\mathrm{ISNR} = \mathrm{PSNR}\left(\mathbf{d}, \hat{\mathbf{d}}\right) - \mathrm{PSNR}\left(\mathbf{d}, \mathbf{g}\right). \quad (31)$$

4.1.1 Additive Gaussian noise

In our first case, we consider additive Gaussian noise with no PSF blur (i.e., $h(n_1, n_2) = \delta(n_1, n_2)$). We consider four different noise standard deviations. The denoising benchmark methods are NLM [15], Globalized NLM (GLIDE-NLM) [37], PLOW [36], BM3D [32], and the single patch AWF [5]. Note that the NLM implementation is from [37], and AWF used is the same as CAWF with no aggregation and $M = 1$.

The PSNR comparison is provided in Table 2, and the SSIM comparison is in Table 3. Note that CAWF provides the highest PSNR results in Table 2 in all but one instance,

Figure 10 All truth images used in simulated data experimental results. (a) Aerial (491 × 434); **(b)** bridge (512 × 512); **(c)** river (Kodak 11) (768 × 512); **(d)** bones (512 × 768 rotated); **(e)** building (768 × 512); and **(f)** gazebo (768 × 512).

Table 2 PSNR comparison for additive Gaussian noise

Image	Method	PSNR (ISNR)			
		$\sigma_\eta = 10$	$\sigma_\eta = 20$	$\sigma_\eta = 30$	$\sigma_\eta = 40$
Aerial	Corrupted	28.13	22.11	18.59	16.09
491 × 434	NLM	30.14 (2.01)	25.56 (3.45)	23.62 (5.03)	22.35 (6.26)
	GLIDE-NLM	30.25 (2.12)	25.82 (3.71)	23.79 (5.19)	22.58 (6.49)
	PLOW	28.76 (0.63)	25.30 (3.19)	23.68 (5.09)	22.64 (6.55)
	BM3D	30.61 (2.48)	26.48 (4.37)	24.38 (5.79)	22.92 (6.82)
	AWF	30.19 (2.06)	26.09 (3.98)	24.05 (5.46)	22.73 (6.63)
	CAWF	*30.67 (2.54)*	*26.55 (4.44)*	*24.50 (5.90)*	*23.13 (7.04)*
Bridge	Corrupted	28.14	22.11	18.59	16.09
512^2	NLM	30.56 (2.42)	26.36 (4.24)	24.69 (6.09)	23.62 (7.52)
	GLIDE-NLM	30.64 (2.51)	26.55 (4.43)	24.82 (6.22)	23.73 (7.64)
	PLOW	30.00 (1.87)	26.70 (4.58)	25.23 (6.63)	24.25 (8.16)
	BM3D	*31.17 (3.04)*	27.27 (5.16)	25.46 (6.87)	24.31 (8.21)
	AWF	30.58 (2.45)	26.87 (4.76)	25.09 (6.50)	23.90 (7.80)
	CAWF	31.11 (2.97)	*27.31 (5.19)*	*25.50 (6.91)*	*24.33 (8.23)*
River	Corrupted	28.13	22.11	18.59	16.09
(Kodak 11)	NLM	29.95 (1.82)	25.46 (3.35)	23.63 (5.04)	22.47 (6.37)
768 × 512	GLIDE-NLM	29.84 (1.71)	25.71 (3.60)	23.78 (5.19)	22.62 (6.53)
	PLOW	28.53 (0.40)	24.70 (2.58)	22.94 (4.35)	22.10 (6.01)
	BM3D	30.37 (2.24)	26.16 (4.05)	24.16 (5.57)	22.86 (6.77)
	AWF	29.96 (1.82)	25.90 (3.78)	23.90 (5.31)	22.61 (6.52)
	CAWF	*30.48 (2.35)*	*26.32 (4.20)*	*24.35 (5.76)*	*23.11 (7.02)*
Bones	Corrupted	28.13	22.11	18.59	16.09
512 × 768	NLM	30.36 (2.23)	26.65 (4.54)	25.50 (6.91)	24.72 (8.63)
	GLIDE-NLM	30.48 (2.35)	26.86 (4.75)	25.50 (6.91)	24.78 (8.68)
	PLOW	29.33 (1.20)	26.54 (4.43)	25.60 (7.01)	24.94 (8.85)
	BM3D	30.85 (2.72)	27.18 (5.07)	25.81 (7.22)	25.05 (8.96)
	AWF	30.46 (2.33)	27.13 (5.02)	25.60 (7.01)	24.51 (8.41)
	CAWF	*30.93 (2.79)*	*27.46 (5.35)*	*26.03 (7.44)*	*25.11 (9.02)*
Building	Corrupted	28.13	22.11	18.59	16.09
768 × 512	NLM	30.61 (2.48)	26.35 (4.24)	24.50 (5.91)	23.30 (7.21)
	GLIDE-NLM	30.32 (2.19)	26.43 (4.32)	24.64 (6.05)	23.45 (7.36)
	PLOW	29.25 (1.12)	25.04 (2.93)	23.32 (4.73)	22.51 (6.42)
	BM3D	30.97 (2.84)	26.91 (4.79)	24.88 (6.29)	23.62 (7.53)
	AWF	30.46 (2.33)	26.51 (4.40)	24.51 (5.92)	23.17 (7.08)
	CAWF	*31.12 (2.98)*	*27.05 (4.94)*	*25.09 (6.50)*	*23.84 (7.74)*
Gazebo	Corrupted	28.13	22.11	18.59	16.09
768 × 512	NLM	31.11 (2.98)	26.71 (4.59)	24.93 (6.34)	23.77 (7.68)
	GLIDE-NLM	31.11 (2.98)	26.97 (4.86)	25.05 (6.46)	23.95 (7.86)
	PLOW	29.82 (1.68)	25.69 (3.58)	24.20 (5.61)	23.46 (7.36)
	BM3D	31.57 (3.44)	27.40 (5.29)	25.48 (6.89)	24.29 (8.19)
	AWF	30.78 (2.64)	26.88 (4.77)	24.91 (6.32)	23.59 (7.50)
	CAWF	*31.60 (3.47)*	*27.49 (5.37)*	*25.55 (6.96)*	*24.30 (8.21)*

Table 3 SSIM for additive Gaussian noise

Image	Method	SSIM			
		$\sigma_\eta = 10$	$\sigma_\eta = 20$	$\sigma_\eta = 30$	$\sigma_\eta = 40$
Aerial	Corrupted	0.9719	0.9046	0.8224	0.7388
491 × 434	NLM	0.9753	0.9058	0.8412	0.7872
	GLIDE-NLM	0.9747	0.9058	0.8427	0.7916
	PLOW	0.9734	0.9252	0.8760	0.8282
	BM3D	0.9783	0.9340	0.8835	0.8287
	AWF	0.9762	0.9304	0.8804	0.8322
	CAWF	*0.9787*	*0.9351*	***0.8859***	*0.8365*
Bridge	Corrupted	0.9542	0.8512	0.7365	0.6305
512^2	NLM	0.9611	0.8651	0.7925	0.7379
	GLIDE-NLM	0.9621	0.8757	0.8010	0.7363
	PLOW	0.9587	0.9051	0.8485	0.7943
	BM3D	*0.9692*	*0.9126*	*0.8539*	0.7963
	AWF	0.9652	0.9061	0.8477	0.7927
	CAWF	0.9676	0.9109	0.8523	*0.7985*
River	Corrupted	0.9512	0.8429	0.7252	0.6196
(Kodak 11)	NLM	0.9525	0.8343	0.7588	0.7042
768 × 512	GLIDE-NLM	0.9548	0.8461	0.7617	0.7007
	PLOW	0.9505	0.8732	0.7921	0.7299
	BM3D	0.9597	0.8792	0.8066	0.7437
	AWF	0.9574	0.8823	0.8142	0.7512
	CAWF	*0.9609*	*0.8873*	*0.8207*	*0.7634*
Bones	Corrupted	0.9259	0.7735	0.6224	0.4979
512 × 768	NLM	0.9329	0.7946	0.7269	0.6821
	GLIDE-NLM	0.9379	0.8163	0.7351	0.6725
	PLOW	0.9240	0.8295	0.7587	0.7047
	BM3D	0.9417	0.8404	0.7641	0.7091
	AWF	0.9389	0.8459	0.7669	0.6970
	CAWF	*0.9439*	*0.8551*	*0.7839*	*0.7277*
Building	Corrupted	0.9153	0.7660	0.6349	0.5318
768 × 512	NLM	0.9395	0.8370	0.7808	0.7386
	GLIDE-NLM	0.9426	0.8416	0.7835	0.7385
	PLOW	0.9379	0.8577	0.7823	0.7262
	BM3D	0.9482	0.8748	0.8141	0.7637
	AWF	0.9398	0.8583	0.7797	0.7072
	CAWF	*0.9502*	*0.8788*	*0.8190*	*0.7665*
Gazebo	Corrupted	0.9130	0.7678	0.6428	0.5435
768 × 512	NLM	0.9561	0.8653	0.8090	0.7649
	GLIDE-NLM	0.9578	0.8712	0.8117	0.7686
	PLOW	0.9525	0.8698	0.8046	0.7559
	BM3D	0.9620	0.8963	*0.8397*	*0.7940*
	AWF	0.9452	0.8740	0.7950	0.7230
	CAWF	*0.9611*	*0.8988*	0.8387	0.7837

with BM3D generally providing the next highest PSNR values. Looking at Table 3, we see that CAWF still performs well, but BM3D does provide a higher SSIM in 5 of 24 scenarios. These results also show that CAWF consistently outperforms AWF. This demonstrates the advantage of using multiple patches within this framework. It is also interesting to note that AWF itself does quite well compared with some of the benchmark methods on these data, especially in the SSIM metric.

Selected regions of interest (ROIs) from images bridge and river for the noise-only case with $\sigma_\eta = 30$ are shown in Figures 11 and 12, respectively. We find that BM3D tends to do a better job in smooth areas, and CAWF generally appears better in high-detail texture areas. Note that more branches on the small trees are visible in the CAWF estimate in Figure 11f, compared with that for BM3D in Figure 11e. Also, the texture in the tree foliage appears to be better preserved with CAWF processing in Figure 12f, compared with that for BM3D in Figure 12e.

The results in Figure 13 show how the CAWF method can produce an estimate of the MSE on a pixel-by-pixel basis. Figure 13a shows an ROI from the image

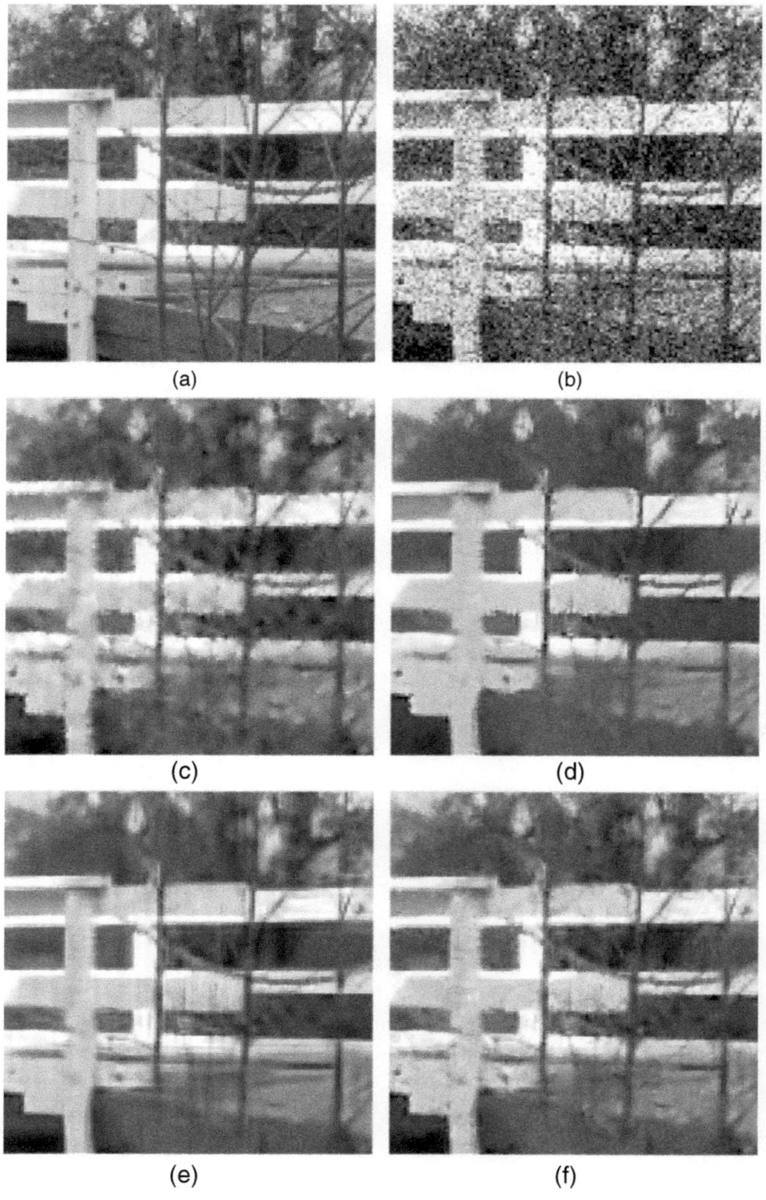

Figure 11 Region of interest from the image bridge with $\sigma_\eta = 30$ Gaussian noise. (a) Truth image, **(b)** noisy image, **(c)** AWF, **(d)** GLIDE-NLM, **(e)** BM3D, and **(f)** CAWF.

aerial with noise of standard deviation 10. The CAWF estimate image is shown in Figure 13b. The estimated MSE, computed according to Equation 26, is shown in Figure 13c. The average squared error over 100 noise realizations is shown in Figure 13d. Aside from some of the small high frequency structures, the estimated MSE is appears similar to the average squared error. The ability to provide an estimate of the MSE is another distinctive feature of the CAWF method among other multi-patch methods.

4.1.2 Gaussian blur plus Gaussian noise

We consider four scenarios of Gaussian blur plus Gaussian noise, and these are listed in Table 4. The benchmark methods in this case must be able to address both blur and noise. We use L0-Abs [47], TVMM [48], BM3DDEB [38], IDD-BM3D [39], and AWF [5]. Note that for IDD-BM3D, the tuning parameters are selected from those used in [39]. In particular, we use the tuning parameters from Scenario 4 in [39], as these produce the highest PSNR values in the current experiments.

Figure 12 Region of interest from the image river with $\sigma_\eta = 30$ Gaussian noise. (a) Truth image, **(b)** noisy image, **(c)** AWF, **(d)** GLIDE-NLM, **(e)** BM3D, and **(f)** CAWF.

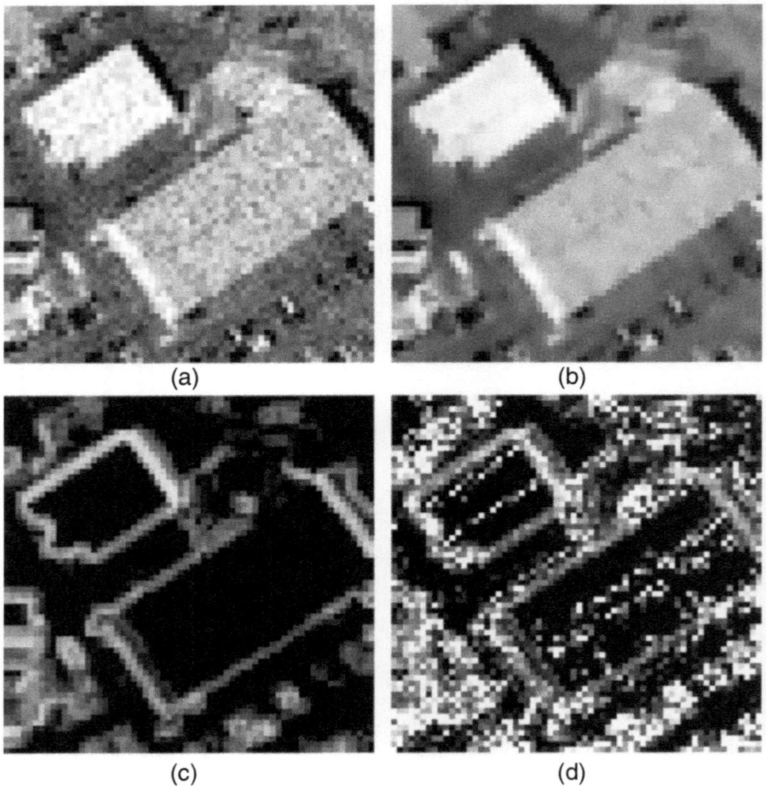

Figure 13 Predicted MSE for the image aerial with $\sigma_\eta = 10$ Gaussian noise. (a) Noisy image, **(b)** non-aggregated CAWF image estimate, **(c)** CAWF estimate of MSE, $\hat{J}_i$, and **(d)** average squared error over 100 noise realizations.

The PSNR comparison is provided in Table 5, and the SSIM comparison is in Table 6. Here, IDD-BM3D provides the highest PSNR in 15 of 24 instances, and CAWF provides the highest in 9 of 24. In terms of SSIM, IDD-BM3D provides the highest values in 5 of 24 instances and CAWF in 19 of 24. We see a similar situation with AWF as we did in the noise-only experiments. The AWF method does well compared to many of the benchmark methods, but CAWF consistently outperforms it. Selected ROIs from the images aerial and bones for Scenario III in Table 4 are shown in Figures 14 and 15, respectively. Again, CAWF appears to do a good job restoring image detail, based on subjective evaluation and SSIM. Note that a larger blur kernel generally demands a larger restoration filter window. The variational benchmark methods, like IDD-BM3D, are not restricted to local processing like CAWF and AWF. Thus, they may have an advantage in high levels of blur. However, the iterative nature of these methods also means that a full parallel implementation may not be possible.

Table 4 Scenarios of Gaussian blur and Gaussian noise

Scenario	PSF	σ_η
I	Gaussian std. = 1.0	10
II	Gaussian std. = 1.5	10
III	Gaussian std. = 1.0	20
IV	Gaussian std. = 1.5	20

4.2 Real data

Real video frames have been acquired of an outdoor natural scene on the campus of the University of Dayton using an Imaging Source 8 bit grayscale camera (DMK 23U618) with Sony ICX618ALA sensor. A short exposure time is used, proving a low SNR. A sequence of 500 frames is acquired for the static scene. This allows us to form a temporal average as a type of reference with which to compare the noise reduction estimates. Since this real noise will have both a signal-dependent and signal-independent component, we apply an Anscombe transform to stabilize the local noise variance prior to applying all denoising methods [49]. After the transform and scaling, an effective constant noise standard deviation of $\sigma_\eta = 9.11$ is estimated and used for all methods.

Table 5 PSNR comparison for Gaussian blur plus Gaussian noise

Image	Method	PSNR (ISNR)			
		Scenario I	Scenario II	Scenario III	Scenario IV
Aerial 491 × 434	Corrupted	22.64	20.79	19.98	18.88
	L0-Abs	24.02 (1.38)	21.88 (1.10)	21.80 (1.82)	20.36 (1.48)
	TVMM	24.18 (1.54)	22.18 (1.40)	21.94 (1.95)	20.49 (1.60)
	BM3DDEB	24.41 (1.76)	22.56 (1.78)	22.54 (2.56)	21.32 (2.44)
	IDD-BM3D	*24.71 (2.06)*	*22.69 (1.90)*	22.71 (2.73)	21.32 (2.44)
	AWF	24.25 (1.61)	22.33 (1.55)	22.64 (2.66)	21.29 (2.41)
	CAWF	24.47 (1.83)	22.56 (1.78)	*22.83 (2.85)*	*21.43 (2.55)*
Bridge 512^2	Corrupted	24.19	22.80	20.76	20.07
	L0-Abs	25.63 (1.44)	24.08 (1.28)	23.52 (2.77)	22.53 (2.47)
	TVMM	25.73 (1.53)	24.27 (1.47)	23.57 (2.82)	22.84 (2.77)
	BM3DDEB	26.02 (1.82)	24.71 (1.91)	24.47 (3.72)	23.62 (3.55)
	IDD-BM3D	*26.23 (2.03)*	*24.79 (2.00)*	*24.52 (3.76)*	*23.63 (3.56)*
	AWF	25.78 (1.59)	24.43 (1.63)	24.39 (3.63)	23.45 (3.38)
	CAWF	25.94 (1.74)	24.56 (1.77)	24.48 (3.72)	23.48 (3.41)
River (Kodak 11) 768 × 512	Corrupted	22.14	20.66	19.69	18.79
	L0-Abs	23.22 (1.08)	21.43 (0.77)	21.45 (1.76)	20.33 (1.54)
	TVMM	23.13 (1.00)	21.52 (0.86)	21.14 (1.45)	20.33 (1.53)
	BM3DDEB	23.24 (1.11)	21.78 (1.12)	21.72 (2.03)	20.90 (2.11)
	IDD-BM3D	*23.63 (1.49)*	21.90 (1.24)	21.95 (2.26)	20.96 (2.17)
	AWF	23.29 (1.15)	21.78 (1.12)	22.06 (2.36)	21.06 (2.27)
	CAWF	23.48 (1.34)	*21.93 (1.28)*	*22.21 (2.52)*	*21.15 (2.36)*
Bones 512 × 768	Corrupted	24.80	23.78	21.01	20.56
	L0-Abs	26.07 (1.27)	25.14 (1.36)	24.65 (3.64)	24.17 (3.62)
	TVMM	25.94 (1.15)	25.11 (1.33)	24.52 (3.51)	24.24 (3.68)
	BM3DDEB	26.33 (1.54)	25.48 (1.71)	25.23 (4.22)	24.79 (4.23)
	IDD-BM3D	*26.46 (1.66)*	*25.51 (1.74)*	25.26 (4.24)	*24.80 (4.24)*
	AWF	26.31 (1.51)	25.37 (1.59)	25.25 (4.24)	24.69 (4.14)
	CAWF	26.44 (1.64)	25.43 (1.66)	*25.28 (4.27)*	24.67 (4.12)
Building 768 × 512	Corrupted	22.60	21.27	19.96	19.18
	L0-Abs	23.88 (1.28)	22.24 (0.97)	22.25 (2.29)	21.22 (2.04)
	TVMM	23.82 (1.22)	22.42 (1.16)	22.11 (2.15)	21.57 (2.39)
	BM3DDEB	23.86 (1.26)	22.53 (1.27)	22.49 (2.53)	21.73 (2.55)
	IDD-BM3D	*24.25 (1.65)*	*22.69 (1.43)*	22.75 (2.79)	21.85 (2.66)
	AWF	23.89 (1.28)	22.52 (1.26)	22.76 (2.80)	21.85 (2.66)
	CAWF	24.05 (1.45)	22.65 (1.38)	*22.89 (2.93)*	*21.92 (2.74)*
Gazebo 768 × 512	Corrupted	23.29	21.90	20.31	19.55
	L0-Abs	24.82 (1.53)	23.11 (1.21)	23.02 (2.71)	21.93 (2.38)
	TVMM	24.80 (1.50)	23.41 (1.50)	23.34 (3.03)	22.17 (2.61)
	BM3DDEB	24.91 (1.61)	23.45 (1.55)	23.39 (3.08)	22.52 (2.97)
	IDD-BM3D	*25.29 (2.00)*	*23.65 (1.75)*	23.68 (3.37)	*22.66 (3.11)*
	AWF	24.83 (1.53)	23.37 (1.47)	23.58 (3.26)	22.58 (3.03)
	CAWF	24.99 (1.69)	23.51 (1.61)	*23.70 (3.39)*	22.66 (3.10)

Table 6 SSIM comparison for Gaussian blur plus Gaussian noise

Image	Method	SSIM			
		Scenario I	Scenario II	Scenario III	Scenario IV
Aerial 491 × 434	Corrupted	0.8891	0.7794	0.8176	0.7117
	L0-Abs	0.9087	0.8050	0.7821	0.6448
	TVMM	0.9215	0.8325	0.7885	0.6604
	BM3DDEB	0.9198	0.8571	0.8400	0.7710
	IDD-BM3D	*0.9292*	0.8591	0.8406	0.7577
	AWF	0.9199	0.8519	0.8517	0.7733
	CAWF	0.9273	*0.8662*	*0.8649*	*0.7957*
Bridge 512^2	Corrupted	0.8836	0.7975	0.7745	0.6929
	L0-Abs	0.8902	0.8044	0.7425	0.6524
	TVMM	0.8995	0.8156	0.7323	0.6689
	BM3DDEB	0.9099	0.8579	0.8294	0.7796
	IDD-BM3D	*0.9173*	0.8573	0.8252	0.7657
	AWF	0.9080	0.8503	0.8344	0.7712
	CAWF	0.9155	*0.8637*	*0.8462*	*0.7892*
River (Kodak 11) 768 × 512	Corrupted	0.8373	0.7198	0.7246	0.6161
	L0-Abs	0.8328	0.7159	0.6670	0.5560
	TVMM	0.8452	0.7219	0.6344	0.5421
	BM3DDEB	0.8487	0.7713	0.7377	0.6745
	IDD-BM3D	0.8672	0.7767	0.7463	0.6666
	AWF	0.8573	0.7736	0.7676	0.6878
	CAWF	*0.8707*	*0.7949*	*0.7837*	*0.7105*
Bones 512 × 768	Corrupted	0.8231	0.7323	0.6681	0.5884
	L0-Abs	0.8051	0.7287	0.6652	0.6181
	TVMM	0.7954	0.7192	0.6427	0.6172
	BM3DDEB	0.8320	0.7768	0.7420	0.7064
	IDD-BM3D	0.8421	0.7753	0.7389	0.6969
	AWF	0.8358	0.7699	0.7526	0.7000
	CAWF	*0.8476*	*0.7866*	*0.7615*	*0.7108*
Building 768 × 512	Corrupted	0.8119	0.7162	0.6609	0.5753
	L0-Abs	0.8356	0.7491	0.7142	0.6397
	TVMM	0.8454	0.7592	0.6936	0.6615
	BM3DDEB	0.8520	0.7913	0.7633	0.7138
	IDD-BM3D	*0.8662*	0.7953	0.7715	0.7130
	AWF	0.8509	0.7848	0.7721	0.7147
	CAWF	0.8607	*0.7974*	*0.7816*	*0.7268*
Gazebo 768 × 512	Corrupted	0.8279	0.7413	0.6819	0.6030
	L0-Abs	0.8651	0.7867	0.7527	0.6788
	TVMM	0.8741	0.8057	0.7785	0.6826
	BM3DDEB	0.8805	0.8249	0.8001	0.7500
	IDD-BM3D	*0.8898*	*0.8282*	0.8064	0.7504
	AWF	0.8796	0.8190	0.8035	0.7481
	CAWF	0.8801	0.8238	*0.8072*	*0.7561*

Figure 16a shows a 500 frame temporal average image. This image represents a near noise-free image of the scene that can be used as a reference. A single noisy frame is shown in Figure 16b. The processed single frame outputs for GLIDE-NLM, AWF, BM3D, and CAWF are shown in Figure 16c-f, respectively. The PSNR values, relative to the temporal average, for observed GLIDE-NLM, AWF, BM3D, and CAWF outputs are 31.78, 36.04, 36.33, 36.47, and 36.65, respectively. The corresponding SSIM values are 0.7525, 0.8929, 0.8975, 0.8973, and 0.9059. These results appear to be consistent with the results obtained with the simulated data.

4.3 Parameter and distance metric sensitivity

In this section, we investigate the sensitivity of the CAWF algorithm to some of the key tuning parameters listed in Table 1 and the distance metric used in the correlation model. We begin with the number of patches M. A plot of PSNR versus the number of similar patches is shown in Figure 17 for CAWF using the image bones with noise

Figure 14 Region of interest from the image aerial for blur and noise Scenario III. (a) Truth image, **(b)** degraded image, **(c)** AWF, **(d)** TVMM, **(e)** IDD-BM3D, and **(f)** CAWF.

only and $\sigma_\eta = 40$. This plot is representative of many of the restoration scenarios. One can see a 'knee' in the curve near $M = 10$. Since increasing M has a significant impact on computational complexity, we have elected to use $M = 10$ for our denoising applications. Note that deconvolution requires a larger window size than denoising. Thus, to manage computational complexity, we compensate with a somewhat lower $M = 8$, as shown in Table 1.

Next, we examine the autocorrelation decay constant, ρ, and the patch similarity decay, α, for the image aerial with additive Gaussian noise. We have evaluated CAWF PSNR for ρ ranging from 0.6 to 0.75, with all other parameter values as listed in Table 1. The maximum change in PSNR as a function of ρ, for noise levels ranging from $\sigma_\eta = 10$ to $\sigma_\eta = 40$, is only 0.13%. Similarly, we have evaluated PSNR values for α ranging from 1.0 to 2.0. The maximum change in PSNR as a function of α is observed to be 0.23%. Thus, we conclude that the CAWF method is not highly sensitive to these tuning parameters within these operating ranges.

Finally, we explore CAWF performance using different distance metrics in the correlation model. Our standard

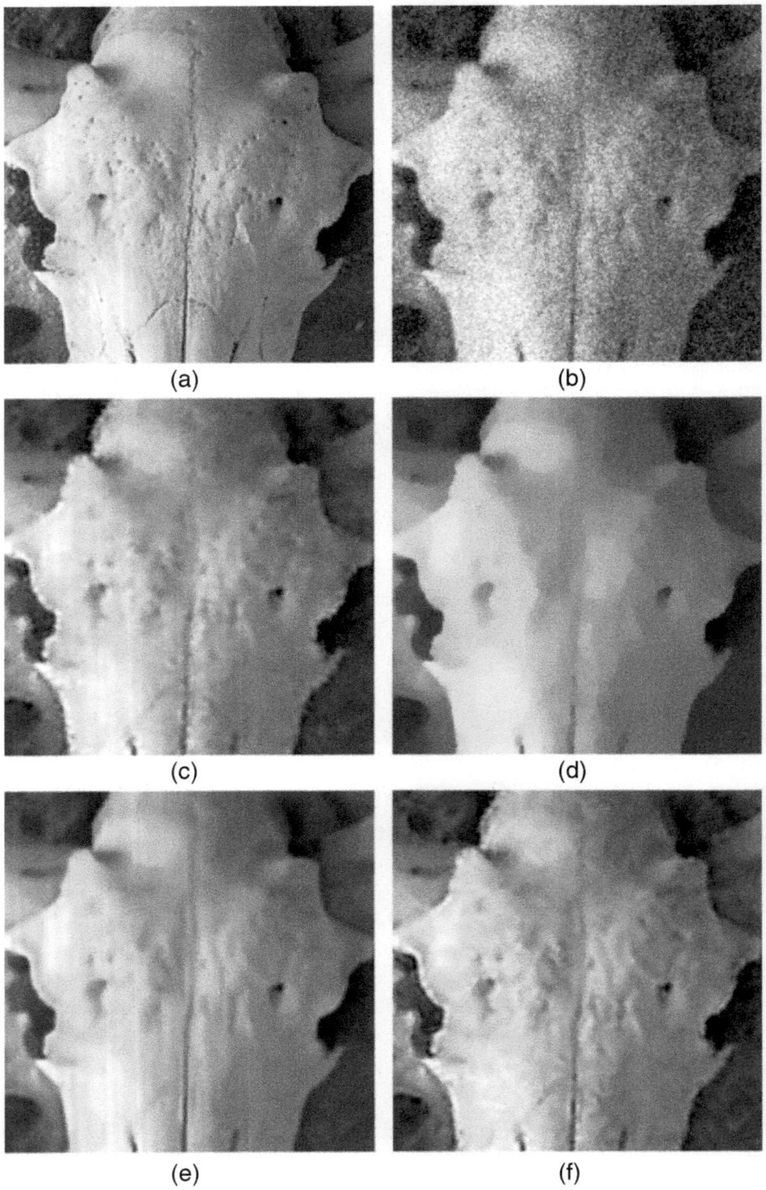

Figure 15 Region of interest from the image bones for blur and noise Scenario III. (a) Truth image, **(b)** degraded image, **(c)** AWF, **(d)** TVMM, **(e)** IDD-BM3D, and **(f)** CAWF.

metric uses the l^2-norm as defined in Equation 13. To test distance metric sensitivity, we compare distances with the l^1-, l^2-, and l^{10}-norms. For aerial with $\sigma_\eta = 10$, the PSNRs are 30.67, 30.67, and 30.60, respectively (with tuned scaling parameters). For aerial with $\sigma_\eta = 40$, the PSNRs are 23.01, 23.13, and 23.05, respectively (also with tuned scaling parameters). As with the other parameters, we do not see a strong sensitivity to the choice of distance metric. However, the l^2-norm generally provides the best results.

5 Conclusions

We have proposed a novel CAWF method for image restoration, which can be thought of an extension of the AWF [5] using multiple patches. For each reference window, M similar patches are identified. The output is formed as a single-pass weighted sum of all of the pixels from the multiple selected patches. Wiener weights are used to provide a minimum MSE estimate for this filter structure. A key aspect of the method is the new spatial-domain multi-patch correlation model, presented

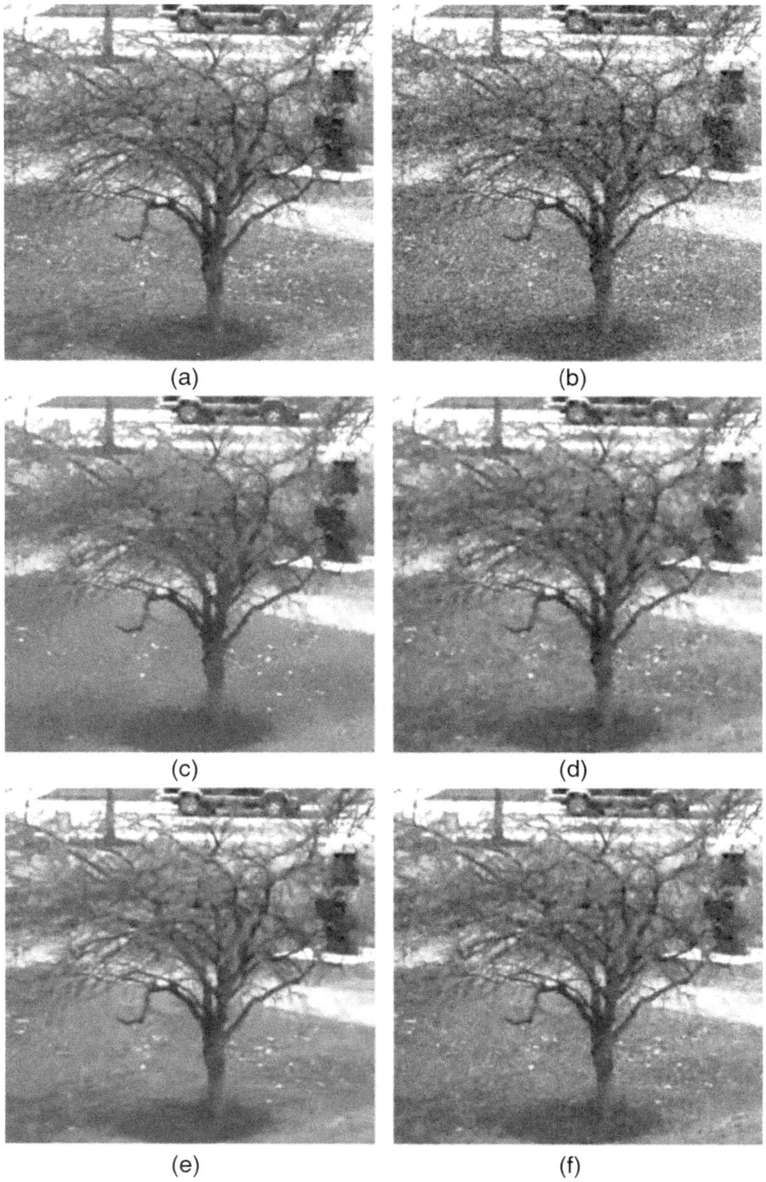

Figure 16 Region of interest from the real image data. (a) 500 frame temporal average image (reference image), **(b)** single observed frame, **(c)** GLIDE-NLM, **(d)** AWF, **(e)** BM3D, and **(f)** CAWF.

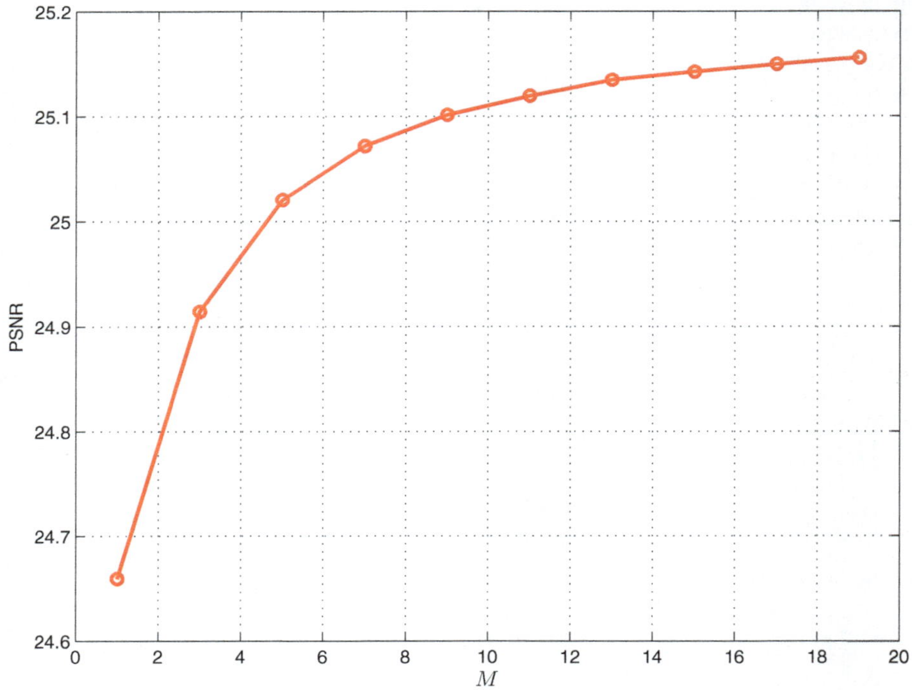

Figure 17 PSNSR versus the number of similar patches *M* for CAWF. Degradation is noise only with Bones image and $\sigma_\eta = 40$.

in Section 3.2. This model attempts to capture the spatial correlation among the samples within a given patch and also the correlations among the patches.

The CAWF is able to jointly perform denoising and deblurring. We believe this type of joint restoration is advantageous, compared with decoupling these operations. The CAWF algorithm is also capable of adapting to local signal and noise variance. Bad or missing pixels can easily be accommodated by leaving them out of the multi-patch observation vector and corresponding correlation statistics. The weights will adapt in a non-trivial way to the missing pixels [5,9,10].

In simulated and real data for Gaussian noise, the CAWF outperforms the benchmark methods in our experiments in Sections 4.1.1 and 4.2, both in PSNR and in SSIM. With blur and noise, CAWF produces the highest SSIM in more cases than the benchmark methods. However, IDD-BM3D does provide a higher PSNR in more instances. Our results show that the CAWF method consistently outperforms the AWF. This clearly demonstrates that incorporating multiple patches within this filter structure is advantageous. From the results in Section 4.3, we also conclude that CAWF performance is not highly sensitive to the tuning parameter values within a given operating range.

We believe the single-pass weighted-sum structure of the CAWF method is conceptually simple and versatile. It is also highly parallel. In principle, each output pixel can be computed in parallel. We have demonstrated that the method provides excellent performance in image restoration with noise and blur and noise. This method may be beneficial in numerous other applications as well, including those where its predecessor, the AWF, is successful [5,9-14]. We think there may also be an opportunity for further improvements in the parametric correlation model that could boost filter performance. Thus, we hope this approach will be of interest to the signal and image processing community.

Competing interests

The authors declare that they have no competing interests.

References

1. G Demoment, Image reconstruction and restoration: overview of common estimation structures and problems. IEEE Trans Acoustics, Speech Signal Process. **37**(12), 2024–36 (1989)
2. MI Sezan, AM Tekalp, Survey of recent developments in digital image restoration. Opt Eng. **29**(5), 393–404 (1990)
3. AK Jain, *Fundamentals of Digital Image Processing*. (Prentice Hall, New Jersey, 1989)
4. SP Ghael, AM Sayeed, RG Baraniuk, in *Proc of SPIE*. Improved wavelet denoising via empirical Wiener filtering, vol. 3169 (SPIE San Diego, 1997), pp. 389–99
5. RC Hardie, A fast image super-resolution algorithm using an adaptive Wiener filter. IEEE Transactions on Image Process. **16**, 2953–64 (Dec. 2007)

6. KE Barner, AM Sarhan, RC Hardie, Partition-based weighted sum filters for image restoration. IEEE Trans Image Process. **8**, 740–745 (May 1999)
7. M Shao, KE Barner, RC Hardie, Partition-based interpolation for image demosaicing and super-resolution reconstruction. Opt Eng. **44**, 107003–1–107003–14 (Oct 2005)
8. B Narayanan, RC Hardie, KE Barner, M Shao, A computationally efficient super-resolution algorithm for video processing using partition filters. IEEE Trans Circuits Syst Video Technol. **17**, 621–34 (May 2007)
9. RC Hardie, KJ Barnard, R Ordonez, Fast super-resolution with affine motion using an adaptive Wiener filter and its application to airborne imaging. Opt Express, 1926208–31 (Dec 2011)
10. RC Hardie, KJ Barnard, Fast super-resolution using an adaptive Wiener filter with robustness to local motion. Opt Express. **20**, 21053–73 (Sep 2012)
11. B Narayanan, RC Hardie, E Balster, Multiframe adaptive Wiener filter super-resolution with JPEG2000-compressed images. EURASIP J Adv Signal Process. **2014**(1), 55 (2014)
12. RC Hardie, DA LeMaster, BM Ratliff, Super-resolution for imagery from integrated microgrid polarimeters. Opt Express. **19**, 12937–60 (Jul 2011)
13. BK Karch, RC Hardie, Adaptive Wiener filter super-resolution of color filter array images. Opt Express, 2118820–41 (Aug 2013)
14. M Rucci, RC Hardie, KJ Barnard, 53. Appl Opt, C1–13 (May 2014)
15. A Buades, B Coll, JM Morel, A review of image denoising algorithms, with a new one. Multiscale Model Simul. **4**, 490–530 (2005)
16. KE Barner, GR Arce, J-H Lin, On the performance of stack filters and vector detection in image restoration. Circuits Syst Signal Process. **11**, No. 1 (Jan 1992)
17. KE Barner, RC Hardie, GR Arce, in *Proceedings of the 1994 CISS*. On the permutation and quantization partitioning of $\mathbf{R}^N$ and the filtering problem (New Jersey, Princeton, Mar 1994)
18. A Buades, B Coll, JM Morel, in *Computer Vision and Pattern Recognition, 2005. CVPR 2005. IEEE Comput Soc Conference on, vol. 2*. A non-local algorithm for image denoising (IEEE, June 2005), pp. 60–65
19. C Kervrann, J Boulanger, Optimal spatial adaptation for patch-based image denoising. IEEE Trans Image Process. **15**(10), 2866–78 (2006)
20. A Buades, B Coll, J-M Morel, Nonlocal image and movie denoising. Int J Comput Vision. **76**, 123–39 (Feb 2008)
21. Y Han, R Chen, Efficient video denoising based on dynamic nonlocal means. Image Vision Comput. **30**, 78–85 (Feb 2012)
22. Tasdizen, Principal neighborhood dictionaries for nonlocal means image denoising. IEEE Trans Image Process. **18**, 2649–60 (July 2009)
23. H Bhujle, S Chaudhuri, Novel speed-up strategies for non-local means denoising with patch and edge patch based dictionaries. IEEE Trans Image Process. **23**, 356–365 (Jan 2014)
24. Y Wu, B Tracey, P Natarajan, JP Noonan, SUSAN controlled decay parameter adaption for non-local means image denoising. Electron Lett. **49**, 807–8 (June 2013)
25. WL Zeng, XB Lu, Region-based non-local means algorithm for noise removal. Electron Lett. **47**, 1125–7 (September 2011)
26. WF Sun, YH Peng, WL Hwang, Modified similarity metric for non-local means algorithm. Electron Lett. **45**, 1307–9 (Dec 2009)
27. C Tomasi, R Manduchi, in *Proceedings of the 1998 IEEE International Conference on Computer Vision, Bombay*. Bilateral filtering for gray and color images (IEEE India, 1998)
28. H Kishan, CS Seelamantula, *Sure-fast bilateral filters. Acoustics, Speech and Signal Processing (ICASSP) 2012 IEEE International Conference on*. (IEEE, Kyoto, 2012), pp. 1129–32
29. W Kesjindatanawaj, S Srisuk, in *Communications and Information Technologies (ISCIT), 2013 13th International Symposium on*. Deciles-based bilateral filtering (IEEE Surat Thani, 2013), pp. 429–33
30. X Changzhen, C Licong, P Yigui, *An adaptive bilateral filtering algorithm and its application in edge detection. Measuring Technology and Mechatronics Automation (ICMTMA), 2010 International Conference on. vol. 1*. (IEEE, Changsha City, 2010), pp. 440–443
31. H Peng, R Rao, SA Dianat, Multispectral image denoising with optimized vector bilateral filter. IEEE Trans Image Process. **23**, 264–73 (Jan 2014)
32. K Dabov, A Foi, V Katkovnik, K Egiazarian, Image denoising by sparse 3-d transform-domain collaborative filtering. IEEE Trans Image Process. **16**, 2080–95 (Aug 2007)
33. K Dabov, A Foi, K Egiazarian, Video denoising by sparse 3d transform-domain collaborative filtering. Proc 15th Eur Signal Process Conference. **1**, 7 (2007)
34. M Maggioni, G Boracchi, A Foi, K Egiazarian, Video denoising, deblocking, and enhancement through separable 4-d nonlocal spatiotemporal transforms. IEEE Trans Image Process. **21**(9), 3952–66 (2012)
35. K Hirakawa, T Parks, Image denoising using total least squares. IEEE Trans Image Process. **15**(9), 2730–42 (2006)
36. P Chatterjee, P Milanfar, Patch-based near-optimal image denoising. IEEE Trans Image Process. **21**, 1635–49 (April 2012)
37. H Talebi, P Milanfar, Global image denoising. IEEE Trans Image Process. **23**, 755–768 (Feb 2014)
38. K Dabov, A Foi, V Katkovnik, K Egiazarian, in *SPIE Electronic Imaging*. Image restoration by sparse 3d transform-domain collaborative filtering, vol. 6812 (San Jose, Jan 2008)
39. A Danielyan, V Katkovnik, K Egiazarian, BM3D frames and variational image deblurring. IEEE Trans Image Process. **21**, 1715–28 (April 2012)
40. K Nasrollahi, TB Moeslund, Super-resolution: a comprehensive survey. Mach Vision Appl. **25**(6), 1423–68 (Aug 2014)
41. M Protter, M Elad, Super-resolution with probabilistic motion estimation. IEEE Trans Image Process. **18**(8), 1899–904 (2009)
42. M Protter, M Elad, H Takeda, P Milanfar, Generalizing the nonlocal-means to super-resolution reconstruction. IEEE Trans Image Process. **18**(1), 36–51 (2009)
43. MH Cheng, HY Chen, JJ Leou, Video super-resolution reconstruction using a mobile search strategy and adaptive patch size. Signal Process. **91**, 1284–97 (2011)
44. B Huhle, T Schairer, P Jenke, W Straber, Fusion of range and color images for denoising and resolution enhancement with a non-local filter. Comput Vision Image Understanding. **114**, 1336–45 (2012)
45. K-W Hung, W-C Siu, Single image super-resolution using iterative Wiener filter. Proc IEEE Int Conference Acoustics, Speech and Signal Process, 1269–72 (2012)
46. Z Wang, A Bovik, H Sheikh, E Simoncelli, Image quality assessment: from error visibility to structural similarity. IEEE Trans Image Process. **13**, 600–12 (April 2004)
47. J Portilla, Image restoration through l0 analysis-based sparse optimization in tight frames. Image Process (ICIP), 2009 16th IEEE Int Conference on, 3909–912 (Nov 2009)
48. J Oliveira, JM Bioucas-Dias, MA Figueire, Adaptive total variation image deblurring: A majorization-minimization approach. Signal Process. **89**, 1683–93 (Sep 2009)
49. M Makitalo, F Foi, Optimal inversion of the generalized Anscombe transformation for Poisson-Gaussian noise. IEEE Trans Image Process. **22**(1), 91–103 (2013)

4

Eigenvector-based initial ranging process for OFDMA uplink systems

Sungeun Lee* and Xiaoli Ma

Abstract

At the contention-based synchronization, e.g., initial ranging process, it is crucial to identify multiple users through ranging codes and estimate the corresponding parameters such as timing offset and frequency offset. This paper presents an improved parameter estimation and pairing algorithm for initial ranging process in orthogonal frequency-division multiple access (OFDMA) uplink systems by exploiting multidimensional harmonic retrieval (HR) technique. Unlike most existing techniques that estimate each parameters independently and associate the ranging codes with the estimated parameters manually, the proposed method estimates multiple user's ranging code, timing and frequency offsets simultaneously, and pair up all the multiple parameters automatically. The simulation results confirm that the proposed technique not only improves the ranging code detection capability and adjusts the acquisition range of the estimators but also increases the maximum number of resolvable users for given samples.

Keywords: Initial ranging; Synchronization; Contention; OFDMA; Eigenvector; Harmonic retrieval; Random access; Uplink; Identifiability

1 Introduction

In general orthogonal frequency-division multiple access (OFDMA) uplink systems, to maintain the orthogonality among multiple users, the signals from all active users should arrive at the base station (BS) synchronously [1]. In order to align the signals of multiple users, contention-based random access procedure, called as initial ranging process, should be performed in the beginning in order to identify multiple users as well as to estimate the corresponding timing and/or frequency offsets for the adjustment/alignment [2,3]. So, code identification as well as multiuser timing estimation are the main tasks of the BS in contention-based synchronization processes.

On the contention-based synchronization process, the same time/frequency resources are shared by multiple users, so the multiple access interference (MAI) limits the performance of ranging detectors. Consequently, it is essential to deal with MAI of ranging subscriber stations (RSSs) to achieve the reliable initial ranging performance.

In the beginning, the existing ranging detection methods simply treat the MAI as noise, which results in the performance degradation as the number of RSSs increases due to the limitation by the amount of MAI [4-6]. To decrease the effect of the accumulated MAI on the ranging process, Ruan et al. [7] proposed a successive multiuser detection (SMUD) algorithm which detects the channel paths of active ranging signals and cancels their interference for further detection. In order to alleviate the high computational complexity of SMUD, the low-complexity method is proposed in [8], and in order to increase the resilience to MAI, a generalized likelihood ratio test (GLRT) approach is also exploited to derive a two-stage interference cancellation scheme [9]. However, it is shown that still SMUD and interference cancellation-based methods are vulnerable for MAI, and the code detection and parameter estimation performance are severely degraded as the number of active RSSs increases due to accumulated residual MAI [10].

In order to reduce the chance undergoing severe amount of MAI and exploiting the frequency selectivity of the fading channels, subchannel-based frame structure is proposed which allocate a smaller number of subcarriers to each ranging opportunity so that most of the RSSs are expected to transmit on disjoint sets of subcarriers with alleviated interference to each other [11-15].

*Correspondence: sungeun@gatech.edu
School of Electrical and Computer Engineering, Georgia Institute of Technology, 777 Atlantic Drive NW, Atlanta, GA 30332-0250, USA

With the improved parameter estimation capability, subspace decomposition-based approaches are exploited to accommodate the detection of multiple RSSs simultaneously [14,15]. However, in order to achieve the code detection and parameter estimation at the same time, the allowable parameter estimation range is strictly limited. In order to bind the detected ranging codes with the estimated timing or frequency offsets, the methods either sacrifice the estimation performance by inserting the code sequence in the estimated parameters [14] or perform exhaustive full search by correlating all sets of ranging codes and offsets at the BS [11,15]. Bacci et al. introduced a game-theoretic approach to derive an energy-efficient method for contention-based synchronization problem [16]. The ranging process is formulated as a noncooperative game for each terminal to determine the transmit power and detection strategy in a selfish way to maximize the detection probability with the minimum energy consumption.

The topic of multidimensional harmonic retrieval (HR) problems are encountered in a variety of signal processing applications including radar, sonar, seismology, communications, MIMO wireless channel, and the different types of schemes are utilized for the application [17-21]. Among these, principal-singular-vector utilization for modal analysis (PUMA) is used as the computationally attractive solution in HR [17]; however, it may not work properly when there are identical frequencies in one or more dimensions. On the other hand, the improved multidimensional folding (IMDF) scheme, which uses eigenvector instead of eigenvalue for estimating frequencies, can resolve identical frequency scenarios by introducing the randomness on the sample data [18,19]. These IMDF algorithms introduced more relaxed identifiability bound, i.e., increased number of resolvable frequencies for the given sample data [19]. In order to improve the performance, the approaches with the use of higher-order singular value decomposition (HOSVD) and/or structured least squares technique are introduced in [20,21]. Instead of using stacking matrix form, the HOSVD method captured the structure inherent in the received data at the expense of a high computational complexity [20,21]. It was shown in [20] that the general computational complexity of the tensor-based approach (HOSVD) is higher than that of the matrix based one (IMDF) even with the computationally efficient subspace estimation methods.

In order to improve the ranging process performance as well as to incorporate the superiority of HR technique with reasonable complexity, the preliminary design utilizing IMDF approach was first introduced in [22]. However, since the single snapshot IMDF approach [18] is simply exploited in [22], the automatic pairing property and multiple resource element utilization were not fully exploited in [22], which results in the limited number of resolvable RSSs as well as the limited detection and estimation performance improvement without proper two-dimensional code design.

Since the multiple snapshot approach is quite well aligned with the inherent OFDMA subchannel allocation concept, we propose an improved ranging technique using IMDF with finite snapshots [19] for the initial ranging process. It enables for all the detected ranging codes, the estimated timing offsets and frequency offsets of multiple users to be automatically paired, so extra ranging code and offset association process, which usually requires high computational complexity, can be avoided.

Even though our approach is based on the IMDF algorithm, but it is not a trivial application of the IMDF algorithm to the ranging process. On the top of the IMDF methodology, our novel contribution is twofold.

First, we properly formulate initial ranging signals into finite snapshots such that automatic pairing property can be fully exploited during the procedure, which brings the improved performance as well as the reduced pairing complexity. Moreover, the maximum number of resolvable RSSs that the BS can distinguish can be also increased by applying multidimensional folding (MDF) algorithm.

Next, with the virtue of the automatic paring property, two-dimensional ranging code is newly proposed. The new code index pair design, ranging code extraction and detection procedure, decision boundary for the circular code index pair, and redefined detection events using the code index pair with *multiset* theory are introduced. According to the new code index pair concept, the enlarged timing and/or frequency offset acquisition range is also derived.

The rest of this paper is organized as follows: Section 2 explains an OFDMA uplink system model in a contention-based synchronization (initial ranging) process. Section 3 explains the proposed harmonic retrieval algorithm with multidimensional folding for extracting the code, timing offset, and frequency offset of active RSSs. In addition, the proposed two-dimensional code design and code detection/evaluation methodology is described. In Section 4, the statistical characteristics of the proposed algorithm, which utilizes IMDF method, is derived to show the identifiability for the number of resolvable RSSs and to confirm the acquisition range of the proposed algorithms on the timing offset and frequency offset. Section 5 evaluates the performance of the proposed method and investigates the comparison with the existing algorithms. Section 6 presents our conclusion.

Throughout this paper, upper (lower) boldface letters will be used for matrices (column vectors). $\mathbf{A}^*$, $\mathbf{A}^T$, $\mathbf{A}^H$, and $\mathbf{A}^\dagger$ denote the conjugate, transpose, Hermitian transpose, and pseudo inverse of $\mathbf{A}$, respectively. We will use $\otimes$ for the Kronecker product, $\odot$ for the Khatri-Rao (column-wise Kronecker) product [23], $\mathbf{I}_n$ for a $n \times n$ identify matrix, $\mathbf{0}_{m\times n}$ for an $m \times n$ zero matrix, $[\mathbf{a}]_n$ for the nth element of $\mathbf{a}$, and $[\mathbf{A}]_{m,n}$ for the (m, n)th element of $\mathbf{A}$, $\mathbf{D}(\mathbf{a})$ for a diagonal matrix with $\mathbf{a}$ as its diagonal.

2 System model

2.1 Ranging process overview

Figure 1 shows the brief signal flow on the initial ranging process. First of all, each RSS catches the downlink synchronization signal to detect its relative frame timing position and to estimate its absolute carrier frequency offset. Since every RSS can compare and compensate its frequency to the reference one from the BS, the coarse frequency synchronization can be completed before the initial ranging trial is started. Therefore, the residual frequency offset error due to the imperfect estimation is only remained for each RSS as shown in Figure 1.

On the other hand, the absolute frame timing of BS is unknown to each RSS since the propagation delay is unknown to each RSS (refer Figure 1). Therefore, each RSS only estimates its relative downlink frame timing position at first. Next, each RSS transmits its own initial ranging sequence, which is locked on the estimated frame timing, to the BS with the network access request. Once BS receives the sequences from multiple RSSs, the BS should identify multiple RSSs and estimate every RSS's timing offset and residual frequency offset independently. Finally, from the given estimate, the BS provides the timing and frequency feedback info to the identified RSSs with the network grant message. Consequently, it is crucial on the initial ranging process for BS to identify as many as RSSs and to estimate their timing and frequency offsets as accurate as possible.

The whole transceiver block diagram of the ranging process is shown in Figure 2, and the detailed procedure on how to construct the ranging sequence with the code and to perform the code detection and timing/frequency estimation is described in the following.

2.2 Resource allocation for ranging process

Let us consider an OFDMA uplink system employing N subcarriers. Among the whole subcarriers, virtual subcarriers are placed at both edges of the spectrum to prevent the spectrum aliasing. Except the virtual subcarriers, the useful subcarriers are grouped by multiple subchannels, and these subchannels are assigned to multiple users (subscribers) for transmission. Typically, each subchannel is divided into Q subbands composed of a set of V adjacent subcarriers [14,15,24]. When one transmission block consists of M consecutive OFDMA symbols, let us define a bunch of consecutive V subcarriers (one subband) and M OFDMA symbols as a *tile*.

Now, among multiple subchannels, let us assume that each RSS only uses a set of R subchannels for ranging with M consecutive OFDMA symbols, and then the total number of subcarriers used for ranging is $N_R = RQV$.

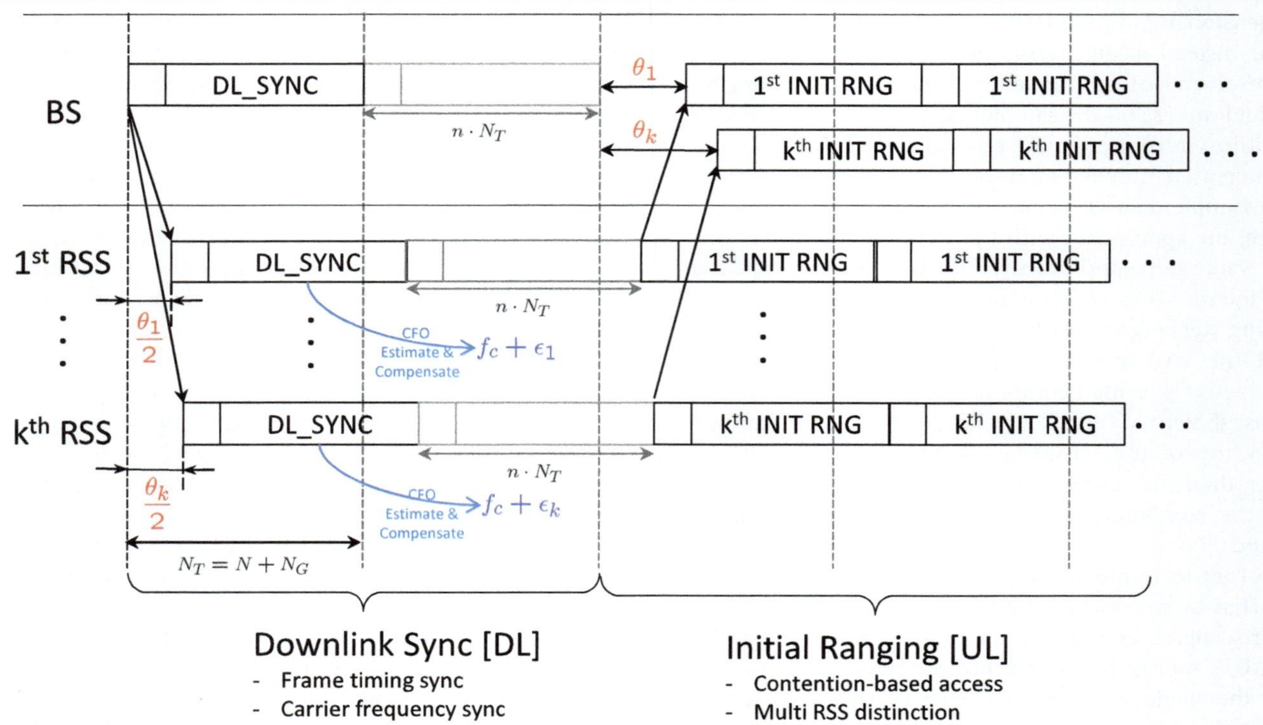

Figure 1 Initial ranging process signal flow and timing.

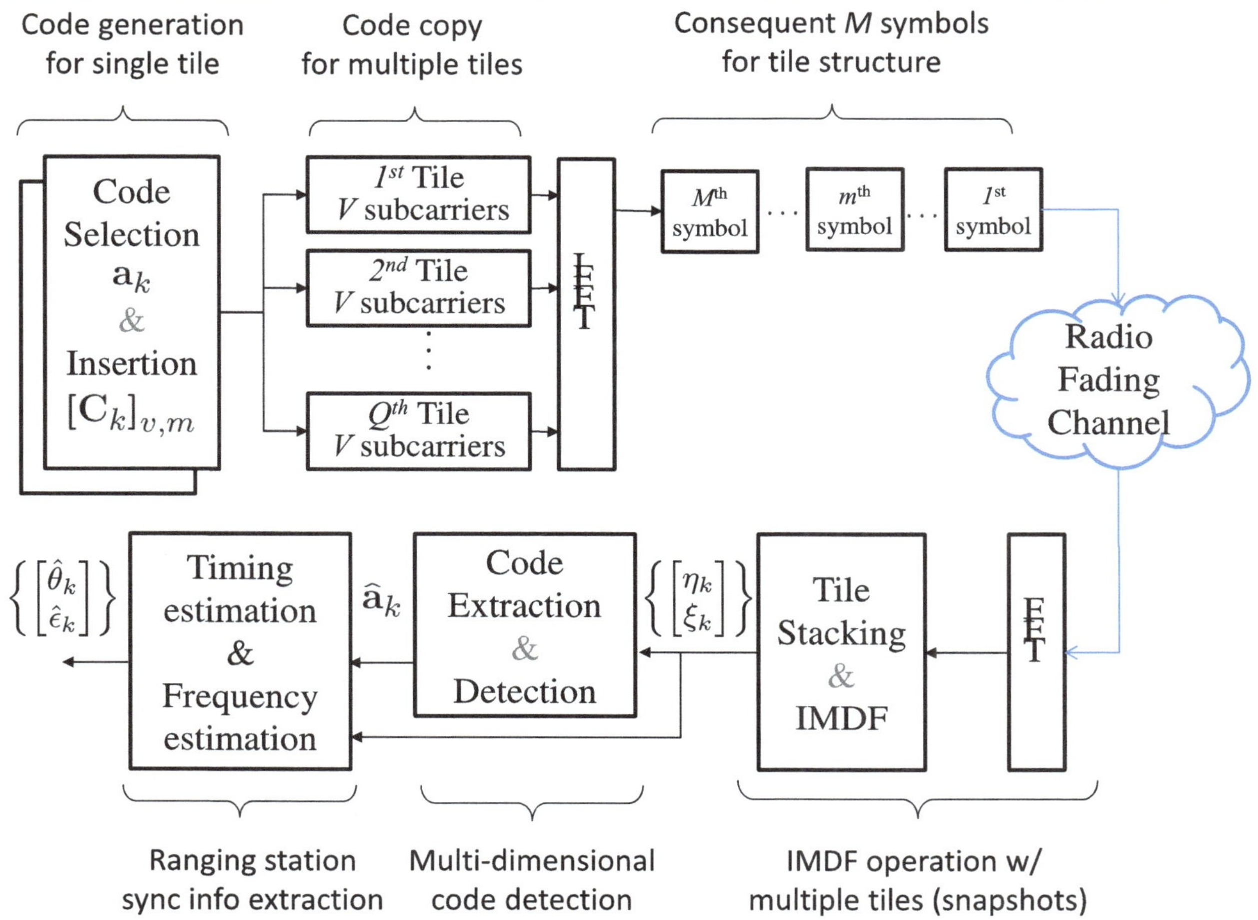

Figure 2 Transceiver block diagram of the initial ranging process.

After the fast Fourier transform (FFT) at the receiver, the received data at the qth tile of the ranging subchannel is written as:

$$[\mathbf{X}_q]_{v,m} = \sum_{k=1}^{K} [\mathbf{C}_k]_{v,m} H_k\left[i_q+v-1\right] e^{-\frac{j2\pi}{N}(i_q+v-1)\theta_k} e^{\frac{j2\pi N_T}{N}(m-1)\epsilon_k} + \left[\mathbf{W}_q\right]_{v,m}, \quad (1)$$

with $\forall q \in [1,Q], \forall v \in [1,V], \forall m \in [1,M]$. In (1), K is the number of active RSSs participating the ranging process, and v and m denote the subcarrier and OFDMA symbol index within the tile, respectively. $H_k[x], x \in [0, N-1]$ is the channel frequency response of the kth RSS on the xth subcarrier position.

Since there is no big frequency offset difference between stations after the coarse frequency offset synchronization, the inter-carrier-interference (ICI) term caused by the residual frequency offset is ignored for the simplicity in (1) whereas the accumulated phase rotation term caused by frequency offset is still included.

In (1), the length of one OFDMA symbol is $N_T = N + N_G$ with N_G guard interval for ranging symbols, and i_q is the starting subcarrier index for the qth tile, and θ_k and ϵ_k denote the timing and frequency offsets of the kth RSS, respectively. $\left[\mathbf{W}_q\right]_{v,m}$ is a complex additive white Gaussian noise (AWGN) with zero mean and variance σ_w^2. $\mathbf{C}_k$ is the code sequence matrix for the kth RSS.

Here, let us assume that $\theta_k \in [0, \theta_{\max}]$ and $\epsilon_k \in [-\epsilon_{\max}, \epsilon_{\max}]$ where $\theta_{\max}$ and $\epsilon_{\max}$ are defined as the maximum allowable timing offset and and absolute frequency offset of the system, respectively.

2.3 Code design

Let us design the code sequence for the kth RSS user as:

$$[\mathbf{C}_k]_{v,m} = e^{j2\pi v \cdot t_k / C_T} e^{j2\pi m \cdot f_k / C_F}. \quad (2)$$

Here, C_T and C_F are defined as the design parameters to determine the two-dimensional code index pair grid. Note that C_T and C_F can be properly chosen according to RSS's timing and frequency distributions. Let us define $\mathbf{a}_k = [t_k\ f_k]^T$ as a two-tuple for the code index pair of the kth user where $t_k \in [0, C_T)$ is the code index at the timing offset domain and $f_k \in [0, C_F)$ is the one at the frequency offset domain, respectively. The tuple $\mathbf{a}_k$ is chosen from the whole code index pair set $\mathcal{A}$, i.e., $\mathbf{a}_k \in \mathcal{A}$.

Since the subband size in one tile is designed to be smaller than the channel coherence bandwidth [14,15], $H_k[x] \approx H_k[x+1]$ within the tile. Therefore, by averaging V subcarriers, the received output can be expressed as an average channel response as follows:

$$[\mathbf{X}_q]_{v,m} \approx \sum_{k=1}^{K} [\mathbf{h}_q]_k e^{j2\pi v\eta_k} e^{j2\pi m\xi_k} + [\mathbf{W}_q]_{v,m}, \quad (3)$$

where $[\mathbf{h}_q]_k$ is the average of the phase-rotated channel response for the kth user at the qth tile written as:

$$[\mathbf{h}_q]_k = \left(\frac{1}{V}\sum_{v=1}^{V} H_k[i_q + v - 1]\right) e^{-j2\pi(i_q-1)\theta_k/N} e^{-j2\pi\frac{N_T}{N}\epsilon_k},$$
$$\forall k \in \{1, 2, \ldots, K\}.$$

In (3), the code sequence of the kth user, $[\mathbf{C}_k]_{v,m}$, is absorbed into the *effective* timing and frequency offsets as η_k and ξ_k, and this *effective* timing and frequency offsets of the kth user are given by:

$$\eta_k = \frac{t_k}{C_T} - \frac{\theta_k}{N}, \quad (4)$$

$$\xi_k = \frac{f_k}{C_F} + \frac{\epsilon_k N_T}{N}. \quad (5)$$

3 MDF estimation using multiple tiles

In order to exploit automatic pairing property of HR, we formulate the received signal as a two-dimensional (2-D) mixture form and then apply MDF estimation method [18] for estimating timing and frequency offsets of RSSs as well as detecting RSSs codes simultaneously.

3.1 2-D mixture model

Let us recall the received symbols at the qth tile is $\mathbf{X}_q \in \mathbb{C}^{V\times M}$. Given (3), $\mathbf{X}_q$ can be written as 2-D mixture matrix form as:

$$\mathbf{X}_q = \mathbf{T}\mathbf{D}(\mathbf{h}_q)\mathbf{F}^T + \mathbf{W}_q, \quad (6)$$

where $\mathbf{T} \in \mathbb{C}^{V\times K}$ and $\mathbf{F} \in \mathbb{C}^{M\times K}$ are the transpose of Vandermonde matrices with the common ratios $e^{j2\pi\eta_k}$ and $e^{j2\pi\xi_k}$ for each column to represent the timing and frequency offsets of K RSSs, respectively.

For simplicity, let us define the (v, m)th element of the matrix $\mathbf{X}_q$ as $x_{q,v,m}$, i.e., $x_{q,v,m} = [\mathbf{X}_q]_{v,m}$. Then, from the given Eq. 6, we can define the sample vector form $\mathbf{x}_q \in \mathbb{C}^{VM\times 1}$, which is:

$$\begin{aligned}\mathbf{x}_q &= \begin{bmatrix} x_{q,1,1} & x_{q,1,2} & \cdots & x_{q,1,M} & x_{q,2,1} & \cdots & x_{q,V,M}\end{bmatrix}^T \\ &= (\mathbf{T} \odot \mathbf{F})\mathbf{h}_q + \mathbf{w}_q,\end{aligned} \quad (7)$$

where $\mathbf{w}_q \in \mathbb{C}^{VM\times 1}$ is the vector form of the noise matrix $\mathbf{W}_q$. Since the received signal is formed as undamped 2-D exponentials for each tile, we can apply MDF algorithm [18] for each tile at the initial ranging process.

3.2 MDF algorithm using multiple tiles

3.2.1 Smoothing operation

As shown in (6) and (7), the observed symbols at each tile can be described as the special form of the multiple harmonic combinations. Based on this structure, we even apply *smoothing operator* $\mathcal{S}(\mathbf{x})$ in order to fully utilize this special characteristics of the observed symbols and improve the estimation performance.

Let us define a set of two-dimensional selection matrices $\mathbf{J}_{v_2 m_2} \in \mathbb{C}^{V_1 M_1\times VM}$ given by:

$$\begin{aligned}\mathbf{J}_{v_2 m_2} &= \bar{\mathbf{J}}_{v_2} \otimes \tilde{\mathbf{J}}_{m_2}, \\ \bar{\mathbf{J}}_{v_2} &= \begin{bmatrix}\mathbf{0}_{V_1\times(v_2-1)} & \mathbf{I}_{V_1} & \mathbf{0}_{V_1\times(V_2-v_2)}\end{bmatrix}, \quad v_2 = 1, \ldots, V_2, \\ \tilde{\mathbf{J}}_{m_2} &= \begin{bmatrix}\mathbf{0}_{M_1\times(m_2-1)} & \mathbf{I}_{M_1} & \mathbf{0}_{M_1\times(M_2-m_2)}\end{bmatrix}, \quad m_2 = 1, \ldots, M_2,\end{aligned} \quad (8)$$

where V_1, V_2 and M_1, M_2 are positive integers satisfying $V_1 + V_2 = V + 1$ and $M_1 + M_2 = M + 1$.

Using (7) and (8), let us define *smoothing operator* $\mathcal{S}(\mathbf{x})$ and the smoothed matrix $\dot{\mathbf{X}}_{q,\mathcal{S}}$ for the sample vector as:

$$\begin{aligned}\dot{\mathbf{X}}_{q,\mathcal{S}} &= \mathcal{S}(\mathbf{x}_q) \\ &= \begin{bmatrix}\mathbf{J}_{11}\mathbf{x}_q & \mathbf{J}_{12}\mathbf{x}_q & \cdots & \mathbf{J}_{1M_1}\mathbf{x}_q & \mathbf{J}_{21}\mathbf{x}_q & \cdots & \mathbf{J}_{V_2M_2}\mathbf{x}_q\end{bmatrix}\end{aligned} \quad (9)$$

where $\dot{\mathbf{X}}_{q,\mathcal{S}} \in \mathbb{C}^{V_1M_1\times V_2M_2}$. Here, we show an explicit expression of $\dot{\mathbf{X}}_{q,\mathcal{S}}$ using $x_{q,v,m}$ in order to figure out the smoothing operator:

$$\dot{\mathbf{X}}_{q,\mathcal{S}} = \begin{bmatrix} \mathbf{X}_{q,1} & \mathbf{X}_{q,2} & \cdots & \mathbf{X}_{q,V_2} \\ \mathbf{X}_{q,2} & \mathbf{X}_{q,3} & \cdots & \mathbf{X}_{q,V_2+1} \\ \vdots & \vdots & \ddots & \vdots \\ \mathbf{X}_{q,V_1} & \mathbf{X}_{q,V_1+1} & \cdots & \mathbf{X}_{q,V}\end{bmatrix}$$

with $\mathbf{X}_{q,v} \in \mathbb{C}^{M_1\times M_2}$:

$$\mathbf{X}_{q,v} = \begin{bmatrix} x_{q,v,1} & x_{q,v,2} & \cdots & x_{q,v,M_2} \\ x_{q,v,2} & x_{q,v,3} & \cdots & x_{q,v,M_2+1} \\ \vdots & \vdots & \ddots & \vdots \\ x_{q,v,M_1} & x_{q,v,M_1+1} & \cdots & x_{q,v,M}\end{bmatrix}.$$

In the absence of noise, it is already verified in [18], Lemma 2 that (9) can be decomposed into harmonic matrices as follows:

$$\dot{\mathbf{X}}_{q,\mathcal{S}} = \mathcal{S}\left(\mathbf{x}_q\right) = \mathbf{G}_1\mathbf{D}(\mathbf{h}_q)\mathbf{G}_2^T \tag{10}$$

where:

$$\mathbf{G}_1 = \mathbf{T}^{(V_1)} \odot \mathbf{F}^{(M_1)}, \tag{11}$$

$$\mathbf{G}_2 = \mathbf{T}^{(V_2)} \odot \mathbf{F}^{(M_2)}, \tag{12}$$

with $V_1 + V_2 = V + 1$ and $M_1 + M_2 = M + 1$. $\mathbf{T}^{(V_i)} \in \mathbb{C}^{V_i \times K}$ and $\mathbf{F}^{(M_i)} \in \mathbb{C}^{M_i \times K}, i = 1, 2$, are the transpose of Vandermonde matrices with the common ratios $e^{j2\pi\eta_k}$ and $e^{j2\pi\xi_k}$.

For the noisy case, the perturbation analysis can be applied to derive the theoretical MSEs of one realization of timing and frequency offset estimation [19], Equations (33) to (47). However, since timing offset θ_k and frequency offset ϵ_k are all unknown random variables in the ranging process, the theoretical performance analysis on the deterministic realization of η_k and ξ_k would be meaningless on the whole. Therefore, for the simplicity, the noiseless case is used on the derivation while the noisy case is revisited on the simulation experiment.

In (10), in order to further explore the data structure, we operate the backward smoothing as well as forward smoothing on the sample vector $\mathbf{x}_q$ in (7). Let us define a backward sample vector $\mathbf{y}_q = \Pi_{VM} \cdot \mathbf{x}_q^*$ where Π_n is an $n \times n$ backward permutation matrix. By using the property of harmonics, it is easily verified that:

$$\begin{aligned} \mathbf{y}_q &= \Pi_{VM} \cdot (\mathbf{T} \odot \mathbf{F})^* \mathbf{h}_q^* \\ &= (\mathbf{T} \odot \mathbf{F})\mathbf{D}(\mathbf{e}_\beta)\mathbf{h}_q^* \\ &= (\mathbf{T} \odot \mathbf{F})\tilde{\mathbf{h}}_q, \end{aligned} \tag{13}$$

where $\tilde{\mathbf{h}}_q = \mathbf{D}\left(\mathbf{e}_\beta^*\right)\mathbf{h}_q^*$ and $\mathbf{e}_\beta = \left[e^{j\beta_1} \quad e^{j\beta_2} \quad \cdots \quad e^{j\beta_K}\right]^T$ with $\beta_k = (V-1)\eta_k + (M-1)\xi_k$ [19]. Since (13) has the same harmonic matrix form as (7), we can use both forward and backward sample vectors to estimate the phase. In the similar way like (9), $\dot{\mathbf{Y}}_{q,\mathcal{S}}$ can be also obtained by backward smoothing operation:

$$\dot{\mathbf{Y}}_{q,\mathcal{S}} = \mathcal{S}\left(\mathbf{y}_q\right) = \mathbf{G}_1\mathbf{D}(\tilde{\mathbf{h}}_q)\mathbf{G}_2^T. \tag{14}$$

3.2.2 Multiple tile combination

Let us define the forward and backward channel matrices as:

$$\mathbf{H} = \left[\mathbf{h}_1\ \mathbf{h}_2 \cdots \mathbf{h}_Q\right], \tag{15}$$

$$\tilde{\mathbf{H}} = \left[\tilde{\mathbf{h}}_1\ \tilde{\mathbf{h}}_2 \cdots \tilde{\mathbf{h}}_Q\right], \tag{16}$$

where $\mathbf{H} \in \mathbb{C}^{K\times Q}$, $\tilde{\mathbf{H}} \in \mathbb{C}^{K\times Q}$. Whereas [22] uses each tile independently for the code detection and timing/frequency offset estimation, now all channel responses and signals from multiple tiles are stacked and utilized together to create just one augmented matrix $\tilde{\mathbf{X}}$. Therefore, it would be more robust for the noisy scenario due to multiple tile usage.

After calculating the smoothed matrices $\dot{\mathbf{X}}_{q,\mathcal{S}}$ and $\dot{\mathbf{Y}}_{q,\mathcal{S}}$ for each tile, we collect all the smoothed matrices in order to perform MDF algorithm [19]:

$$\tilde{\mathbf{X}} = \left[\dot{\mathbf{X}}_{1,\mathcal{S}} \quad \dot{\mathbf{X}}_{2,\mathcal{S}} \cdots \dot{\mathbf{X}}_{Q,\mathcal{S}} \quad \dot{\mathbf{Y}}_{Q,\mathcal{S}} \quad \dot{\mathbf{Y}}_{Q-1,\mathcal{S}} \cdots \dot{\mathbf{Y}}_{1,\mathcal{S}}\right] \tag{17}$$

$$= \mathbf{G}_1\left(\mathbf{G}_2 \odot \mathbf{B}\right)^T \tag{18}$$

$$= \mathbf{G}_1\hat{\mathbf{G}}_2^T \tag{19}$$

where $\tilde{\mathbf{X}} \in \mathbb{C}^{V_1M_1\times 2QV_2M_2}$ and $\mathbf{B} \in \mathbb{C}^{2Q\times K}$ as follows:

$$\mathbf{B} = \begin{bmatrix} \mathbf{H}^T \\ \Pi_Q\tilde{\mathbf{H}}^T \end{bmatrix}.$$

3.2.3 Eigenvector-based estimation

$\tilde{\mathbf{X}}$ is of rank K almost surely. Therefore, the singular value decomposition (SVD) of $\tilde{\mathbf{X}}$ yields:

$$\tilde{\mathbf{X}} = \mathbf{U}_s\Sigma_s\mathbf{V}_s^H \tag{20}$$

where $\mathbf{U}_s$ has K columns that together span the column space of $\tilde{\mathbf{X}}$. Since the same space is spanned by the columns of $\mathbf{G}_1$ from (19), there exists $K \times K$ nonsingular matrix $\mathbf{L}^{-1}$ such that $\mathbf{U}_s = \mathbf{G}_1\mathbf{L}^{-1}$.

$\mathbf{U}_s$ can be partitioned into two submatrices $\mathbf{U}_1$ and $\mathbf{U}_2$ (refer (46) and (47) in Appendix). Then, $\mathbf{L}$ can be obtained from the eigenvalue decomposition of $\mathbf{U}_1^\dagger\mathbf{U}_2$ up to column scaling and permutation ambiguity, i.e., $\mathbf{L}_{sp} = \mathbf{L}\boldsymbol{\Lambda}\boldsymbol{\Delta}$ where $\mathbf{L}_{sp}$ is a practically obtained eigenvector including column scaling and permutation ambiguity, $\boldsymbol{\Lambda}$ is a nonsingular diagonal column scaling matrix, and $\boldsymbol{\Delta}$ is a permutation matrix [18,19]. This part is maybe most computationally complex part of the whole calculation, and the major computational load complexity to acquire $\mathbf{L}_{sp}$ is calculated as $\mathcal{O}(2k_tV_1M_1V_2M_2K)$ [19-21].

Once we obtain $\mathbf{L}_{sp}$, we can retrieve $\mathbf{P} \in \mathbb{C}^{(V_1-1)M_1\times K}$ up to column scaling and permutation ambiguity as:

$$\mathbf{P}_{sp} = \mathbf{U_s}\mathbf{L} = \mathbf{P}\boldsymbol{\Lambda}\boldsymbol{\Delta}, \tag{21}$$

where $\mathbf{P}_{sp}$ is a practically calculated matrix which consists of Khatri-Rao product of two Vandermonde matrices with are K columns up to column scaling and permutation ambiguity. Here, the original stacking matrix $\mathbf{P}$ is defined as:

$$\mathbf{P} = \mathbf{T}^{(V_1-1)} \odot \mathbf{F}^{(M_1)} \tag{22}$$

$$= \begin{bmatrix} \alpha_{\eta_1}\alpha_{\xi_1} & \alpha_{\eta_2}\alpha_{\xi_2} & \cdots & \alpha_{\eta_K}\alpha_{\xi_K} \\ \alpha_{\eta_1}\alpha_{\xi_1}^2 & \alpha_{\eta_2}\alpha_{\xi_2}^2 & \cdots & \alpha_{\eta_K}\alpha_{\xi_K}^2 \\ \vdots & \vdots & \ddots & \vdots \\ \alpha_{\eta_1}\alpha_{\xi_1}^{M_1} & \alpha_{\eta_2}\alpha_{\xi_2}^{M_1} & \cdots & \alpha_{\eta_K}\alpha_{\xi_K}^{M_1} \\ \alpha_{\eta_1}^2\alpha_{\xi_1} & \alpha_{\eta_2}^2\alpha_{\xi_2} & \cdots & \alpha_{\eta_K}^2\alpha_{\xi_K} \\ \vdots & \vdots & \ddots & \vdots \\ \alpha_{\eta_1}^{V_1-1}\alpha_{\xi_1}^{M_1} & \alpha_{\eta_2}^{V_1-1}\alpha_{\xi_2}^{M_1} & \cdots & \alpha_{\eta_K}^{V_1-1}\alpha_{\xi_K}^{M_1} \end{bmatrix}. \tag{23}$$

where $\alpha_{\eta_k} = e^{j2\pi\eta_k}$ and $\alpha_{\xi_k} = e^{j2\pi\xi_k}$, i.e., $[\mathbf{P}]_{p,k} = \alpha_{\eta_k}^{\lceil \frac{p}{M_1} \rceil} \alpha_{\xi_k}^{M_1-((M_1-p) \bmod M_1)}$. In $\mathbf{P}$, the phase rotation components caused by timing offset and frequency offset appear in the same column of the $\mathbf{P}$. In other words, for a fixed kth column, $e^{j2\pi\eta_k}$ and $e^{j2\pi\xi_k}$ appear simultaneously in the same column of $\mathbf{P}$. Consequently, thanks to this combined structure, we can estimate K RSS's timing offset and frequency offset at the same time by dividing suitably chosen elements of the $\mathbf{P}$.

The detailed procedure on how to obtain the above $\mathbf{P}_{sp}$ and $\mathbf{L}_{sp}$ from $\mathbf{U}_s$ is described in Appendix.

3.3 Effective timing offset and frequency offset estimation

Even though $\mathbf{P}_{sp}$ contains the column scaling ambiguity $\mathbf{\Lambda}$, it is immaterial for estimating the timing and frequency offset of each user. It is because the timing and frequency offset of each user is obtained by phase differential term. Therefore, the scaling effect on the matrix is diminished. In addition, each timing offset with frequency offset is even automatically paired without heavy computational search for the pairing.

The effective timing offset on the kth column can be estimated as:

$$\widehat{e^{j2\pi\eta_k}} = \frac{1}{(V_1-2)M_1} \sum_{p=1}^{(V_1-2)M_1} \frac{[\mathbf{P}_{sp}]_{p+M_1,k}}{[\mathbf{P}_{sp}]_{p,k}}, \quad \forall k = 1, \ldots, K, \tag{24}$$

and the effective timing offset from the kth column can be obtained by:

$$\hat{\eta}_k = \frac{1}{2\pi} Im \left(\ln \widehat{e^{j2\pi\eta_k}} \right). \tag{25}$$

In the similar way, the effective frequency offset on the kth column can be estimated as:

$$\widehat{e^{j2\pi\xi_k}} = \frac{1}{(V_1-1)(M_1-1)} \sum_{\substack{p=1,\\ \bmod (p,M_1)\neq 0}}^{(V_1-1)M_1} \frac{[\mathbf{P}_{sp}]_{p+1,k}}{[\mathbf{P}_{sp}]_{p,k}}, \quad \forall k = 1, \ldots, K, \tag{26}$$

and the effective frequency offset from the kth column can be obtained by:

$$\hat{\xi}_k = \frac{1}{2\pi} Im \left(\ln \widehat{e^{j2\pi\xi_k}} \right). \tag{27}$$

Note that still we have the the permutation ambiguity issue on the estimation, i.e., unknown $\mathbf{\Delta}$. Since the user (column) identification along with timing and frequency estimation should be performed accordingly in the ranging process, therefore it is crucial to identify each column correctly and to remove the permutation ambiguity clearly. This can be done by the proposed code extraction and code detection process.

3.4 Code extraction

Recall the *effective* timing and frequency offsets are composed of the code index as well as actual timing and frequency offsets as shown in (4) and (5). Since $\hat{\eta}_k$ and $\hat{\xi}_k$ are automatically paired, the paired ranging code, timing offset, and frequency offset of the kth RSS can be detected and estimated simultaneously by using the effective timing and frequency offsets.

Followed by (4) and (5), let us define the extracted code index pair $\mathbf{b}_k$ as:

$$\mathbf{b}_k = \begin{bmatrix} C_T \cdot \hat{\eta}_k + \Delta T \\ C_F \cdot \hat{\epsilon}_k \end{bmatrix}, \tag{28}$$

where $\Delta T = \frac{\theta_{\max}}{2} \frac{C_T}{N}$. By plugging (4) and (5) in (28), $\mathbf{b}_k$ is derived as follows:

$$\mathbf{b}_k = \begin{bmatrix} t_k + \left(\frac{\theta_{\max}}{2} - \theta_k \right) \frac{C_T}{N} \\ f_k + \epsilon_k \frac{C_F N_T}{N} \end{bmatrix}, \tag{29}$$

where $\left(\frac{\theta_{\max}}{2} - \theta_k \right) \frac{C_T}{N} \in \left[-\frac{\theta_{\max}}{2} \frac{C_T}{N}, \frac{\theta_{\max}}{2} \frac{C_T}{N} \right]$ and $\epsilon_k \frac{C_F N_T}{N} \in \left[-\epsilon_{\max} \frac{C_F N_T}{N}, \epsilon_{\max} \frac{C_F N_T}{N} \right]$. Since the timing offset θ_k only has the positive values, by introducing the bias ΔT on the extracted code index pair, the disturbance due to timing offset can be spread out around the original code index point t_k. Consequently, by observing $\mathbf{b}_k$, we can retrieve the code index pair (t_k, f_k) accordingly.

3.5 Code detection

Now, the new code detection methodology and the decision boundary/criterion for the circularly repeated code index pair are introduced using *multiset* theory. In addition, the code detection events are newly redefined for the multidimensional code, e.g., *phantom code, collided code, missing code, false-alarmed code, and correctly detected code* in this section.

Figure 3 illustrates the example of code index pair set $\mathcal{A}$, decision boundary, and decision region $\mathcal{R}_k$ for $C_T = C_F = 5$. As discussed before, from the automatic pairing property of HR, it is possible to build up the two-dimensional code index pair.

From $\mathbf{b}_k$, the code index pair $\hat{\mathbf{a}}_k$ can be extracted as follows:

$$\hat{\mathbf{a}}_k = \mathbf{a}_i \quad \text{s.t. } \mathbf{b}_k \in \mathcal{R}_i, \forall i \in [1, |\mathcal{A}|], \tag{30}$$

where $\mathcal{R}_i$ is the decision region for the code index pair $\mathbf{a}_i$ $\forall i \in [1, |\mathcal{A}|]$. Since the code index is circularly repeated per design parameter C_T and C_F, the modified maximum likelihood approach which reflects the repeated lattice

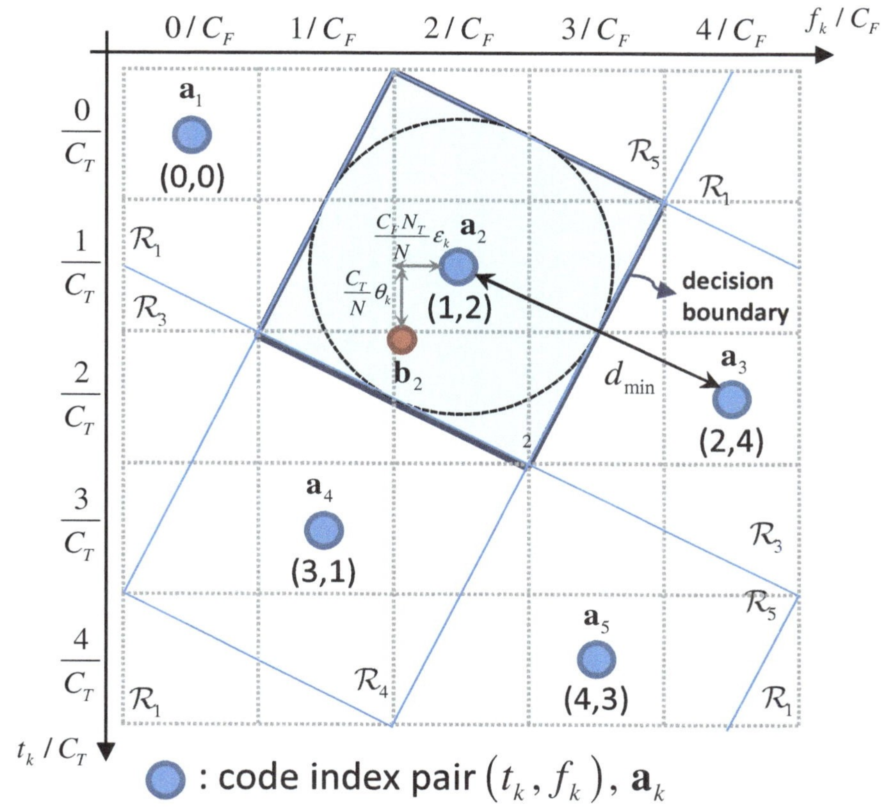

Figure 3 Code tuple example. $\mathcal{A} = \{\mathbf{a}_k\}$ with $C_T = C_F = 5$.

structure can be applied for extracting the code index pair as follows:

$$\kappa(k) = \arg_i \min_{\forall \mathbf{a}_i \in \mathcal{A}, z_t \in \mathbb{Z}, z_f \in \mathbb{Z}} \left\| \mathbf{b}_k - \mathbf{a}_i + \begin{bmatrix} z_t C_T \\ z_f C_F \end{bmatrix} \right\|_2^2, \quad (31)$$

$$\hat{\mathbf{a}}_k = \mathbf{a}_{\kappa(k)} \quad (32)$$

where $\kappa(k)$ is the detected code index for the kth received tuple (kth extracted column), $\mathbb{Z}$ is the set of all integers, and $\|\mathbf{x}\|_2^2$ is the square of the two-norm. Once all of K code index pair is detected, and then, it should be investigated from the detected code index pairs whether or not it contains the phantom code index pair which means the index pair is duplicated more than one time.

In order to redefine the code detection events using code tuple, let us assume $\mathcal{A}_T$ and $\mathcal{A}_R$ as the transmitted and the detected code index pair *multisets* (or bag), respectively, i.e., $\mathbf{a}_k \in \mathcal{A}_T$ and $\hat{\mathbf{a}}_k \in \mathcal{A}_R$. In contrast to set concept, the members of the multiset are allowed to appear more than once. Here is an example to figure out the different events on the detection procedure:

$$\mathcal{A}_T = \{\mathbf{a}_1, \ \mathbf{a}_1, \ \mathbf{a}_2, \ \mathbf{a}_3, \ \mathbf{a}_4, \ \mathbf{a}_5\}, \quad (33)$$

$$\mathcal{A}_R = \{\mathbf{a}_1, \ \mathbf{a}_1, \ \mathbf{a}_2, \ \mathbf{a}_2, \ \mathbf{a}_3, \ \mathbf{a}_6\} = \{\hat{\mathbf{a}}_k\}. \quad (34)$$

This illustrates the case that six active RSSs are participated on ranging process, and two of the RSSs pick up the same code index pair $\mathbf{a}_1$ and transmit ($\mathcal{A}_T$), and the BS detects the six code index pairs from the received tuple ($\mathcal{A}_R$). The multiset indicator function of a subset of a multiset $\mathcal{A}$ is:

$$\mathbf{1}_{\mathcal{A}} : \mathbf{x} \to \mathbb{Z}_+ \quad (35)$$

defined by:

$$\mathbf{1}_{\mathcal{A}}(\mathbf{x}) = \begin{cases} \mu(\mathbf{x}) & \text{if } \mathbf{x} \in \mathcal{A}, \\ 0 & \text{if } \mathbf{x} \notin \mathcal{A}, \end{cases} \quad (36)$$

where $\mathbb{Z}_+$ is the nonnegative integer set and $\mu(\mathbf{x})$ denotes the number of occurrences of $\mathbf{x}$ in the set. Now, let us take a look at different events on the code detection procedure.

3.5.1 Phantom code

Let us define phantom code that one code index pair is observed more than once at $\mathcal{A}_R$. The index set of the phantom code is defined as:

$$\mathcal{G}_{ph} = \left\{ \kappa(k) \,\middle|\, \mathbf{1}_{\mathcal{A}_R}(\hat{\mathbf{a}}_k) > 1, \forall k \in [1, K] \right\}, \tag{37}$$

with $\kappa(k)$ defined in (31). Since $\mathbf{a}_1$ and $\mathbf{a}_2$ are detected more than once in (34), here, $\mathcal{G}_{ph} = \{1, 2\}$. The phantom code can be detected at the receiver, so this code index can be opted out during the detection procedure.

3.5.2 Collided code

Collided code is defined such that one code index pair is utilized more than once at $\mathcal{A}_T$, which is the case when more than two RSSs choose the same code for the ranging. The index set of the collided code is defined as:

$$\mathcal{G}_{co} = \left\{ k \,\middle|\, \mathbf{1}_{\mathcal{A}_T}(\mathbf{a}_k) > 1, \forall k \in [1, K] \right\}. \tag{38}$$

From the example, $\mathcal{G}_{co}$ is defined as $\mathcal{G}_{co} = \{1\}$. Typically, it is hard to detect the collided code at the receiver side, but with the virtue of automatic pairing property in HR, if each RSS has different timing offsets and/or frequency offsets, it has a chance to remove this collided code as phantom code at the receiver side (like the example). Therefore, HR can avoid the false detection of the collided code effectively.

3.5.3 Missing code

Let us redefine missing code such that the transmitted code index pair is not detected at the receiver side except collided code, which is denoted as:

$$\mathcal{G}_{m} = \left\{ k \,\middle|\, \mathbf{1}_{\mathcal{A}_T}(\mathbf{a}_k) = 1, \mathbf{1}_{\mathcal{A}_T \setminus \mathcal{A}_R}(\mathbf{a}_k) > 0, \forall k \in [1, K] \right\}. \tag{39}$$

From the example, the missing code is shown as $\mathcal{G}_m = \{4, 5\}$.

3.5.4 False-alarmed code

False-alarmed code means that the wrong code index pair, which is never transmitted, is detected at the receiver side. There is no way to classify the false-alarmed code at the receiver, so it usually degrades the detection and estimation performance of the system. The false-alarmed code can be calculated as:

$$\mathcal{G}_{f} = \left\{ \kappa(k) \,\middle|\, \mathbf{1}_{\mathcal{A}_T}(\hat{\mathbf{a}}_k) = 0, \mathbf{1}_{\mathcal{A}_R \setminus \mathcal{A}_T}(\hat{\mathbf{a}}_k) > 0, \forall k \in [1, K] \right\}. \tag{40}$$

Here, $\mathcal{G}_f = \{6\}$.

3.5.5 Correctly detected code

After excluding the phantom code, all the remained codes are treated as detected codes including the false-alarmed code. However, in order for the accurate evaluation of the code detection performance, let us newly define the correctly detected code which only counts the code truly transmitted and accurately received as:

$$\mathcal{G}_{d} = \left\{ k \,\middle|\, \mathbf{1}_{\mathcal{A}_T \times \mathcal{A}_R}(\hat{\mathbf{a}}_k) = \mathbf{1}_{\mathcal{A}_T}(\hat{\mathbf{a}}_k) \cdot \mathbf{1}_{\mathcal{A}_R}(\hat{\mathbf{a}}_k) = 1, \forall k \in [1, K] \right\}. \tag{41}$$

From the example, the correctly detected code is calculated as $\mathcal{G}_d = \{3\}$.

On the whole, the index set of detected codes is described as $\{3\}$ while the index multiset of misdetected codes including both $\mathcal{G}_{co}$ and $\mathcal{G}_m$ is $\{1, 1, 2, 4, 5\}$. Finally, the index set of false-detected codes is defined as $\{6\}$. Note $\{1\}$ appears twice on the misdetected multiset since $\mathcal{A}_T$ already carries the two collided code.

3.6 Actual timing and frequency estimation

Among K extracted code index pairs, the phantom code $\mathcal{G}_{ph}$ is firstly excluded, and only remaining code index pairs, $\hat{\mathbf{a}}_{k^*} = \left[t_{k^*}\ f_{k^*}\right]^T, \forall \kappa(k^*) \in \mathcal{G}_d \cup \mathcal{G}_f$, participate the actual timing and frequency estimation. So, the timing and frequency offsets are calculated only for the $\kappa(k^*)$th RSSs $\forall \kappa(k^*) \in \mathcal{G}_d \cup \mathcal{G}_f$. Based on (4) and (25), the estimated actual timing offset for the $\kappa(k^*)$th RSS from the k^*th received tuple is given by:

$$\hat{\theta}_{\kappa(k^*)} = N\left(\frac{t_{k^*}}{C_T} - \hat{\eta}_{k^*}\right), \tag{42}$$

and it would be compared with θ_{k^*} if it is successfully identified without false-alarm ($k^* \in \mathcal{G}_d$) and the offset is accurate estimated.

In the same way, the estimated actual frequency offset for the kth RSS is written by:

$$\hat{\epsilon}_{\kappa(k^*)} = \frac{N}{N_T}\left(\hat{\xi}_{k^*} - \frac{f_{k^*}}{C_F}\right), \tag{43}$$

and it would be also compared with ϵ_{k^*} if it is successfully identified without false-alarm ($k^* \in \mathcal{G}_d$).

4 Statistical characteristics

4.1 Identifiability

When the estimator uses the timing offset dimension as a primary dimension, the maximum number of RSSs, which is almost surely uniquely identifiable, is derived as [19], Equation (31):

$$K_{\text{uid}} = \min\left\{(V_1 - 1)M_1, 2QV_2M_2\right\}. \tag{44}$$

Actually, (44) can be easily induced from (19) since the matrix size of $\mathbf{G}_1$ and $\hat{\mathbf{G}}_2$ would determine the overall rank.

By adjusting the smoothing factors V_1, V_2, M_1, and M_2, the code detection and timing and frequency offset estimation performance can be improved as well as K_{uid} can be increased.

4.2 Acquisition range

Basically, the transmitted code index pair $\mathbf{a}_{k^*}$ is twisted by the timing and frequency offsets. So, large timing and/or frequency offsets can cause the misdetection and/or false detection. Therefore, for the reliable code detection, it is essential to limit the allowable timing and/or frequency offset ranges.

In principle, once $\mathbf{b}_{k^*}$ resides on the correct decision region, i.e., $\mathbf{b}_{k^*} \in \mathcal{R}_{\kappa(k^*)}$, the timing and frequency offsets of the $\kappa(k^*)$th RSS can be successfully recovered from the k^*th received tuple, (refer Figure 3), and depending on the individual θ_{k^*}, ξ_{k^*}, and $\mathcal{R}_{\kappa(k^*)}$, the maximum allowable timing and frequency offset ranges, $\theta_{\max}$ and $\epsilon_{\max}$, can be changed. Meanwhile, from (4) and (5), the guaranteed acquisition range not to make ambiguity at the code detection without any noise term can be derived as:

$$\left(\frac{C_T}{2N}\theta_{\max}\right)^2 + \left(\frac{C_F}{N}N_T\epsilon_{\max}\right)^2 < \left(\frac{d_{\min}}{2}\right)^2, \tag{45}$$

where $\theta_{\max}$ and $\epsilon_{\max}$ stand for the maximum allowable timing and frequency offsets, respectively. So, from the inequality, it is available to adjust the maximum ranges of timing offset and frequency offset accordingly by enlarging one of the allowable ranges while shrinking another ranges at the same time. Remark that even though the acquisition range is derived as a lower bound by taking only a portion of the decision region $\mathcal{R}$ in (45), the maximum allowable ranges $\theta_{\max}$ and $\epsilon_{\max}$ are larger than the one introduced in MUSIC [15] and ESPRIT [14]. For example, with the same $C_T = C_F = 5$, the maximum ranges of MUSIC and ESPRIT were defined as $\theta_{\max} < \frac{N}{C_T}$ and $\epsilon_{\max} < \frac{N}{2N_T C_F}$ while the ranges of the proposed HR can be picked up as $\theta_{\max} < \frac{\sqrt{5}}{2} \cdot \frac{N}{C_T}$ and $\epsilon_{\max} < \frac{\sqrt{5}}{2} \cdot \frac{N}{2N_T C_F}$ (whose range is enlarged by more than 10%) when we just treat the importance of both timing offset and frequency offset ranges equivalently with $d_{\min} = \sqrt{5}$ from the Figure 3.

5 Simulation results

The overall system model and simulation parameters are based on the IEEE 802.16m and 3GPP LTE environments in [2,3]. The total bandwidth is 10 MHz, and $N = 1024$. Let us assume that each ranging subchannel is composed of $Q = 3$ tiles with $V = 6$ consecutive subcarriers, while $M = 5$ OFDMA blocks are presented in ranging subchannels. In order to corporate with wide bandwidth, extended vehicular A (EVA) channel model is applied to evaluate the performance [25].

The timing and frequency offset ranges are designed for $\theta_k \in [0, 114)$ and $\epsilon_k \in [-0.02, 0.02]$ which correspond to the system more than 1 km cell radius considering the round-trip delay, e.g., dense small cell, and 135 km/h Doppler frequency at $f_c = 2.4$ GHz. Note that all the simulation results reflect the ICI caused by residual frequency offset even though it is neglected in (1) for the simplicity.

The code design parameters C_T and C_F are set as $C_T = C_F = 7$ for the proposed HR algorithm.

To verify the superiority of the proposed algorithm in terms of the code detection and offset estimation performance, the ranging algorithms using MUSIC [15] and ESPRIT [14] are also compared under a common simulation setup. In addition, the ranging algorithm exploiting Zadoff-Chu sequence structure [2,26] is also compared as a reference under almost the same environment. Here, to evaluate the code detection performance solely, no collided code is assumed at the transmitter for all the algorithms, i.e., $\mathcal{G}_{co} = \{\}$ like [14,15], which means all transmitted code index pair is exclusive to each other. The overall simulation parameters and frame structure are shown in Table 1, Table 2, and Figure 4.

Table 1 Common parameters for the simulation, [2,3]

Parameters	HR, MUSIC, ESPRIT	Zadoff-Chu sequence
Sampling frequency	15.36 MHz	15.36 MHz
Total number of subcarriers	1,024	2,048
Number of OFDM symbols (M)	5	2
Each ranging symbol length	$T_{RP_{HR}} = 75$ μs	$T_{RCP_{ZC}} = 100$ μs, $T_{RP_{ZC}} = 133$ μs
Total ranging symbol length	$5.675N_T = 425.625$ μs	$5.5N_T = 412.5$ μs
Total number of ranging subchannels (R)	4	1
Number of tiles/ranging subchannel (Q)	3	1
Number of subcarriers/tile (V)	6	144
Total number of ranging subcarriers (RQV)	72	144
Timing offset range	[0, 114] samples	[0, 114] samples
Frequency offset range	[−300, 300] Hz	[−300, 300] Hz

Table 2 Contention-based initial ranging capability with the number of RSSs

Parameters	MUSIC	ESPRIT	HR	Zadoff-Chu sequence
Number of uniquely identifiable RSSs per subchannel, K_{uid}	4	4	16	N/A
Total number of uniquely identifiable RSSs, $R \cdot K_{uid}$	16	16	64	64
Total number of code sequences for RSSs, $R \cdot \lvert\mathcal{A}\rvert$	16	16	28	28
Number of active RSSs per subchannel, $K = \lvert\mathcal{A}_t\rvert$	3, 6	3, 6	3, 6	N/A
Total number of active RSSs, $R \cdot K$	12, 24	12, 24	12, 24	12, 24

5.1 Smoothing factor decision and identifiability comparison

Let us recall that the smoothing operator $\mathcal{S}(\mathbf{x})$ enlarges the observed data matrix $\mathbf{X}_q \in \mathbb{C}^{V \times M}$ to the smoothed data matrix $\dot{\mathbf{X}}_{q,\mathcal{S}} \in \mathbb{C}^{V_1M_1 \times V_2M_2}$ by choosing appropriate smoothing factors V_1, V_2 and M_1, M_2 with the condition $V_1 + V_2 = V + 1$ and $M_1 + M_2 = M + 1$. Therefore, the size of smoothed data matrix $\dot{\mathbf{X}}_{q,\mathcal{S}}$ depends on the smoothing factors. In fact, this smoothing factor could affect the code detection and offset estimation performance as well as the maximum number of distinguishable RSSs.

As shown in (44), K_{uid} is determined by (V_1, V_2) and (M_1, M_2) pair values, and Figure 5a shows the variation of this K_{uid} according to V_1 and M_1 for given $Q = 3$, $V = 6$, and $M = 5$. First, we investigated most popularly obtained number in K_{uid} distribution, which is $K_{uid} = 12$ as shown in Figure 5a. Then, from the given four cases with $K_{uid} = 12$, i.e., $\{(V_1, M_1) \,|(4,4), (5,3), (6,4), (5,5)\}$, the smoothing factor combination which shows better performance on the whole is picked up, which is $(V_1, M_1) = (4,4)$.

In addition, Figure 5b shows the trend on the maximum number of identifiable RSSs according to the number of subcarriers for fixed number of OFDM symbols $M = 5$. As seen in the figure, the maximum number of RSSs which MUSIC and/or ESPRIT can support is limited by M even though we allocate more subcarriers for the ranging whereas the number of RSSs for HR can keep increasing without limitation. This addresses that the proposed HR technique can be popularly exploited on the scenarios such that high volume of users and/or devices should be separated at the same time at the dense small cell.

5.2 Misdetection probability

Figure 6 describes the misdetection probability for four code detection algorithms with respect to signal-to-noise ratio (SNR) with the number of active RSSs $R \cdot K = 12$ and 24. The notation for each code detector and offset estimator is expressed in the Table 3.

As seen in the figures, the proposed HRCD shows good P_{miss} performance. It is because automatic pairing property in HR improves the performance of code indices detection. Even though ESPRIT code detector (ECD) also exploits two independent estimates of the code indices to make a decision, the misdetection frequently occurs for ECD due to the discrepant code indices for two independent estimations. In addition, it is shown in Figure 6a,b that ZCCD has very poor performance regardless of SNR. It is because the orthogonality property of Zadoff-Chu sequence can sustain only with small number of RSSs. Different multipath fading channels as well as asynchronous timing/frequency offsets among multiple RSSs deteriorate the ZCCD detection performance dramatically.

Note that in Figure 6b, MUSIC and ESPRIT misdetection performance cannot be evaluated for $R \cdot K = 24$ since it exceeds the maximum number of RSSs which MUSIC and/or ESPRIT can support, which is the limiting factor of MUSIC and ESPRIT algorithms.

5.3 False detection probability

Figure 7 depicts the false detection probability of different algorithms with RSSs $R \cdot K = 12$ and 24. The performance trend is almost the same as the misdetection probability. Since the false detection probability of MCD and ZCCD

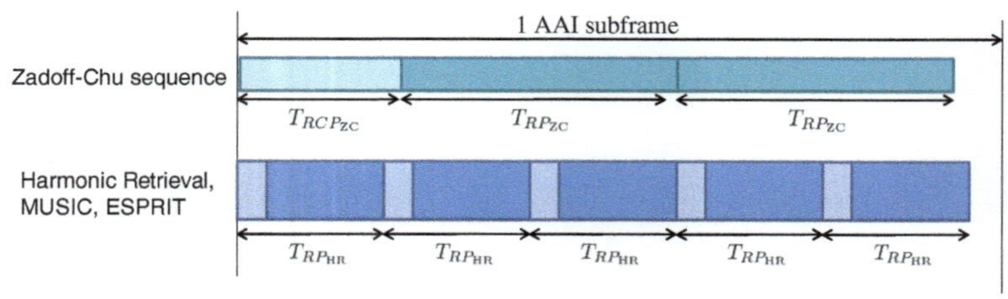

Figure 4 Transmitted frame structure. Two different frame structures are described, harmonic retrieval (HR) and Zadoff-Chu (ZC) sequence algorithm. MUSIC and ESPRIT use the same frame structure as HR.

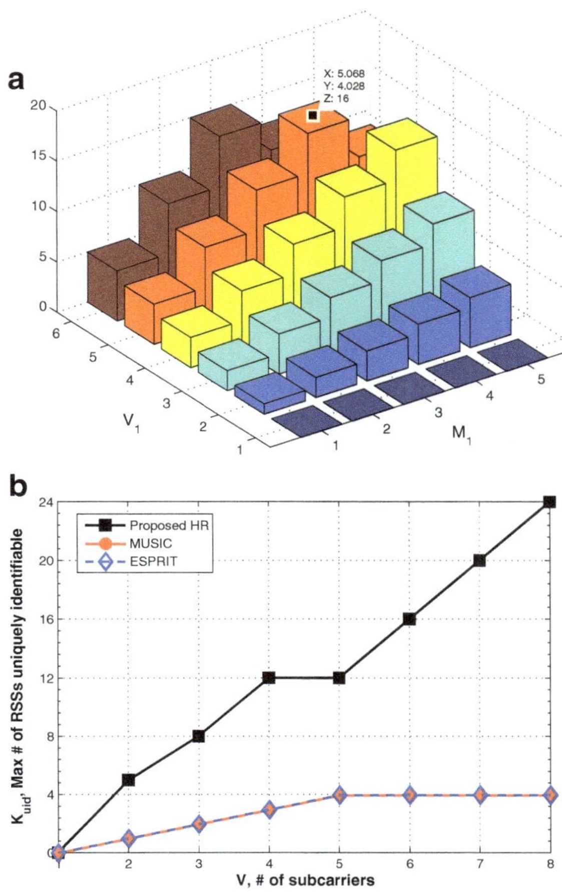

Figure 5 The number of identifiable RSSs, K_{uid} with (a) different smoothing factor configuration or (b) different algorithms.

is very poor, we expect that the overall estimation performance can be also degraded by the false detection accordingly.

5.4 Timing and frequency offset estimation accuracy

Figures 8 and 9 show the error variance performances of the timing and frequency offset estimation given the correctly detected codes, respectively, i.e., $\mathrm{Var}\left(\hat{\theta}_{\kappa(k^*)} - \theta_{k^*} \,\middle|\, \kappa(k^*) \in \mathcal{G}_d\right)$ and $\mathrm{Var}\left(\hat{\epsilon}_{\kappa(k^*)} - \epsilon_{k^*} \,\middle|\, \kappa(k^*) \in \mathcal{G}_d\right)$. The performance is evaluated with different numbers of active RSS users in the system. Note that the performance evaluation on the timing and frequency estimation is only performed for the corrected detected code index pairs, i.e., $\kappa(k^*) \in \mathcal{G}_d$. Unlike other methods, the proposed algorithm with HR can link the k^*th received tuple to the code index $\kappa(k^*)$ directly without any brute-force code index search effort due to automatic pairing property.

As seen in the Figure 8, HRTE outperforms other timing estimators over the whole SNR ranges regardless of the number of active users. Since it is crucial to obtain the timing misalignment information of each RSS on the initial ranging process, these curve confirm that the proposed HRTE can sustain the great performance even with high contention volume.

In Figure 9, it is shown that HRFE outperforms MFE and EFE over the whole range, and HRFE becomes better than ZCFE with SNR increase. Even though HRFE shows worse performance on low SNR ranges compared to ZCFE, but please remark that this estimation error variance is only evaluated for the successfully corrected code sets, i.e., $\mathrm{Var}\left(\hat{\epsilon}_{\kappa(k^*)} - \epsilon_{k^*} \,\middle|\, \kappa(k^*) \in \mathcal{G}_d\right)$. Therefore, if we consider the poor false-alarm and missing probability performance of ZCCD shown in Figure 7 (more than 10% of false-alarm and missing), the overall ZCFE performance will be severely degraded on the whole.

On the whole, the performance of other algorithms undergo severe performance degradation because wrongly detected code index disturbs timing and frequency offset estimation. However, the proposed HRTE and HRFE algorithm more robust to the wrongly detected code effect.

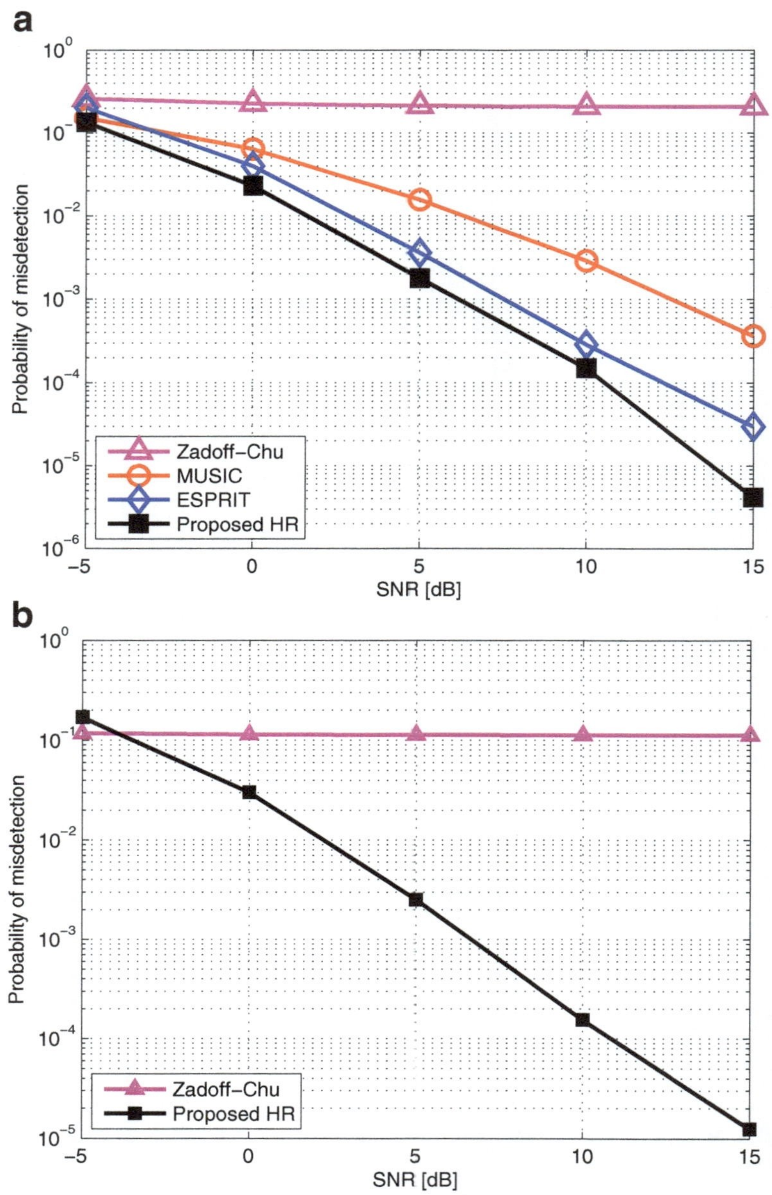

Figure 6 Misdetection probability, P_{miss} with (a) $R \cdot K = 12$ or (b) $R \cdot K = 24$ RSSs.

Table 3 Code detection and offset estimation notation

Main algorithm	Code detector (CD)	Timing offset estimator (TE)	Frequency offset estimator (FE)
Zadoff-Chu (ZC)	ZCCD	ZCTE	ZCFE
MUSIC	MCD	AHTE (*adhoc* TE)	MFE
ESPRIT	ECD	ETE	EFE
Harmonic retrieval (HR)	HRCD	HRTE	HRFE

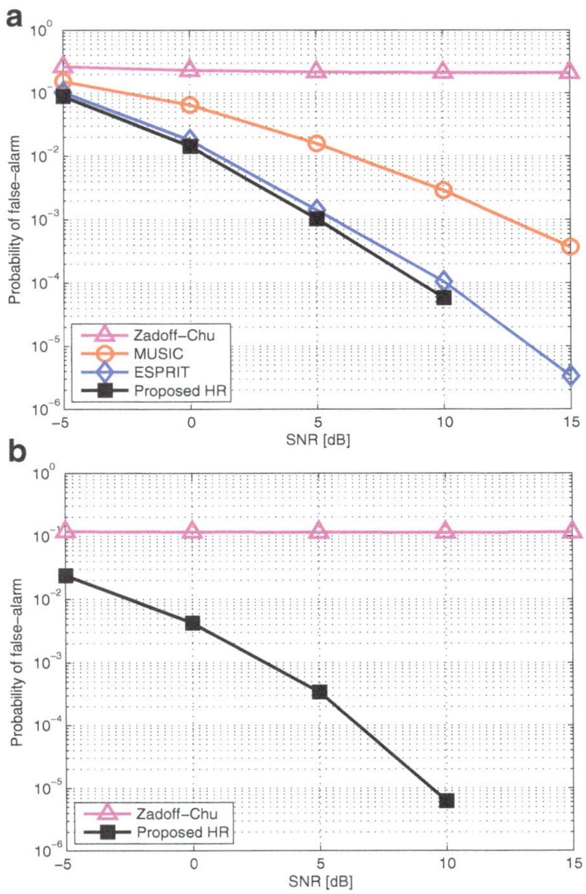

Figure 7 False detection probability, P_{false} with (a) $R \cdot K = 12$ or (b) $R \cdot K = 24$ RSSs.

6 Conclusions

First, the proposed method utilizing IMDF scheme supports the flexible subcarrier allocation usage. Basically, the proposed algorithm can be applied for the subcarrier allocation based on small *tile* and/or resource block structure whereas Zadoff-Chu sequence algorithm requires only the subband allocation, which should consist of consecutive subcarriers. Since the proposed algorithm also supports frequency noncontiguous combination of multiple tiles, actually it would even bring the performance improvement on the proposed scheme due to the enriched frequency diversity experience on the sample data. In addition, *tile* concept is compliant to the state-of-the-art wireless cellular systems, e.g., IEEE 802.16m and 3GPP LTE-Advanced [2,3], so this scheme can be adopted directly to the practical system without any painful remedy on the frame and resource allocation structure.

Next, the proposed method using IMDF scheme fully utilizes the resources to distinguish multiple RSSs as many as possible. In the practical scenario which only a few OFDMA symbols are utilized in the ranging process, the proposed HR algorithm can still increase K_{uid} by adjusting the number of tiles Q and the number of subcarriers V. Note that the number of RSSs in MUSIC and ESPRIT is easily restricted by the limited number of OFDMA symbols. Consequently, it confirms that the HR algorithm can smartly utilize the given data matrix to distinguish many RSSs simultaneously.

From the simulation results, it is shown that the proposed method outperforms the other algorithms in terms of the code detection and offset estimation while supporting more number of RSSs. Especially, the newly designed and introduced two-dimensional code index pair with the aid of automatic pairing property enables for the algorithm to increase the minimum Euclidean distance between two code index pairs and improve the acquisition range and detection/estimation performance accordingly. Consequently, this HR method can be proper for the dense network ranging process such as small cell, Machine-type communications, device-to-device communications.

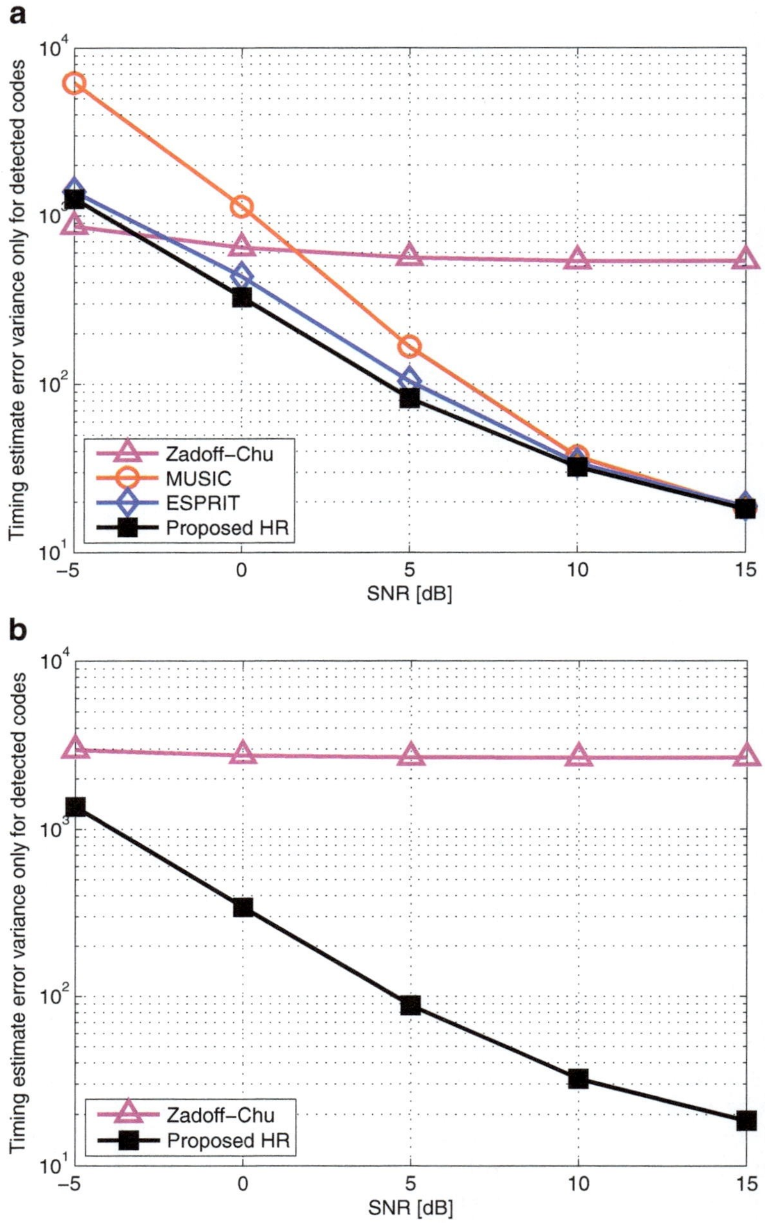

Figure 8 Timing estimation error variance, Var$\left(\hat{\theta}_{\kappa(k^*)} - \theta_{k^*} \,\middle|\, \kappa(k^*) \in \mathcal{G}_d\right)$, with (a) $R \cdot K = 12$ or (b) $R \cdot K = 24$ RSSs.

Appendix

Let us define two selection matrices $\mathbf{S}_1$ and $\mathbf{S}_2$ as:

$$\mathbf{S}_1 = \mathbf{S}_{V_1,1} \otimes \mathbf{I}_{M_1}$$
$$\mathbf{S}_2 = \mathbf{S}_{V_1,2} \otimes \mathbf{I}_{M_1}$$

where $\mathbf{S}_{V_1,1} = \begin{bmatrix}\mathbf{I}_{V_1-1} & \mathbf{0}_{V_1-1}\end{bmatrix}$ and $\mathbf{S}_{V_1,2} = \begin{bmatrix}\mathbf{0}_{V_1-1} & \mathbf{I}_{V_1-1}\end{bmatrix}$.

Then, for the given $\mathbf{U}_s$ from (20), we obtain two equal-sized matrices $\mathbf{U}_1$ and $\mathbf{U}_2$ by:

$$\mathbf{U}_1 = \mathbf{S}_1 \mathbf{U}_s, \tag{46}$$

$$\mathbf{U}_2 = \mathbf{S}_2 \mathbf{U}_s. \tag{47}$$

According to the property of Khatri-Rao product: $(\mathbf{A} \otimes \mathbf{B})(\mathbf{C} \odot \mathbf{D}) = \mathbf{AC} \odot \mathbf{BD}$, by plugging (11) and $\mathbf{U_s} = \mathbf{G}_1 \mathbf{L}^{-1}$ into (46) and (47), it can be verified that:

$$\mathbf{U}_1 = \mathbf{PL}^{-1},$$
$$\mathbf{U}_2 = \mathbf{PD}(\mathbf{e}_\eta)\mathbf{L}^{-1},$$

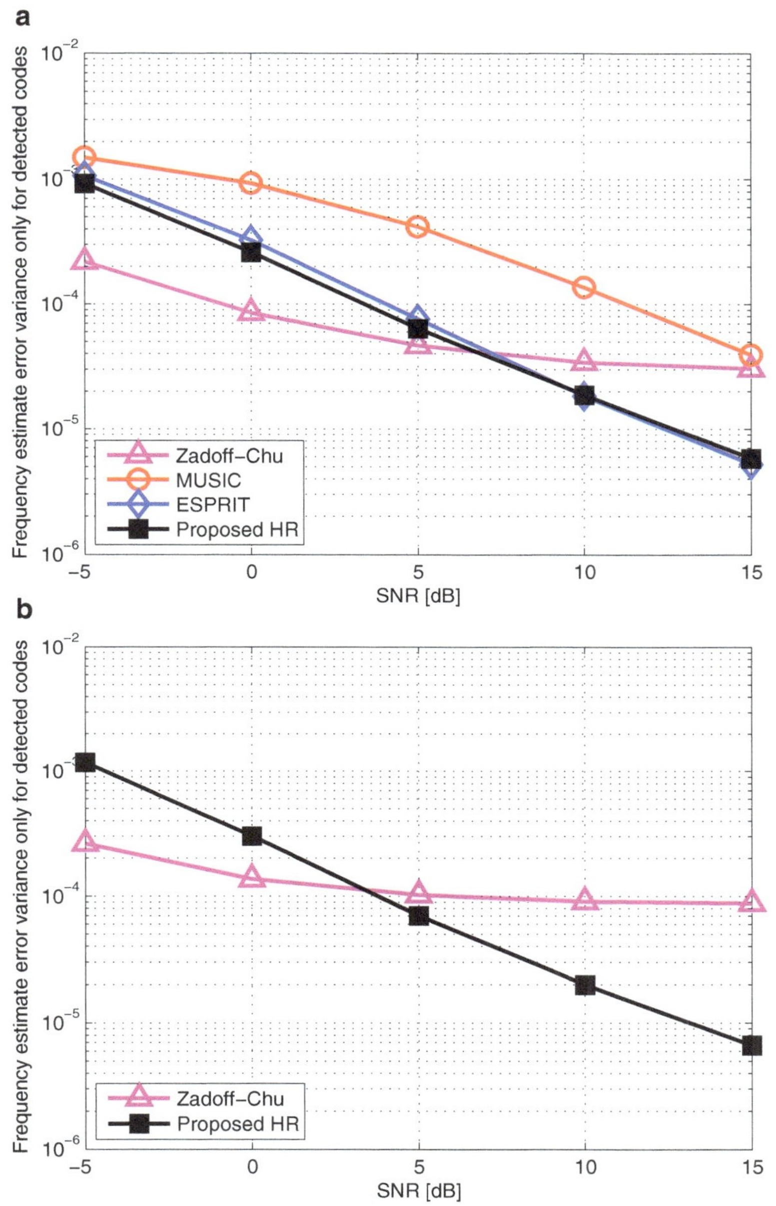

Figure 9 **Frequency estimation error variance, Var$\left(\hat{\epsilon}_{\kappa(k^*)} - \epsilon_{k^*} \,\middle|\, \kappa(k^*) \in \mathcal{G}_d\right)$, with (a) $R \cdot K = 12$ or (b) $R \cdot K = 24$ RSSs.**

where $\mathbf{P} = \mathbf{T}^{(V_1-1)} \odot \mathbf{F}^{(M_1)}$ and $\mathbf{e}_\eta = \left[e^{j2\pi\eta_1}, e^{j2\pi\eta_2}, \ldots, e^{j2\pi\eta_K}\right]^T$. Assuming that the elements of $\mathbf{e}_\eta$ are distinct, $\mathbf{L}$ can be obtained from the eigenvalue decomposition of $\mathbf{U}_1^\dagger \mathbf{U}_2$ up to column scaling and permutation ambiguity [18], i.e., the eigenvalue decomposition of $\mathbf{U}_1^\dagger \mathbf{U}_2$ gives:

$$\mathbf{U}_1^\dagger \mathbf{U}_2 = \mathbf{L}_{sp} \mathbf{D}(\mathbf{e}_\eta) \mathbf{L}_{sp}^{-1} \tag{48}$$

where $\mathbf{L}_{sp} = \mathbf{L}\mathbf{\Lambda}\mathbf{\Delta}$ and $\mathbf{\Lambda}$ is a nonsingular diagonal column scaling matrix and $\mathbf{\Delta}$ is a permutation matrix. Once we obtain $\mathbf{L}_{sp}$, we can retrieve $\mathbf{P}$ up to column scaling and permutation ambiguity:

$$\mathbf{P}_{sp} = \mathbf{U}_s \mathbf{L} = \mathbf{P}\mathbf{\Lambda}\mathbf{\Delta}. \tag{49}$$

Competing interests

The authors declare that they have no competing interests.

Received: 15 April 2014 Accepted: 24 October 2014
Published: 16 January 2015

References

1. S Lee, H Yoo, M Park, B Park, D Hong, in *Proc. IEEE VTC 2005-Spring*. Timing synchronization using phase difference between subcarriers for OFDMA uplink systems over frequency selective fading channels (Stockholm, Sweden, May 30–Jun. 1 2005), pp. 1258–1262
2. Air Interface for Broadband Wireless Access Systems Amendment 3: Advanced Air Interface. IEEE. Std. 802.16m-2011, (2011)
3. Physical channels and modulation, 3GPP Std. TS 36.211 V12.0.0, (2013)
4. DH Lee, in *Proc. IEEE and IFIP International Conference in central Asia on Internet*. OFDMA uplink ranging for IEEE 802.16e using modified generalized chirp-like polyphase sequences, (2005)
5. HA Mahmoud, H Arslan, MK Ozdemir, in *Proc. IEEE MILCOM 2006*. Initial ranging for WiMAX (802.16e) OFDMA, (2006), pp. 1–7
6. X Fu, H Minn, in *Proc. IEEE GLOBECOM 2004*, vol. 6. Initial uplink synchronization and power control (ranging process) for OFDMA systems, (2004), pp. 3999–4003
7. M Ruan, MC Reed, Z Shi, Successive multiuser detection and interference cancelation for contention based OFDMA ranging channel [transactions letters]. IEEE Trans. Wireless Commun. **9**(2), 481–487 (2010). [Online]. Available: http://ieeexplore.ieee.org/stamp/stamp.jsp?arnumber=5403522
8. C-L Lin, S-L Su, in *Proc. WCSP 2011*. A differential successive multiuser ranging detection for OFDMA systems, (2011), pp. 1–5
9. L Sanguinetti, M Morelli, An initial ranging scheme for the IEEE 802.16 OFDMA uplink. IEEE Trans. Wireless Commun. **11**(9), 3204–3215 (2012). [Online]. Available: http://ieeexplore.ieee.org/stamp/stamp.jsp?arnumber=6247449
10. X Yujie Xia, G Ren, Precoding-based initial ranging signal for SMUD algorithm in OFDMA systems. Electronics Lett. **48**(18), 1167–1168 (2012). [Online]. Available: http://ieeexplore.ieee.org/stamp/stamp.jsp?arnumber=6294580
11. X Fu, Y Li, H Minn, A new ranging method for OFDMA systems. IEEE Trans. Wireless Commun. **6**(2), 659–669 (2007)
12. XJ Zhuang, K Baum, V Nangia, M Cudak, Ranging improvement for 802.16e OFDMA PHY (2004). IEEE, Tech. Rep. IEEE C802.16e-04/143r1,
13. J Zeng, H Minn, A novel OFDMA ranging method exploiting multiuser diversity. IEEE Trans. Commun. **58**(3), 945–955 (2010)
14. L Sanguinetti, M Morelli, HV Poor, An ESPRIT-based approach for initial ranging in OFDMA systems. IEEE Trans. Commun. **57**(11), 3225–3229 (2009)
15. M Morelli, L Sanguinetti, HV Poor, A robust ranging scheme for OFDMA-based networks. IEEE Trans Commun. **57**(8), 2441–2452 (2009)
16. G Bacci, L Sanguinetti, M Luise, HV Poor, A game-theoretic approach for energy-efficient contention-based synchronization in OFDMA systems. IEEE Trans. Signal Process. **61**(5), 1258–1271 (2013). [Online]. Available:, http://ieeexplore.ieee.org/stamp/stamp.jsp?arnumber=6373741
17. W Sun, H So, Accurate and computationally efficient tensor-based subspace approach for multidimensional harmonic retrieval. IEEE Trans. Signal Process. **60**(10), 5077–5088 (2012). [Online]. Available: http://ieeexplore.ieee.org/stamp/stamp.jsp?arnumber=6222372
18. J Liu, X Liu, An eigenvector-based approach for multidimensional frequency estimation with improved identifiability. IEEE Trans. Signal Process. **54**(12), 4543–4556 (2006)
19. J Liu, X Liu, X Ma, Multidimensional frequency estimation with finite snapshots in the presence of identical frequencies. IEEE Trans. Signal Process. **55**(11), 5179–5194 (2007)
20. M Haardt, F Roemer, G Del Galdo, Higher-order SVD-based subspace estimation to improve the parameter estimation accuracy in multidimensional harmonic retrieval problems. IEEE Trans. Signal Process. **56**(7), 3198–3213 (2008). [Online]. Available: http://ieeexplore.ieee.org/stamp/stamp.jsp?arnumber=4545266
21. W Sun, H So, F Chan, L Huang, Tensor approach for eigenvector-based multi-dimensional harmonic retrieval. IEEE Trans. Signal Process. **61**(13), 3378–3388 (2013). [Online]. Available: http://ieeexplore.ieee.org/stamp/stamp.jsp?arnumber=6506111
22. S Lee, X Ma, in *Proc. ASILOMAR 2010*. An improved synchronization scheme for OFDMA systems with initial ranging transmissions (Pacific Grove, USA, Nov. 7–10 2010), pp. 150–154
23. CG Khatri, CR Rao, Solutions to some functional equations and their applications to characterization of probability distributions. Sankhyā: Ind. J. Stat. Series A. **30**(2), 167–180 (1968). [Online]. Available: http://www.jstor.org/stable/25049527
24. Air Interface for Fixed and Mobile Broadband Wireless Access Systems Amendment 2: Physical and Medium Access Control Layers for Combined Fixed and Mobile Operation in Licensed Bands and Corrigendum 1. IEEE Std. 802.16e-2005, (2006)
25. User equipment (UE) radio transmission and reception. 3GPP Std. TS 36.101 V12.2.0, (2013)
26. H Lee, JS Kwak, H Cho, Y-H Kwon, S Moon, X Wang, Y-S Choi, H Yoo, S-H Park, C-W Huang, P-H Kuo, Y-T Hsieh, P-A Ting, Z Yan-Xju, Proposed AWD text on the ranging structures for non-synchronized AMSs. IEEE, Tech. Rep. IEEE C802.16m-09/1092r3, (2009)

5

A new frequency approach for light flicker evaluation in electric power systems

Luigi Feola*, Roberto Langella and Alfredo Testa

Abstract

In this paper, a new analytical estimator for light flicker in frequency domain, which is able to take into account also the frequency components neglected by the classical methods proposed in literature, is proposed. The analytical solutions proposed apply for any generic stationary signal affected by interharmonic distortion. The light flicker analytical estimator proposed is applied to numerous numerical case studies with the goal of showing i) the correctness and the improvements of the analytical approach proposed with respect to the other methods proposed in literature and ii) the accuracy of the results compared to those obtained by means of the classical International Electrotechnical Commission (IEC) flickermeter. The usefulness of the proposed analytical approach is that it can be included in signal processing tools for interharmonic penetration studies for the integration of renewable energy sources in future smart grids.

Keywords: Light flicker; IEC flickermeter; Interharmonic; Distributed energy resources; Power quality; Smart grids

1 Introduction

Light flicker (LF) phenomenon is still considered one of the most important power quality (PQ) problems due to its ability to be directly perceived by customers, producing complaints from them.

LF is caused by the modulation of the supply fundamental voltage, which produces modulated light emissions whose severity, in terms of annoying effects on humans, depends on modulation amplitudes and frequencies as well as on lamp technologies [1]. LF is commonly measured by means of the International Electrotechnical Commission (IEC) flickermeter [2] that, for historical reasons, was designed and tested only with reference to voltage amplitude modulation (AM), which was the first source of LF identified and referring only to 60-W incandescent bulbs, which were the most diffused lamps all over the word at that time.

Today, incandescent lamps are going to be banned, in particular in Europe, Australia and North America, but the IEC flickermeter is still the only instrument used also because international standards are based on it.

The main drawbacks of the IEC flickermeter are as follows: i) it is based on the incandescent bulb model; ii) it requires 10 min of time domain signals to give the short-term flicker sensation index output, P_{st}; and iii) the output data cannot be used to study LF propagation effects in distribution and transmission networks.

Basic literature demonstrates the perfect equivalence of amplitude modulation to the summation of interharmonic tones of proper amplitudes and phase angles superimposed to the fundamental [3].

Starting from the beginning of the last decade, several papers aimed to model the IEC flickermeter in the frequency domain have been written [4-13]. Some of them [4,7,10,11] are pure frequency domain methods. Some others [6,9,12] are hybrid time-frequency domain methods.

Mayordomo et al. obtained very accurate analytical formulas that were applied to the voltages of DC and AC electrical arc furnace (EAF) measurements. In [13], the analytical formulas have been used to evaluate the propagation in the network of flicker produced by rapidly varying loads. In [11], a spectral decomposition-based approach is proposed to estimate LF caused by EAFs where

* Correspondence: luigi.feola@unina2.it
Department of Industrial and Information Engineering, Second University of Naples, DIII-SUN, Via Roma, 29, 81031 Aversa, Caserta, Italy

the system frequency deviates significantly due to the EAF operation. Both methods start from the discrete Fourier transform (DFT) performed over 200 ms, which is perfectly compatible with the IEC Standard 61000-4-7 [14] that defines harmonic and interharmonic measurement techniques.

All of the above-mentioned methods are based on simplified assumptions essentially based on the concept that, due to the design specifications of the filters of the IEC flickermeter, the interharmonic components below 15 Hz and above 85 Hz can be neglected, with reference to 50 Hz systems. This assumption is demonstrated to be valid when the interharmonic source is an EAF which mainly produces modulation of the fundamental voltage in the frequency range from 0 to 20 Hz, that is to say modulations produced by interharmonics in the frequency range from 20 to 80 Hz. Recent studies have demonstrated that modern distributed energy resources, in particular wind turbines, are able to produce interharmonics in a wide range of frequency from DC to some kilohertz [15]. Moreover, in [16-18], it was demonstrated that interharmonic components produced by adjustable speed drives can cover all the frequency range from DC.

In this paper, the above-mentioned simplified assumption is overcome, leading to analytical solutions, of different complexities, able to take into account also the frequency components neglected by the classical methods. The analytical solutions proposed apply for any generic stationary signal affected by interharmonic distortion. The LF analytical estimator proposed is applied to numerous numerical case studies with the goal of showing i) the correctness and the improvements of the analytical approach proposed with respect to the other methods proposed in literature and ii) the accuracy of the results compared to those obtained by means of the classical IEC flickermeter. The usefulness of the proposed analytical approach is that it can be included in signal processing tools for interharmonic penetration studies for the integration of renewable energy sources in future smart grids.

2 Analytical assessment of IEC flickermeter response due to interharmonics

In this section, the behaviour of the IEC flickermeter in terms of instantaneous flicker sensation (PU), that is the output of the block 4 of the IEC flickermeter (Figure 1), is analytically assessed.

The analytical solutions proposed apply for any generic stationary signal affected by interharmonic distortion. The generic signal is decomposed into N interharmonic pairs. Each pair is constituted of two tones in symmetrical frequency positions with respect to the fundamental frequency. Moreover, each of two components of the pair has generic amplitude and generic phase angle. Obviously, the case of single interharmonic components can be easily obtained assuming the amplitude of one of the two components of the pair equal to zero.

Here, explicit reference is made to 50-Hz systems, but the considerations developed may also be applied to 60-Hz systems by changing the constants and parameters.

2.1 Single pair

A normalized voltage with a couple of superimposed interharmonic tones ($N = 1$) in symmetrical angular frequency positions (lower and upper) with respect to the fundamental signal ($\omega_{1\_L} = \omega_1 - \Delta\omega_1$ and $\omega_{1\_U} = \omega_1 + \Delta\omega_1$) can be expressed as:

$$\frac{u(t)}{\hat{u}} = a_0 cos(\omega_1 t + \varphi_1) + a_{1\_L} cos[(\omega_1 - \Delta\omega_1)t + \varphi_{1\_L}] + a_{1\_U} cos[(\omega_1 + \Delta\omega_1)t + \varphi_{1\_U}], \quad (1)$$

with $\hat{u}$ representing the half cycle rms value processed through a first-order filter with a time constant of 27.3 s; a_0, ω_1 and φ_1, respectively, representing the relative amplitude, the angular frequency and the phase angle of the fundamental tone; a_{1_L} and a_{1_U} representing the relative amplitudes of the interharmonic tones and φ_{1_L} and φ_{2_L} representing their phase angles, respectively.

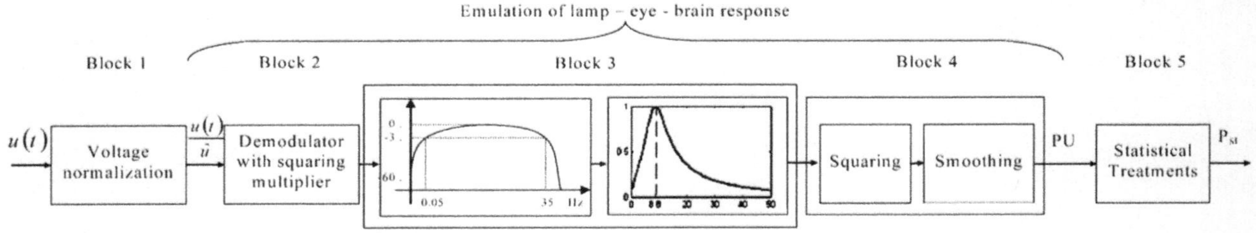

Figure 1 Simplified scheme of the IEC flickermeter.

The squaring of (1) (block 2 in Figure 1) yields a demodulated signal that is very close to the luminous flux Φ:

$$\left[\frac{u(t)}{\hat{u}}\right]^2 = \frac{1}{2}\left(a_0^2 + a_{1\_\perp}^2 + a_{1\_U}^2\right)$$
$$+\frac{1}{2}\{a_0^2\cos(2\omega_1 t + 2\varphi_1) + a_{1\_\perp}^2\cos[2(\omega_1 - \Delta\omega_1)t + 2\varphi_{1\_\perp}]$$
$$+a_{1\_U}^2\cos[2(\omega_1 + \Delta\omega_1)t + 2\varphi_{1\_U}]\}$$
$$+a_0 a_{1\_\perp}\{\cos(\Delta\omega_1 t - \varphi_{1\_\perp} + \varphi_1) + \cos[(2\omega_1 - \Delta\omega_1)t + \varphi_{1\_\perp} + \varphi_1]\}$$
$$+a_0 a_{1\_U}\{\cos(\Delta\omega_1 t + \varphi_{1\_U} + \varphi_1) + \cos[(2\omega_1 + \Delta\omega_1)t + \varphi_{1\_U} + \varphi_1]\}$$
$$+a_{1\_\perp}a_{1\_U}[\cos(2\Delta\omega_1 t - \varphi_{1\_\perp} + \varphi_{1\_U}) + \cos(2\omega_1 t + \varphi_{1\_\perp} + \varphi_{1\_U})], \quad (2)$$

Block 3 is composed of a cascade of two filters. The first is used to eliminate the DC component and the component at twice the fundamental frequency present at the output of the demodulator. This filter is composed of a cascade of a high-pass filter (HPF) of first order with a cutoff frequency of 0.05 Hz and a Butterworth low-pass filter (LPF) of the sixth order with a cutoff frequency of 35 Hz for 230 V/50 Hz systems. The second is used to weigh the voltage fluctuation according to the lamp-eye-brain sensitivity. The transfer functions and the Bode diagrams for these filters are reported in (3) and (4) and in Figure 2 together with their product $G(s) = X(s)\ F(s)$.

$$X(s) = \frac{s}{s + \omega_{HPF}}\frac{n_0}{s^6 + d_5 s^5 + d_4 s^4 + d_3 s^3 + d_2 s^2 + d_1 s + d_0}, \quad (3)$$

$$F(s) = \frac{k\omega_1 s}{s^2 + 2\lambda s + \omega_1^2}\frac{1 + s/\omega_2}{(1 + s/\omega_3)(1 + s/\omega_4)}, \quad (4)$$

being the coefficient of $X(s)$ and $F(s)$ shown in Table 1 and in Table 2, respectively [2].

Under the hypothesis that block 3 gates the DC component and all the components at a frequency certainly higher than 100 Hz (frequencies greater than the cutoff frequency of 35 Hz) and considering $G(s) = X(s)\ F(s)$ the output of the block 3 is:

$$\Delta\Phi \cong H\Big\{ |G_{\Delta\omega_1}|a_0 a_{1\_\perp}\cos(\Delta\omega_1 t - \varphi_{1\_\perp} + \varphi_1 + \angle G_{\Delta\omega_1})$$
$$+ |G_{\Delta\omega_1}|a_0 a_{1\_U}\cos(\Delta\omega_1 t + \varphi_{1\_U} - \varphi_1 + \angle G_{\Delta\omega_1})$$
$$+ |G_{2\Delta\omega_1}|a_{1\_\perp}a_{1\_U}\cos(2\Delta\omega_1 t - \varphi_{1\_\perp} + \varphi_{1\_U} + \angle G_{2\Delta\omega_1})$$
$$+ |G_{2\omega_1-\Delta\omega_1}|a_0 a_{1\_\perp}\cos[(2\omega_1 - \Delta\omega_1)t + \varphi_{1\_\perp} + \varphi_1 + \angle G_{2\omega_1-\Delta\omega_1}]$$
$$+ |G_{2\omega_1-2\Delta\omega_1}|\frac{a_{1\_\perp}^2}{2}\cos[2(\omega_1 - \Delta\omega_1)t + 2\varphi_{1\_\perp} + \angle G_{2\omega_1-2\Delta\omega}]\Big\}, \quad (5)$$

where $H = \sqrt{1238400}$ is a gain factor related to the normalization of the weighting curve, and four different magnitude and phase gains $G_{\Delta\omega_1}, G_{2\Delta\omega_1}, G_{2\omega_1-\Delta\omega_1}$ and $G_{2\omega_1-2\Delta\omega_1}$ are introduced for the five components depending on their angular frequency distances.

Block 4 of the flickermeter squares and filters the input signal. The filtering is obtained by means of a low-pass filter with a cutoff frequency of 0.5305 Hz and transfer function equal to:

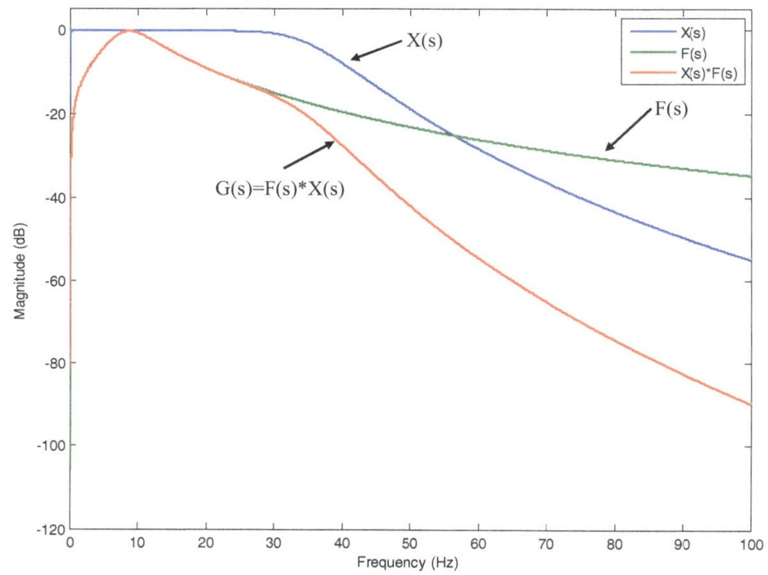

Figure 2 Magnitude Bode diagrams of block 3 filters.

Table 1 Design parameter of *X*(*s*)

ωHPF	n_0	d_0	d_1	d_2	d_3	d_4	d_5
2π0.05	1.13 10^14	1.13 10^14	1.99 10^12	1.75 10^10	9.72 10^7	3.61 10^5	849.7

$$Y(s) = \frac{2\pi 0.5305}{s + 2\pi 0.5305}. \quad (6)$$

This filter has the goal to attenuate all the sinusoidal components, leaving the continuous component unaffected. Although the bandwidth of the filter is very narrow, not all the sinusoidal components may be void. For this reason, the instantaneous flicker sensation due to the single interharmonic pair can be written as:

$$PU_1(t) = PU_{1\_DC} + PU_{1\_AC}(t), \quad (7)$$

where PU_{1_DC} and $PU_{1\_AC}(t)$ are the DC and the residual AC component of the instantaneous flicker sensation, respectively.

With reference only to the DC component of PU_1 and applying at the block 4 the signal (5), it is possible to demonstrate that:

$$PU_{1\_DC} = PU_{1\_DC1} + PU_{1\_DC2}, \quad (8)$$

where PU_{1_DC1} and PU_{1_DC2} are equal, respectively, to:

$$PU_{1\_DC1} \cong \frac{1}{2} H^2 |G_{\Delta\omega_1}|^2 a_0^2 \times \left[a_{1\_L}^2 + a_{1\_U}^2 + 2a_{1\_L}a_{1\_U}\cos(\varphi_{1\_L} + \varphi_{1\_U} - 2\varphi_1)\right], \quad (9)$$

$$PU_{1\_DC2} \cong \frac{1}{2} H^2 \left[|G_{2\Delta\omega_1}|^2 a_{1\_L}^2 a_{1\_U}^2 + |G_{2\omega_1+\Delta\omega_1}|^2 a_0^2 a_{1\_L}^2 + |G_{2\omega_1+2\Delta\omega_1}|^2 \frac{a_{1\_L}^4}{4} \right]. \quad (10)$$

As it will be shown in a more comprehensive manner in the following sections, the first component (9) assumes prevalent values for all the values of $\Delta\omega_1$ while the second (10) assumes values increasingly larger how close is the frequency of the lower interharmonic tone to zero.

It is worth noting that (9) corresponds to the analogous expressions reported in [7] and in [10]. The difference is that in cited references, the filter *X*(*s*) was considered ideal, that is to say that *G*(*s*) = *X*(*s*)*F*(*s*) = *F*(*s*); so (9) considers the red curve of Figure 2 instead of the green one considered in [7] and in [10].

Table 2 Design parameter of *F*(*s*)

k	λ	ω_1	ω_2	ω_3	ω_4
1.74802	2π 4.05981	2π 9.15494	2π 2.27979	2π 1.22535	2π 21.9

With reference only to the residual AC component of PU_1 and applying at the block 4 the signal (5), it is possible to demonstrate that:

$$PU_{1\_AC}(t) = PU_{1\_AC1}(t) + PU_{1\_AC2}(t), \quad (11)$$

where PU_{1_AC1} and PU_{1_AC2}, considering only the component of major interest, are equal, respectively, to:

$$PU_{1\_AC1}(t) \cong PU_{1\_AC1\_\Delta\omega_1}(t) + PU_{1\_AC1\_2\Delta\omega_1}(t) + PU_{1\_AC1\_3\Delta\omega_1}(t) + PU_{1\_AC1\_4\Delta\omega_1}(t), \quad (12)$$

$$PU_{1\_AC2}(t) \cong PU_{1\_AC2\_(2\omega_1-2\Delta\omega_1)}(t) + PU_{1\_AC2\_(4\omega_1-4\Delta\omega_1)}(t). \quad (13)$$

PU_{1_AC1} and PU_{1_AC2} are oscillating components due to the summation of sinusoidal signals with different angular frequencies (the angular frequency of each summand is indicated in the subscript). In particular, PU_{1_AC1} assumes higher values how close $\Delta\omega_1$ is to zero, while PU_{1_AC2} assumes higher values how close $\Delta\omega_1$ is to the fundamental angular frequency.

For the sake of brevity, the analytical assessment of the summands of (12) and (13) are reported extensively in Appendix A.

2.2 Two pairs

A normalized voltage with two pairs of superimposed interharmonic tones in symmetrical frequency positions ($\Delta\omega_1$ and $\Delta\omega_2$) with respect to the fundamental signal can be expressed as:

$$\begin{aligned} u(t)\hat{u} = {} & a_0\cos(\omega_1 t + \varphi_1) + a_{1\_L}\cos[(\omega_1 - \Delta\omega_1)t + \varphi_{1\_L}] \\ & + a_{1\_U}\cos[(\omega_1 + \Delta\omega_1)t + \varphi_{1\_U}] \\ & + a_{2\_L}\cos[(\omega_1 - \Delta\omega_2)t + \varphi_{2\_L}] \\ & + a_{2\_U}\cos[(\omega_1 + \Delta\omega_2)t + \varphi_{2\_U}], \end{aligned} \quad (14)$$

where a_{1_L} and a_{1_U} and φ_{1_L} and φ_{1_U} represent the relative amplitudes and the phase angles of the first interharmonic pair, respectively, and a_{2_L} and a_{2_U} and φ_{2_L} and φ_{2_U} of the second interharmonic pair, respectively.

Using the same procedure used for the single pair of interharmonic tones (shown in Section 2.1) and neglecting the effect on the AC component of the instantaneous flicker sensation of the interaction between the two interharmonic pairs, it is possible to demonstrate that:

$$\begin{aligned}\mathrm{PU}_{1+2} &= \mathrm{PU}_1 + \mathrm{PU}_2 + \mathrm{PU}_{12\_DC} = \mathrm{PU}_{1\_DC} + \mathrm{PU}_{1\_AC} \\ &\quad + \mathrm{PU}_{2\_DC} + \mathrm{PU}_{2\_AC} + \mathrm{PU}_{12\_DC} \\ &= (\mathrm{PU}_{1\_DC} + \mathrm{PU}_{2\_DC} + \mathrm{PU}_{12\_DC}) \\ &\quad + (\mathrm{PU}_{1\_AC} + \mathrm{PU}_{2\_AC}) \\ &= \mathrm{PU}_{1+2\_DC} + \mathrm{PU}_{1+2\_AC},\end{aligned} \tag{15}$$

where PU_1 and PU_2 are obtained by means of (7) for the interharmonic pair tones and PU_{12_DC} is the DC component of instantaneous flicker sensation due to the interaction between the two interharmonic pairs. For sake of brevity, the analytical assessment of PU_{12_DC} is reported in Appendix B. Again, as in (7), two different components, one DC (PU_{1+2_DC}) and the other AC (PU_{1+2_AC}), are defined.

From the previous formulas, it should be noted that in case of two interharmonic couples superimposed to the fundamental signal, the PU is not only due to the sum of the contributions to the instantaneous flicker sensation of the single interharmonic pair, but there is also a component due to the combination effect of the interharmonic pairs. The entity of this effect will be evaluated in the following sections.

2.3 *N* pairs

Starting from the analytical assessments of Section 2.2, it is possible to generalize the analytical assessment to the case of N pairs of interharmonic tones. In fact, from (15), it is possible to obtain:

$$\begin{aligned}\mathrm{PU}_{\mathrm{TOT}} &= \sum_{j=0}^{\infty}\mathrm{PU}_j + \sum_{j=0}^{\infty}\sum_{k=j+1}^{\infty}\mathrm{PU}_{kj\_DC} \\ &= \sum_{j=0}^{\infty}(\mathrm{PU}_{j\_DC} + \mathrm{PU}_{j\_AC}) + \sum_{j=0}^{\infty}\sum_{k=j+1}^{\infty}\mathrm{PU}_{kj\_DC} \\ &= \sum_{j=0}^{\infty}\left(\mathrm{PU}_{j\_DC1} + \sum_{k=j+1}^{\infty}\mathrm{PU}_{kj\_DC}\right) + \sum_{j=0}^{\infty}\mathrm{PU}_{j\_AC} \\ &= \mathrm{PU}_{\mathrm{TOT\_DC}} + \mathrm{PU}_{\mathrm{TOT\_AC}},\end{aligned} \tag{16}$$

where $\mathrm{PU}_{\mathrm{TOT\_DC}}$ and $\mathrm{PU}_{\mathrm{TOT\_AC}}$ are the DC and the AC components, respectively, of the instantaneous flicker sensation due to N symmetric interharmonic pairs superimposed to the fundamental signal.

3 Numerical case studies

In this section, different numerical case studies are shown to validate the analytical assessment presented in Section 2. The results obtained by a numerical implementation of the IEC flickermeter (according with the standard IEC 61000-4-15) and the results obtained by the formulas of the analytical flickermeter presented in this paper, both implemented in MATLAB, are compared. The block diagram of the numerical case studies is shown in Figure 3.

The steps performed to obtain the results of each numerical case study are as follows:

1. A time domain signal with a length of 10 min with different characteristics, according with the case study, is generated.
2. The signal is analysed by means of the IEC flickermeter, which returns both the reference of the instantaneous flicker sensation and of the short-term flicker severity value, $\mathrm{PU}_{\mathrm{ref}}$ and $P_{\mathrm{st\_ref}}$, respectively.
3. A DFT is performed only on the Fourier period of the signal-generated [1] since the signal is stationary in terms of fundamental and interharmonic signals in all the 10 min.
4. The output of the DFT is analysed by means of the analytical formulas of Section 2, according with the case study, to obtain the value of instantaneous flicker sensation (PU).
5. The instantaneous flicker sensation is used as input to the statistical evaluation (like that described in the Standard IEC 61000-4-15) to obtain the short-term flicker severity (P_{st}).
6. The values of $\mathrm{PU}_{\mathrm{ref}}$, $P_{\mathrm{st\_ref}}$, PU and P_{st} so obtained have been post-processed to calculate the error of the analytical estimation.

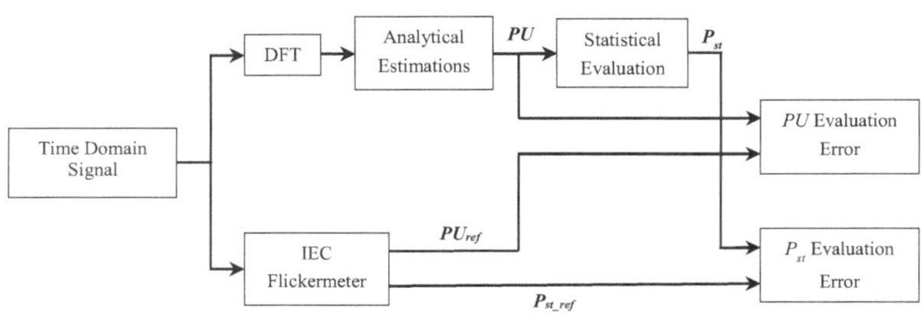

Figure 3 Block diagram of the numerical case studies.

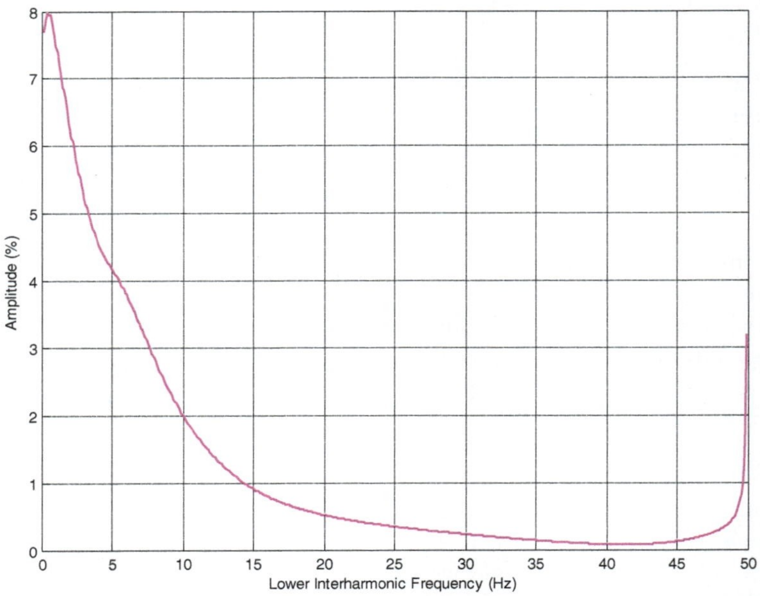

Figure 4 Amplitudes of pairs of symmetric interharmonics causing unitary P_{st} versus the pair lower interharmonic frequency.

3.1 Single pair producing AM

The interharmonic pairs have modulation frequencies varying between 0.1 and 49.9 Hz with steps of 0.1 Hz. For each modulation frequency, the amplitude of the pair taken from Figure 4, where the interharmonic amplitudes of pairs of symmetric interharmonic tones superimposed to the fundamental causing a P_{st} equal to 1 versus the lower interharmonic frequency of the pair, is shown. Furthermore, it should be noted that since the Fourier period of the signal generated is 10 s, a DFT with a spectral resolution of 0.1 Hz is used.

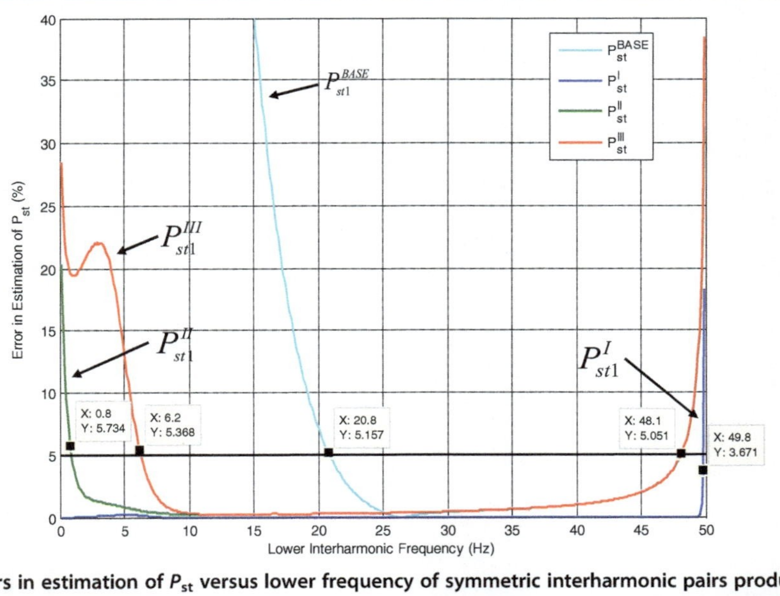

Figure 5 Percentage errors in estimation of P_{st} versus lower frequency of symmetric interharmonic pairs producing AM and unitary P_{st}. P^{BASE}_{st1}, P^{I}_{st1}, P^{II}_{st1} and P^{III}_{st1} indicate the short-term flicker severity index considering, respectively, the analytical estimation proposed in [7] (cyan), the AC and the DC component of PU_1 (17) (blue), only the DC component of PU_1 (18) (green) and only the PU_{1_DC1} component of PU_1 (19) (red).

Three different analytical estimations of the instantaneous Flicker sensation, with decreasing complexity, have been considered:

$$\text{I) } PU_1^{I} = PU_1 = PU_{1\_DC1} + PU_{1\_DC2} + PU_{1\_AC}; \quad (17)$$

$$\text{II) } PU_1^{II} = PU_{1\_DC} = PU_{1\_DC1} + PU_{1\_DC2}; \quad (18)$$

$$\text{III) } PU_1^{III} = PU_{1\_DC1}. \quad (19)$$

Obviously, the corresponding values of $P_{st1}^{I}, P_{st1}^{II}$ and P_{st1}^{III} have been calculated. Moreover, the results obtained implementing the analytical assessment proposed in [7] are reported and referred to as P_{st1}^{BASE}.

In Figure 5, the percentage errors (- P_{st1}^{BASE} - cyan curve, - PU_1^{I} - blue curve, - PU_1^{II} - green curve, - PU_1^{III} - red curve) versus the lower interharmonic frequency of the symmetric interharmonic pair are shown.

From Figure 5, it is possible to observe the following:

- The blue curve $\left(P_{st1}^{I}\right)$, in all the frequency range analysed, gives the best results and only for the pair with the frequency of 49.9 Hz shows an error greater than 5% (black line). For this frequency, the error is caused mainly by the transient behaviour of the filters contained in the IEC flickermeter, which becomes non-negligible for interharmonic frequencies very close to the fundamental. This transient behaviour is not taken into account by pure frequency domain methods differently from hybrid methods [12].
- The green curve $\left(P_{st1}^{II}\right)$ and the red curve $\left(P_{st1}^{III}\right)$ are virtually indistinguishable for frequencies greater than 15 Hz. Below this frequency (as mentioned in Section 2), the impact of PU_{1_DC2} on the total value of PU_1 is non-negligible, and for this reason, in this frequency range, the error committed by the analytical estimation P_{st1}^{II} is less than that produced by P_{st1}^{III}.
- The P_{st1}^{II} estimation makes an error greater than 5% (black line) for the pairs with frequencies higher than 48 Hz and lower than 0.8 Hz. The reason of these errors (in addition to the previously mentioned transient behaviour of the IEC flickermeter for interharmonic frequencies very close to the fundamental) is due to the AC component of PU_1, which is more consistent for interharmonic tones with $\Delta\omega_1$ both close to the fundamental frequency or close to zero.
- The P_{st1}^{III} estimation makes an error greater than 5% (black line) for frequencies lower than 6.2 Hz and higher than 48 Hz. The reasons of the different trend of the red curve with respect to the other two curves are the same as the ones mentioned in the previous point.
- The P_{st1}^{BASE} (cyan curve) makes an error almost equal to the P_{st1}^{II} and the P_{st1}^{III} for frequencies higher than 25 Hz, but for lower frequencies, as a result of the simplifying assumptions, the error diverges, rapidly reaching the 40% already for a frequency of 15 Hz.

3.2 Two pairs producing AM

The amplitude of the single pair of interharmonic tones has been chosen to produce singly a P_{st} equal to 1 (Figure 4). The frequency modulation of the two pairs, which are defined $\Delta\omega_1$ and $\Delta\omega_2$ in (14), has been chosen to vary between 1 and 49 Hz, with steps of 1 Hz, and all their possible combinations have been evaluated. Since the Fourier period of the input signal is equal to 1 s, a DFT with a spectral resolution of 1 Hz is used.

Three different analytical estimations of the cumulative P_{st} (P_{st1+2}) have been considered, all based on (15):

$$\text{I) } PU_{1+2}^{I} = PU_{1+2} = PU_{1+2\_DC} + PU_{1+2\_AC}. \quad (20)$$

$$\text{II) } PU_{1+2}^{II} = PU_{1+2\_DC}. \quad (21)$$

$$\text{III) } PU_{1+2}^{III} = PU_1 + PU_2. \quad (22)$$

Obviously, the corresponding values of $P_{st1+2}^{I}, P_{st1+2}^{II}$ and P_{st1+2}^{III} have been calculated. Moreover, the results obtained implementing the analytical assessment proposed in [7] are reported and referred to as P_{st1+2}^{BASE}.

In Figures 6, 7, 8 and 9, the percentage errors for the analytical estimation, $PU_{1+2}^{I}, PU_{1+2}^{II}, PU_{1+2}^{III}$ and P_{st1+2}^{BASE} are shown.

From these figures, it should be noted that

- The x-axis and the y-axis represent the frequency of lower interharmonic tone of the first and of the second interharmonic pair, respectively.
- The most complete analytical estimation P_{st1+2}^{I} (Figure 6), for the most part of the cases analysed, makes a mistake lower than 5% (light green squares). The cases with an error greater than the 5% occur essentially when the interaction between the two interharmonic pairs produces one or more oscillating component of the PU_{1+2} with an angular frequency that the filter $Y(s)$ (6) is not able to eliminate, which is not evaluated analytically.

These events occur when

i) $\Delta\omega_1 \approx \Delta\omega_2$ (main diagonal);
ii) Both the lower interharmonic tones of the two pairs are lower than 5 Hz (in the top left corner);

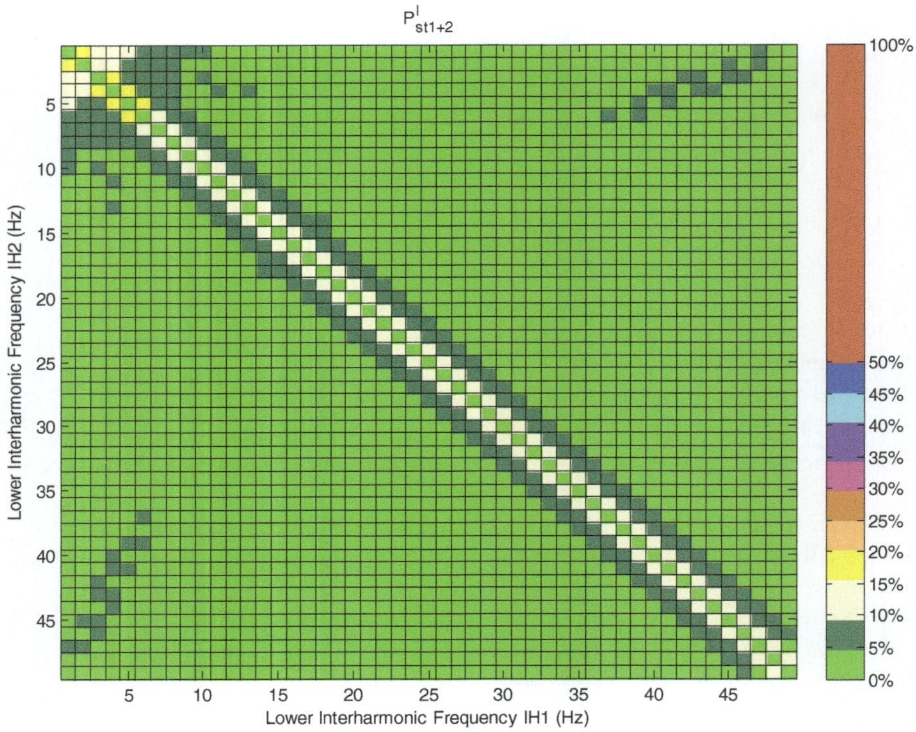

Figure 6 **Percentage errors in the evaluation of P_{st} of two pairs of interharmonic tones producing AM for P^{I}_{st1+2} (20).**

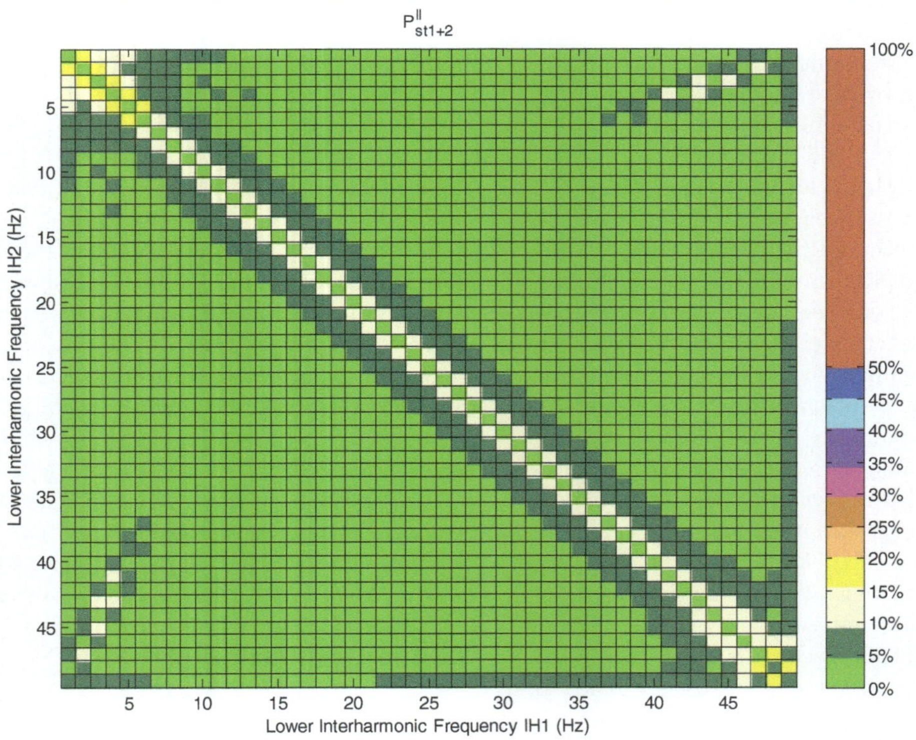

Figure 7 **Percentage errors in the evaluation of P_{st} of two pairs of interharmonic tones producing AM for PU^{II}_{1+2} (21).**

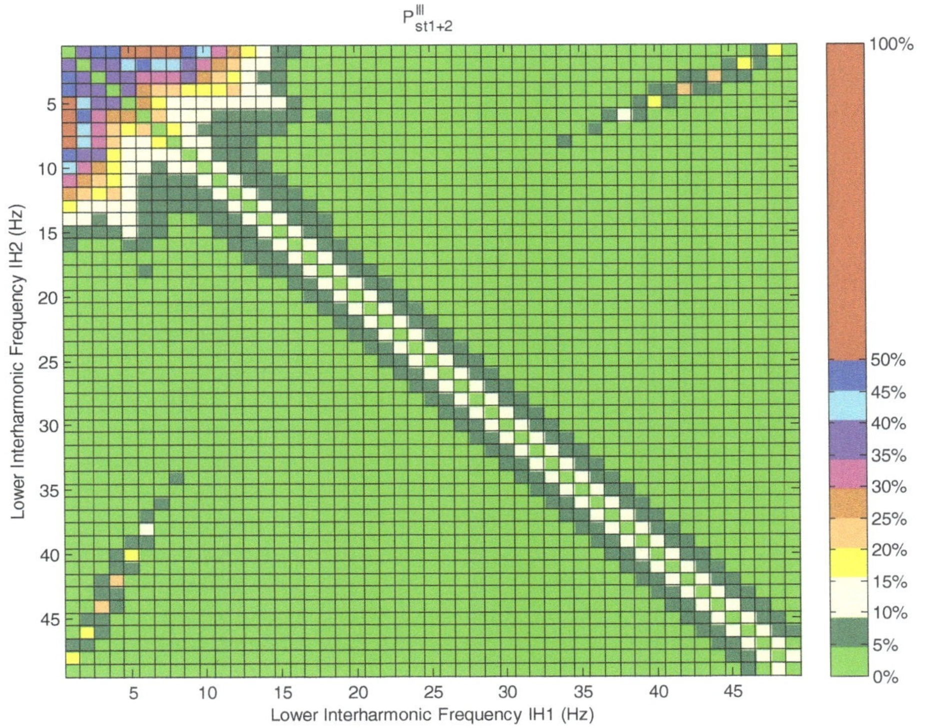

Figure 8 **Percentage errors in the evaluation of P_{st} of two pairs of interharmonic tones producing AM for PU^{III}_{1+2} (22).**

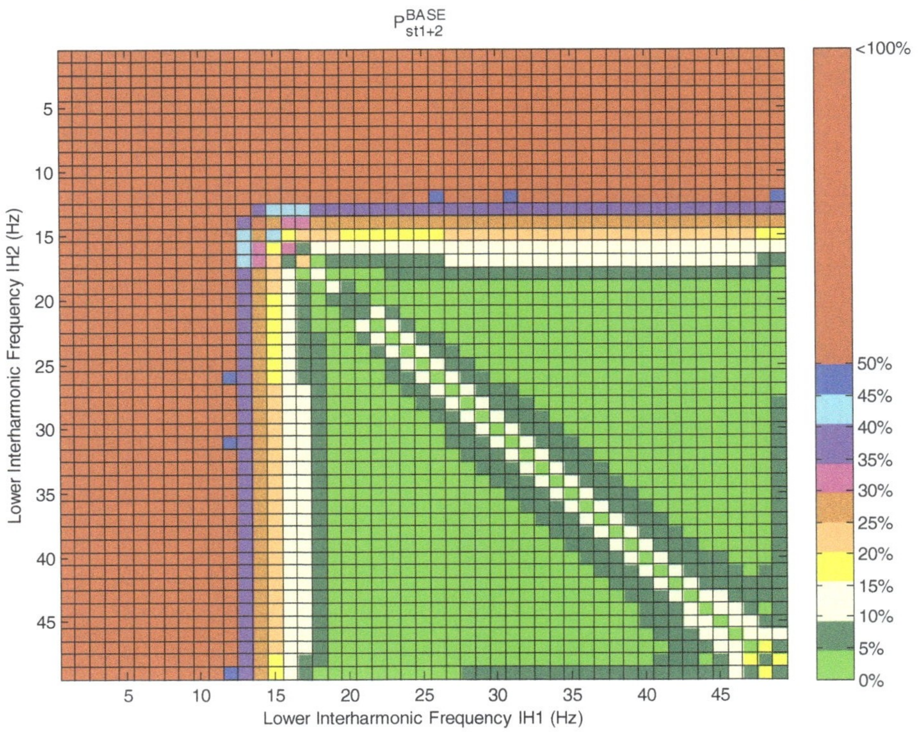

Figure 9 **Percentage errors in the evaluation of P_{st} of two pairs of interharmonic tones producing AM for P^{BASE}_{st1+2}.**

iii) $\Delta\omega_1 \approx 2(\omega_1 - \Delta\omega_2)$ (in the top right corner);
iv) $\Delta\omega_2 \approx 2(\omega_1 - \Delta\omega_1)$ (in the bottom left corner).

- Comparing the results obtained for $P^{\mathrm{I}}_{\mathrm{st1+2}}$ (Figure 6) and for $P^{\mathrm{II}}_{\mathrm{st1+2}}$ (Figure 7), it should be noted that neglecting the effects of the AC components produced by the single interharmonic pair increases the number of non-light green squares when

i) $\Delta\omega_1 \approx \Delta\omega_2$ (main diagonal);
ii) $\Delta\omega_1 \approx 2(\omega_1 - \Delta\omega_2)$ (in the top right corner) and when $\Delta\omega_2 \approx 2(\omega_1 - \Delta\omega_1)$ (in the bottom left corner);
iii) $\Delta\omega_1 = 1$ (last column to the right) and when $\Delta\omega_2 = 1$ (last column to the bottom).

- Comparing the results obtained for $P^{\mathrm{I}}_{\mathrm{st1+2}}$ (Figure 6) and for $P^{\mathrm{III}}_{\mathrm{st1+2}}$ (Figure 8), it should be noted that neglecting the effects of the interactions between the two interharmonic pairs on the DC component of the PU_{1+2} are present with a non-negligible entity when

i) Both the pairs have the lower interharmonic tone frequency lower than 15 Hz (in the top left corner);
ii) $\Delta\omega_1 = 2(\omega_1 - \Delta\omega_2)$ (in the top right corner) or when $\Delta\omega_2 \approx 2(\omega_1 - \Delta\omega_1)$ (in the bottom left corner).

- In similar mode to the case of an interharmonic pair shown in Figure 5, the analytical assessment $P^{\mathrm{BASE}}_{\mathrm{st1+2}}$ (Figure 9) makes errors very high (sometimes even more than 100%) when one or both the pairs have the lower interharmonic tone with frequencies lower than 20 Hz.
- Finally, it is worthwhile to note that from the implementation point of view on a general PQ instrument, the proposed analytical approach requires only some manipulations, of different complexities depending on the level of approximation desired (see (20), (21) and (22)), of the spectra which are already evaluated by the PQ instrument for the harmonic and interharmonic analysis. On the other hand, the digital signal processing of the conventional IEC flickermeter requires the implementation of blocks 1 to 4 independently from the spectral analysis.

4 Conclusions

In this paper, a new analytical estimator for LF in the frequency domain, which is able to take into account also the frequency components neglected by the classical methods proposed in literature, has been proposed. The analytical solutions proposed apply for any generic stationary signal affected by interharmonic distortion. The LF analytical estimator proposed has been applied to numerous numerical case studies with the goal of showing i) the correctness and the improvements of the analytical approach proposed in respect with the other method proposed in literature and ii) the accuracy of the results compared to those obtained by means of the classical IEC flickermeter. The usefulness of the proposed analytical approach is that it can be included in signal processing tools for interharmonic penetration studies for the integration of renewable energy sources in future smart grids.

The main outcomes of the paper are as follows:

- In the presence of interharmonic tones in the frequency range from DC to 15 Hz and from 85 to 100 Hz, the simplified assumptions made by classical methods proposed in literature can lead to very inaccurate results.
- The analytical formulas can be used to perform interharmonic penetration studies in transmission and distribution networks.

Future development of the research will be aimed to generalize the methodology adapted to interharmonic components at frequencies higher than 100 Hz which is proven to affect modern lighting systems different from incandescent bulbs.

5 Appendix A Analytical assessment of $\mathrm{PU}_{1\_\mathrm{AC}}(t)$

With reference to (11), it is possible to demonstrate that the summands of (12) are equal to the following:

$$
\begin{aligned}
PU_{1\_\mathrm{AC1}\_\Delta\omega_1}(t) \cong \frac{H^2|Y_{\Delta\omega_1}|}{2}\{&2|G_{\Delta\omega_1}||G_{2\Delta\omega_1}|a_0 a_{1\_\mathrm{L}}^2 a_{1\_\mathrm{U}}\cos(\Delta\omega_1 t\\
&+\varphi_{1\_\mathrm{U}}-\varphi_1-\angle G_{\Delta\omega_1}+\angle G_{2\Delta\omega_1}+\angle Y_{\Delta\omega_1})\\
&+2|G_{\Delta\omega_1}||G_{2\Delta\omega_1}|a_0 a_{1\_\mathrm{L}} a_{1\_\mathrm{U}}^2\cos(\Delta\omega_1 t-\varphi_{1\_\mathrm{L}}\\
&+\varphi_1-\angle G_{\Delta\omega_1}+\angle G_{2\Delta\omega_1}+\angle Y_{\Delta\omega_1})\\
&+|G_{2\omega_1-2\Delta\omega_1}||G_{2\omega_1-\Delta\omega_1}|a_0 a_{1\_\mathrm{L}}^3\cos(\Delta\omega_1 t-\varphi_{1\_L}\\
&+\varphi_1-\angle G_{2\omega_1-2\Delta\omega_1}+\angle G_{2\omega_1-\Delta\omega_1}+\angle Y_{\Delta\omega_1})\},
\end{aligned}
\tag{23}
$$

$$
\begin{aligned}
\mathrm{PU}_{1\_\mathrm{AC1}\_2\Delta\omega}(t) \cong \frac{H^2}{2}|Y_{2\Delta\omega_1}||G_{\Delta\omega_1}|^2 a_0^2[&a_{1\_\mathrm{L}}^2\cos(2\Delta\omega_1 t-2\varphi_{1\_\mathrm{L}}\\
&+2\varphi_1+2\angle G_{\Delta\omega_1}+\angle Y_{2\Delta\omega_1})+a_{1\_\mathrm{U}}^2\cos(2\Delta\omega_1 t\\
&+2\varphi_{1\_U}+2\varphi_1-2\angle G_{\Delta\omega_1}+\angle Y_{2\Delta\omega_1})\\
&+2a_{1\_\mathrm{L}}a_{1\_\mathrm{U}}\cos(2\Delta\omega_1 t-\varphi_{1\_\mathrm{L}}+\varphi_{1\_U}+2\angle G_{\Delta\omega_1}\\
&+\angle Y_{2\Delta\omega_1})],
\end{aligned}
\tag{24}
$$

$$
\begin{aligned}
\mathrm{PU}_{1\_\mathrm{AC1}\_3\Delta\omega_1}(t) \cong H^2&|Y_{3\Delta\omega_1}||G_{\Delta\omega_1}||G_{2\Delta\omega_1}|a_0\\
&\times[a_{1\_\mathrm{L}}^2 a_{1\_\mathrm{U}}\cos(3\Delta\omega_1 t-2\varphi_{1\_\mathrm{L}}+\varphi_{1\_\mathrm{U}}-\varphi_1\\
&+\angle G_{\Delta\omega_1}+\angle G_{2\Delta\omega_1}+\angle Y_{3\Delta\omega_1})\\
&+a_{1\_\mathrm{L}}a_{1\_\mathrm{U}}^2\cos(3\Delta\omega_1 t+2\varphi_{1\_\mathrm{U}}-\varphi_{1\_\mathrm{L}}-\varphi_1\\
&+\angle G_{\Delta\omega_1}+\angle G_{2\Delta\omega_1}+\angle Y_{3\Delta\omega_1})],
\end{aligned}
\tag{25}
$$

$$\mathrm{PU}_{1\_AC1\_4\Delta\omega_1}(t) \cong \frac{H^2}{2}|Y_{4\Delta\omega_1}||G_{2\Delta\omega_1}|^2 a_{1\_L}^2 a_{1\_U}^2 \cos(4\Delta\omega_1 t - 2\varphi_{1\_L} + 2\varphi_{1\_U} + 2\angle G_{2\Delta\omega_1} + \angle Y_{4\Delta\omega_1}), \quad (26)$$

and that the summands of (13) are equal to the following:

$$\begin{aligned}\mathrm{PU}_{1\_AC2\_2\omega_1-2\Delta\omega_1}(t) \cong\ & H^2|Y_{2\omega_1-2\Delta\omega_1}||G_{\Delta\omega_1}||G_{2\omega_1-\Delta\omega_1}|a_0^2 \\ & \{a_{1\_L}^2 \cos[2(\omega_1-\Delta\omega_1)t + 2\varphi_{1\_L} - \angle G_{\Delta\omega_1} \\ & + \angle G_{2\omega_1-\Delta\omega_1} + \angle Y_{2\omega_1-2\Delta\omega_1}] \\ & + a_{1\_L}a_{1\_U}\cos[2(\omega_1-\Delta\omega_1)t + \varphi_{1\_L} - \varphi_{1\_U} \\ & + 2\varphi_1 - \angle G_{\Delta\omega_1} + \angle G_{2\omega_1-\Delta\omega_1} + \angle Y_{2\omega_1-2\Delta\omega_1}]\}\end{aligned} \quad (27)$$

$$\mathrm{PU}_{1\_AC2\_4\omega1-4\Delta\omega_1}(t) \cong \frac{H^2}{8}|Y_{4\omega_1-4\Delta\omega_1}||G_{2\omega_1-2\Delta\omega_1}|^2 a_{1\_L}^4 \cos[4(\omega_1-\Delta\omega_1)t + 4\varphi_{1\_L} + 2\angle G_{2\omega_1-2\Delta\omega_1} + \angle Y_{4\omega_1-4\Delta\omega_1}], \quad (28)$$

6 Appendix B Analytical Assessment of PU_{12_DC}

With reference to (15), PU_{12_DC} can be expressed as:

$$\mathrm{PU}_{12\_DC} = \mathrm{PU}_{12\_DC1} + \mathrm{PU}_{12\_DC2}. \quad (29)$$

It is possible to demonstrate that PU_{12_DC1} is equal to:

$$\begin{aligned}\mathrm{PU}_{12\_DC1} = \frac{1}{2}H^2\{ & |G_{2\omega_1-\Delta\omega_1-\Delta\omega_2}|^2 a_{1\_L}^2 a_{2\_L}^2 \quad (30) \\ & + |G_{2\omega_1-\Delta\omega_1+\Delta\omega_2}|^2 a_{1\_L}^2 a_{2\_U}^2 \\ & + |G_{2\omega_1+\Delta\omega_1-\Delta\omega_2}|^2 a_{1\_U}^2 a_{2\_L}^2 \\ & + |G_{\Delta\omega_1+\Delta\omega_2}|^2 [a_{1\_U}^2 a_{2\_L}^2 + a_{1\_L}^2 a_{2\_U}^2 \\ & + 2a_{1\_L}a_{1\_U}a_{2\_L}a_{2\_U}\cos(\varphi_{1\_L} \\ & + \varphi_{1\_U} - \varphi_{2\_L} - \varphi_{2\_L})] \\ & + |G_{\Delta\omega_1-\Delta\omega_2}|^2 [a_{1\_L}^2 a_{2\_L}^2 + a_{1\_U}^2 a_{2\_U}^2 \\ & + 2a_{1\_L}a_{1\_U}a_{2\_L}a_{2\_U}\cos(\varphi_{1\_L} \\ & + \varphi_{1\_U} - \varphi_{2\_L} - \varphi_{2\_L})]\},\end{aligned}$$

and that PU_{12_DC2}, considering $\Delta\omega_1 > \Delta\omega_2$, assumes different values according to the relationship between $\Delta\omega_1$ and $\Delta\omega_2$:

- For $3\Delta\omega_1 - \Delta\omega_2 = 2\omega_1$, it is equal to:

$$\begin{aligned}\mathrm{PU}_{12\_DC2} = \frac{1}{2}H^2\{ & |G_{\Delta\omega_1-\Delta\omega_2}|^2[a_{1\_L}^3 a_{2\_L}\cos(3\varphi_{1\_L} - \varphi_{2\_L}) \\ & + a_{1\_L}^2 a_{1\_U}a_{2\_U}\cos(2\varphi_{1\_L} - \varphi_{1\_U} + \varphi_{2\_U})] \\ & + 2|G_{2\Delta\omega_1}|^2 a_{1\_L}a_{1\_U}^2 a_{2\_L}\cos(-\varphi_{1\_L} - \varphi_{2\_L})\},\end{aligned} \quad (31)$$

- For $\Delta\omega_2 + 2\Delta\omega_1 = 2\omega_1$, it is equal to:

$$\begin{aligned}\mathrm{PU}_{12\_DC2} = \frac{1}{2}H^2\{ & |G_{\Delta\omega_2}|^2[a_{2\_L}a_{1\_L}^2 a\cos(-\varphi_{2\_L} - 2\varphi_{1\_L} \\ & + \varphi) + a_{2\_U}a_{1\_L}^2 a\cos(\varphi_{2\_U} - 2\varphi_{1\_L} + \varphi)] \\ & + 2|G_{\Delta\omega_1+\Delta\omega_2}|^2[a_{1\_L}a_{1\_U}a_{2\_L}a\cos(\varphi_{1\_L} - \varphi_{1\_U} \\ & + \varphi_{2\_L} + \varphi) + a_{1\_L}^2 a_{2\_U}a\cos(2\varphi_{1\_L} - \varphi_{2\_U} + \varphi)] \\ & + 2|G_{\Delta\omega_1}|^2[a_{2\_L}a_{1\_L}^2 a\cos(-\varphi_{2\_L} - 2\varphi_{1\_L} + \varphi) \\ & + a_{1\_L}a_{2\_L}a_{1\_U}a\cos(-\varphi_{1\_L} - \varphi_{2\_L} + \varphi_{1\_U} - \varphi)] \\ & + 2|G_{2\Delta\omega_1}|^2 a_{2\_L}a_{1\_L}a_{1\_U}a\cos(-\varphi_{2\_L} - \varphi_{1\_L} \\ & + \varphi_{1\_U} - \varphi)\},\end{aligned} \quad (32)$$

- For $\Delta\omega_1 = 2\Delta\omega_2$, it is equal to:

$$\begin{aligned}\mathrm{PU}_{12\_DC2} = \frac{1}{2}H^2\{ & 2|G_{\Delta\omega_1}|^2[a_{1\_L}a_{2\_L}a_{2\_U}a\cos(-\varphi_{1\_L} + \varphi_{2\_L} - \varphi_{2\_U} \\ & + \varphi) + a_{1\_U}a_{2\_L}a_{2\_U}a\cos(\varphi_{1\_U} + \varphi_{2\_L} - \varphi_{2\_U} - \varphi)] \\ & + 2|G_{\Delta\omega_2}|^2[a_{1\_L}a_{2\_L}^2 a\cos(+\varphi_{1\_L} - 2\varphi_{2\_L} + \varphi) \\ & + a_{1\_U}a_{2\_L}a_{2\_U}a\cos(-\varphi_{1\_U} - \varphi_{2\_L} + \varphi_{2\_U} \\ & + \varphi) + a_{1\_L}a_{2\_L}a_{2\_U}a\cos(\varphi_{1\_L} - \varphi_{2\_L} + \varphi_{2\_U} - \varphi) \\ & + a_{1\_U}a_{2\_U}^2 a\cos(-\varphi_{1\_U} + 2\varphi_{2\_U} - \varphi)] \\ & + 2|G_{2\omega_1-\Delta\omega_1+\Delta\omega_2}|^2 a_{1\_L}a_{2\_L}a_{2\_U}a\cos(-\varphi_{1\_L} \\ & + \varphi_{2\_L} - \varphi_{2\_U} + \varphi) + |G_{2\omega_1-2\Delta\omega_2}|^2 a_{1\_L}a_{2\_L}^2 a \\ & \cos(-\varphi_{1\_L} + 2\varphi_{2\_L} - \varphi)\}.\end{aligned} \quad (33)$$

Competing interests

The authors declare that they have no competing interests.

Acknowledgements

The research activity discussed in this paper has been partially supported by the Project PON01_02582 "Command, control, protection and supervision integrated system for production, transmission and distribution (ColAdMin integrated SCADA) of renewable and non renewable electrical energy, with field-device-interface, for rational use of electrical power" funded by the Italian Ministry for the Instruction, University and Research.

References

1. IEEE Task, Force on harmonics modeling and simulation, interharmonics: theory and modeling. IEEE Trans. Power Deliv. **22**(4), 2335–2348 (2007). doi:10.1109/TPWRD.2007.905505
2. IEC Standard 61000-4-15. Flickermeter—functional and design specifications (IEC, Geneva, Switzerland, Edition 2.0, 2010-07)
3. R Langella, A Testa, Amplitude and phase modulation effects of waveform distortion in power systems. J. Electrical Power Qual. Util. **XIII**(1), 25–32 (2007)
4. D Gallo, R Langella, A Testa, *Light Flicker Prediction Based on Voltage Spectral Analysis* (Paper presented at the IEEE Porto Power Tech Conference, Porto, Portogallo, 2001), pp. 10–13
5. R Langella, A Testa, Power System Subharmonics, ed. by IEEE (Paper invited at the IEEE Power Engineering Society General Meeting 2005, S. Francisco, USAm, 2005)
6. T Keppler, NR Watson, S Chen, J Arrilaga, Digital flickermeter realisations in the time and frequency domains, ed. by Australasian Committee for Power Engineering (Paper presented at the Australasian Power Engineering Conference, Perth, Australia, September, 2001)

7. A Hernandez, JG Mayordomo, R Asensi, LF Beites, A new frequency domain approach for flicker evaluation of arc furnaces. IEEE Trans. Power Deliv. **18**(2), 631–638 (2003). doi:10.1109/TPWRD.2003.809733
8. CJ Wu, TH Fu, Effective voltage flicker calculation algorithm using indirect demodulation method. IEE P-Gener. Transm. D. **150**(4), 493–500 (2003). doi:10.1049/ip-gtd:20030302
9. T Keppler, NR Watson, J Arrilaga, S Chen, Theoretical assessment of light flicker caused by sub- and interharmonic frequencies. IEEE Trans. Power Deliv. **18**(1), 329–333 (2003). doi:10.1109/TPWRD.2002.806690
10. D Gallo, R Langella, C Landi, A Testa, On the use of the flickermeter to limit low-frequency interharmonic voltages. IEEE Trans. Power Deliv. **23**(4), 1720–1727 (2008). doi:10.1109/TPWRD.2008.2002842
11. N Kose, O Salor, New spectral decomposition based approach for flicker evaluation of electric arc furnaces. IET Gener. Transm. Dis. **3**(4), 393–411 (2009). doi:10.1049/iet-gtd.2008.0479
12. GW Chang, C Cheng-I, H Ya-Lun, A digital implementation of flickermeter in the hybrid time and frequency domains. IEEE Trans. Power Deliv. **24**(3), 1475–1482 (2009). doi:10.1109/TPWRD.2009.2022673
13. A Hernández, JG Mayordomo, R Asensi, LF Beites, A method based on interharmonics for flicker propagation applied to arc furnaces. IEEE Trans. Power Deliv. **20**(3), 2334–2342 (2005). doi:10.1109/TPWRD.2005.848677
14. IEC standard 61000-4-7: General guide on harmonics and interharmonics measurements, for power supply systems and equipment connected thereto (IEC, Geneva, Switzerland, Edition 2.0, 2002-08)
15. K Yang, MHJ Bollen, EO Anders Larsson, M Wahlberg, Measurements of harmonic emission versus active power from wind turbines. Electr. Pow. Syst. Res. **108**, 304–314 (2014). doi:10.1016/j.epsr.2013.11.025
16. F De Rosa, R Langella, A Sollazzo, A Testa, On the interharmonic components generated by adjustable speed drives. IEEE Trans. Power Deliv. **20**(4), 2535–2543 (2005). doi:10.1109/TPWRD.2005.852313
17. R Carbone, F De Rosa, R Langella, A Testa, A new approach for the computation of harmonics and interharmonics produced by line-commutated AC/DC/AC converters. IEEE Trans. Power Deliv. **20**(3), 2227–2234 (2005). doi:10.1109/TPWRD.2005.848448
18. R Langella, A Sollazzo, A Testa, A new approach for the computation of harmonics and interharmonics produced by AC/DC/AC conversion systems with PWM inverters. Eur. T. Electr. Power **20**(1), 68–82 (2010). doi:10.1002/etep.400

6

An optimised twin precision multiplier for ASIC environment

Rosi Asirvatham* and Seshasayanan Ramachandran

Abstract

In this paper, we present the performance of twin precision technique in reduced computation modified booth (RCMB) multiplier to achieve double throughput, and an algorithm is proposed. Twin precision technique is the efficient way to obtain double throughput in the multipliers. We describe how to apply twin precision technique to RCMB multipliers. Implementation of twin precision in RCMB multiplier requires lesser changes to be made in partial product array for obtaining double throughput. Multiplexers usually do the signal selection for N and $N/2$ bit multiplication. In RCMB multiplier, $[N/2] + 1$ partial product are reduced to $N/2$ rows. Our idea of implementing twin precision technique to RCMB results in less utilisation of multiplexers of about $[N/2] + 3$ which gave a way for optimization in the twin precision (TP) multiplier. Thereby, we have achieved the drastic reduction in multiplexer utilisation of about 40% to 50% (for $N = 8$ to 128) compared to the existing twin precision modified booth multiplier. In our proposed optimised TP modified booth multiplier this reduction in multiplexers gave a way for overall reduction in area, power and delay. Lesser utilisation of multiplexer results in the area reduction of about 5% to 18%, delay of 5% to 20% and a considerable reduction in power of 8% to 32% were noticed in the proposed TP booth multiplier for N = 8 to 128. Our proposed optimised TP multiplier is implemented in FFT complex multiplication which is taken as an application case study and achieves better performance (area, delay and power) compare to prior TP multiplier. All our evaluation are made using cadence RTL compiler using TSMC 180 nm library.

Keywords: Twin precision; Multiplexer; Throughput; RCMB; FPGA; ASIC

1 Introduction

Multiplication is an influential arithmetic operation in processor and digital signal-processing application, and thus it plays a foremost role in digital computation. In multiplication, the processing delay is directly proportional to the critical path. In order to design an efficient multiplier, limits such as multiplier speed, power and area have to be thought thoroughly. In this paper, we narrowed our research work towards achieving double throughput in signed multipliers with lesser hardware complexity. Twin precision (TP) is the technique that can be exploited to obtain the dual output [1] in multipliers, and this technique was implemented for both signed and unsigned multipliers [2].

Achieving double throughput multipliers in application-specific integrated circuit (ASIC) environment is a challenging task where re-programmability is not possible like in field-programmable gate array (FPGA). Two technologies such as FPGA and ASIC have their own pros and cons [3]. One of the biggest advantages considered in FPGA is re-programmability. Modern FPGA has many variable precision embedded multipliers [4]. Though FPGA has variable precision embedded multipliers, blindly we cannot say it is more advantageous than ASIC multipliers. Because further optimization in FPGA embedded multipliers are not possible. In ASIC, re-programmability is not possible but increased throughput in multipliers is achieved efficiently by twin precision technique.

Initially, double throughput in very large scale integration (VLSI) multipliers is achieved by using several multipliers and at least two share the same route,

* Correspondence: arosyme@gmail.com
Department of Electronics and Communication Engineering, College of Engineering, Guindy, Chennai 600025, India

which is adopted in [5,6]. However, these methods have several disadvantages like increased multiplier area and high fan-out that increases the overall delay and area. This method uses multiplexers to connect the active multiplier to the output. We can say twin precision is a subset of subword parallelism (SWP) if the lower precision multiplications stop within two levels, i.e., if a 16 bit multiplier performs two 8-bit multiplication, then it is TP multiplication. SWP is capable of performing four 4-bit multiplications or two 8-bit multiplications in the same multiplier architecture, and we referred it as multiple SWP. A subword is a lower precision of data contained within a word. By exploiting SWP in signed multipliers, multiple lower precision multiplications can be performed. So unlike TP, the throughput wont stop within two. But the other case, we have to consider here is when multiple (above two) SWP is performed on modified booth multiplier many changes are to be made for obtaining MB algorithm in all levels of lower precision multiplication. While obtaining multiple (above two) SWP in signed multiplier which adopts modified booth algorithm, steps like sign extension and inversion of bits and carry suppression are to be made in all levels of lower precision operation. And multiplexers are needed at all lower precision multiplication to select appropriate partial products because sign extension and inversion of the most significant bits (MSB) in partial product rows will vary for multiple SWP and full precision multiplication. Though SWP offers more flexibility like multiple throughputs, their hardware complexity increases with an increase in multiplexer utilisation for implementing a signed multiplication algorithm like modified booth (MB). Much architecture based on SWP has been implemented in [7-9]. These architectures are specially designed for media processing.

In [2], double throughput is realised effectively by adopting TP technique in MB multiplier. MB algorithm [10] is a widely used signed algorithm since it has reduced partial product row. The possibility of combining N and $N/2$ bit (b) multiplication in the same N b tree multiplier is called as TP multiplier where N is the bit width of the multiplier. Here we can split the partial product bits of the N b multiplier in such a way that $N/2$ b multiplication can be performed in the least significant part (LSP) of the multiplier in parallel with another $N/2$ b multiplication in the most significant part (MSP). And this is done in the partial product reduction tree without inclusion of any additional logic as explained in [1]. Multiplexers (muxes) are generally employed in TP multiplier to select appropriate partial product for N and $N/2$ b multiplication.

While implementing TP technique in the signed multipliers like MB [2], multiplexers are involved for the selection of partial products during N and $N/2$ b multiplication. This addition of multiplexers give rise to an area overhead, and thereby significant delay is added in the TP multiplier. By making the calculation of multiplexer utilisation, the TP implementation in MB multiplier [2] utilises $N + 3$ multiplexers to select the partial products. The unwanted partial products that are not in use for $N/2$ bit multiplication are set to zero by modifying two input AND gate which produces partial products to three input AND gate and the third input is the control signal. Since the area overhead in TP multiplier is caused by the multiplexers, we aimed at optimising the TP multiplier by reducing the multiplexer utilisation. Multiplexer will get reduced only when the changes to be made in $N/2$ b multiplication are lesser. Our goal of obtaining optimised TP MB multiplier is achieved when implementing the TP technique in reduced computation modified booth (RCMB) that consists of $[N/2]$ rows and not $[N/2] + 1$ rows as in modified booth algorithm and uses the simple sign extension prevention scheme which requires lesser changes to be made in TP implementation, i.e., for $N/2$ b multiplication.

In this paper, our goal of obtaining optimised double throughput signed multiplier has achieved without inclusion of complex logic, which reflects in reduced area, delay and power. We have implemented twin precision in an efficient manner with less hardware constraint compared to previous implementation, and a suitable algorithm is proposed. Implementing TP in RCMB [11] has achieved reduced area, delay and power compared to prior TP technique applied in MB algorithm. We have analysed that implementing TP technique in RCMB results in better performance, and it has been discussed in the rest of our paper.

2 TP implementation in modified booth

MB algorithm is a widely used algorithm for signed multipliers, and it holds an advantage of producing half the number of partial products. In this algorithm, X is multiplied by Y using modified booth encoding (MBE) scheme [12]. This encoding scheme groups Y into three bits as shown in Figure 1 and encodes it into one of the following {−2, −1, 0, 1, 2}, and the MB decoder generate partial products by multiplying encoded signal with the multiplicand Y. In the TP implementation made in MB, [2] adopts modified booth encoder and decoder are shown in Figure 2(a) & (b). Different methodologies of implementing MB recoding logic are explained in [13,14].

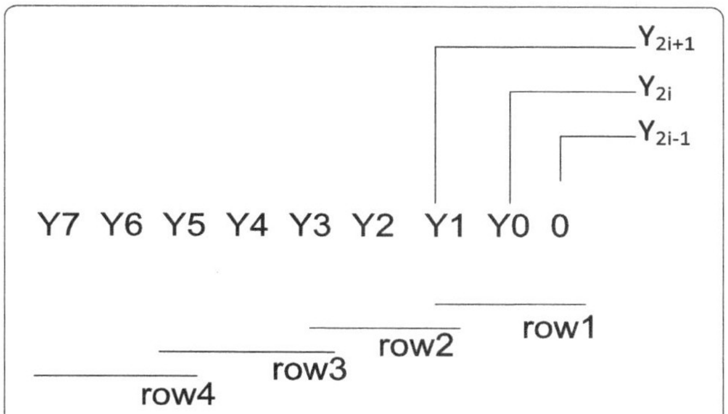

Figure 1 Encoding logic of 8-bit modified booth.

To multiply the encoded {−2, −1} with multiplicand X, two's complement representation of partial products is in need to indicate the change in sign bit. A '1' should be inserted in the least significant bit (LSB) of each partial product row for sign change which is a normal procedure in two's complement representation. While doing so, an irregular partial product array will be obtained, and this can be overwhelmed by the idea proposed in [15]. The idea is to precompute the two LSB positions of recoded partial product row by the insertion of 1 during the sign change. Actually, precomputation calculates the addition of LSB with potential '1' whose sum is the new LSB and carry is inserted in the second least significant bit position. Precomputation works according to the Equation 1 and its carry is calculated as given by Equation 2, which is adopted in [2].

$$\text{PLSB}i = X_0(Y_{2i-1} \oplus Y_{2i}) \tag{1}$$

$$a_i = Y_{2i+1}(Y_{2i-1} + Y_{2i} + X_{LSB} + Y_{2i} + X_{LSB} + Y_{2i-1}) \tag{2}$$

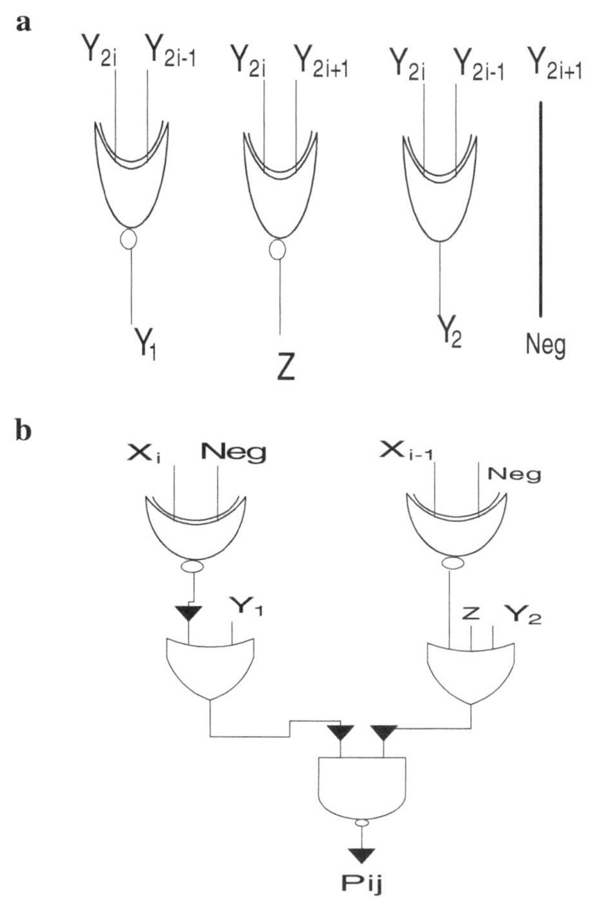

Figure 2 Modified booth encoder (a) and modified booth decoder (b).

TP technique facilitates the increase in computational throughput by allowing narrow width operation in parallel. Double throughput was obtained by the TP technique and was implemented in signed algorithm such as Baugh Woolley (BW) and MB [2]. Normally, TP multiplier either performs N b multiplication or $N/2$ b multiplication. If TP multiplier performs N bit multiplication, then its output sum (S) bits are from S0 to S15 as demonstrated in Figure 3, and when $N/2$ bit multiplication is performed, the output sum bits are from S0 to S7 and S8 to S15, i.e., twin output is taken as illustrated in Figure 4 and $N = 8$ for the illustrated example.

Implementation of TP technique is not similar for BW and MB algorithm. Figure 3 illustrates the MB multiplication for 8 × 8 which uses the sign extension scheme proposed by Fadavi [16] and uses recoding logic proposed in [15]. Achieving double throughput in MB algorithm requires appropriate selection of partial product signals in N and $N/2$ b multiplication, i.e., looking after sign extension prevention in LSB and MSB multiplication is to be taken care when $N/2$ b multiplications are performed. Simply we can say the steps performed for N b multiplication has to be carried over for $N/2$ b multiplication. Apart from booth encoding and decoding, the steps (changes) to be performed in $N/2$ LSB and MSB multiplication for MB algorithm [2] are precomputed LSB (PLSB) and its carry ai has to be calculated separately. Likewise inversion of MSB in each partial product row and sign bit pattern 1 s and 0 s will differ for $N/2$ LSB and MSB multiplication. Multiplexers will make these changes in N b partial product array by selecting appropriate partial products for N and $N/2$ b multiplication.

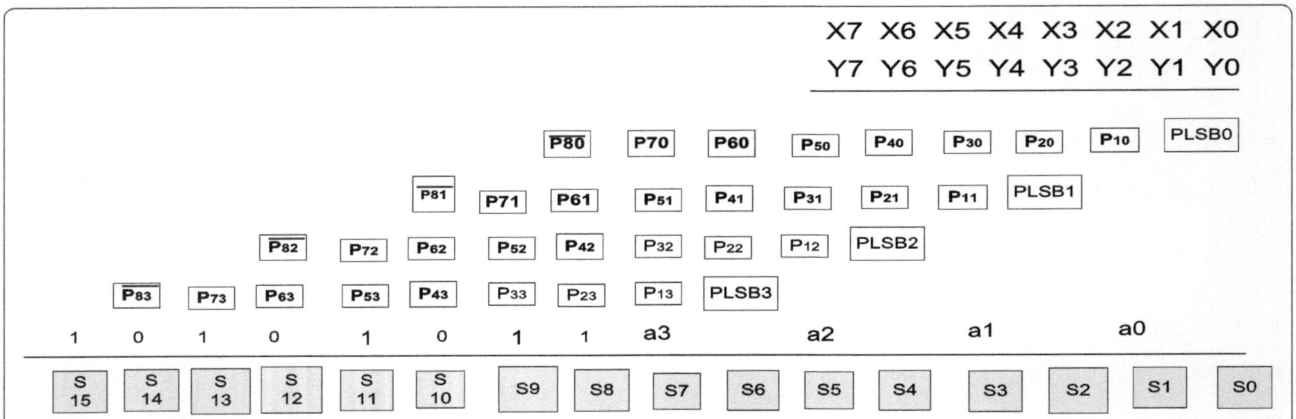

Figure 3 Signed 8-bit multiplication using modified booth algorithm-*N* bit multiplication.

Since the partial product generation of MB multiplication is based on the encoding and decoding logic, it is not possible to use the results of full precision (N b multiplication) for narrow width ($N/2$ b) MB multiplication. So changes are to be made in the recoded N b partial product array to obtain double throughput. Though the MB algorithm reduces N number of partial product rows to $N/2$ rows, the inclusion of sign bit 1 s in the last row which is due to negative decoding of MBE makes partial product strength to $[N/2] + 1$. In [2] $(N/2 + 1)$th row is due to sign bit and potential carry (ai, $i = 0$ to $(N/2 - 1)$) of PLSB. The twin precision technique implemented in MB multiplication [2] consisting of $[N/2] + 1$ rows which has been illustrated in Figure 4 and double throughput is achieved, and Figures 3, 4, 5, and 6 are inspired from [2].

To obtain TP in modified booth, the partial products that are shaded (grey colour) in Figure 4 are needed when multiplier performs $N/2$ b multiplication whereas the non-shaded partial products are set to zero. The partial products are set to zero by connecting the MB decoder output to two input AND gate and the other input of the AND is the control signal. Based on this control signal, either MB decoded signal or zero will appear as the output. Multiplexers are needed to select the appropriate partial products for N and $N/2$ b multiplications. Encoding scheme for $N/2$ b multiplication is shown in Figure 5.

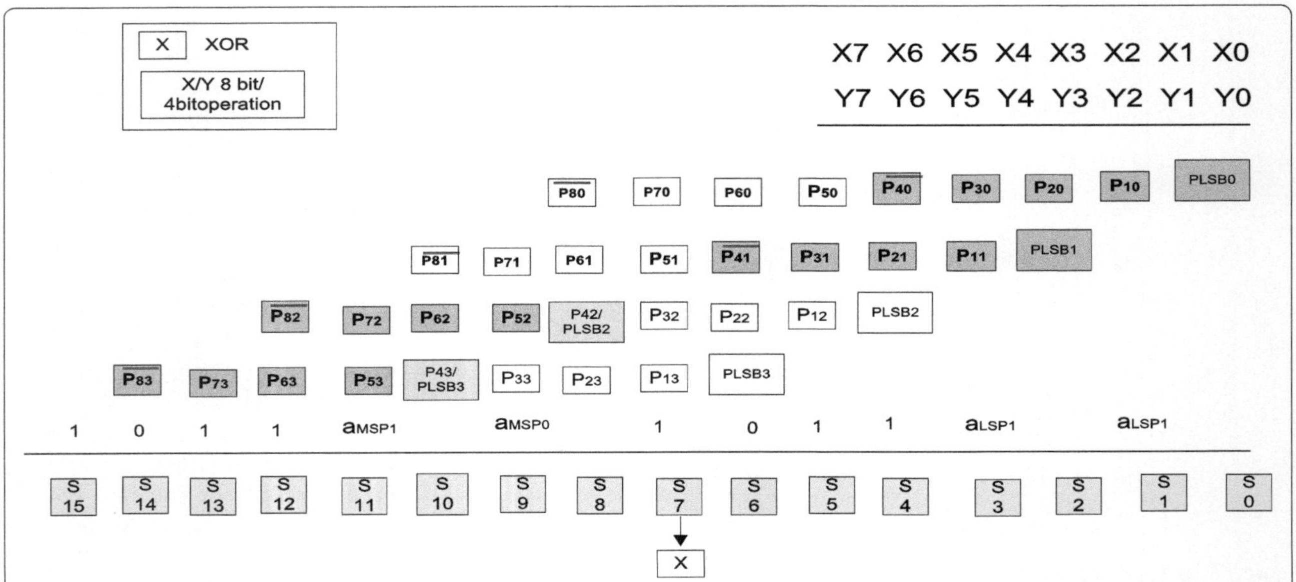

Figure 4 Implementation of twin precision in modified booth multiplier-*N*/2 multiplication.

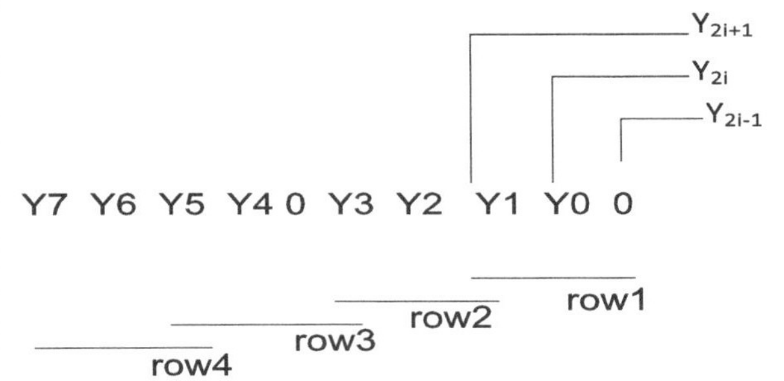

Figure 5 MB encoding scheme for *N*/2 multiplication.

By comparing Figures 3 and 4, the changes to be made to obtain twin precision are

- Partial products – P40 and P41 during normal 8 × 8 multiplications (Figure 3) need to describe sign extension prevention for *N*/2 LSB multiplication in MSB (Figure 4).
- Partial products – P42 and P43 in normal (*N* bit) multiplication has to be replenished as PLSB2 and PLSB3 for *N*/2 MSB bit multiplication.
- aMSP0 and aMSP1 are to be added for *N*/2 MSB multiplication.
- The pattern 1 s and 0 s are different for *N* and *N*/2 bit multiplication.

In Figure 6, the partial products are mapped to high-performance multiplier (HPM) reduction tree [17] and the shaded regions indicate that they are involved in *N*/2 b multiplication. Another issue that has to be looked through in *N*/2 b multiplication is that the multiplication in the LSB should not interfere with MSB multiplication, and so an encoding logic in Figure 5 has been followed. The 1 s and 0 s pattern in the last row of the partial product array is for the sign extension prevention and the most significant '1' is to invert the S2*N*-1 bit. So while performing *N*/2 LSB multiplication, the most significant '1' present in the S7 column produces a carry, and when it propagates, it will affect the *N*/2 MSB multiplication result. In order to avoid this problem, output of the S7 is passed to an exclusive OR (XOR) gate and the inversion is made, and it is illustrated in Figure 6 where the most significant '1' is made to zero and S7 output is passed to the XOR gate.

Apart from setting the unwanted partial products to zero in *N*/2 b multiplication, the selection of partial products for *N* and *N*/2 b multiplication are done by multiplexers. These multiplexers are added to the TP multiplier design to get the dual output, and thereby a significant delay is added. This delay increases linearly as the bit width (*N*) of multiplier increases because for a higher bit width, more multiplexers are utilised. The number of multiplexers depends upon the changes (steps) or selection of appropriate signals in each column to produce sum bit (S). For an 8 × 8 MB multiplier which adopts sign extension scheme [16] and recoding logic [15], the changes that have to be made for implementing TP in each column to

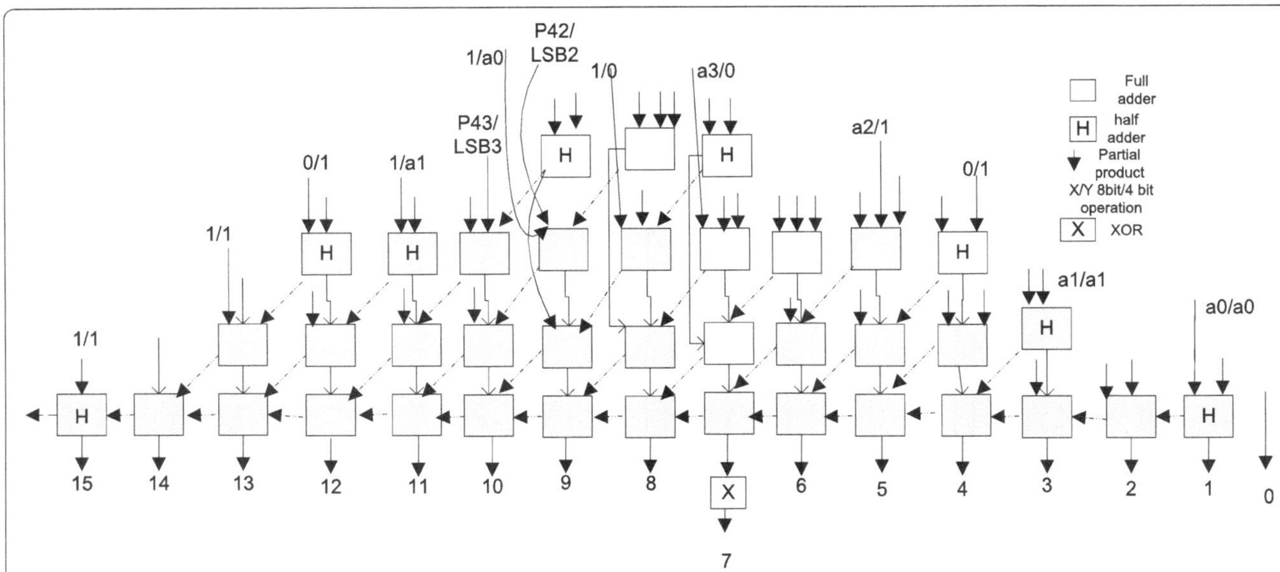

Figure 6 TP implementation of MB in HPM.

produce sum bits S0 to S15 are elaborated in Table 1. The signal selection x/y in Table 1 represents partial product (P) for N bit/ P for $N/2$ bit multiplication.

By comparing Figures 3 and 4, the changes needed to perform in each column of 8 × 8 MB multiplication for TP are tabulated in Table 1. Depending on the multiplication performed (N or $N/2$ b), the appropriate changes that have to be made in the partial product array are executed by multiplexers. In N b multiplication (Figure 3), the S4 column usually has no sign bit, but when $N/2$ LSB multiplication (Figure 4) is performed, '1' has to be included as sign bit and the inversion of MSB bit (P40) has to be performed. So in the S4 column, two multiplexers are required to make these two changes. As explained earlier, the MSB bit of each partial product row has to be inverted so P41 bit has to be inverted for $N/2$ LSB multiplication . In the seventh column, the potential carry (a3) in N bit partial product array has to be replaced as sign bit 1 for $N/2$ LSB multiplication. Likewise the changes such as precomputed LSB (PLSB2, PLSB3), the potential carry (aMSP0, aMSP1), and sign bit pattern 1 s and 0 s made for $N/2$ MSB multiplication are tabulated in Table 1.

Totally, for an 8 × 8 TP MB multiplication, we need 11 multiplexers. For various bit width, the multiplexer (mux) utilisation is calculated. From the analysis made in the MUX utilisation for implementation of TP in MB [2], it is clear that this method inquires $N + 3$ multiplexers for selection of appropriate partial products, where N is the bit width of multiplier.

Table 1 Changes needed for twin precision

Sum bit	Twin 2009 Signal selection
S0	No change
S1	No change
S2	No change
S3	No change
S4	1/0 and P40/$\bar{P}$40
S5	1/a2
S6	P41/$\bar{P}$41
S7	0/a3
S8	PLSB2/P42 and 0/1
S9	aMSP0/1
S10	PLSB3/P43
S11	aMSP1/1
S12	1/0
S13	No change
S14	No change
S15	No change

3 Implementation of TP in RCMB

In this study, we have designed an optimised TP multiplier. Optimization is achieved by implementing TP technique in RCMB and an algorithm is proposed which is applicable for $N = 8 \times$ (multiples of eight). Implementation of TP in RCMB reduces the mux utilisation to ($N/2 + 3$). Muxes are normally employed in TP multiplier for partial product selection in N and $N/2$ b multiplication. In prior work [2], the implementation of TP involves $N + 3$ multiplexers for partial product selection in TP multiplier. By applying our proposed algorithm, the TP can be obtained in the RCMB multiplier.

RCMB algorithm [11] is the classic twos complement that uses a radix-4 MBE scheme. Partial product row in RCMB (Figure 7) consists of partial product (Pi,j) that are generated based on booth encoding and decoding, negk signals ($k = 0$ to $((N/2) - 1)$) are added in the LSB position of each partial product row for generating twos complement and 1 s in the leftmost part of partial product rows that are for sign extension in twos complement representation. Figure 1 in [11] describes the gate level diagram of partial product generation. The maximum height of the partial product in MB is ($N/2 + 1$). This is reduced to $N/2$ rows in RCMB algorithm which is the biggest advantage. The neg($N/2 - 1$) bit which actually lies in the ($N/2 + 1$)th row of partial product is added to the MSB part of first row partial product represented as qij, i.e., in Figure 7, we have $\bar{q}$90 q90 q80 q70 q60 = 0 0, $\bar{P}$80 P70 P60 + 0 1 1 0 neg3 (refer to Figure 6 in [11]). And by this way, the height of partial product is reduced to $N/2$. Finally the partial products are mapped to half and full adders in HMM tree [17] structure.

In our proposed algorithm, we have assumed the partial product rows of RCMB as reduced computation matrix set (R_s). This set is divided equally and named it as upper and lower bound as shown in Figure 8. This set consists of partial product Pij, negk signals ($k = 0$ to $((N/2) - 1)$) and 1 s for sign extension. This set consists of i rows and j columns where $i = (1$ to $N/2)$ and $j = (1$ to $2N)$ and represented as $R_s\{i,j\}$. The N b procedure (steps) has to be followed for $N/2$ b multiplication. The changes or steps to be performed for $N/2$ (LSB and MSB) bit multiplication in N bit architecture are inversion of MSB in each partial product row which has to be made for twos complement method, addition of 1 s for sign extension prevention, generation of negk

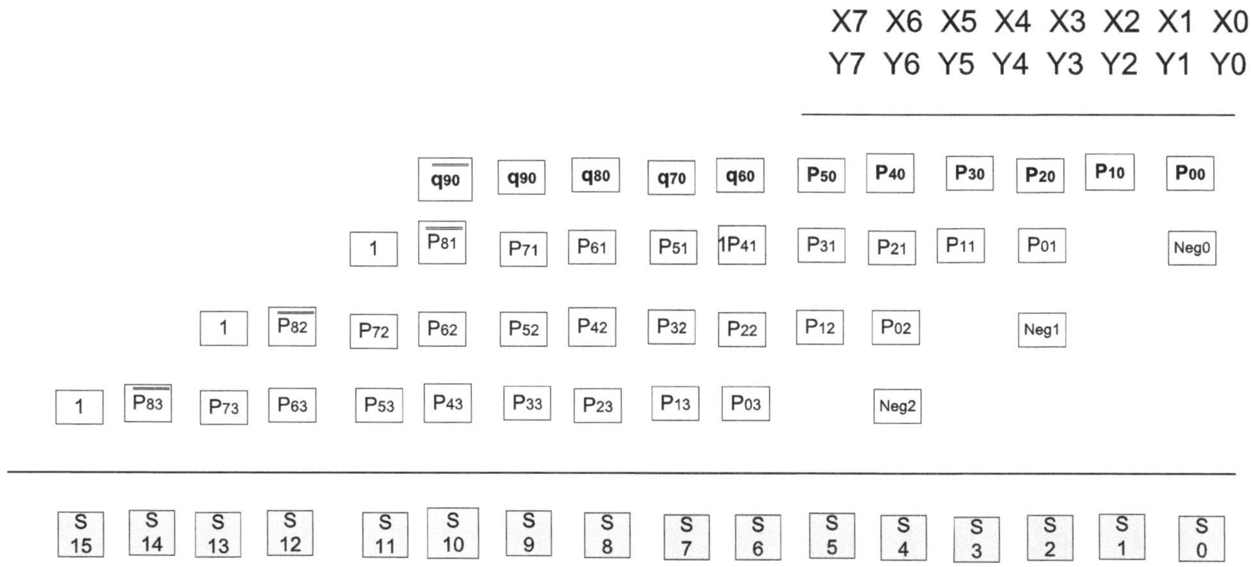

Figure 7 8 × 8 multiplication using reduced computation modified booth (RCMB).

signals in the LSB position for generating twos complement. While performing $N/2$ LSB multiplication, the changes are to be in the upper bound, and for $N/2$ MSB multiplication, changes are made in the lower bound. Multiplexers are employed in the corresponding places of N b architecture where the changes are to be made, and depending upon the multiplication, appropriate partial products are selected. Each bit (element) in R_s is denoted as Bx, and $x = 1$ to $N + 3$ denotes relative bit position in each row.

In this paper, we have implemented TP technique in RCMB multiplier which has $N/2$ partial rows. The sign extension prevention scheme proposed in [18] is used in the RCMB. In our proposed algorithm, much of the work has to be done in the upper bound compared to the lower bound. Inversion of most significant bit in partial product rows and adding of sign bit 1 s are two major steps followed to perform twin precision. In prior work of implementing twin precision technique [2] to obtain double throughput, more changes (steps) are to be performed in partial product array than proposed implementation of TP in RCMB. This in turn increases the need of multiplexers. Therefore, the area overhead for the implementation of twin precision in MB algorithm is higher than our proposed method. While obtaining twin precision in RCMB algorithm, the $N/2$ rows will be increased to $(N/2) + 1$ that uses sign extension prevention scheme in [18], and MB recoding logic proposed [19]. When the TP technique is implemented in RCMB, the multiplexer utilisation reduces drastically, and its impact on area, delay and power is analysed.

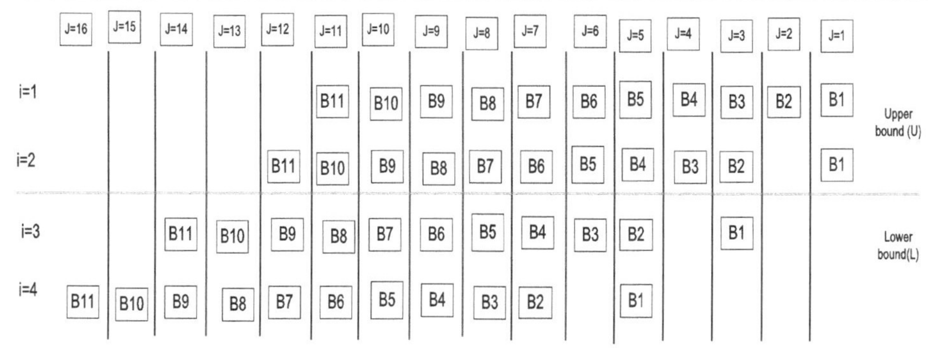

Figure 8 Reduced computation matrix RS for 8 × 8 multiplication.

3.1 Algorithm for obtaining twin precision in RCMB

1. $R_s\{i,j\}$ - reduced computation matrix set

   $i = 1\ to \frac{N}{2}\ ; j = 1\ to\ 2N;$
   N-bit width

2. For $N = 8$, upper bound $= U\{i_1,i_2\}$; lower bound $= L\{i_3,i_4\}$
3. $(Pi, j \in (i, j))$; Partial product is an element of i and j
4. $B_T = N + 3$; B_T - total bits (BT) in a row
5. Inversion and addition of sign bit 1s are needed for TP

   (i) Inversion of MSB bit while performing $N/2$ LSB multiplication is given as $B[X_i]$. X_i value gives relative bit position in the ith row so that inversion of MSB can be made correctly

   $$\text{for}\begin{cases} i = 1 & X = X_i = B_T - C_N \\ i = 2\ to \frac{N}{4} & X = X_i = X_{(i-1)} + 2 \end{cases}$$

   $B[X_i]$-bit (B) position (X) in the corresponding i row
   $C_N = \frac{N}{2} + 2$
   C_N - constant which depends on N value

   (iia) Addition of 1s in corresponding columns (j) are made for following iterations

   Iteration $(I) = 1$ to $(N/4) + 1$

   $$\text{for}\begin{cases} I = 1 & j_{ou(I)} = j_{ou(1)} = \left(\frac{N}{2}\right) + 1 \\ I = 2 & j_{ou(I)} = j_{ou(2)} = \left(\frac{N}{2}\right) + 2 \\ I = 3\ to\ (N/4) + 1 & j_{ou(I)} = j_{ou(I-1)} + 2 \end{cases}$$

   $J_{ou(I)}$ ones (O) in upper (U) bound in corresponding column (j) for corresponding iteration (I)

   (iib)

   $$OB_{L(j)} = \left(\frac{N}{2} + 1 + N\right)$$

   $OB_{L(j)}$ - addition of one in the lower bound in the corresponding column j

6. Negative bits (negk) have to be inserted in the LSB part of each partial product row for $N/2$ MSB multiplication. For this, we need to know the column (j) with respect to row (i). k varies from 0 to $(N/2 - 1)$.

   $$\text{negk} \rightarrow j_{(i)}$$

   $$\text{for}\begin{cases} i = \frac{N}{4} + 1, k = N/4\ j_{(i)} = N + 1 \\ i = \left(\frac{N}{4} + 2\ to\ N/2\right), k = \left(\left(\frac{N}{4} + 1\right)\ to\ \left(\frac{N}{2} - 1\right)\right)\ j_{(i)} = j_{(i-1)} + 2 \end{cases}$$

   i = row, j = column
   negk bits where $k = 0$ to $(N/2) - 1$

The total bits in any row in R_s will be equal and Equation 3 gives the formula to calculate total bits in a row. We have generalised the partial product (Pij), 1 s that are for twos complement implementation and negk bit in each row as a bit (B).

$$B_T = N + 3 \tag{3}$$

Proof

$$\text{For}\, N = 8 \quad B_T = 8 + 3 = 11$$

$B_T(i)$ - total bits (BT) in a row

For an 8 × 8 RCMB multiplier illustrated in Figure 8, $B_T = 11$ and this is equal for all the rows. When we perform $N/2$ LSB multiplication in N b tree, the most significant bit in each partial product row has to be inverted (negated). To perform this, we need to know the bit position (Xi) in the upper bound with respect to corresponding row (i) and given as $B[Xi]$ in Equation 4, where B is the bit and X denotes the bit position. In the lower bound, there is no need for inversion of bits for $N/2$ MSB multiplication because during N bit multiplication itself, the inversion was made. The inversion of bits that has to be implemented in the upper bound in corresponding bit position of ith row is described in Equation 4.

For rows $i = 1$ to $N/4$, the inversion of bits $B[Xi]$ are done. For initial condition $i = 1$, $[Xi]$ is the difference of total bits $B_T(i)$ in row and constant C_N.

$$for\begin{cases} i = 1 & X = X_i = X_1 = B_T - C_N \\ i = 2\,to\,\dfrac{N}{4} & X = X_i = X_{(i-1)} + 2 \end{cases} \tag{4}$$

$$C_N = \frac{N}{2} + 2$$

$B[Xi]$ - bit position in the corresponding i row

C_N - constant, which depends on N value

Proof

For $N = 8$

$$i = 1 \quad X = X_i = B_T - C_N$$
$$X_1 = 11 - 6 = 5$$
$$i = 2\,to\,\frac{N}{4}\, X = X_i = X_{(i-1)} + 2$$
$$X = X_i = X_2 = 5 + 2 = 7$$

Moreover, for the succeeding values of i that is up to $N/4$, the inversion bit position is the sum of previous iteration value and 2. For $N = 8$, the values of $X_1 = 5$ and $X_2 = 7$, so the fifth and seventh bit has to be inverted in the first and second row as implemented in Figure 9. The changes are made in the upper bound for $N/2$ LSB multiplication and in the lower bound for $N/2$ MSB multiplication.

Next step in the algorithm to perform TP is the addition of sign bit 1 s for sign extension in the upper and lower bound and has to be included in the suitable column. For this, much work has to be performed in the upper bound compared to the lower bound. Because for N b multiplication, RCMB algorithm already has sign extension bit 1 s in the most significant bit position of the lower bound. So in our algorithm for obtaining TP, addition of 1 s for sign extension in corresponding column (j) is performed in the upper bound according to Equation (5). The addition of sign bit 1 s and 0 s pattern that is used for sign extension prevention will vary for N and $N/2$ b multiplication. It is not necessary to find the bit position for the insertion of 1 s as we do for inversion of bits to get the correct sum bit (S). While performing inversion (negation) of bits in the upper bound, bit position is needed for the corresponding i because if the inversion is made in the wrong bit position, then we will wind up with incorrect sum bit and thereby will end up with the erroneous output. But in the case of adding 1 s, it is just an addition of extra signal, so finding out the corresponding column (j) is sufficient. Total 1 s to be added in upper bound for $N/2$ LSB multiplication depends on iteration (I).

$$for\begin{cases} I = 1 & j_{ou(I)} = j_{ou(1)} = \left(\dfrac{N}{2}\right) + 1 \\ I = 2 & j_{ou(I)} = j_{ou(2)} = \left(\dfrac{N}{2}\right) + 2 \\ I = 3\,to\,(N/4) + 1 & j_{ou(I)} = j_{ou(I-1)} + 2 \end{cases} \tag{5}$$

$j_{ou(I)}$ - ones (O) in the upper (U) bound for iteration (I)

Proof

$N = 8$ Iteration $I = 1$ to 3

$$I = 1 \qquad j_{ou(1)} = \left(\frac{8}{2}\right) + 1 = 5$$
$$I = 2 \qquad j_{ou(2)} = \left(\frac{8}{2}\right) + 2 = 6$$
$$I = 3 \qquad j_{ou(3)} = j_{ou(3-1)} + 2 = 6 + 2 = 8$$

So in the fifth, sixth and eighth columns, the 1 s are inserted

Iteration (I) varies from 1 to ($N/4 + 1$). For addition of 1 s in the upper bound, the iteration (I) lies from 1 to ($N/4$) + 1 and for a first two iteration, i.e., $i = 1$ and 2, the column where 1 has to be added is $\left(\frac{N}{2}\right) + 1$ and $\left(\frac{N}{2}\right) + 2$. And for the subsequent values of I, it is the sum of previous iteration value and 2. For $N = 8$, the $I = 1$ to 3, and according to Equation 5, the columns are calculated for the insertion of 1 s. From proof, it is clear that the 1 s are added in the fifth, sixth and eighth columns. In Figure 9, the '1' that is inserted in the fifth column is joined with partial product represented as P02/1 (Figure 10, Table 2) which implies

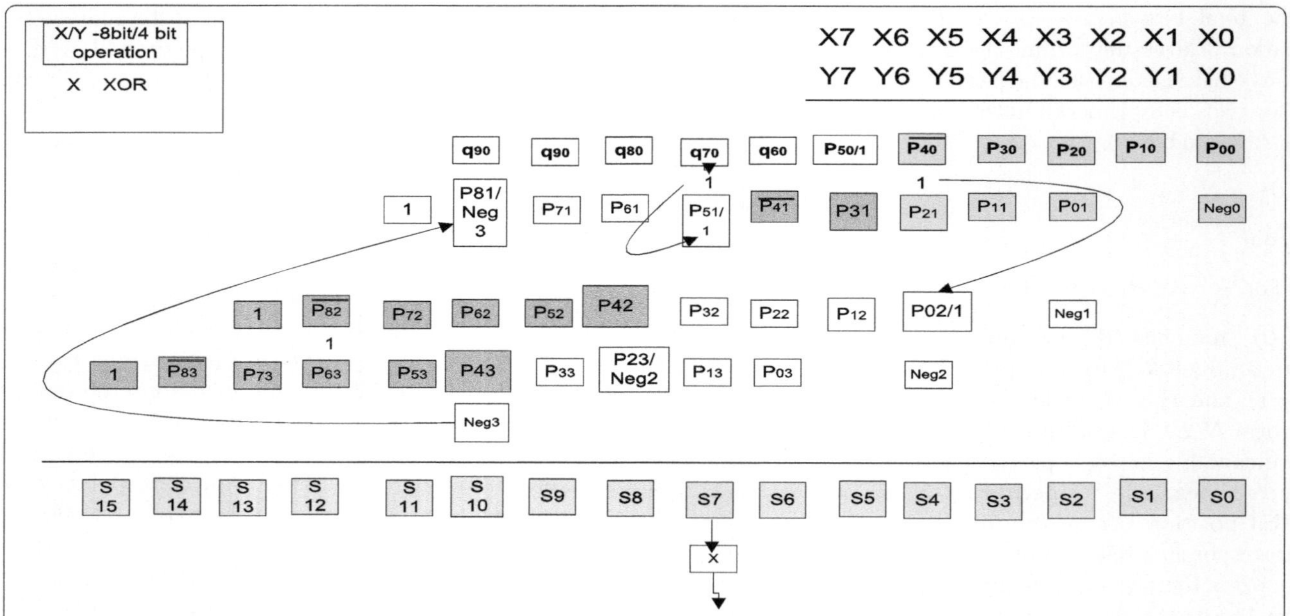

Figure 9 Implementation of our proposed algorithm in RCMB to obtain twin output.

during N bit multiplication P02 is used whereas for $N/2$ LSB multiplication, the '1' is utilised. And this selection is made by multiplexers. Likewise '1' in the sixth column is joined with P50. '1' that is inserted in the eighth column is to invert the $SN\text{-}1$ bit. If 1 is added in the eighth column, it generates a carry and affects the $N/2$ MSB multiplication. To avoid this problem, the MSB bit of $N/2$ LSB multiplication is passed through an XOR gate after the final addition and the inversion of $SN\text{-}1$ bit made.

In the lower bound, 1 has to be added in only one column for TP implementation as per Equation 6. Because in N b multiplication of RCMB, the sign bit 1 s are already present for each partial product row, so there is no need of extra effort to insert 1 s in the lower bound for $N/2$ MSB multiplication.

$$OB_{L(j)} = \left(\frac{N}{2} + 1 + N\right) \tag{6}$$

Proof
$N = 8$

$$OB_{L(j)} = \left(\frac{N}{2} + 1 + N\right)$$
$$= (8/2 + 1 + 8) = 13$$

In thirteenth column, 1 is added

$OB_{L(j)}$ - one (O) in the lower bound (BL) in corresponding column (j)

From the proof of Equation 6, it is clear 1 has to be added in the 13th column. All the steps in the algorithm are made in RCMB multiplier and mapped in HPM

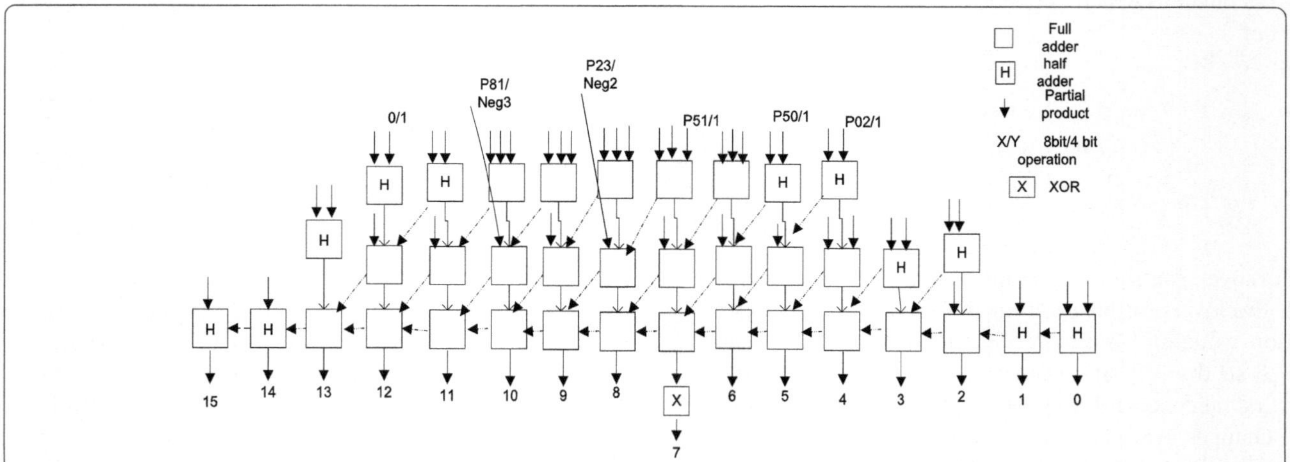

Figure 10 RCMB in HPM.

Table 2 Changes to be made to obtain TP

Sum bit	Twin 2009 [2]	Proposed twin signal selection
S15	No change	No change
S14	No change	No change
S13	No change	No change
S12	1/0	1
S11	aMSP1/1	No change
S10	PLSB3/P43	P81/neg3
S9	aMSP0/1	No change
S8	PLSB2/P42 and 0/1	*P*23/neg2
S7	0/a3	*P*51/1
S6	P41/$\overline{P41}$	P41/$\overline{P41}$
S5	1/a2	*P*31/1
S4	1/0 and *P*40/$\overline{P40}$	*P*02/1 and *P*40/$\overline{P40}$
S3	No change	No change
S2	No change	No change
S1	No change	No change
S0	No change	No change

reduction tree [17] shown in Figure 10. For N b multiplication, the sum (S) bit is from S0 to S15, and for $N/2$ LSB and MSB multiplication, sum bits are S0 to S7 and S8 to S15.

In RCMB [11], the $N/2+1$ rows are reduced to $N/2$ rows by the addition of last row sign bit, i.e., (neg(($N/2$) – 1)) bit to MSB part of the first row partial product. In our chosen example in Figure 7, it is neg3 bit which leads to $(N/2)+1$ rows added to the first row whose partial product are represented as *qij*. We exploited TP technique in RCMB multiplier and suitable algorithm is proposed. In our proposed TP implementation in RCMB when we perform $N/2$ MSB multiplication, the first row ($i=3$) of lower bound is not modified like N bit multiplication because the complexity of the design will get increased.

For N bit multiplication the neg3 bit is added to the first row whereas for $N/2$ bit MSB multiplication, the sign bit (neg3) is not added to the first row ($i=3$). But for $N/2$ MSB multiplication, P23 is replaced by the sign bit neg3 by multiplexer. This way of replacing the partial product as neg bit for $N/2$ multiplication maintains the partial product array to ($N/2$) rows without the first row modification for $N/2$ MSB multiplication.

Proof

$N=8$

$$negk \rightarrow j_{(i)}$$

$$i=\frac{N}{4}+1, k=N/4 \quad j_{(i)}=N+1$$

$i=3$, $k=2$ $j_{(3)}=8+1=9$ neg2 → 9 (neg2 bit is added to ninth column)

$$i=\left(\frac{N}{4}+2,\frac{N}{4}+3\, to\, N/2\right), k=\left(\left(\frac{N}{4}+1\right) to \left(\frac{N}{2}-1\right)\right)$$

$$j_{(i)}=j_{(i-1)}+2$$

$$i=N/2+2=4, k=N/4+1=3$$

$j_{(4)}=j_{(4-1)}+2=9+2=11$ neg3 → 11 (neg3 bit is added to eleventh column)

Note: $i=1$ to $N/2$ (first step in the 'Algorithm for obtaining twin precision in RCMB' section). For $N=8$, i varies from 1 to 4. Since i value ends with 4, it is not necessary to find $j_{(i)}$ for values of i after $N/4+2$ (i.e.,) for $i=N/4+3=5$ where we do not have $i=5$ (fifth row) in 8 × 8 ($N=8$) RCMB multiplier.

The negk ($k=0$ to $(N/2-1)$) bits are usually used to generate twos complement in RCMB algorithm. These bits are added to the LSB part in each partial product row. During N bit multiplication, negk bits are added in the LSB part of each partial product row. So for $N/2$ LSB multiplication, we do not face any problem in adding the neg(0 to $(N/4-1)$) bits in the LSB part, i.e., in the upper bound. But for $N/2$ MSB multiplication, the neg($N/4$ to $(N/2-1)$) bits have to be added in the corresponding LSB part of each partial product row in the lower bound. By Equation 7, this can be performed by adding neg($N/4$ to $(N/2-1)$) bits to the corresponding column. From proof of Equation 7, it is clear the neg2 and neg3 are added in the 9th and 11th columns. In Figure 9, the neg2 bit is joined with *P*23, i.e., for $N/2$ MSB multiplication; instead of *P*23, the neg2 bit will be selected by the multiplexer. Likewise in the tenth column, neg3 bit is joined with *P*81.

And by implementing TP technique in RCMB, we found that our approach requires less mux utilisation compared to previous implementation [2]. Figure 9 depicts the implementation of all the above steps for N and $N/2$ bit multiplication. When the multiplier performs

$$\text{for}\begin{cases} i=\frac{N}{4}+1, k=N/4 & j_{(i)}=N+1 \\ i=\left(\frac{N}{4}+2,\frac{N}{4}+3\, to\, N/2\right), k=\left(\left(\frac{N}{4}+1\right) to \left(\frac{N}{2}-1\right)\right) & j_{(i)}=j_{(i-1)}+2 \end{cases} \quad (7)$$

$N/2$ b multiplication, the unwanted partial products that are not shaded in Figure 9 are set to zero and the multiplexers are implemented where ever the changes are required. The columns where the changes have to be made in RCMB to obtain twin output using our algorithm are S4, S5, S7, S8, S10 and S12 as charted in Table 2. And by consolidating these changes, we can formulate the mux utilisation for our approach.

Here in this study, we have chosen 8 × 8 RCMB multiplier to explain our idea. From the consolidated changes, it is revealed that we entail $(N/2) + 3$ multiplexers and verified for various bit width multiples of eight as outlined in Table 2. The earlier work [2] utilises $N + 3$ multiplexers for the selection of appropriate signals. Since multiplier structure is the interconnection of full adders and half adders, the adders involved in the critical path have to wait for the previous carry signal to produce output sum bit S. Implementation of our proposed algorithm for obtaining TP in RCMB results in less multiplexers utilisation which has its impacts on area, delay and power that are analysed in the Results and discussion section. Because of this drastic reduction in multiplexers, the area, delay and power of TP multiplier are reduced compared to prior work of implementing TP in modified booth algorithm [2].

4 SWP

A subword is a lower precision unit of data contained within a word. When SWP concept is adopted in multiplier, then multiple lower precision multiplications are possible. By applying SWP to signed multiplication like modified booth, the MB algorithm has to be performed for all the lower precision multiplication. While doing so, the number of multiplexers increases for selecting appropriate partial products. And this increases the hardware complexity of the design than TP multiplier. In this section, we have explained multiple SWP in RCMB multiplier for $N = 16$, i.e., four 4-bit multiplications or two 8-bit multiplications ($N/2$ or $N/4$) are performed which is shown in Figure 11. When two 8-bit multiplications are performed, i.e., only $N/2$ operation is performed and it is called as TP multiplication. This is the reason we say TP is a subset of SWP. The 'Algorithm for obtaining twin precision in RCMB' section describes the operation of TP in RCMB, and we need $(N/2 + 3)$ multiplexers for selection of partial products. Apart from this, we need to calculate mux utilisation for four 4-bit operations in multiple SWP.

The bounded regions in Figure 11 that are green in colour illustrate four 4-bit ($N/4$) multiplications and red bounded region shows TP, i.e., $N/2$ multiplications. Multiple SWP multiplication means either $N/2$ or $N/4$ bit multiplication is possible. While doing four 4-bit multiplications, the first lower precision multiplication will perform for inputs $x0$ to $x3$ and $y0$ to $y3$, second precision multiplication inputs are from $x4$ to $x7$ and $y4$ to $y7$, third is from $x8$ to $x11$ and $y8$ to $y11$ and fourth is from $x12$ to $x15$ and $y12$ to $y15$. For all these four lower precision multiplication, RCMB algorithm is applied. In this algorithm, all the MSB bit of partial product row has to be inverted for achieving twos complement, 1 s are added for sign extension and negk bits are added in the LSB of each partial product row. For first lower precision multiplication, the first two rows (only green bounded region) of partial product in Figure 11 for $N =$ 16 are taken into consideration. Since the inputs are $x0$ to $x3$ and $y0$ to $y3$, the MSB bit for this multiplication is P30 and P31 has to be inverted to achieve twos

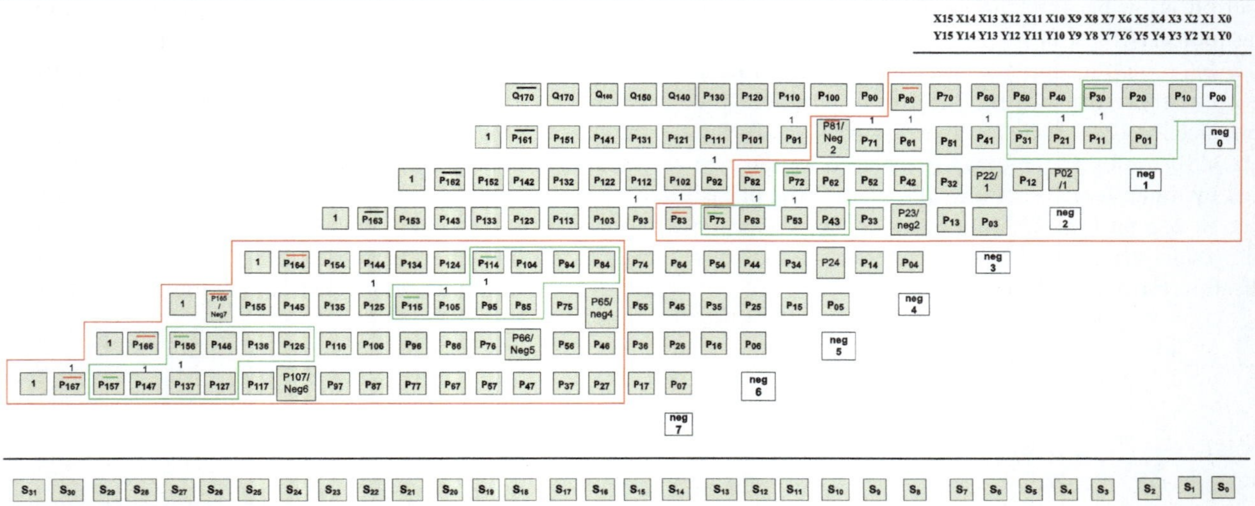

Figure 11 Multiple SWP in RCMB multiplier for $N = 16$.

complement. Multiplexer will select either $P30$ or $P\bar{3}0$ and $P31$ or $P\bar{3}1$. Sign bit '1' has to be added in fourth, fifth and seventh columns which is for sign extension in RCMB algorithm. In the fourth column, none of the partial product is available to merge sign bit '1', so this leads to extra half adder during the addition of partial products to produce sum bit (S) and this sign bit will be 0 for $N/2$ and N bit operation. Whereas in fifth and seventh column sign bit 1 is merged with P02/1, P22/1, i.e., mux is employed in the places of P02 and P22 which selects either P02 or 1 depends on the operation. neg0 and neg1 are added in the LSB. This leads to utilisation of four multiplexers in the first lower precision region bounded in the green colour to select the partial products for N/4 operation.

Likewise for second lower precision multiplication, the algorithm has to be applied in the third and fourth rows. The MSB bit such as $P72$, $P73$ are to be inverted and sign bit 1 has to be added in the 12th, 13th and 15th columns. In the 12th column, sign bit 1 has been inserted for $N/2$ operation so this can be used for $N/4$ bit operation and we can avoid mux for this operation. In the 13th and 15th columns, sign bit 1 has been added for second lower precision multiplication and it is merged with partial products which are not involved in $N/4$ bit operation. neg2 and neg3 bits have to be added in the LSB position, i.e., in columns 9 and 11. $P23$ and $P81$ will be replaced as neg2 and neg3 by mux which is represented as $P23$/neg2 and $P81$/neg3 in Figure 11. Since we have inserted mux in these places for $N/2$ bit operation, it is not necessary to add additional mux. Totally, four muxes are added in the second lower precision region. Likewise in the third and forth lower precisions, the algorithm is applied which utilises four muxes for each multiplication. Apart from mux utilisation of $N/2$ bit operation (twin precision) by making the above changes, we need 16 muxes in $N/4$ operation for $N = 16$. So to perform multiple SWP with either $N/2$ or $N/4$ bit RCMB multiplication, we need 27 muxes in the multiplier architecture. This mux calculation for multiple SWP is made by addition of mux count in the proposed TP multiplier as tabulated in Table 3 and mux required for four 4-bit multiplications. This results in an increase in area than our proposed twin multiplier. Table 4 shows mux count for proposed twin and multiple SWP. Throughput in the signed multiplier like MB, RCMB can be increased beyond two in multiple SWP. But if the throughput is increased beyond two, then mux count will also increase as shown in Table 4. The unwanted partial products (Pij) that are not required for $N/4$ bit operation are made zero. This is done by passing unwanted Pij to two input AND gate, and other input will act as control signal and for $N/4$ bit operation; more partial products are made zero than $N/2$ operation. And due to these reasons, area overhead will be higher than proposed twin. Hereby we conclude hardware complexity is more for multiple SWP while performing signed multiplication based on algorithms like MB and RCMB. We have calculated mux count for multiple SWP for $N = 16$, 32 in Table 4. Likewise for higher bit width mux utilisation will be higher than proposed twin.

Table 3 Mux utilisation

S number	Bit width	Twin 2009 MUX ($N + 3$)	Proposed twin mux ($N/2$) + 3
1	8	11	7
2	16	19	11
3	32	35	19
4	64	67	35
5	128	131	67

5 Results and discussion

In this paper, we have implemented TP technique in RCMB and a suitable algorithm is proposed for obtaining double throughput, which is applicable for all bit width that are multiples of eight. Our approach of implementing TP in RCMB gives better performance compared to [2]. Our implementation of the proposed algorithm yields less mux utilisation, and its impacts on area, delay and power compared to prior work of implementing twin precision in MB algorithm are analysed. In the twin precision implementation, changes to be made for N and $N/2$ b multiplication are selected using multiplexers which increase the design complexity of the multiplier. Our proposed method decreases this complexity by utilising less multiplexers. Less multiplexers are utilised because when TP implementation is made in RCMB according to our proposed algorithm, lesser changes are to be made in partial product array for $N/2$ bit multiplication. We have achieved nearly 40% to 48% (for $N = 8$ to 128) of less multiplexers compare to previous work [2].

To our knowledge, TP implementation to signed multipliers is made in [2] and further optimization in TP multiplier is not done. Compared to prior work [2], our proposed algorithm implementation requires less change for performing $N/2$ multiplication. These

Table 4 Mux utilisation for SWP and TP

S number	Bit width	Proposed twin mux	Mux count for multiple SWP-($N/2 + N/4$)
1	16	11	27
2	32	19	52

changes or selection of appropriate partial products are typically selected using multiplexers. Table 1 illustrates the signal selection that has to be made for $N/2$ multiplication in MB. TP implementation in MB contains $(N/2)+1$ partial product rows and adopts sign extension scheme presented in [16]. TP implementation in MB which adopts the sign extension scheme in [16] requires that more changes are to be made for $N/2$ bit multiplication. Multiplexer utilisation is directly related to the changes to be made in partial product row for $N/2$ b multiplication, and for every change, the multiplexers are to be deployed to select required partial product for N and $N/2$ b multiplication.

For twin precision implementation in modified booth algorithm [2], the changes (steps) that have to be made for $N/2$ b multiplication apart from booth encoding and decoding are PLSB, and its potential carry (most significant part (aMSP)) has to be performed separately for $N/2$ LSB and MSB multiplication. And also sign extension has to be made separately for $N/2$ b multiplication. So these three steps have to be performed in $N/2$ LSB and MSB multiplication. In RCMB [11] apart from booth encoding and decoding, the two steps to be performed for $N/2$ b multiplication are neg bit added in the LSB of each partial product row to generate twos complement and 1 s added for sign extension. RCMB does not require PLSB and potential carry (less complexity) so this makes RCMB more suitable for TP implementation and utilises less multiplexers. For an 8×8 TP multiplier, twin 2009 [2] requires 11 multiplexers, and our proposed method requires only 7 multiplexer as consolidated in Table 3. Reduction in multiplexer occurs because only two steps have to be performed in RCMB $N/2$ b multiplication compared to [2] which performs three steps in $N/2$ b multiplication.

Our proposed method has been tested for various bit widths that are multiples of eight and results are compared with prior work [2] and tabulated in Table 4. From the analysis made for various bit widths, it is inferred that our method needs $(N/2)+3$ multiplexers. Due to reduction in multiplexers, overall area, delay and power are reduced for TP multiplier. For the previous implementation of TP technique [2], the multiplexer utilisation has been formulated as $N+3$. This reduction in multiplexer utilisation in our proposed work reduces the design complexity of TP multiplier.

Figure 12 shows a graphical representation of Table 5. This table shows the comparison chart of implementation results of TP in MB (twin 2009) and RCMB (proposed twin) with non-TP multiplier. Figure 12a illustrates the utilisation of multiplexer (mux) in the implementation of TP for the proposed and existing method. A drastic decrease in the mux utilisation is much noticeable for higher bit width, i.e., from bit width of 16. Figure 12b,c,d portrays the changes in area, delay and power.

Cadence RTL compiler with TSMC library 180 nm is used to synthesized and analyse the cell area, delay and power. All the evaluation including power consumption is taken after post place and route synthesis of cadence digital flow. For higher bit width of multiplier, the reduction in area, delay and power are more noticeable for proposed twin compared with twin 2009 [2]. TP multiplier usually requires muxes for selection of partial products, so in Table 5, non-TP multiplier (TPM) results in less area compared to twin 2009 [2] and proposed twin. Usually, non-TP multiplier requires less area because of the absence of multiplexers, but the TP multiplier produces double output which is not possible in non-TP multiplier. Though non-TP multiplier produces less delay compared with TP multiplier, it cannot produce N or two $N/2$ b output at a time. And also increase in delay for TP is only 15% when compared to non-TP multiplier. While performing $N/2$ b multiplication in TP multiplier, the unwanted partial products are made to zero and this is the reason for achieving lesser power in TP multiplier compared to non-TP.

From Table 5, it is inferred that our TP implementation in RCMB achieves reduction in delay of about 5% to 18%, area of about 5% to 20% and power reduction of 8% to 32% due to reduction in mux utilisation. And the mux utilisation is drastically reduced up to 40% than the previous method [2]. When $N/2$ multiplication is performed, i.e., when multiplier performs narrow width operation, a significant reduction in power is achieved. Most of the partial products are made zero in $N/2$ b multiplication, so it leads to overall power reduction.

From the results in Figure 11, it is inferred that our implementation gives better performance compared with the twin 2009 [2] methodology.

In hardware architectures, obtaining reconfigurable architecture in ASIC is more challenging. Through TP technique, double throughput is achieved in signed multipliers. In this study, we have proposed an optimised twin precision multiplier. To prove the effectiveness of the proposed multiplier, we have implemented the twin 2009 [2] multiplier and the proposed optimised TP multiplier in the FFT complex multiplication and analysis were made. Apart from the effective hardware utilisation, our proposed TP multiplier yields reduced area, power and delay compared to the twin 2009 [2] TP multiplier.

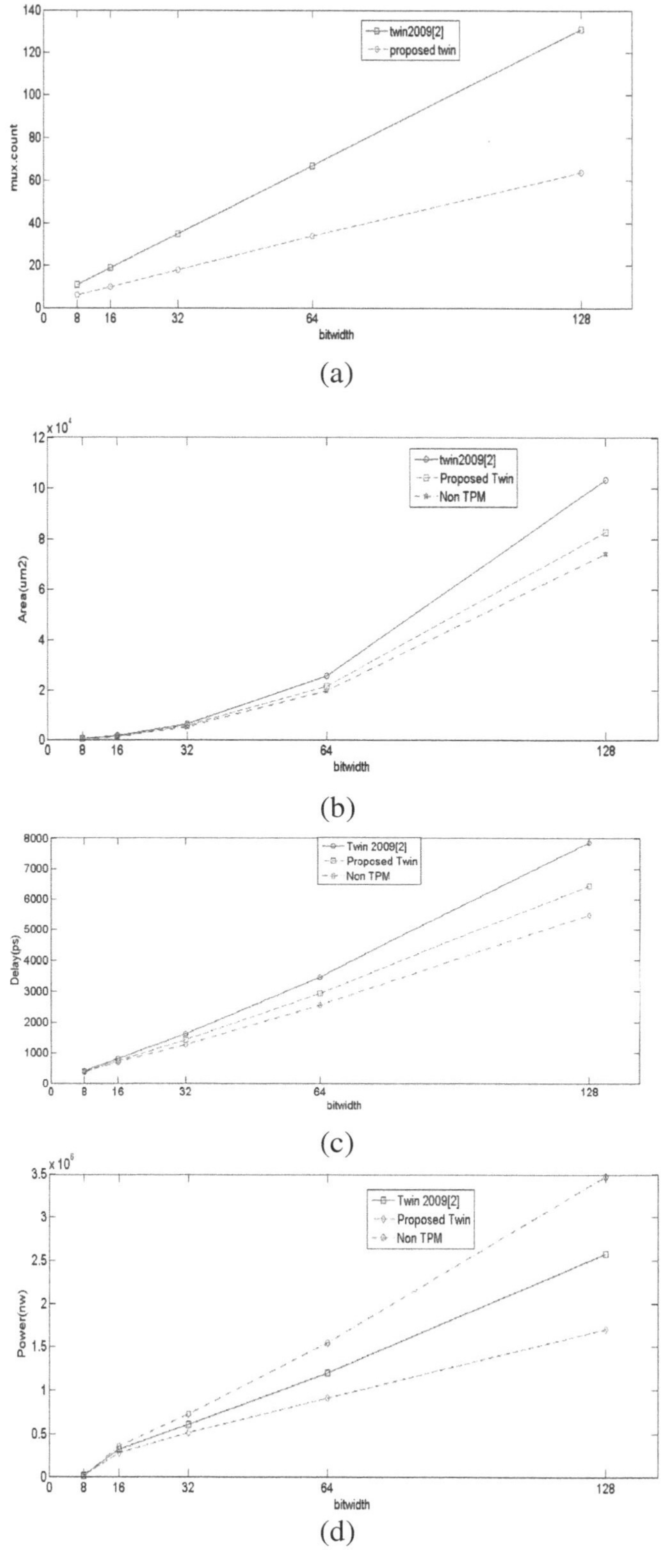

Figure 12 Graphical representation of Table 5. Comparison of existing and proposed TP; **(a)** bit width vs mux, **(b)** bit width vs area, **(c)** bit width vs delay and **(d)** bit width vs power.

Table 5 Comparison chart of existing TP multiplier, proposed TP multiplier and non-TP

	Non-TPM					Twin 2009 [2]					Proposed twin				
Bit width	8	16	32	64	128	8	16	32	64	128	8	16	32	64	128
Mux	-	-	-	-	-	11	19	35	67	131	7	11	19	35	67
Area (μm^2) (cells)	398	1,385	5,247	19,706	74,390	444	1,603	6,412	25,780	103,320	412	1,458		216,552	90,836
Delay (ps)	359	692.27	1,273.8	2,553.5	5,481.65	394.6	818.5	1,627.4	3,453.8	7,865.3	378.2	7438	1,415.4	2,935.05	6,449
Power (nW)	13,051.8	362,256.4	731,984.4	1,548,441.2	3,478,721.9	12,198	320,581	609,987	1,200,342	2,576,831	11,222	280,212	5,123,891	912,259	1,700,708

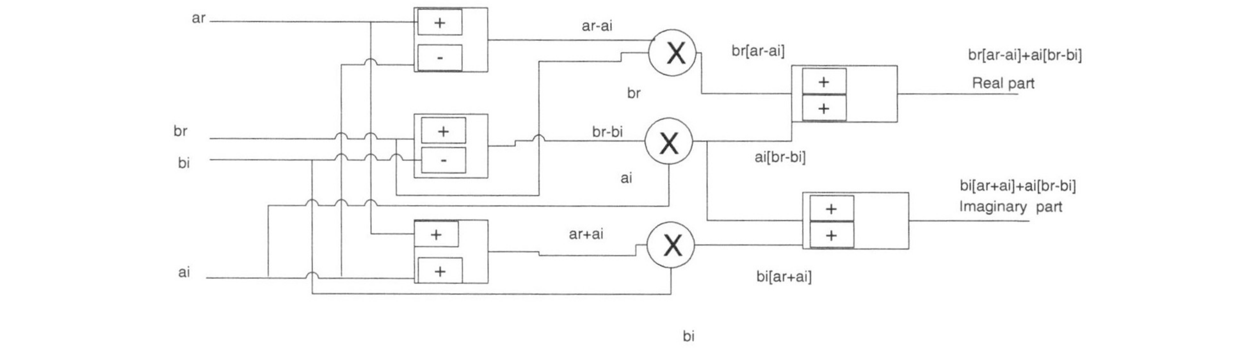

Figure 13 Complex multiplication in FFT.

The complex multiplier in FFT [20] is realised using four real multipliers, one adder and one subtractor as per Equation 8 as shown in Figure 13. In VLSI implementation, the complex multiplier occupies majority chip area. So according to Equation 9, complex multiplication can be realised by only three multipliers.

$$(a_r + ja_i)(b_r + jb_i) = (a_r b_r - a_i b_i) + j(a_i b_r + a_r b_i) \tag{8}$$

$$(a_r + ja_i)(b_r + jb_i) = \{b_r(a_r - a_i) + a_i(b_r - b_i)\} + j\{b_i(a_r + a_i) + a_i(b_r - b_i)\} \tag{9}$$

First, our analysis is made by replacing three multipliers in the complex multiplication in Figure 13 as TP multiplier. Second, the experiment was conducted with a 32-bit TP multiplier. So in a single 32-bit multiplier, two 16-bit multiplications ($N/2$ multiplication) or one 32-bit multiplication (N bit multiplication) can be performed. During N bit multiplication, the data width is 32 bit and twiddle factor is also 32 bit whereas in $N/2$ multiplication, data width is 16 bit and twiddle factor is 16 bit. When the three 32-bit multiplier in Figure 13 is replaced by TP multiplier as per TP logic, it can perform one 32-bit multiplication or two 16-bit multiplications.

While performing $N/2$ operation since the data and twiddle factor bit width are 16, the $b_r(a_r - a_i)$ and $a_i(b_r - b_i)$ can be performed in one single TP multiplier and in another multiplier $a_i(b_r - b_i)$, and $b_i(a_r + a_i)$ is performed and one multiplier can be left free. So this gives a way to overall power reduction when $N/2$ multiplication is performed or multiplier is operated in narrow width. The results are tabulated in Table 6. All the results are synthesized in cadence RTL compiler with TSMC 180 nm library. The inferences from the results in Table 6 are the percentage of improvement in area is 16%, delay of 17% and power improvement of 22%. Our proposed method gives better performance because it utilises less multiplexers compared to the existing TP multiplier.

6 Conclusions

Double throughput in ASIC environment is achieved effectively by implementing TP technique in RCMB, and a suitable algorithm is proposed. Our implementation requires less changes in the partial product array to acquire TP. Depending upon the multiplication either N or $N/2$ bit, selection of appropriate partial products in TP multiplier is done using multiplexers. Since our implementation utilised less multiplexers of about $(N/2) + 3$, the overall delay, area and power are reduced compared to prior implementation of TP technique in MB algorithm. Our proposed TP implementations consume 40% to 50% (for $N = 8$ to 128%) of less multiplexers. And our implementation gives better performance (area, delay and power) compared to prior implementation of TP in MB algorithm. To test the efficiency of the system, our proposed TP multiplier is implemented in FFT complex multiplication and its results gives better performance for our approach.

Table 6 TP in complex multiplication

Parameter	Twin 2009 [2] multiplier	Proposed TP multiplier (proposed twin)
Data width	32	32
Area (μm^2)	20,186.47	16,956.99
Delay (ps)	5,123	4,252
Power (nW)	1,867,221.23	1,456,432.67

Competing interests

The authors declare that they have no competing interests.

Authors' information

A. Rosi received her Bachelor of Engineering in Electronics and Communication and Master of Engineering degree in Applied Electonics from Anna University, India, in the year 2007 and 2009, respectively. Since 2010, she has been a senior project assistant in the project funded by the Ministry of Earth Science - India. She is currently pursuing Ph.D in VLSI Architecture at Anna University, India.

Dr. R. Seshasayanan was born in the year 1958 in India and received his B.E. degree from the College of Engineering, M.E. degree from Anna University in the year 1980 and 1983, respectively. He received his Ph.D from Anna University. He is presently working as an Associate Professor in the Department of Electronics and Communication, Anna University, and his area of interests are Low Power VLSI Design and Reconfigurable Architectures for Image processing. He is actively involved in various projects funded by the Ministry of Earth Science and Ministry of Defence - India.

References

1. M Själander, H Eriksson, P Larsson-Edefors, An efficient twin- precision multiplier, in *Proc. 22nd IEEE Int. Conf. Comput. Des*, 2004, pp. 30–33
2. M Själander, P Larsson-Edefors, Multiplication Acceleration Through Twin Precision, in *IEEE Transactions on Very Large Scale Integration (VLSI) Systems, vol. 17*, 9th edn., 2009
3. Differences between FPGA and ASIC. http://www.xilinx.com/fpga/asic.htm.
4. About ASIC and FPGA. http://www.altera.com
5. GH Loh, Exploiting data-width locality to increase superscalar execution bandwidth, in *Proc. 35th Int. Symp. Microarchitecture*, 2002, pp. 395–405
6. D Brooks, M Martonosi, Dynamically exploiting narrow width operands to improve processor power and performance, in *Proc. 5th Int. Symp. High Perform. Comput. Arch*, 1999, pp. 13–22
7. S Krithivasan, MJ Schulte, Multiplier Architectures for Media Processing, in *IEEE Conference*, 2003
8. S Khan, E Casseau, D Menard, Reconfigurable SWP operator for multimedia processing, in *IEEE International Conference on Application-Specific Systems, Architectures and Processors*, 2009
9. A Danysh, D Tan, Architecture and Implementation of a vector/SIMD multiply-accumulate unit, in *IEEE Transaction on Computers, vol. 54*, 2005, pp. 284–293
10. AD Booth, A signed binary multiplication technique, in *Quarterly J. Mechanical and Applied Math, vol. 4*, 1951, pp. 236–240
11. F Lamberti, N Andrikos, E Antelo, P Montuschi, Reducing the Computation Time in (Short Bit-Width) Two's Complement Multipliers, in *IEEE Transactions On Computers, vol. 60*, 2nd edn., 2011
12. OL MacSorley, High speed arithmetic in binary computers. Proc. Inst. Radio Eng. **49**(1), 67–97 (1961)
13. Z Huang, MD Ercegovac, High-Performance Low-Power Left-to-Right Array Multiplier Design. IEEE Trans. Comput. **54**(3), 272–283 (2005)
14. R Zimmermann, DQ Tran, Optimized Synthesis of Sum-of- Products, in *Proc. Conf. Record of the 37th Asilomar Conf. Signals, Systems and Computers, vol. 1*, 2003, pp. 867–872
15. W-C Yeh, C-W Jen, High-speed Booth encoded parallel multiplier design. IEEE Trans. Comput. **49**(7), 692–701 (2000)
16. J Fadavi-Ardekani, M N Booth encoded multiplier generator using optimized Wallace trees, in *IEEE Trans. Very Large Scale Integr. (VLSI) Syst, vol. 1*, 2nd edn., 1993, pp. 120–125
17. H Eriksson, P Larsson-Edefors, M Sheeran, M Själander, D Jo-hansson, M Schölin, Multiplier reduction tree with logarithmic logic depth and regular connectivity, in *Proc. IEEE Int. Symp. Circuits Syst. (ISCAS)*, 2006, pp. 4–8
18. MD Ercegovac, T Lang, *Digital Arithmetic*. Morgan Kaufmann Publishers - An Imprint of Elsevier, 2004
19. JY Kang, JL Gaudiot, A Fast and Well-Structured Multiplier, in *Proc. Euromicro Symp. Digital System Design*, 2004, pp. 508–515
20. YS Algnabi, R Teymourzadeh, M Othman and S Islam. FPGA Implementation of Pipeline Digit-Slicing Multiplier-Less Radix 22 DIF SDF Butterfly for Fast Fourier Transform Structure, Institute of MicroEngineering and Nanoelectronics IMEN, VLSI Design Department, Malaysia, in *The 5th European conference on antennas andpropagation (EUCAP2011)*, pp 4168–4172.

Nonlinear joint transmit-receive processing for coordinated multi-cell systems: centralized and decentralized

Zhirui Hu[1,2*], Chunyan Feng[1], Tiankui Zhang[1], Qin Niu[1] and Yue Chen[3]

Abstract

This paper proposes a nonlinear joint transmit-receive (tx-rx) processing scheme for downlink-coordinated multi-cell systems with multi-stream multi-antenna users. The nonlinear joint tx-rx processing is formulated as an optimization problem to maximize the minimum signal-to-interference noise ratio (SINR) of streams to guarantee the fairness among streams of each user. Nonlinear Tomlinson-Harashima precoding (THP) is applied at transmitters, and linear receive processing is applied at receivers, to eliminate the inter-user interference and inter-stream interference. We consider multi-cell systems under two coordinated modes: centralized and decentralized, corresponding to systems with high- and low-capacity backhaul links, respectively. For the *centralized coordinated* mode, transmit and receive processing matrices are jointly determined by the central processing unit based on the global channel state information (CSI) shared by base stations (BSs). For the *decentralized coordinated* mode, transmit and receive processing matrices are computed independently based on the local CSI at each BS. In correspondence, we propose both a centralized and a decentralized algorithm to solve the optimization problem under the two modes, respectively. Feasibility and computational complexity of the proposed algorithms are also analyzed. Simulation results prove that the proposed nonlinear joint tx-rx processing scheme can achieve user fairness by equalizing the bit error rate (BER) among streams of each user and the proposed scheme outperforms the existing linear joint tx-rx processing. Moreover, consistent with previous research results, performance of the proposed centralized nonlinear joint tx-rx processing scheme is proved to be better than that of the decentralized nonlinear joint tx-rx processing.

Keywords: Coordinated multi-cell; Centralized coordinated; Decentralized coordinated; Joint transmit-receive processing; Nonlinear precoding; Tomlinson-Harashima precoding (THP)

1 Introduction

Coordinated multi-cell is a promising technology to reduce inter-cell interference and increase user data rate, which has been considered as one of the potential technologies for LTE Advanced [1,2]. To fully utilize the advantage of coordinated multi-cell technology, it is essential to manage the multi-user interference (MUI) within the coordinated area appropriately as it is directly related to the achievable spectrum efficiency [3]. Precoding is a well-known technique for MUI mitigation in multi-user multiple-input multiple-output (MU-MIMO) systems [4,5]. The joint transmit-receive (tx-rx) processing can be used to further improve the downlink performance of MU-MIMO systems by optimizing the transmit precoding and receive filter matrices jointly. According to the processing of the transmit precoding, the joint tx-rx processing technology can be divided into two types, linear and nonlinear schemes.

The coordinated multi-cell technology can be implemented in a centralized or decentralized mode based on the backhaul capacity of the systems. The centralized coordinated mode can achieve higher data rate at the cost of high-capacity backhaul links in order to enable base stations (BSs) to share their channel state information (CSI) (defined as local CSI) and data. Hence, the centralized

* Correspondence: huzhirui74@163.com
[1]Beijing Key Laboratory of Network System Architecture and Convergence, Beijing University of Posts and Telecommunications, Beijing 100876, China
[2]School of Communication Engineering, Hangzhou Dianzi University, Hangzhou 310018, China
Full list of author information is available at the end of the article

approach is limited to systems with sufficient backhaul capacity [6,7]. The decentralized coordinated mode does not require BSs to share their local CSI, and the precoding or tx-rx processing is conducted at each BS [8]. This approach has less requirement on backhaul link capacity at some loss on the data rate in comparison to the centralized coordinated mode.

In recent years, relevant works on joint tx-rx processing in coordinated multi-cell systems have been widely studied under either centralized [9-19] or decentralized mode [20-23]. For designing the nonlinear joint tx-rx processing, many different optimal objectives have been considered, such as minimizing the sum mean square error (S-MSE) or maximizing the SINR; yet, the fairness among the streams of each user has not been solved for the coordinated multi-cell systems with multi-stream multi-antenna users.

1.1 Prior art

Linear joint tx-rx processing algorithms have been widely studied for coordinated multi-cell systems under centralized mode [9-14]. In [9], block diagonalization (BD) precoding was designed to maximize the weighted sum rate of all users. The tx-rx processing optimization with the criterion of minimizing the S-MSE was presented in [10-12], and the authors of [13] proposed a weighted S-MSE minimization algorithm by considering the channel gain as the weight factor. In [14], the energy efficiency was considered in the tx-rx processing design. A new criterion of maximizing weighted sum energy efficiency was formulated, and the optimization problem was solved by an iterative algorithm. For the decentralized coordinated mode, D. Gesbert and R. Holakouei, et al. studied the decentralized linear precoding techniques for the system with single-antenna users recently [20-22]. In [20], a distributed precoding scheme based on zero-forcing (ZF) criterion (defined as DZF) and several centralized power allocation approaches was proposed. In [21,22], a characterization of the optimal linear precoding strategy was derived. Distributed virtual SINR (DV-SINR) precoding approaches, where each BS balances the ratio between signal gain at the intended user and the interference caused by other users, had been proposed for the particular case of two users in [21] and generalized for multi-user in [22]. The DV-SINR scheme was illustrated to satisfy the optimal precoding characterization and outperform DZF.

Compared with the linear joint tx-rx processing schemes, the nonlinear joint tx-rx processing schemes are more complex but can obtain more system gain, which have gained much attention recently. Most research about the nonlinear precoding focus on Tomlinson-Harashima precoding (THP), as it can achieve approximate performance with the optimal dirty paper coding but has a much lower complexity [5]. For the centralized coordinated mode, the tx-rx processing scheme was designed to minimize the S-MSE in [15] and maximize the SINR in [16], wherein both should be solved by an iterative method, resulting in high computational complexity. The schemes with low complexity were proposed and derived a closed-form solution based on minimum average bit error rate (BER) in [17], minimum mean square error (MMSE) in [18], or ZF criterion in [19]. In [18], the receive processing matrix was firstly computed by CSI. Then,the transmit processing matrix and receive weight coefficient were computed based on MMSE. In [19], the algorithm decomposed the MU-MIMO channel into parallel independent single user MIMO (SU-MIMO) channels, and then, closed-form expressions of transmit and receive processing matrices were derived to optimize the performance of each user. The above research works on nonlinear tx-rx processing were all developed for the centralized coordinated mode. The relevant works for the decentralized coordinated mode are relatively fewer. A decentralized nonlinear precoding, ZF-THP, was proposed in [23] but can only be applied for the system with a single user. To the best of our knowledge, for the system with multi-stream multi-antenna users, the tx-rx processing solutions under decentralized coordination mode have not been addressed in the literature.

Previous work did not consider fairness among streams of each user in the coordinated multi-cell system with multi-stream multi-antenna users. It is essential to study the fairness for nonlinear scheme, as unfairness is an inherent character of THP and the worst performance determines the whole performance of the user [24].

1.2 Contributions

In this paper, a nonlinear joint tx-rx processing scheme is proposed to improve fairness among streams of each user with multi-antenna. The nonlinear joint tx-rx processing is formulated as an optimization problem to maximize the minimum SINR of streams. The performance of the proposed scheme is evaluated under both centralized and decentralized coordinated modes. Two algorithms for solving the optimization problem are derived.

The main work of this paper can be summarized as follows.

- Nonlinear joint tx-rx processing scheme is developed for a coordinated multi-cell system with multi-stream multi-antenna users under two coordinated modes, centralized and decentralized mode.
- Two algorithms, the centralized and the decentralized algorithms, are proposed to solve the

optimization problem, and both of them derive the closed-form solutions.

- The algorithms guarantee the fairness among the streams of each user, which not only boost the performance of each user, but bring much convenience to the modulation/demodulation and coding/decoding procedures.

The remainder of this paper is organized as follows. Section 2 presents the coordinated multi-cell system model. The proposed nonlinear joint tx-rx processing scheme is described in detail in Section 3. A performance analysis of the proposed algorithms is developed in Section 4. Simulation results and conclusions are presented in Section 5 and Section 6, respectively.

1.3 Notation

We use uppercase boldface letters to denote matrices and lowercase boldface to denote vectors. The operators $(\cdot)^T, (\cdot)^H, (\cdot)^\dagger, \mathrm{E}(\cdot)$, and $\mathrm{Tr}(\cdot)$ stand for transpose, Hermitian, Moore-Penrose pseudo-inverse, expectation, and the trace of a matrix, respectively. diag$(\cdot)$ and blockdiag$(\cdot)$ denote diagonal and block diagonal matrix. $\mathbf{I}$ and $\mathbf{0}$ are the identity and the all-zero matrix, respectively, with appropriate dimensions. $\|\cdot\|_F$ represents the Frobenius norm of a matrix. $[\cdot]_{i:j,k:l}$ denotes the submatrix comprised of row i through row j and column k through column l of a matrix.

2 System model

Consider a downlink coordinated multi-cell system, where N BSs cooperatively serve K users. Each BS and user is equipped with n_t and n_r antennas, respectively. All BSs share user data and cooperatively transmit the data to an intended user. Each BS transmits $L = \sum_{k=1}^{K} l_k$ data streams to K users, where l_k is the number of transmitted data streams for user k.

We assume that BSs' transmit power for every user is P. Therefore, the total transmit power of BSs is KP. Denote $\mathbf{x}^k = \left[\mathbf{x}_1^{k^T}, \cdots, \mathbf{x}_N^{k^T}\right]^{\mathrm{T}}$, where $\mathbf{x}_n^k$ denotes the preprocessed signal transmitted by the nth BS for user k, satisfying $\mathrm{Tr}\left\{\mathbf{x}^k\mathbf{x}^{k^H}\right\} = P$. The received signal of the kth user is:

$$\mathbf{y}^k = \sum_{n=1}^{N}\mathbf{H}_n^k\mathbf{x}_n^k + \sum_{t=1,t\neq k}^{K}\sum_{n=1}^{N}\mathbf{H}_n^k\mathbf{x}_n^t + \mathbf{n}^k = \mathbf{H}^k\mathbf{x}^k + \sum_{t=1,t\neq k}^{K}\mathbf{H}^k\mathbf{x}^t + \mathbf{n}^k \quad (1)$$

where $\mathbf{H}^k = \left[\mathbf{H}_1^k, \cdots, \mathbf{H}_N^k\right]$ is the global CSI between BSs and the kth user and $\mathbf{H}_n^k \in C^{n_r\times n_t}$ denotes the local CSI between the nth BS and the kth user, whose entries are independent and identically distributed (i.i.d.) complex Gaussian variables with zero mean and unit variance. In Equation 1, the second term on the right-hand side is MUI, and $\mathbf{n}^k \sim CN(\mathbf{0}, \sigma^2\mathbf{I}_{n_r})$ is the additive white Gaussian noise variable.

Each user decodes the desired data by multiplying with the receive processing matrix. The received data of the kth user is given as:

$$\tilde{\mathbf{y}}^k = \mathbf{R}^k\mathbf{y}^k = \mathbf{R}^k\sum_{n=1}^{N}\mathbf{H}_n^k\mathbf{x}_n^k + \mathbf{R}^k\sum_{t=1,t\neq k}^{K}\sum_{n=1}^{N}\mathbf{H}_n^k\mathbf{x}_n^t + \tilde{\mathbf{n}}^k = \mathbf{R}^k\mathbf{H}^k\mathbf{x}^k + \sum_{t=1,t\neq k}^{K}\mathbf{R}^k\mathbf{H}^k\mathbf{x}^t + \tilde{\mathbf{n}}^k \quad (2)$$

where $\mathbf{R}^k \in C^{l_k\times n_r}$ denotes the receive processing matrix of the kth user. $\tilde{\mathbf{n}}^k = \mathbf{R}^k\mathbf{n}^k$ is the equivalent received noise vector at the kth user.

Let $\mathbf{y} = \left(\mathbf{y}^{1^T}, \cdots, \mathbf{y}^{K^T}\right)^T$ represent the received signal of the K users. Equation 2 can be expressed as:

$$\tilde{\mathbf{y}} = \mathbf{R}\mathbf{y} = \mathbf{R}\sum_{n=1}^{N}\mathbf{H}_n\mathbf{x}_n + \tilde{\mathbf{n}} = \mathbf{R}\mathbf{H}\mathbf{x} + \tilde{\mathbf{n}} \quad (3)$$

where $\mathbf{R} = \mathrm{blockdiag}(\mathbf{R}^1, \cdots, \mathbf{R}^K)$ is a $L \times Kn_r$ matrix. $\mathbf{H} = \left[\mathbf{H}^{1^T}, \cdots, \mathbf{H}^{K^T}\right]^{\mathrm{T}} \in C^{Kn_r\times N_t}$ is the global CSI between BSs and K users. $\mathbf{H}_n = \left[\mathbf{H}_n^{1^T}, \cdots, \mathbf{H}_n^{K^T}\right]^T \in C^{Kn_r\times n_t}$ denotes the nth local CSI between the nth BS and K users. $\mathbf{x} = \sum_{k=1}^{K}\mathbf{x}^k$ denotes the transmit signal at BSs, and $\mathbf{x}_n = \sum_{k=1}^{K}\mathbf{x}_n^k$ is the transmit signal at the nth BS. $\tilde{\mathbf{n}} = \left(\tilde{\mathbf{n}}^{1^T}, \cdots, \tilde{\mathbf{n}}^{K^T}\right)^T$ is the combination of the receive noise at the K users.

Define $\mathbf{\Lambda}(k,t) = \mathbf{R}^k\sum_{n=1}^{N}\mathbf{H}_n^k\mathbf{x}_n^t = \mathbf{R}^k\mathbf{H}^k\mathbf{x}^t$. The rate of the kth user is given by

$$r^k = \log_2\left|\mathbf{I} + \frac{\mathbf{\Lambda}(k,k)\mathbf{\Lambda}(k,k)^H}{\tilde{\mathbf{n}}^k\tilde{\mathbf{n}}^{k^H} + \sum_{t=1,t\neq k}^{K}\mathbf{\Lambda}(k,t)\mathbf{\Lambda}(k,t)^H}\right| \quad (4)$$

Then, the system sum rate can be obtained by $r = \sum_{k=1}^{K} r^k$.

The coverage of N-coordinated BSs is defined as one coordinated area. We mainly focus on the interference within the coordinated area. The interference from other coordinated areas is ignored in this paper, which can be eliminated by inter-cell interference coordination technology [25] or interference alignment technology [26].

For the centralized coordinated mode, it is assumed that all BSs exchange their local CSI, and the tx-rx processing matrices are jointly designed at the central processing unit. The system can be seen as a virtual MU-MIMO system with $N_t = Nn_t$ transmit antennas. On the contrary, for the decentralized coordinated mode, BSs do not share their CSI, and every BS only has knowledge of local CSI between itself and K users. Therefore, the tx-rx processing matrices are independently designed at each BS.

3 Nonlinear joint transmit-receive processing algorithm

In this section, we present nonlinear joint tx-rx processing algorithms for a coordinated multi-cell system under two different coordinated modes. The algorithm structure is firstly shown. Then, we formulate the optimization problem, aiming at maximizing the minimum SINR of streams to guarantee the fairness among the streams of each user. Finally, the algorithms for different coordinated modes are proposed.

3.1 Algorithm structure

The structure of the proposed algorithm is shown in Figure 1. In the proposed algorithms, nonlinear preprocessing is applied at transmitters; meanwhile, linear processing is applied at each receiver.

At the nth($n = 1,\ldots,N$) transmitter, $\mathbf{s} = \left(\mathbf{s}^{1^T}, \cdots, \mathbf{s}^{K^T}\right)^T \in C^{L\times 1}$ denotes the modulated data vector satisfying $E\{\mathbf{ss}^H\} = \mathbf{I}$, where $\mathbf{s}^k$ is comprised of the l_k data streams for the kth user. In THP, feedback matrix $\mathbf{B}_n$ is a unit lower triangular matrix,

$$\mathbf{B}_n = \begin{bmatrix} \mathbf{B}_n^{1,1} & \mathbf{0} & \cdots & \mathbf{0} \\ \mathbf{B}_n^{2,1} & \mathbf{B}_n^{2,2} & \ddots & \vdots \\ \vdots & \ddots & \ddots & \mathbf{0} \\ \mathbf{B}_n^{K,1} & \cdots & \mathbf{B}_n^{K,K-1,} & \mathbf{B}_n^{K,K} \end{bmatrix} \tag{5}$$

where $\mathbf{B}_n^{k,k}$ is a unit lower triangular matrix with $l_k \times l_k$ size. $\mathbf{u}_n$ is the output data of THP. Therefore, the lth data stream of $\mathbf{u}_n$ is interfered by the first (l-1) data streams; in other words, the lth($l = 2, \cdots, L$) element u_n^l in $\mathbf{u}_n$ is a linear combination of $s^j(j \le l)$. Assume M-ary square constellation is employed to $\mathbf{s}$. To ensure that the real and the imaginary parts of u_n^l are constrained into $\left(-\sqrt{M}, \sqrt{M}\right]$, modulo $2\sqrt{M}$ operation $\mathrm{mod}_{2\sqrt{M}}(\cdot)$ is introduced. The output data of THP is expressed as:

$$\begin{aligned} \mathbf{u}_n &= \mathrm{mod}_{2\sqrt{M}}[(\mathbf{I}-\mathbf{B}_n)\mathbf{u}_n + \mathbf{s}] \\ &= (\mathbf{I}-\mathbf{B}_n)\mathbf{u}_n + \mathbf{s} + \mathbf{d}_n \end{aligned} \tag{6}$$

where $\mathbf{d}_n = 2\sqrt{M}\left(\mathbf{z}_I + j\mathbf{z}_Q\right)$, $\mathbf{z}_I$ and $\mathbf{z}_Q$ are both integers. Define $\mathbf{v}_n = \mathbf{s} + \mathbf{d}_n$, and then, $\mathbf{u}_n$ is written as:

$$\mathbf{u}_n = \mathbf{B}_n^{-1}\mathbf{v}_n \tag{7}$$

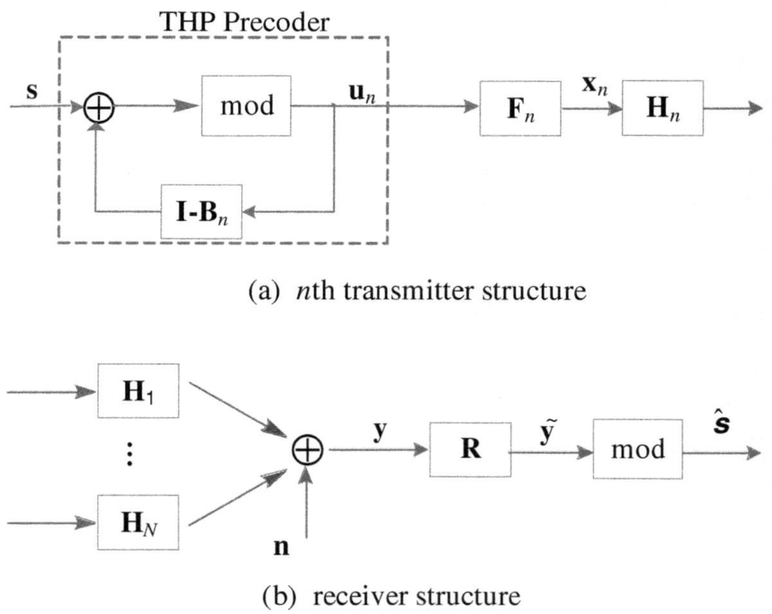

Figure 1 Structure of the transmit-receive processing: (a) transmitter structure; (b) receiver structure.

There is a power enhancement of $\tau = M/(M-1)$ due to THP, i.e., $E\{\mathbf{u}_n\mathbf{u}_n^H\} = \tau\mathbf{I}$ [27]. The transmit signal at the nth BS is:

$$\mathbf{x}_n = \mathbf{F}_n\mathbf{u}_n = \sum_{k=1}^{K}\mathbf{F}_n^k\mathbf{x}_n^k \tag{8}$$

where $\mathbf{F}_n = \left[\mathbf{F}_n^1, \cdots, \mathbf{F}_n^K\right] \in C^{n_t \times L}$ is the transmit processing matrix.

At the receivers, the received signal in Equation 3 can be rewritten as:

$$\tilde{\mathbf{y}} = \mathbf{R}\sum_{n=1}^{N}\mathbf{H}_n\mathbf{F}_n\mathbf{B}_n^{-1}\mathbf{v}_n + \tilde{\mathbf{n}} = \mathbf{RHFC}_{\mathbf{B}^{-1}}\mathbf{v} + \tilde{\mathbf{n}} \tag{9}$$

where $\mathbf{F} = \text{diag}(\mathbf{F}_1, \cdots, \mathbf{F}_N)$ is a block diagonal matrix with $N_t \times NL$ size and $\mathbf{C}_{\mathbf{B}^{-1}} = \left[\mathbf{B}_1^{-1^T}, \cdots, \mathbf{B}_N^{-1^T}\right]^T \in C^{NL \times L}$. The user data will finally be obtained by modulo operation and demodulation. Obviously, the received noise power of l_k streams of the kth user is:

$$\left(\sigma_{n,1}^2, \cdots, \sigma_{n,l_k}^2\right) = \sigma^2\text{diag}\left(\mathbf{R}^k\mathbf{R}^{k^H}\right) \tag{10}$$

3.2 Problem formulation

From Equation 1, it is noticed that the received signal of every user is influenced by MUI. In order to liberate every user from MUI, the relative matrices in this algorithm are designed to satisfy ZF criterion:

$$\mathbf{RHFC}_{\mathbf{B}^{-1}} = \mathbf{W} \tag{11}$$

where $\mathbf{W} = \text{diag}\left(\sqrt{P/(\tau l_1)}\mathbf{I}_{l_1}, \cdots, \sqrt{P/(\tau l_K)}\mathbf{I}_{l_K}\right)$. The SINR of the kth user can be obtained as:

$$\left(\rho_1^k, \cdots, \rho_{l_k}^k\right) = \frac{P}{\tau l_k}\left(\frac{1}{\sigma_{n,1}^2}, \cdots, \frac{1}{\sigma_{n,l_k}^2}\right) \tag{12}$$

In order to guarantee the fairness among streams of each user, we investigate the tx-rx processing matrices design to maximize the minimum SINR for each stream of each user, which is formulated as follows:

$$\begin{aligned} \max_{\mathbf{B},\mathbf{F},\mathbf{G}} \quad & \min_{k} \quad \left(\rho_1^k, \cdots, \rho_{l_k}^k\right) \\ s.t. \quad & \mathbf{RHFC}_{\mathbf{B}^{-1}} = \mathbf{W} \qquad (a) \\ & \mathbf{S}_i\mathbf{B}_n\mathbf{e}_i = \mathbf{0}_i, i = 1, \cdots, L \quad (b) \\ & Tr\left\{\mathbf{F}^k\mathbf{F}^{k^H}\right\} = P/\tau \qquad (c) \end{aligned} \tag{13}$$

for $k = 1, \cdots, K, n = 1, \cdots, N$. (a) denotes ZF criterion. (b) denotes that $\mathbf{B}_n$ is the unit lower triangular matrix, where $\mathbf{S}_i = [\mathbf{I}_i, \mathbf{0}_{i \times (L-i)}]$ and $\mathbf{e}_i$ is the ith column of $\mathbf{I}_L$. (c) is used to guarantee the power constraint and $\mathbf{F}^k = \left[\mathbf{F}_1^{k^H}, \cdots, \mathbf{F}_N^{k^H}\right]^H$.

3.3 Centralized algorithm

In Equation 13, the relative matrices are entangled with each other. To solve this problem, we start from the ZF constraint. Every BS is assumed to have the same feedback matrix, denoted as $\mathbf{B}$, which will be determined at the central processing unit based on the global CSI. (a) in Equation 13 can be rewritten as

$$\mathbf{RHF}' = \mathbf{WB} \tag{14}$$

where $\mathbf{F}' = \left[\mathbf{F}_1^T, \cdots, \mathbf{F}_N^T\right]^T$. As $\mathbf{B}$ is a unit lower triangular matrix, the left side of Equation 14 should satisfy:

$$\mathbf{R}^t\mathbf{H}^t\mathbf{F}^k = \mathbf{0} \quad (t < k) \tag{15}$$

which reveals that $\mathbf{F}^k = \left[\mathbf{F}_1^{k^T}, \cdots, \mathbf{F}_N^{k^T}\right]^T \in C^{N_t \times l_k}(k > 1)$ lies in the null space of $\breve{\mathbf{H}}^k = \left[\bar{\mathbf{H}}^{1^T}, \cdots, \bar{\mathbf{H}}^{(k-1)^T}\right]^T$, where $\bar{\mathbf{H}}^i = \mathbf{R}^i\mathbf{H}^i$ is the equivalent CSI of the ith user. $\mathbf{F}^k$ can be found by doing singular value decomposition (SVD) on $\breve{\mathbf{H}}^k$:

$$\breve{\mathbf{H}}^k = \mathbf{U}^k\left[\boldsymbol{\Sigma}^k \quad \mathbf{0}\right]\left[\mathbf{V}_1^k \quad \mathbf{V}_0^k\right]^H \tag{16}$$

We assume that $\mathbf{F}^k$ is represented as $\mathbf{F}^k = \bar{\mathbf{F}}^k\bar{\bar{\mathbf{F}}}^k$. Then, $\bar{\mathbf{F}}^k = \mathbf{V}_0^k \in C^{N_t \times \left[N_t - \sum_{i=1}^{k-1}l_i\right]}$ is named as the transmit space matrix and $\bar{\bar{\mathbf{F}}}^k$ is the transmit diversity matrix with $\left[N_t - \sum_{i=1}^{k-1}l_i\right] \times l_k$ size.

The above analysis is suitable for the kth$(k > 1)$ user. Since the first user is not limited by Equation 15, we use $\bar{\mathbf{F}}^1 = \mathbf{I}_{N_t}$. $\bar{\mathbf{F}}^k$ can be formed as:

$$\bar{\mathbf{F}}^k = \begin{cases} \mathbf{I}_{N_t}, & k = 1 \\ \mathbf{V}_0^k, & 1 < k \le K \end{cases} \tag{17}$$

With $\bar{\mathbf{F}}^k$ and THP, the proposed algorithm decomposes the MU-MIMO channel into parallel independent SU-MIMO channels [19]. We can comprehend this as follows: for the kth user, $\bar{\mathbf{F}}^k$ is designed to avoid the interference from users $(k+1, \cdots, K)$. Meanwhile, THP is used to eliminate the interference from the first $(k-1)$ users. Therefore, user k will not suffer from MUI. For any user $k(k = 1, \cdots, K)$, $\mathbf{B}^{k,k}, \bar{\bar{\mathbf{F}}}^k$ and $\mathbf{R}^k$ satisfy:

$$\mathbf{R}^k\mathbf{H}^k\bar{\mathbf{F}}^k\bar{\bar{\mathbf{F}}}^k = \sqrt{P/(\tau l_k)}\mathbf{B}^{k,k}, \quad k = 1, \cdots, K \tag{18}$$

where $\mathbf{B}^{k,k}$ is a unit lower triangular matrix with $l_k \times l_k$ size. Therefore, $\mathbf{B}^{k,k}, \bar{\bar{\mathbf{F}}}^k$ and $\mathbf{R}^k(k = 1, \cdots, K)$ can be designed separately. For the kth user, Equation 13 can be reduced to:

$$\begin{aligned} \max_{\mathbf{R}^k, \mathbf{F}_2^k, \mathbf{B}^{k,k}} \quad & \min \quad \left(\rho_1^k, \cdots, \rho_{l_k}^k\right) \\ s.t. \quad & \mathbf{R}^k \mathbf{H}^k \bar{\mathbf{F}}^k \bar{\bar{\mathbf{F}}}^k = \sqrt{P/(\tau l_k)} \mathbf{B}^{k,k} \quad (a) \\ & \mathbf{S}_i^k \mathbf{B}^{k,k} \mathbf{e}_i^k = \mathbf{0}_i, i = 1, \cdots, l_k \quad (b) \\ & \boldsymbol{Tr}\left\{\bar{\bar{\mathbf{F}}}^k \bar{\bar{\mathbf{F}}}^{k^H}\right\} = P/\tau \quad (c) \end{aligned} \tag{19}$$

where $\mathbf{S}_i^k = \left[\mathbf{I}_i, \mathbf{0}_{i \times (l_k - i)}\right]$, and $\mathbf{e}_i^k$ is the ith column of $\mathbf{I}_{l_k}$.

The optimal solution of Equation 19 can be obtained from the generalized triangular decomposition of $\mathbf{H}^k \bar{\mathbf{F}}^k$ [19,28]:

$$\mathbf{H}^k \bar{\mathbf{F}}^k = \mathbf{Q}^k \mathbf{D}^k \mathbf{P}^{k^H} \tag{20}$$

where $\mathbf{Q}^k \in C^{n_r \times S}$ and $\mathbf{P}^k \in C^{S \times \left[N_t - \sum_{i=1}^{k-1} l_i\right]}$ have orthonormal columns, and S is the rank of $\mathbf{H}^k \bar{\mathbf{F}}^k$. $\mathbf{D}^k \in C^{S \times S}$ is a lower triangular matrix, the diagonal elements of which satisfy:

$$d_{i,i}^k = \begin{cases} \bar{\lambda}^k, & i = 1, \cdots, l_k \\ \lambda_i^k, & i = l_k + 1, \cdots, S \end{cases}, \tag{21}$$

where λ_i^k is the ith largest positive singular value of $\mathbf{H}^k \bar{\mathbf{F}}^k$, and $\bar{\lambda}^k = \left(\prod_{j=1}^{l_k} \lambda_j^k\right)^{1/l_k}$. Define $\mathbf{\Lambda}^k = \left(\bar{\lambda}^k\right)^{-1} \mathbf{I}_{l_k}$. Then $\mathbf{R}^k$, $\bar{\bar{\mathbf{F}}}^k$ and $\mathbf{B}^{k,k}$ are given by:

$$\begin{aligned} \mathbf{R}^k &= \mathbf{\Lambda}^k \left[\mathbf{Q}^k\right]^{\mathrm{H}}_{:,1:l_k} \\ \mathbf{B}^{k,k} &= \mathbf{\Lambda}^k \left[\mathbf{D}^k\right]_{1:l_k,1:l_k} \\ \bar{\bar{\mathbf{F}}}^k &= \sqrt{P/(\tau l_k)} [\mathbf{P}]_{:,1:l_k} \end{aligned} \tag{22}$$

Based on Equation 10 and Equation 22, the received noise power of every stream of the kth user is $\sigma^2 \left(\bar{\lambda}^k\right)^{-2}$. Therefore, the SINR of the kth user is:

$$\rho_1^k = \rho_2^k = \cdots = \rho_{l_k}^k = \frac{p}{\tau l_k} \cdot \frac{\left(\bar{\lambda}^k\right)^2}{\sigma^2} \tag{23}$$

It can be seen that every stream of the kth user can achieve equal SINR.

Note that for the computation of the transmit space matrix of the kth user $\bar{\mathbf{F}}^k$, we need to know the receive processing matrices of the first $(k - 1)$ users $\mathbf{R}^t (t < k)$ and that, for the computation of the receive processing matrix of the kth user $\mathbf{R}^k$, we need to know the transmit space matrix of the kth user. Therefore, $\bar{\mathbf{F}}^k$ and $\mathbf{R}^k$ are designed step-by-step, which starts by computing the transmit space matrix and the receive processing matrix of the first user, then computes the matrices for the second user by utilizing the receive processing matrix of the first user and so on.

All of the matrices are designed at the central processing unit, and the receive processing matrices are transmitted to each user by downlink channel. The procedure of the proposed centralized algorithm is summarized in Table 1.

3.4 Decentralized algorithm

In this scenario, as BSs do not exchange their local CSI, each BS independently preprocesses the user data with the local CSI of itself. The data processed by each BS cannot be obtained by other BSs. In order to ensure that the user's receive signal is not interfered by MUI, relative matrices at each BS should satisfy the ZF criterion. Therefore, Equation 11 is reduced to:

$$\mathbf{R}\mathbf{H}_n \mathbf{F}_n \mathbf{B}_n^{-1} = \mathbf{W}_n, \quad n = 1, \cdots, N \tag{24}$$

where $\mathbf{W}_n = \mathrm{diag}\left(p_n^1 \mathbf{I}_{l_1}, \cdots, p_n^K \mathbf{I}_{l_K}\right)$ satisfies $\sum_{n=1}^{N} \mathbf{W}_n = \mathbf{W}$.

The receive processing matrix $\mathbf{R}^k (k = 1, \cdots, K)$ of each user is related to the transmit signals from N BSs. If $\mathbf{R}^k$ is computed at BSs, each BS can only decide it dependently as the local CSI of each BS is not exchanged. Generally, $\mathbf{R}^k$ derived at different BSs has different values, which is unreasonable. Otherwise, for each user, frequently interactive information with all coordinated BSs is required. It will largely increase the system computational complexity. Therefore, we firstly compute $\mathbf{R}^k (k = 1, \cdots, K)$ at users. Denote $\mathbf{H}^k = \left[\mathbf{U}_1^k \ \mathbf{U}_0^k\right] \mathbf{\Sigma}^k \mathbf{V}^{k^H}$ as the SVD of $\mathbf{H}^k$, where $\mathbf{U}_1^k \in C^{n_r \times l_k}$. Then $\mathbf{R}^k$ can be obtained by $\mathbf{R}^k = \mathbf{G}^k \left(\mathbf{U}_1^k\right)^H$, where $\mathbf{G}^k$ is a diagonal matrix for normalizing the received signal and will be determined at the BSs. For frequency division duplex system, user k can only feedback the equivalent local CSI $\bar{\mathbf{H}}_n^k = \mathbf{U}_1^k \mathbf{H}_n^k$ to the nth BS. Therefore, Equation 10 can be rewritten as:

$$\left(\sigma_{n,1}^2, \cdots, \sigma_{n,l_k}^2\right) = \sigma^2 \mathrm{diag}\left(\mathbf{G}^k \mathbf{G}^{k^H}\right) \tag{25}$$

Table 1 Centralized nonlinear joint tx-rx processing algorithm

1	for $k = 1{:}K$
2	Compute $\bar{\mathbf{F}}^k$ using Equation 17
3	Obtain $\mathbf{R}^k$, $\bar{\bar{\mathbf{F}}}^k$, and $\mathbf{B}^{k,k}$ using Equation 22
4	Compute $\mathbf{F}^k$ by $\mathbf{F}^k = \bar{\mathbf{F}}^k \bar{\bar{\mathbf{F}}}^k$
5	end
6	Compute $\mathbf{B}$ by $\mathbf{B} = \mathbf{W}^{-1} \mathbf{R}\mathbf{H}\mathbf{F}'$

Assume that every BS has equal transmit power $p = P/N$. Based on the above analysis, Equation 13 is equivalent to the following optimization problem:

$$\begin{aligned}
&\max_{\mathbf{B},\mathbf{F},\mathbf{G}} \quad \min_{k} \quad \left(\rho_1^k, \cdots, \rho_{l_k}^k\right) \\
&s.t. \quad \mathbf{G}\bar{\mathbf{H}}_n\mathbf{F}_n\mathbf{B}_n^{-1} = \mathbf{W}_n \qquad (a) \\
&\qquad \mathbf{G} = diag\{g_1, \cdots, g_L\} \qquad (b) \\
&\qquad \mathbf{S}_i\mathbf{B}_n\mathbf{e}_i = \mathbf{0}_i, i = 1, \cdots, L \qquad (c) \\
&\qquad \boldsymbol{Tr}\left\{\mathbf{F}_n^k\mathbf{F}_n^{k^H}\right\} = p/\tau \qquad (d) \\
&\qquad \sum\nolimits_{n=1}^{N} \mathbf{W}_n = \mathbf{W} \qquad (e)
\end{aligned} \tag{26}$$

for $k = 1, \cdots, K, n = 1, \cdots, N$. In (a), $\bar{\mathbf{H}}_n = \left[\bar{\mathbf{H}}_n^{1^T}, \cdots, \bar{\mathbf{H}}_n^{K^T}\right]^T$. In (b), $g_j (j = 1, 2, \cdots, L)$ are diagonal elements of $\mathbf{G}$. $(d)(e)$ are used to guarantee the power constraint.

In Equation 26, the relative matrices are entangled with each other. Similarly, to solve this problem, we start from the ZF constraint. Take the nth BS for example. (a) in Equation 26 can be rewritten as:

$$\bar{\mathbf{H}}_n\mathbf{F}_n = \mathbf{G}^{-1}\mathbf{W}_n\mathbf{B}_n \tag{27}$$

As $\mathbf{G}$ and $\mathbf{W}_n$ are diagonal matrices and $\mathbf{B}_n$ is a unit lower triangular matrix, the left side of Equation 27 is a lower triangular matrix, i.e.,

$$\bar{\mathbf{H}}_n^t\mathbf{F}_n^k = 0 \quad (t < k) \tag{28}$$

which reveals that $\mathbf{F}_n^k (k > 1)$ lies in the null space of $\breve{\mathbf{H}}_n^k = \left[\bar{\mathbf{H}}_n^{1^T}, \cdots, \bar{\mathbf{H}}_n^{(k-1)^T}\right]^T$. It can be found by doing SVD on $\breve{\mathbf{H}}_n^k$:

$$\breve{\mathbf{H}}_n^k = \mathbf{U}_n^k\left[\boldsymbol{\Sigma}_n^k \quad \mathbf{0}\right]\left[\mathbf{V}_{n1}^k \quad \mathbf{V}_{n0}^k\right]^H \tag{29}$$

We assume that $\mathbf{F}_n^k$ is represented as $\mathbf{F}_n^k = \bar{\mathbf{F}}_n^k\bar{\bar{\mathbf{F}}}_n^k$. We define $\bar{\mathbf{F}}_n^k = \mathbf{V}_{n0}^k \in C^{n_t \times \left[n_t - \sum_{i=1}^{k-1} l_i\right]}$, and $\bar{\bar{\mathbf{F}}}_n^k$ is a $\left[n_t - \sum_{i=1}^{k-1} l_i\right] \times l_k$ matrix.

The above analysis is suitable for the kth$(k > 1)$ user. Since the first user is not limited by Equation 28, we use $\bar{\mathbf{F}}_n^k = \mathbf{I}_{n_t}$. $\bar{\mathbf{F}}_n^k$ can be achieved by:

$$\bar{\mathbf{F}}_n^k = \begin{cases} \mathbf{I}_{n_t}, & k = 1 \\ \mathbf{V}_{n0}^k, & 1 < k \leq K \end{cases} \tag{30}$$

Similarly, with $\mathbf{F}_{n1}$ and THP, the algorithm decomposes the MU-MIMO channel into parallel independent SU-MIMO channels. Define $\bar{\bar{\mathbf{H}}}_n^k = \bar{\mathbf{H}}_n^k\bar{\mathbf{F}}_n^k$. For any user k $(k = 1, \cdots, K)$, $\mathbf{B}_n^{k,k}$, $\bar{\bar{\mathbf{F}}}_n^k$ and $\mathbf{G}^k$ satisfy:

$$\mathbf{G}^k\bar{\bar{\mathbf{H}}}_n^k\bar{\bar{\mathbf{F}}}_n^k = p_n^k\mathbf{B}_n^{k,k} \tag{31}$$

Therefore, $\mathbf{B}_n^{k,k}$, $\bar{\bar{\mathbf{F}}}_n^k$ and $\mathbf{G}^k (k = 1, \cdots, K)$ can be designed separately. For the kth user, Equation 26 can be reduced to:

$$\begin{aligned}
&\max_{\mathbf{B}_n^{k,k},\mathbf{F}_{n2}^k,\mathbf{G}^k} \quad \min_{k} \quad \left(\rho_1^k, \cdots, \rho_{l_k}^k\right) \\
&s.t. \quad \mathbf{G}^k\bar{\bar{\mathbf{H}}}_n^k\bar{\bar{\mathbf{F}}}_n^k = p_n^k\mathbf{B}_n^{k,k} \qquad (a) \\
&\qquad \mathbf{G}^k = diag\left\{g_1^k, \cdots, g_{l_k}^k\right\} \qquad (b) \\
&\qquad \mathbf{S}_i^k\mathbf{B}_n^{k,k}\mathbf{e}_i^k = \mathbf{0}_i, i = 1, \cdots, l_k \qquad (c) \\
&\qquad \boldsymbol{Tr}\left\{\bar{\bar{\mathbf{F}}}_n^k\bar{\bar{\mathbf{F}}}_n^{k^H}\right\} = p/\tau \qquad (d) \\
&\qquad \sum\nolimits_{n=1}^{N} p_n^k = \sqrt{P/(\tau l_k)}, \qquad (e)
\end{aligned} \tag{32}$$

where $\mathbf{S}_i^k = \left[\mathbf{I}_i, \mathbf{0}_{i\times(l_k-i)}\right]$, and $\mathbf{e}_i^k$ is the ith column of $\mathbf{I}_{l_k}$. (d) is obtained because $\bar{\mathbf{F}}_n^{k^H}\bar{\mathbf{F}}_n^k = \mathbf{I}$.

The optimal solution of Equation 32 is obtained when all l_k streams attain equal SINR [29]. According to Equation 12 and Equation 25, it is equivalent to possess equal value for diagonal elements of $\mathbf{G}^k$, expressed as $\mathbf{G}^k = \alpha^k\mathbf{I}$. Equation 32 is equivalent to:

$$\begin{aligned}
&\min_{\mathbf{B}_n^{k,k},\mathbf{F}_{n2}^k,\mathbf{G}^k} \quad \left|\alpha^k\right|^2 \\
&s.t. \quad \gamma_n\bar{\bar{\mathbf{H}}}_n^k\bar{\bar{\mathbf{F}}}_n^k = \mathbf{B}_n^{k,k}, \qquad (a) \\
&\qquad \mathbf{S}_i^k\mathbf{B}_n^{k,k}\mathbf{e}_i^k = \mathbf{0}_i, i = 1, \cdots, l_k \qquad (b) \\
&\qquad \boldsymbol{Tr}\left\{\bar{\bar{\mathbf{F}}}_n^k\bar{\bar{\mathbf{F}}}_n^{k^H}\right\} = p/\tau \qquad (c) \\
&\qquad \sum\nolimits_{n=1}^{N} p_n^k = \sqrt{P/(\tau l_k)}, \qquad (d)
\end{aligned} \tag{33}$$

where $\gamma_n = \alpha^k/p_n^k$. The constrain condition (a) in Equation 33 can be rewritten as $\bar{\bar{\mathbf{F}}}_n^k = \gamma_n^{-1}\bar{\bar{\mathbf{H}}}_n^{k^\dagger}\mathbf{B}_n^{k,k}$. Combining with (c), $\gamma_n^2 p/\tau = \boldsymbol{Tr}\left\{\bar{\bar{\mathbf{H}}}_n^{k^\dagger}\mathbf{B}_n^{k,k}\left(\bar{\bar{\mathbf{H}}}_n^{k^\dagger}\mathbf{B}_n^{k,k}\right)^H\right\}$ is obtained. The problem of Equation 33 can be rewritten as:

$$\begin{aligned}
&\min_{\mathbf{B}_n^{k,k}} \quad \left\|\bar{\bar{\mathbf{H}}}_n^{k^\dagger}\mathbf{B}_n^{k,k}\right\|_F^2 \\
&s.t. \quad \mathbf{S}_i^k\mathbf{B}_n^{k,k}\mathbf{e}_i^k = \mathbf{0}_i, i = 1, \cdots, l_k.
\end{aligned} \tag{34}$$

Actually, $\mathbf{B}_n^{k,k}\mathbf{e}_i^k$ denotes the ith column of $\mathbf{B}_n^{k,k}$. The objective of Equation 34 is equivalent to minimizing $\left\|\bar{\bar{\mathbf{H}}}_n^{k^\dagger}\mathbf{B}_n^{k,k}\mathbf{e}_i^k\right\|^2$ for any $i (i = 1, \cdots, l_k)$.

Let $\mathbf{L}_n^i (i = 1, \cdots, l_k)$ represent the ith column of $\bar{\bar{\mathbf{H}}}_n^{k^\dagger}$. $\mathbf{Y}_n^i$ is comprised of $(i+1, \cdots, l_k)$ columns of $\bar{\bar{\mathbf{H}}}_n^{k^\dagger}$, i.e., $\mathbf{Y}_n^i = \left[\mathbf{L}_n^{i+1}, \cdots, \mathbf{L}_n^{l_k}\right]$. Then, we can attain:

$$\left\| \bar{\bar{\mathbf{H}}}_n^{k^\dagger} \mathbf{B}_n^{k,k} \mathbf{e}_i^k \right\|^2 = \left\| \left[\mathbf{L}_n^i, \mathbf{Y}_n^i\right] \begin{bmatrix} 1 \\ \mathbf{b}_n^i \end{bmatrix} \right\|^2 \tag{35}$$

By differentiating of Equation 35 with respect to $\mathbf{b}_n^i$ and setting the result to zero, $\mathbf{b}_n^i$ is achieved by:

$$\mathbf{b}_n^i = -\left(\mathbf{Y}^{i^H} \mathbf{Y}^i\right)^{-1} \mathbf{Y}^{i^H} \mathbf{L}_n^i, i = 1, \cdots, l_k - 1 \tag{36}$$

Therefore, the ith column of $\mathbf{B}_n^{k,k}$ is obtained by:

$$\mathbf{B}_n^{k,k} \mathbf{e}_i^k = \begin{cases} \left[1, \mathbf{b}_n^{i^T}\right]^T, & i = 1 \\ \left[0_{1 \times (i-1)}, 1, \mathbf{b}_n^{i^T}\right]^T, & i = 2, \cdots, l_k - 1 \\ \mathbf{e}_{l_k}^k, & i = l_k \end{cases} \tag{37}$$

Then, $\mathbf{B}_n^{k,k}$ is obtained by combining all columns $\mathbf{B}_n^{k,k} \mathbf{e}_i^k$ $(i = 1, \cdots, l_k)$. Therefore, we can derive $\gamma_n^2 = \tau \left\| \bar{\bar{\mathbf{H}}}_n^{k^\dagger} \mathbf{B}_n^{k,k} \right\|_F^2 / p$. By combining (a) and (d) in Equation 33, $\mathbf{G}^k = \alpha^k \mathbf{I}$ is obtained, where:

$$\alpha^k = \frac{\sqrt{P/(\tau l_k)}}{\sum_{n=1}^{N} 1/\gamma_n} \tag{38}$$

Finally, $\mathbf{B}_n$ is determined by $\mathbf{B}_n = \mathbf{W}_n^{-1} \mathbf{R} \mathbf{H}_n \mathbf{F}_n$.

In this algorithm, $\mathbf{R}^k (k = 1, \cdots, K)$ are designed at the users, and other matrices are designed at the BSs. The design of matrices is independently performed at every BS. $\left\| \bar{\bar{\mathbf{H}}}_n^{k^\dagger} \mathbf{B}_n^{k,k} \right\|_F^2 (n = 1, \cdots, N)$ should be transmitted to the kth$(k = 1, \cdots, K)$ user by downlink channel for achieving $\mathbf{G}^k$ at the kth user. The procedure of the proposed decentralized algorithm is described in Table 2.

3.5 Remark 1 (applicability)

It should be noted that the proposed two algorithms are also suitable for the system with a single-data stream transmitted for each user. Moreover, the proposed two algorithms both are applicable to the noncoordinated system. However, the centralized scheme is suggested to apply for the noncoordinated system, as the decentralized scheme is a suboptimal solution in this situation.

4 Performance analysis

From Equation 23 and Equation 38, it is noted that both of the proposed two algorithms can achieve equal SINR for every stream of the user. They guarantee the balance

Table 2 Decentralized nonlinear joint tx-rx processing algorithm

for $k = 1{:}K$
Doing SVD on $\mathbf{H}^k$. Obtain $\mathbf{U}_1^k = \left[\mathbf{U}^k\right]_{:,1:l_k}^H$
end
for $n = 1{:}N$
for $k = 1{:}K$
Compute $\bar{\mathbf{F}}_n^k$ using Equation 30
Obtain l_k columns of $\mathbf{B}_n^{k,k}$ using Equation 37
Compute γ_n by $\gamma_n^2 = \tau \left\| \bar{\bar{\mathbf{H}}}_n^{k^\dagger} \mathbf{B}_n^{k,k} \right\|_F^2 / p$
Compute $\bar{\bar{\mathbf{F}}}_n^k$ by $\bar{\bar{\mathbf{F}}}_n^k = \gamma_n^{-1} \bar{\bar{\mathbf{H}}}_n^{k^\dagger} \mathbf{B}_n^{k,k}$
Compute $\mathbf{G}^k = \alpha^k \mathbf{I}$ using Equation 38 and derive $\mathbf{R}^k$ by $\mathbf{R}^k = \mathbf{G}^k \left(\mathbf{U}_1^k\right)^H$
end
Compute $\mathbf{B}_n$ by $\mathbf{B}_n = \mathbf{W}_n^{-1} \mathbf{R} \mathbf{H}_n \mathbf{F}_n$
end

performance among streams of each user, which bring much convenience to the modulation/demodulation and coding/decoding procedures. In this section, we analyze the feasibility and the computational complexity of the proposed two algorithms.

4.1 Feasibility analysis

In the MIMO system, in order to distinguish every transmit stream, the constraint that the number of transmit data streams is no more than the number of transmit and receive antennas should be satisfied. For the centralized coordinated mode and the decentralized coordinated mode, the constraint on the number of transmit data streams is specified as follows:

Lemma 1: For the centralized coordinated mode, the number of transmit data streams are bounded by $L \le N_t$, $l_k \le n_r$; for the decentralized coordinated mode, the number of transmit data streams are bounded by $L \le n_t$, $l_k \le n_r$.

In the proposed centralized algorithm, the design of the transmit space matrix $\bar{\mathbf{F}}^k (k = 1, \cdots, K)$ requires $N_t - \sum_{i=1}^{k-1} l_i > 0$. Furthermore, to guarantee that the optimization problem Equation 19 has solutions, $S \ge l_k$ is required. As the entries of $\mathbf{H}^k \bar{\mathbf{F}}^k$ are zero-mean complex Gaussian variables, the rank of $\mathbf{H}^k \bar{\mathbf{F}}^k$ is $S = \min\left(n_r, N_t - \sum_{i=1}^{k-1} l_i\right)$ with a probability of 1. Therefore, $S \ge l_k$ is the necessary condition to carry out the algorithm. Base on *Lemma 1*, the necessary condition is satisfied to the centralized coordinated system. Therefore, the proposed centralized algorithm is feasible.

In the proposed decentralized algorithm, $n_t - \sum_{i=1}^{k-1} l_i > 0$ is required to guarantee the existence of the transmit space matrix $\bar{\mathbf{F}}_n^k (k = 1, \cdots, K, n = 1, \cdots, N)$. Moreover, the solution of optimization problem Equation 34 requires $n_t - \sum_{i=1}^{k-1} l_i \geq l_k$, which is satisfied in the decentralized coordinated system. Therefore, the proposed decentralized algorithm is feasible.

4.2 Computational complexity

For simplicity, the number of float point operations is used to measure the computational complexity of the proposed algorithms.

In the proposed centralized algorithm, the design of the relative matrices for the kth user includes the following: a onetime multiplication of a $l_{k-1} \times n_r$ matrix and a $n_r \times N_t$ matrix, the complexity of which is $O(l_{k-1} n_r N_t)$; a onetime computation of the null space of a $\sum_{i=1}^{k-1} l_i \times N_t$ matrix with $O\left(\left(\sum_{i=1}^{k-1} l_i\right)^2 N_t\right)$ complexity; a onetime multiplication of a $n_r \times N_t$ matrix and a $N_t \times \left[N_t - \sum_{i=1}^{k-1} l_i\right]$ matrix, the complexity of which is $O\left(n_r N_t \left[N_t - \sum_{i=1}^{k-1} l_i\right]\right)$; and a onetime computation of the singular value of a $n_r \times \left[N_t - \sum_{i=1}^{k-1} l_i\right]$ matrix with $O\left(n_r^2 \left[N_t - \sum_{i=1}^{k-1} l_i\right]\right)$ complexity. Therefore, the complexity of the relative matrices designed for the kth user is $O\left(\left(\sum_{i=1}^{k-1} l_i\right)^2 N_t + n_r^2 N_t + n_r N_t^2\right)$.

In the proposed decentralized algorithm, every BS has the same computational complexity. For any BS, the design of the relative matrices for the kth user includes the following: a onetime computation of the singular vector of a $n_r \times n_t$ matrix with $O(n_r^2 n_t)$ complexity; onetime multiplications of a $l_k \times n_r$ matrix and a $n_r \times n_t$ matrix, the complexity of which is $O(l_k n_r n_t)$; a onetime computation of the null space of a $\sum_{i=1}^{k-1} l_i \times n_t$ matrix with $O\left(\left(\sum_{i=1}^{k-1} l_i\right)^2 n_t\right)$ complexity; a onetime multiplication of a $l_k \times n_t$ matrix and a $n_t \times \left[n_t - \sum_{i=1}^{k-1} l_i\right]$ matrix, the complexity of which is $O\left(l_k n_t \left[n_t - \sum_{i=1}^{k-1} l_i\right]\right)$; and l_k-times computation of the Moore-Penrose pseudo-inverse of a $\left[n_t - \sum_{i=1}^{k-1} l_i\right] \times [l_k - i]$ $(i = 1, \cdots, l_k)$ matrix, the complexity of which is $O\left(\left[n_t - \sum_{i=1}^{k-1} l_i\right] \sum_{i=1}^{l_k} [l_k - i]^2\right)$. The complexity of other scalar computations can be ignored. Therefore, the complexity of the relative matrices designed for the kth user is $O\left(\left(\sum_{i=1}^{k-1} l_i\right)^2 N_t + n_r^2 N_t + N l_k n_t^2 + l_k^3 N_t\right)$.

Assume that the data streams for every user is equal, i.e., $l_1 = \cdots = l_K = l$. Thus, the complexity of the proposed centralized algorithm is $O\left(KL^2 N_t + K n_r^2 N_t + K n_r N_t^2\right)$, and the decentralized algorithm is $O\left(KL^2 N_t + K n_r^2 N_t + N L n_t^2 + K l^3 N_t\right)$.

4.3 Remark 2 (backhaul latency effect)

For centralized coordinated mode, tx-rx processing matrices are jointly computed at the central processing unit and then reported to every BS through the backhaul link. The existing backhaul latency can affect the system performance. We ignore the backhaul latency effect in the paper and will study it in the future work.

5 Numerical results and discussions

This section presents some simulation results to evaluate the BER performance of the proposed two algorithms. We compare them with the following algorithms: the interference-free algorithm, the joint transit-receive processing algorithm proposed in [19], and the centralized BD (CBD) and decentralized BD (DBD). As the traditional BD cannot be directly applied in a decentralized manner, here in DBD, receive processing matrix is derived firstly based on the same method for receive processing matrix in the proposed decentralized algorithm, then the precoding matrix is derived based on the ZF criterion. For the system with a single stream transmitted for each user, i.e., $l_k = 1$ $(k = 1,\ldots,K)$ system, we also compare the proposed decentralized algorithm with DZF [20] and DV-SINR [22]. Flat Rayleigh fading channels are considered in simulations. The elements of the channels are i.i.d. complex Gaussian variables with zero mean and unit variance. In this simulation, a 64-QAM modulation scheme is employed in the simulation. The signal to noise ratio (SNR) is defined as $\mathrm{SNR} = P/(SM\sigma^2)$, where $M = 4$ is the signal constellation size and S is the average number of the data streams transmitted for each user.

5.1 Balance BER performance among streams of each user

Figure 2 verifies the balance performance among the streams of each user in the proposed algorithms. We consider a 3-cell coordinated system with $n_t = 6$ transmit antennas and $K = 3$ users each equipped with $n_r = 3$ receive antennas. There are $l_k = 2$ $(k = 1,\ldots,K)$ data streams transmitted for each user. The BER performance of two streams of the first user and the third user are shown. It can be seen that the two streams of any user achieve the approximately equal BER, not only for the centralized algorithm, but also for the decentralized algorithm. The simulation results are in accordance with the theoretical analysis.

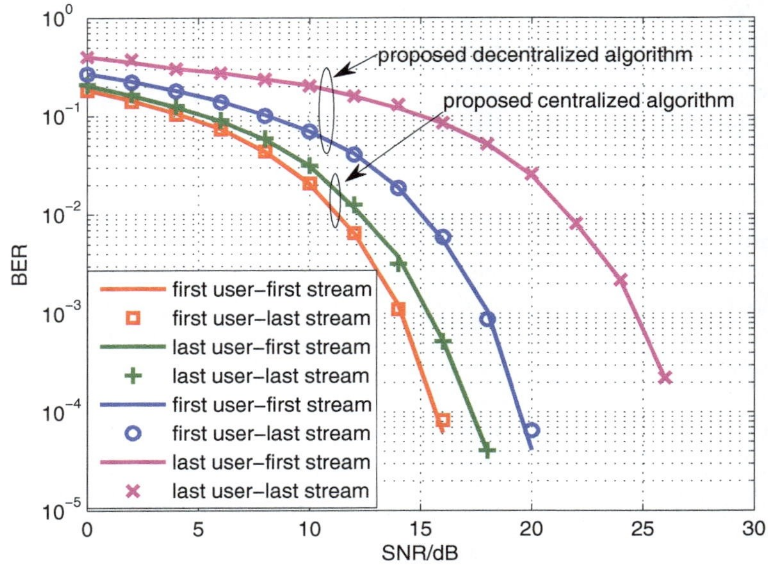

Figure 2 BER performance of each stream in the proposed two algorithms with $N = 3$, $K = 3$, $n_t = 6$, $n_r = 3$, $l_k = 2$ ($k = 1,\ldots,K$).

5.2 BER performance comparison of different algorithms

Figure 3 presents the BER performance comparison of the six algorithms. A 2-cell coordinated system with $n_t = 6$ transmit antennas and $K = 3$ users each equipped with $n_r = 5$ antennas is considered. The number of the data streams transmitted to each user is set to 2, i.e., $l_k = 2$ ($k = 1,\ldots,K$). It is noticed that in the interference-free algorithm, only a single user is served by BSs. On the whole, centralized algorithms have better performance than decentralized algorithms, at the cost of information exchange among BSs. For the proposed algorithms, the centralized algorithm achieves about 7-dB gain related to the decentralized algorithm at BER = 10^{-3}. Compared with the existing algorithms, when BER = 10^{-3}, the proposed centralized algorithm has an approximately 5-dB gain to the algorithm in [19] and a 10-dB gain to CBD. Also, about a 10-dB gain is achieved by the proposed decentralized algorithm related to DBD at BER = 10^{-2}.

In Figure 4, we consider the BER performance of a 3-cell coordinated system, with $n_t = 6$ transmit antennas

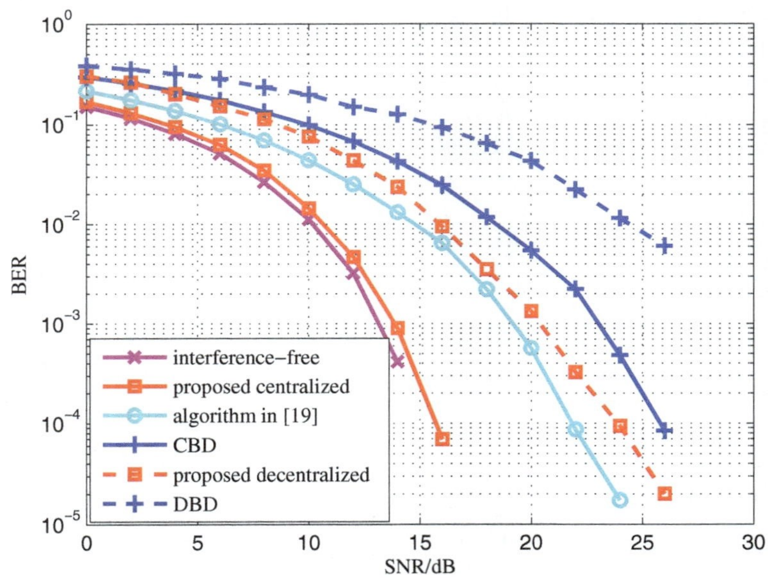

Figure 3 BER performance of centralized and decentralized algorithms with $N = 2$, $K = 3$, $n_t = 6$, $n_r = 5$, $l_k = 2$ ($k = 1,\ldots,K$).

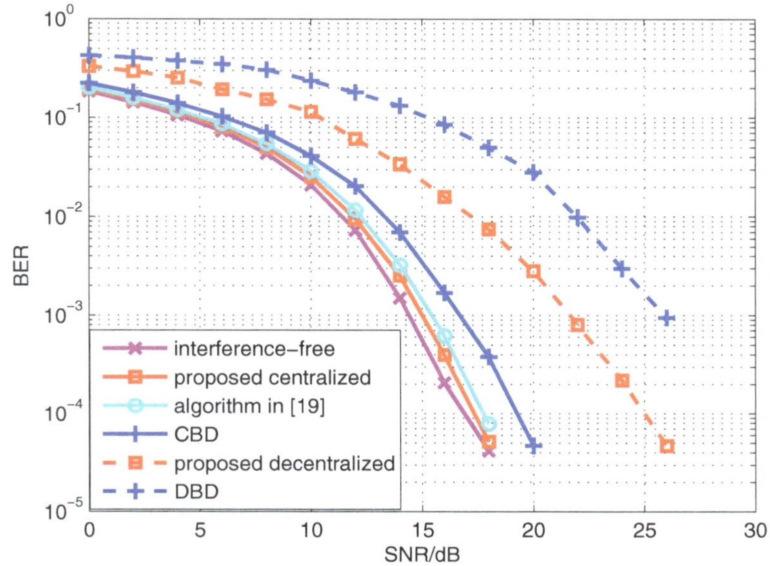

Figure 4 BER performance of centralized and decentralized algorithms with $N = 3$, $K = 3$, $n_t = 6$, $n_r = 3$, $l_k = 2$ ($k = 1,\ldots,K$).

and $K = 3$ users each equipped with $n_r = 3$ antennas. The number of the data streams transmitted to each user is set to 2, i.e., $l_k = 2$ ($k = 1,\ldots,K$). As mentioned in Figure 3, in the interference-free algorithm, only a single user is served by BSs. As can be seen from Figure 4, centralized algorithms have better BER performance than decentralized algorithms. The proposed centralized algorithm has a lower BER than the algorithm in [19] and CBD, and the proposed decentralized algorithm achieves better performance than DBD. Compared with Figure 3, the performance gains among algorithms are different, as they are related with system configuration.

In Figure 5, the performance of the proposed decentralized algorithm for the system with a single stream transmitted for each user, i.e., $l_k = 1$ ($k = 1,\ldots,K$), is verified and compared with the existing decentralized algorithms, DZF [20], and DV-SINR [22] in BER. A 3-cell coordinated system with $n_t = 6$ transmit antennas and $K = 5$ users each equipped with $n_r = 3$ antennas is considered. As can be seen from Figure 5, the proposed decentralized

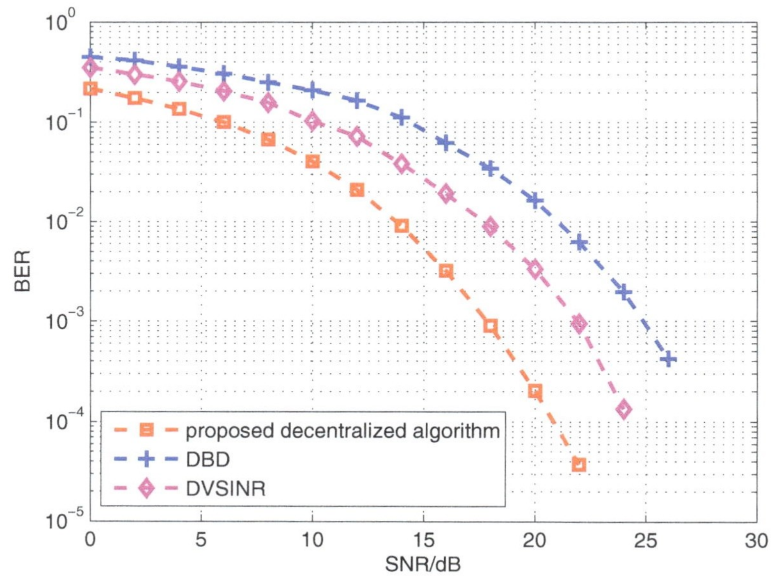

Figure 5 BER performance of different decentralized algorithms with single-stream users, $N = 3$, $K = 5$, $n_t = 6$, $n_r = 3$, $l_k = 1$ ($k = 1,\ldots,K$).

algorithm has a lower BER than other algorithms. When $BER = 10^{-3}$, it can achieve an approximately 7-dB gain compared with DZF, and a 5-dB gain compared with DV-SINR.

5.3 The effect of the receive antennas and user's number to centralized algorithms

Figure 6 illustrates the effect of the number of receive antennas to the proposed centralized algorithm, the algorithm in [19], and the CBD. We consider a 3-cell coordinated system with $n_t = 6$ transmit antennas and $K = 3$ users. The number of the data streams transmitted to each user is set to 2, i.e., $l_k = 2$ $(k = 1,...,K)$. As can be seen from Figure 6, the performance difference between the proposed centralized algorithm and the algorithm in [19] is increased with the number of the receive antennas. In the proposed centralized algorithm, the receive processing matrix is considered into the MU-MIMO channel decomposition. Compared with the algorithm in [19], the decomposed SU-MIMO channels have larger dimensions, which increases the system diversity gain and improves the system performance. With a larger number of the receive antennas, the decomposed SU-MIMO channels have the same dimension in the proposed centralized algorithm but have smaller dimensions in the algorithm in [19]. Therefore, with increased number of the receive antennas, the proposed centralized algorithm can achieve more performance gain than the algorithm in [19].

In Figure 7, the effect of the number of users to the proposed centralized algorithm, the algorithm in [19] and CBD is illustrated. We consider a 3-cell coordinated system with $n_t = 6$ transmit antennas and $n_r = 3$ receive antennas. The number of the data streams transmitted to each user is set to 2, i.e., $l_k = 2$ $(k = 1,...,K)$. As can be seen from Figure 7, the increased number of users enlarges the performance differences among the algorithms. The tx-rx processing matrices of each user, in CBD, are used to eliminate the interference of all other users. Differently, in the proposed centralized algorithm and the algorithm in [19], they are used to eliminate the interference of part of the other users, bringing in more space dimensions for the diversity gain.

5.4 BER performance of the proposed algorithms in a noncoordinated system

In Figure 8, we illustrate the performance of the proposed algorithms for the noncoordinated system with $N = 1$ and compare them with CBD. In this situation, CBD is equivalent to the traditional BD in a single-cell MIMO system. A MU-MIMO system, in which there are $n_t = 8$ transmit antennas and $K = 4$ users each equipped with $n_r = 2$ antennas, is considered. The number of the data streams transmitted to each user is set to 2, i.e., $l_k = 2$ $(k = 1,...,K)$. It is shown that the proposed algorithms can achieve better BER performance than BD, and that the proposed decentralized algorithm is only a suboptimal scheme for the noncoordinated system, as part of the receive processing matrix is not jointly derived with the transmit processing matrix. In this situation, the proposed centralized algorithm is verified to achieve lower BER than the proposed decentralized algorithm. It exhibits an approximately 6-dB gain over the decentralized scheme at $BER = 10^{-2}$.

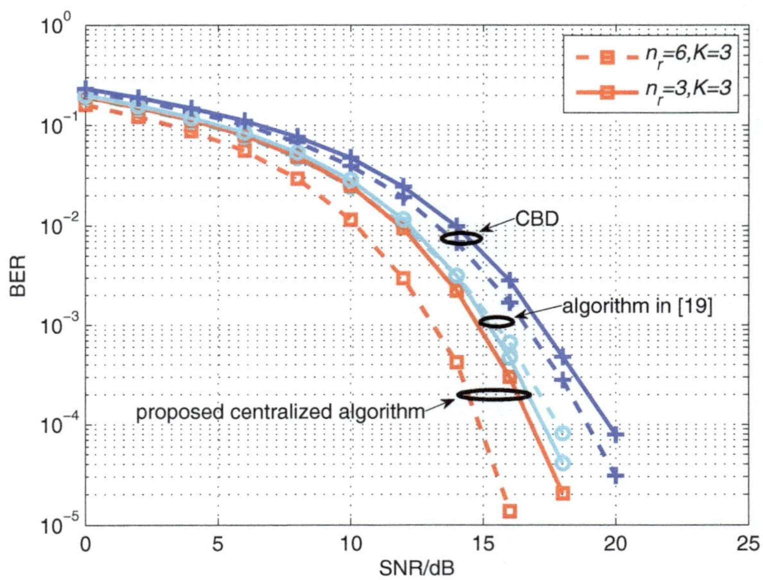

Figure 6 The effect of the number of receive antennas n_r to different centralized algorithms with $N = 3$, $n_t = 6$, $l_k = 2$ $(k = 1,...,K)$.

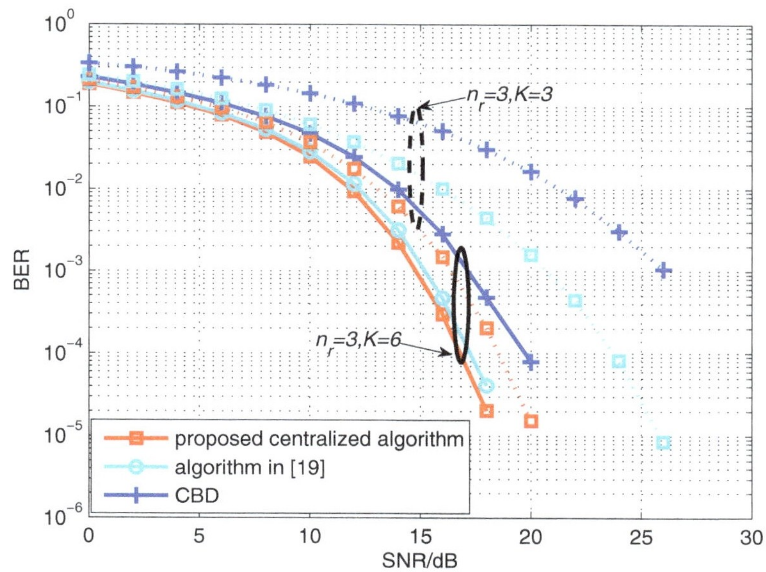

Figure 7 **The effect of the number of users** K **to different centralized algorithms with** $N = 3$, $n_t = 6$, $l_k = 2$ ($k = 1,\ldots,K$).

6 Conclusions

Nonlinear joint tx-rx processing technology for a coordinated multi-cell system with multi-stream multi-antenna users has been studied. The capacity of the backhaul link determines different coordinated modes among BSs, including centralized and decentralized coordinated. The proposed centralized algorithm is proposed to derive the tx-rx processing matrices jointly at the central processing unit. The proposed decentralized algorithm allows each BS design to transmit precoding in a decentralized manner, which alleviates the demand on the backhaul capacity. The analysis and simulation results show that the centralized algorithm achieves better performance than the decentralized algorithm. And, the proposed algorithms achieve better performance than the existing joint tx-rx processing algorithms and the decentralized linear precodings.

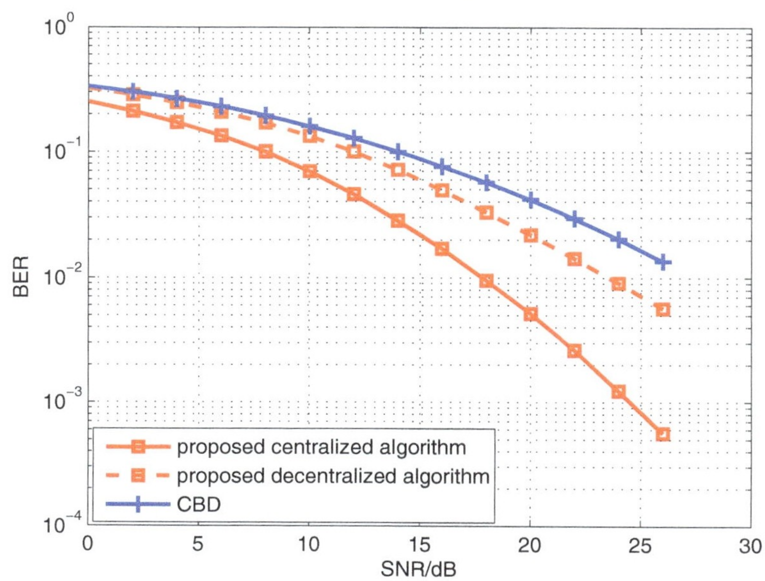

Figure 8 **BER performance of the proposed algorithms in noncoordinated multi-cell system with** $N = 1$, $K = 4$, $n_t = 8$, $n_r = 2$, $l_k = 2$ ($k = 1,\ldots,K$).

Competing interests

The authors declare that they have no competing interests.

Acknowledgements

This work is supported by the Special Funding for Beijing Common Construction Project and the Beijing Natural Science Foundation (4144079).

Author details

[1]Beijing Key Laboratory of Network System Architecture and Convergence, Beijing University of Posts and Telecommunications, Beijing 100876, China. [2]School of Communication Engineering, Hangzhou Dianzi University, Hangzhou 310018, China. [3]School of Electronic Engineering and Computer Science, Queen Mary University of London, London E1 4NS, UK.

References

1. M Sawahashi, Y Kishiyama, A Morimoto, D Nishikawa, M Tano, Coordinated multipoint transmission/reception techniques for LTE-advanced [Coordinated and Distributed MIMO]. IEEE Wireless Commun. **17**(3), 26–34 (2010)
2. D Lee, H Seo, B Clerckx, E Hardouin, D Mazzarese, S Nagata, K Sayana, Coordinated multipoint transmission and reception in LTE-advanced: deployment scenarios and operational challenges. IEEE Commun. Mag. **50**(2), 148–155 (2012)
3. D Gesbert, S Hanly, H Huang, S ShamaiShitz, O Simeone, W Yu, Multi-cell MIMO cooperative networks: a new look at interference. IEEE J. Sel. Areas Commun. **28**(9), 1380–1408 (2010)
4. K Karakayali, GJ Foschini, RA Valenzuela, R Yates, *On the maximum common rate achievable in a coordinated network. Proceedings of the IEEE International Conference Communications* (IEEE, Istanbul, 2006), pp. 4333–4338
5. Z Keke, RC de Lamare, M Haardt, Multi-branch Tomlinson-Harashimaprecoding design for MU-MIMO systems: theory and algorithms. IEEE Trans Commun. **62**(3), 939–951 (2014)
6. S Jing, D Tse, J Soriaga, J Hou, J Smee, R Padovani, Multicell downlink capacity with coordinated processing. EURASIP J. Wireless Commun.Netw **2008**, 586878 (2008)
7. S Liyan, Y Chenyang, H Shengqian, The value of channel prediction in CoMP systems with large backhaul latency. IEEE Trans Commun. **61**(11), 4577–4590 (2013)
8. Papadogiannis, E Hardouin, D Gesbert, Decentralisingmulticell cooperative processing: a novel robust framework. EURASIP J. Wireless Commun.Netw **2009**, 890685 (2009)
9. R Zhang, Cooperative multi-cell block diagonalization with per-basestation power constraints. IEEE J. Sel. Areas Commun. **28**, 1435–1445 (2010)
10. S Shi, M Schubert, N Vucic, H Boche, *MMSE optimization with per-base-station power constraints for network MIMO systems. Proceedings of the IEEE International Conference Communications* (IEEE, Beijing, 2008), pp. 4106–4110
11. J Zhang, Y Wu, S Zhou, J Wang, Joint linear transmitter and receiver design for the downlink of multiuser MIMO systems. IEEE Commun Lett **9**(11), 991–993 (2005)
12. RC de Lamare, Adaptive and iterative multi-branch MMSE decision feedback detection algorithms for multi-antenna systems. IEEE Trans. Wirel. Commun. **12**(10), 5294–5308 (2013)
13. H Park, SH Park, HB Kong, I Lee, Weighted sum MSE minimization under per-BS power constraint for network MIMO systems. IEEE Commun. Lett **16**(3), 360–363 (2012)
14. S He, Y Huang, L Yang, B Ottersten, Coordinated multicell multiuser precoding for maximizing weighted sum energy efficiency. IEEE Trans. Signal Process. **62**(3), 741–751 (2014)
15. M Wei, C Xiang, Z Ming, W Jing, Joint stream-wise THP transceiver design for the multiuser MIMO downlink. IEICE Trans. Commun. **92**(1), 209–218 (2009)
16. W Hardjawana, B Vucetic, Y Li, Multi-user cooperative base station systems with joint precoding and beamforming. IEEE J Sel Top Signal Process **3**(6), 1079–1093 (2009)
17. S Adão, H Reza, G Atílio, Power allocation strategies for distributed precodedmulticell based systems. EURASIP J. Wireless Commun. Netw **2011**, 1 (2011)
18. Y Sun, M Wu, M Zhao, C Xu, *Transceiver designs using non-linear precoding for multiuser MIMO systems with limited feedback. Proceedings of the IEEE VehicularTechnology Conference* (IEEE, Dresden, 2013), pp. 1–5
19. L Sun, M Lei, *Adaptive joint nonlinear transmit-receive processing for multi-cell MIMO networks. Proceedings of the IEEE Globe Communications Conference* (IEEE, Anaheim, 2012), pp. 3766–3771
20. R Holakouei, A Silva, A Gameiro, *Distributed versus centralized zero-forcing precoding for multicell OFDM systems. Proceedings of the IEEE Globe Communications ConferenceWorkshops* (IEEE, Houston, 2011), pp. 188–193
21. R Zakhour, D Gesbert, Distributed multicell-MISO precoding using the layered virtual SINR framework. IEEE Trans. Wireless Commun. **9**(8), 2444–2448 (2010)
22. E Bjornson, R Zakhour, D Gesbert, B Ottersten, Cooperative multicell precoding: rate region characterization and distributed strategies with instantaneous and statistical CSI. IEEE Trans. Signal Processing **58**(8), 4298–4310 (2010)
23. X Zhao, H Xu, X Yang, *Performance enhancement for CoMP based on power allocation and a modified ZF-THP. Proceedings of the IEEE Personal Indoor and Mobile Radio Communications* (IEEE, Sydney, 2012), pp. 2309–2313
24. I Krikidis, B Ottersten, Diversity fairness in Tomlinson–Harashimaprecoded multiuser MIMO through retransmission. IEEE Signal Process Lett. **20**(4), 375–378 (2013)
25. G Boudreau, J Panicker, N Guo, R Chang, N Wang, S Vrzic, Interference coordination and cancellation for 4G networks. IEEE Commun. Mag. **47**(4), 74–81 (2009)
26. J Tang, S Lambotharan, Interference alignment techniques for MIMO multi-cell interfering broadcast channels. IEEE Trans. Commun. **61**(1), 164–175 (2013)
27. R.F.H.Fisher, Precoding and signal shaping for digital transmission (John Wiley & Sons Ltd, 2002)
28. Y Jiang, W Hager, J Li, The generalized triangular decomposition. Math. Comput. **77**, 1037–1056 (2007)
29. A Wiesel, E Yonina, CS Shlomo, Linear precoding via conic optimization for fixed MIMO receivers. IEEE Trans. Signal Process. **54**(1), 161–176 (2006)

8

Rapid algorithm prototyping and implementation for power quality measurement

Krzysztof Kołek* and Krzysztof Piątek

Abstract

This article presents a Model-Based Design (MBD) approach to rapidly implement power quality (PQ) metering algorithms. Power supply quality is a very important aspect of modern power systems and will become even more important in future smart grids. In this case, maintaining the PQ parameters at the desired level will require efficient implementation methods of the metering algorithms. Currently, the development of new, advanced PQ metering algorithms requires new hardware with adequate computational capability and time intensive, cost-ineffective manual implementations. An alternative, considered here, is an MBD approach. The MBD approach focuses on the modelling and validation of the model by simulation, which is well-supported by a Computer-Aided Engineering (CAE) packages. This paper presents two algorithms utilized in modern PQ meters: a phase-locked loop based on an Enhanced Phase Locked Loop (EPLL), and the flicker measurement according to the IEC 61000-4-15 standard. The algorithms were chosen because of their complexity and non-trivial development. They were first modelled in the MATLAB/Simulink package, then tested and validated in a simulation environment. The models, in the form of Simulink diagrams, were next used to automatically generate C code. The code was compiled and executed in real-time on the Zynq Xilinx platform that combines a reconfigurable Field Programmable Gate Array (FPGA) with a dual-core processor. The MBD development of PQ algorithms, automatic code generation, and compilation form a rapid algorithm prototyping and implementation path for PQ measurements. The main advantage of this approach is the ability to focus on the design, validation, and testing stages while skipping over implementation issues. The code generation process renders production-ready code that can be easily used on the target hardware. This is especially important when standards for PQ measurement are in constant development, and the PQ issues in emerging smart grids will require tools for rapid development and implementation of such algorithms.

Keywords: Automatic code generation; Flicker; PLL; Power quality; Rapid prototyping; Simulink; Synchronisation

1 Introduction

Smart Grids offer an opportunity for effective power system management and utilisation of distributed and renewable energy resources. Efficient utilisation of the grid includes maintaining power supply (voltage) quality and emission (current) quality according to the specified limits. These limits can be set by the regulator, provided by standards or grid codes, or even agreed as part of an energy supply contract. The requirement of measuring the power quality (PQ) parameters has increased the interest in applications using signal processing techniques for PQ measurement. Comprehensive overviews of this problem are presented in [1,2]. These books describe classical methods of PQ analysis as well as the application of advanced techniques like neural networks, wavelet transform, and machine learning methods. It is clear that there is a need for increasingly complex PQ metering algorithms.

On the other hand, PQ metering is defined by standards, e.g. EN 61000-4-30. These standards define voltage quality measurement and provide implementation details needed to build a PQ meter. Consequently, algorithms for a PQ meter include not only signal processing techniques, but also data aggregation, gapless measurement, and synchronization with line voltage waveforms or time references, etc.

Because further requirements for PQ are constantly developing, there is a need for a rapid development platform for new algorithms. A rapid development framework

* Correspondence: kko@agh.edu.pl
AGH University of Science and Technology, al. A.Mickiewicza 30, Kraków 30-059, Poland

should enable fast modelling, testing, and hardware implementation. The framework would be useful for preparing a reference PQ meter to test algorithms for PQ measurement or event classification in the field, prepare test procedures for upcoming certification processes, or test new ideas for PQ metering. This framework should enable rapid modelling, facilitate model validation and testing, as well as enable automatic code generation for different hardware platforms, including code generation for microprocessors and reconfigurable FPGA devices.

The conventional V-model of software development includes specification, design, implementation, integration, testing, and verification and validation stages. This implementation process is performed manually and is time intensive. The complexity of PQ measurement algorithms and the requirement for short development cycles motivate the investigation of a more efficient, reliable, and cost-effective alternative. Model-Based Design (MBD), adapted to the development of measurement algorithms, is here considered as an option. The MBD approach to the development of measurement algorithms consists of five steps:

1. Modelling input test signals and defining reference responses to the test inputs
2. Synthesis of measurement algorithms (Computer-Aided Engineering (CAE) tools are heavily used at this stage.)
3. Simulation of measurement algorithms excited by the test sequences
4. Tuning the algorithms to achieve compliance of the simulated responses with reference responses
5. Synthesis of the algorithms on a target hardware platform

The MBD approach focuses on modelling and simulation, hence most of the development time is independent of target hardware. Moreover, some CAE packages automatically convert the models into a source code library, which simplifies and accelerates the synthesis stage. The automatic code generation method reduces the number of errors that appear during manual coding.

Several attempts have been made to design rapid development methods for power systems algorithms. In [3], an analysis of CAE tools such as MATLAB/Simulink and dSpace/TargetLink describes how they perform fundamental steps such as physical modelling, simulations, control algorithms design and testing, hardware evaluation, parameter optimization, and finally code generation for real-time software implementations. As an example, the development of a controller for a low power single-phase transformerless inverter was presented. The development, verification, and implementation stages of automatic code generation to implement over/under-power protection functions and MATLAB/Simulink active power filter control blocks are given in [4] and [5], respectively. The active power filter allows the rapid on-line tuning of the filter's parameters, selection of control algorithms, and can be applied to generate code for microcontrollers. Kirubakaran et al. in [6] presented a Digital Signal Processing (DSP) controller for a fuel-cell based system. MATLAB/Simulink was used to generate Pulse Width Modulation (PWM) signals for the TMS320F2812 DSP. The same DSP was applied to implement a unity power factor PWM rectifier and shunt active power filter, as described in [7]. The automatic code generation in the MATLAB/Simulink environment for the fixed-point DSP was presented. A slightly different development path is given in [8]. Simulink diagrams are automatically converted to C code, afterward converted to assembly language, and finally transferred to a DSP for execution.

SCADA systems are considered to be the target platform for algorithms generated in Simulink. In [9], a Simulink implementation of a grid load forecast algorithm was given. The Simulink diagram was applied to automatically generate OPen Connectivity (OPC) Data Access and Modbus/TCP servers. The servers were "embedded" into a SCADA for real-time load prediction.

More specifically, PQ algorithms have been investigated. McGranaghan and Santoso in [10] presented a summary of the problems encountered during both offline and on-line analyses of PQ algorithms. In [11], a discrete model of the IEC 61000-4-15 flickermeter was presented. The full discrete flickermeter channel, developed as Simulink diagram, is shown, however the aim of this model is only simulation, and hence the implementation path is not considered. An implementation method for a Simulink algorithm that conforms the IEEE 1459-2000 standard for measurement of electric power quantities is presented in [12]. The MATLAB xPC Target toolbox was used to run the code generated automatically from the Simulink diagram on a PC computer. The methods defined by the IEC 61000-4-30 standard were investigated in [13]. The modelling and analysis of the algorithms were carried out in a MATLAB/Simulink environment. The algorithms were then automatically executed on a dSPACE DS 1104 board.

In addition, studies on the compliance of automatically generated code with safety standards were performed. Krizan et al. in [14] dealt with the usability of automatically generated C code in critical applications. Compliance with the DO-178C and DO-331 standards were presented. DO-178C and DO-331 address the entire software development cycle, from functional specification to software verification. The DO-178C and DO-331 are documents used by certification authorities such as the Federal Aviation Administration, European Aviation Safety Agency and Transport Canada to approve software for aerospace systems. In [15], the authors used the IEC 61508 standard and investigated options for running automatically generated code on safety PLCs. Stages for code generation, evaluation, and conversion before the code was applied in safety systems were presented.

Finally, some effort has been made to generate automatic code for reconfigurable FPGA circuits and their applications in power systems. Mekonnen et al. in [16] presented the simulation of a complete full-bridge PWM converter for a solar grid-tie inverter that also provided the benefits of auto VHDL-code generation for an RMS-based digital controller. In [17], a rapid prototyping platform for photovoltaic power inverters was presented. The hardware-in-the-loop approach was used to develop and validate PWM signals for different converter topologies. The control strategies are developed as Simulink diagrams, and the goal of the platform is the automatic VHDL code generation for Xilinx FPGAs.

The scope of this paper covers the development of algorithms for PQ metering while following the MBD approach. The algorithms developed and validated first as models are then transferred to C code automatically. The MATLAB/Simulink CAE plays a central role during the development. The generated code is executed on a hardware platform and real-time processing results are compared to the measurement generated by a commercial reference device. The implementation path for two example algorithms is presented. The algorithms are chosen because of their importance in PQ metering. The authors' intention is to show not the novelty of the algorithms but rather the rapid implementation issues.

1.1 Rapid prototyping

1.1.1 Rapid prototyping approach

The main advantage of the rapid prototyping approach is the ability to focus on the problem while omitting the details of its implementation. Development is carried out in an environment that offers a high level of abstraction, and the implementation on a hardware platform is performed by automatically generated code in the target programming language. The differences with respect to the classical software development process can be shown using the example of a filter design. The classical approach requires the selection of the filter structure, the calculation of parameters, and, finally, the implementation of the filter in a programming language. The rapid prototyping path is reduced to giving the type and parameters of the filter bandwidth. The source code of the filter procedures is then generated automatically. This approach results in ability to rapidly test and verify the proposed strategy while at the same time reducing the number of programming errors.

The MATLAB/Simulink stages of design and implementation of a measurement algorithm are shown in Figure 1. A data processing algorithm is developed as a Simulink diagram and then excited by test data vectors until the results are compliant with the reference results. At this point in the process, the design consists entirely of simulations without any references to the hardware on which the designed algorithm is intended to run.

The implementation on target hardware requires the replacement of simulation inputs and outputs by real signals. The interface can be established by controlling A/D and D/A converters, using files or databases, or over standard communication protocols such as Modbus, OPC Data Access, or OPC Unified Architecture. Modern processors map the A/Ds and D/As as memory, and communication protocol buffers also are allocated in memory, hence, without loss of generality, it can be assumed that in each case, communication is performed by reading and writing to memory locations. To switch from simulation to execution by a microprocessor, only the simulated inputs and outputs need to be swapped for memory access blocks.

At this stage, the C code can be automatically generated. The *MATLAB Coder* [18] and *Simulink Coder* [19] packages are required to translate a Simulink diagram into equivalent C procedures. The generated code contains three sets of procedures: initialisation procedures executed once when the calculations begin, solver procedures that calculate the model at each sampling period, and termination procedures. Usually, the metering algorithms do not have a specific operating time, so the termination procedures are never called.

The generated code is functionally equivalent to the Simulink diagram. The solver procedures read data from the defined memory location, carry out the calculations, and store the results to the given memory addresses. In the case of a PQ metering application, the Simulink diagrams implement calculations for root mean square values (RMS), spectrum, asymmetry, and flicker, resulting in the automatic generation of a PQ measurement library.

The C procedures, generated from Simulink diagrams, are used to create tasks in a real-time operating system (RTOS). The data comes directly from A/D converters with a sampling frequency of 10.24 kHz. The A/Ds measure phase voltages and line currents, and the A/D results are stored in First In-First Out (FIFO) queues that trigger the execution of the tasks. The FIFO is a buffer where samples are stored until the following tasks become ready for execution, synchronizing the data source to the data consumer, and preventing any measurement points from being lost. The tasks read voltages and currents by accessing the FIFO memory and perform calculations for every incoming data sample. Inter-task communication is performed over memory locations, as was defined in the Simulink diagrams.

An alternative to the C-code generation path, the HDL Coder [20] package converts Simulink diagrams into Verilog or VHDL descriptions (see [16] and [17] as examples).

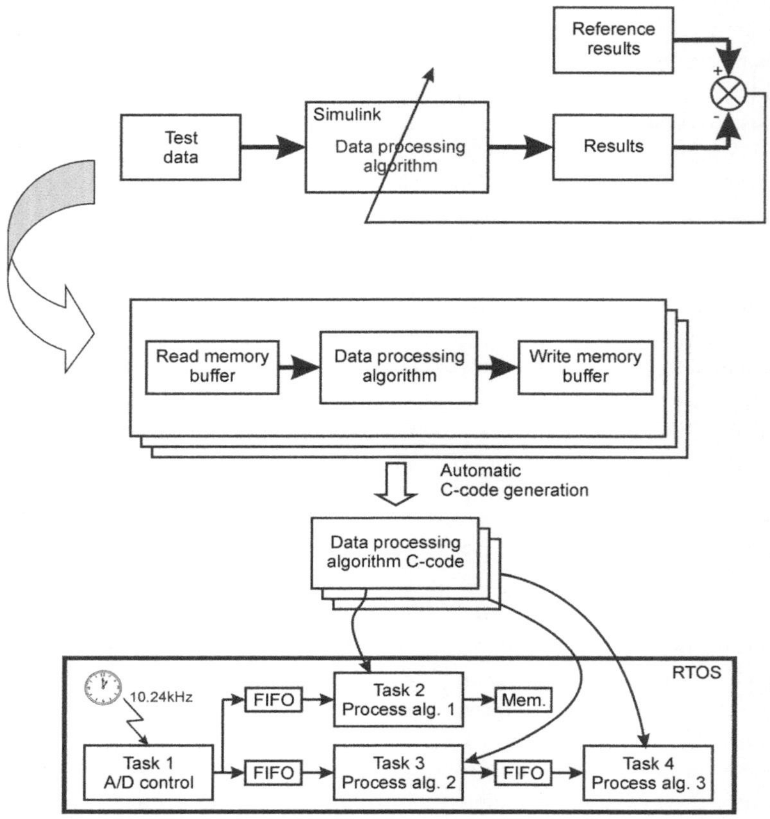

Figure 1 Stages of rapid prototyping approach.

HDL code is synthesizable and can be implemented as an FPGA module. However, unlike C-code procedures, HDL modules may not perform floating point calculations, and the algorithms must be converted into a fixed-point representation. The *Fixed-Point Designer* [21] toolbox supports this conversion, however, the conversion is usually time-consuming and sometimes may not even be possible.

1.1.2 Hardware platform

The Mars Starter Kit evaluation board from Enclustra was selected as the test hardware platform [22]. The board contains Zynq integrated circuits from Xilinx. The Zynq units contain within the single integrated circuit a dual-core ARM Cortex A9 processor equipped with the double precision floating point extension, reconfigurable FPGA fabric equipped with gates, registers, and RAM resources as well as communication cores (see Figure 2). An external A/D converter board is connected to measure grid voltages and currents. The board is equipped with Analog Devices AD7980 1MSPS, 16-bit, successive approximation analogue-to-digital converters. The A/Ds are controlled by an IP (intellectual property) core implemented in the FPGA. The core is developed in the VHDL language and operates as a component of the FPGA, as a kind of simple coprocessor. The A/D converters require the sequence of control signals to be generated with accuracy on the order of a few dozen of nanoseconds. Such accuracy cannot be achieved in programming, and has to be implemented as a silicon circuit.

One ARM processor runs the FreeRTOS real-time operating kernel. FreeRTOS tasks are created directly from the code generated automatically from Simulink diagrams. The second ARM processor runs a Linux distribution and is responsible mainly for implementing communication with the device via Ethernet.

The laboratory prototype of the flickermeter is shown in Figure 3. It consists of two boards—one equipped with the Zynq, memory, interfaces, and connectors to the communication channels. The second one operates as an interface to the analogue inputs.

The prototype has compact dimensions, but also has enough computational power to implement the chosen measurement algorithms. The worst case CPU utilisation of the ARM 1 processor is 7% when running the synchronisation and flicker calculation algorithms. The

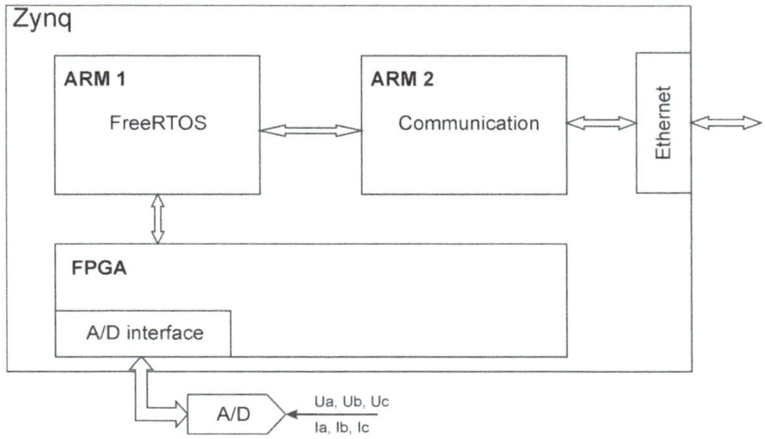

Figure 2 Zynq hardware platform.

average ARM 2 utilisation is 5% when running the Modbus/TCP server in the Linux environment.

1.2 Case studies

1.2.1 Synchronization with line voltage

A significant part of the PQ metering process is synchronisation with the line voltage. The PQ metering standards, namely EN 61000-4-30 and EN 61000-4-7, define strict requirements in this area. In particular, all quantities should be measured within 10 cycle intervals for 50 Hz systems (for 60 Hz systems 12 cycles are required). In addition, RMS measurements for event detection and flicker reference values must be measured by means of a sliding one cycle interval that moves in half-cycle steps. Consequently, correct detection of a cycle is crucial. A typical phase locked loop (PLL) could be used for this task; however, it can be prone to voltage disturbances (e.g. harmonic distortion) or events (e.g. voltage dips). For three phase systems, the asymmetry could also affect a typical PLL system. Therefore, a robust implementation of the PLL is an important task in order to make a PQ meter operate correctly.

The PLL technique is a subject of great interest, and many PLL architectures and applications have been presented (e.g. [23-26]). The Enhanced PLL (EPLL) contains an adaptive notch filter and provides a higher degree of insensitivity to variable-frequency input signals. A detailed description of the EPLL structure is given in [23,27-29]. This structure has been selected to demonstrate the proposed rapid implementation method.

Figure 4 presents the Simulink diagram of the EPLL. The *Reference sin generator* block is the source of test data (see Figure 1). The *RefSin* output generates the sine wave for which amplitude, phase, and frequency are disturbed. The output of the EPLL is a sine signal denoted as *u(t)*. The EPLL also outputs the detected amplitude, frequency, and phase, marked respectively as *A(t)*, *w0(t)*, and *phase(t)*. The *Compare* scope acquires the simulation

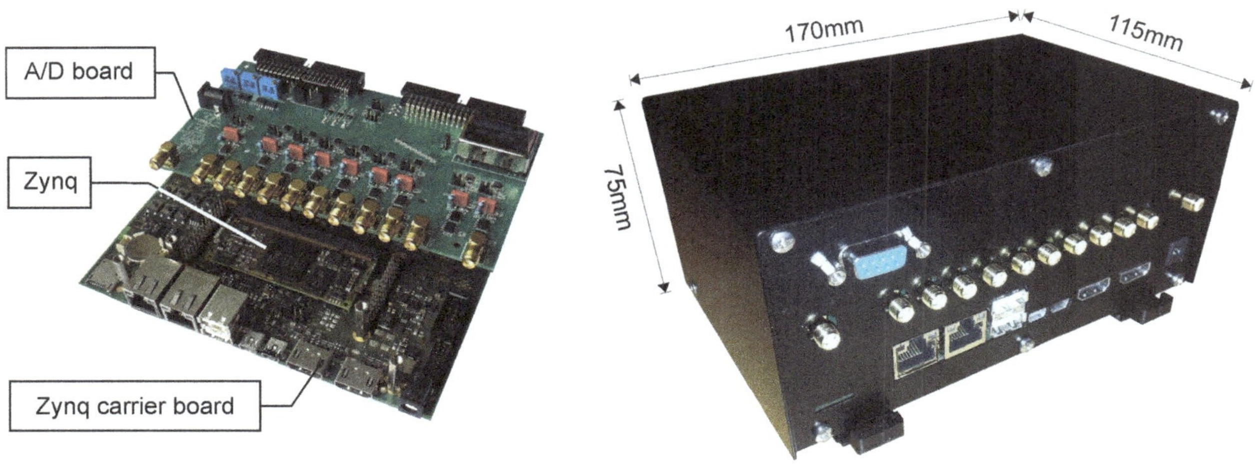

Figure 3 Laboratory flickermeter prototype.

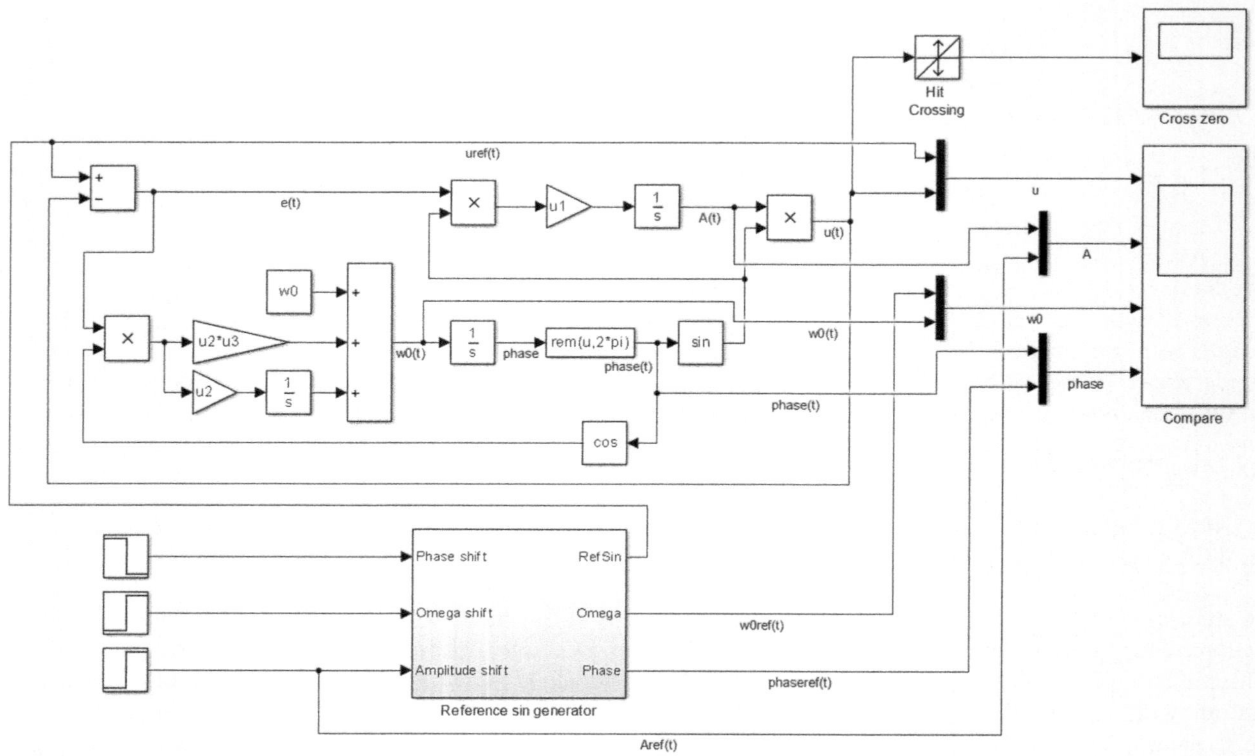

Figure 4 EPLL - simulation diagram.

results and is used to compare the response of the EPLL to the reference signals.

The *Hit Crossing* block detects the zero crossing points of the sine signal generated by the EPLL. It is used for synchronising the PQ measurement algorithms to the power signal. Correct detection of the amplitude, frequency, and phase by the EPLL indicate the synchronous operation of the loop. The difference between the reference sine signal *uref(t)* and the EPLL output *u(t)* is used to tune the loop. However, only the zero crossing signal is used for the measurement algorithms.

The EPLL tuning requires calculation of the *u1, u2,* and *u3* gains (see Figure 4). The parameter values were determined by numerical optimization of the loop response to the test excitation. The response to a step change in amplitude, frequency, and phase was tested. The following cost sub-functions were applied:

$$J = \begin{bmatrix} \sqrt{\sum_{i=0}^{N-1} (u(iT_0) - u_{\text{ref}}(iT_0))^2} \\ \sqrt{\sum_{i=0}^{N-1} (A(iT_0) - A_{\text{ref}}(iT_0))^2} \\ \sqrt{\sum_{i=0}^{N-1} (w(iT_0) - w_{\text{ref}}(iT_0))^2} \\ \sqrt{\sum_{i=0}^{N-1} (\text{phase}(iT_0) - \text{phase}_{\text{ref}}(iT_0))^2} \end{bmatrix}$$

where T_0 is the sampling period, u_{ref} and u are respectively input and output sine signals of the loop, A_{ref} and A are respectively reference and measured amplitudes, w_{ref} and w are respectively reference and measured pulsation values, $phase_{ref}$ and $phase$ are respectively reference and measured phase values, and N is the number of samples during the numerical optimisation experiment. The sampling frequency was set to 10.24 kHz. The optimisation procedure can be defined as:

$$[u_{1\text{opt}}, u_{2\text{opt}}, u_{3\text{opt}}] = \min_{u_1, u_2, u_3} W * J$$

where W is a four element row vector of non-negative constants. The optimisation is performed in the MATLAB environment by calling the *fminsearch* function [30].

The selection of the W vector elements determines the optimal $u_{1\text{opt}}, u_{2\text{opt}}$ and $u_{3\text{opt}}$ values and finally the behaviour of the EPLL. For the flicker measurement, the EPLL is used to synchronise the algorithm to the grid signal when the loop output signal $u(iT_0)$ follows the input signal $u_{\text{ref}}(iT_0)$. The optimal loop parameters are calculated for W such that only the first element is a non-negative value and the remaining elements are set to zero. During the optimisation, the reference sine signal is disturbed by step phase, frequency, and amplitude changes. The calculated optimal values are:

$$[u_{1opt}, u_{2opt}, u_{3opt}] = [3\,919.07,\ \ 147.7206,\ \ 208.1230]$$

For test purposes, the optimised loop is applied not to the simulated signal but to synchronise to the real grid signal measured in the university power laboratory. The total harmonic distortion (THD) of the signal is 2.95% and, additionally, the signal is disturbed by a $-\pi/6$ rad step phase change, 5% frequency step, and 30% amplitude sag. The disturbances of the phase, frequency, and amplitude appear respectively at 60, 120, and 180 ms. The results are shown in Figure 5. The upper diagram presents input test signal u_{ref} and the sinusoidal response of loop u. The bottom left picture shows the difference between u_{ref} and u. The time period between the moments when the u_{ref} and u signals cross the zero voltage level is given in the bottom right diagram.

The miscellaneous sets of vector W values enable us to switch the EPLL into amplitude, frequency, or phase measurement modes and exemplify the flexibility of the proposed approach.

After simulation and optimisation phases, the test signal sources are replaced by memory access drivers. The updated diagram is given in Figure 6. The input driver *Ua* and the output drivers *Cross zero, EPLL sin, Amplitude, Omega,* and *Phase* are created as Simulink level-2 C-code S-functions [31]. The S-functions access memory buffers and operate as interfaces between the Simulink blocks and to the remaining flickermeter blocks.

The diagram is used to automatically generate code. The generation procedures call the input S-function to read the input signal from the memory hexadecimal address FCA0000, run the EPLL algorithm, and call the output S-functions to store the results to memory locations from EA000000 to EA000010. When running the generated software in real time, it is essential that the execution takes place with the same sampling period as during code generation.

The steps presented during the EPLL implementation can be applied to other algorithms. The next section presents the rapid development path of the flicker measurement algorithm. However this approach can also be applied to the development of the remaining PQ algorithms, and even to the development of control strategies.

1.2.2 Flicker implementation

A nontrivial example of rapid prototyping and implementation of a metering problem is the flickermeter. Light flicker measurement and the flickermeter are described in detail in standard EN 61000-4-15. Basically, the meter consists of several blocks (see Figure 7(a)).

The flickermeter is described in the standard as an analogue device, i.e. all signal processing blocks are

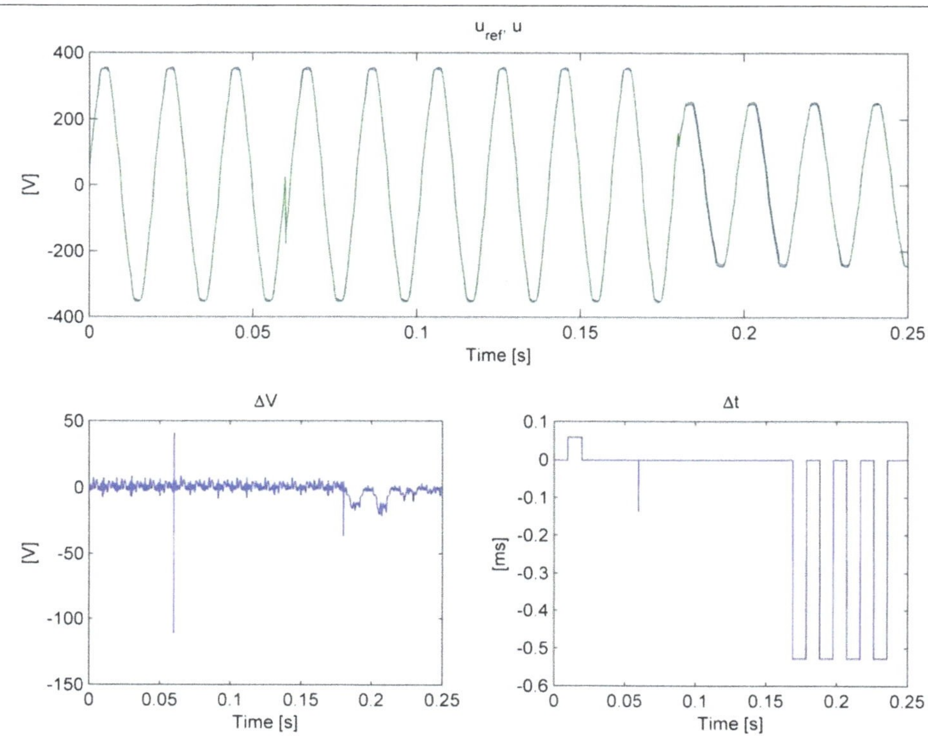

Figure 5 EPLL synchronisation to a real signal.

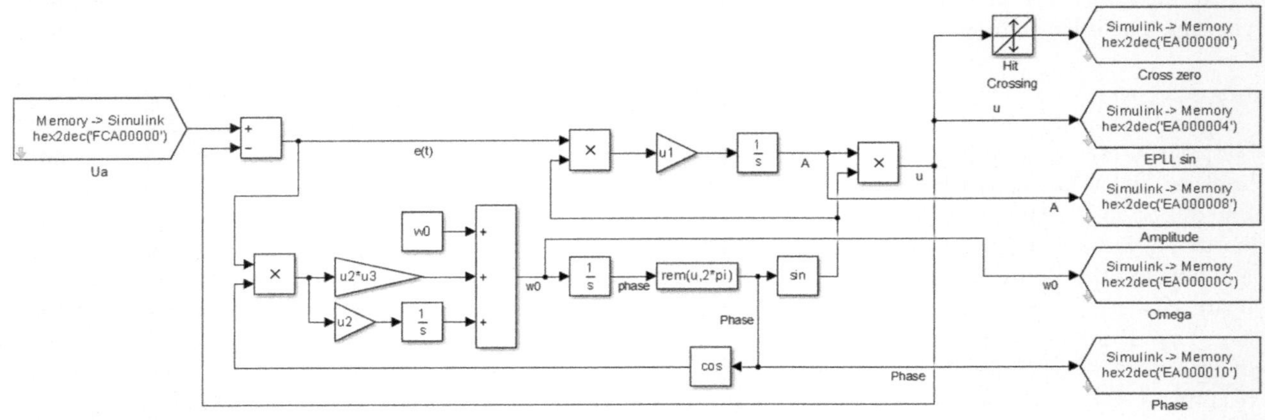

Figure 6 **EPLL - diagram prepared for code generation.**

defined in the analogue domain. The reason for this is that at the time of the standard creation, the flickermeter was intended to be single device built using standard electronic parts. Nowadays, the flicker severity measurement is a function of a digital meter (PQ analyser) but the algorithms for computing short-term flicker P_{ST} and long-term flicker P_{LT} must follow the description in the standard. In state-of-the-art devices, all the operations on the input signal are done in the digital domain. Consequently, the problem of implementing a flickermeter in a digital processor requires discretisation of the analogue blocks and implementation of the discrete transfer functions. The considered blocks are (see Figure 7(a)):

1. Input gain control block: in the standard, the block operates as a voltage adaptor and signal conditioning unit and provides initial scaling. In the digital model, the block provides only the scaling, which is done by dividing the input signal by the reference signal. The reference signal is the half-period RMS value processed by the first order digital filter with a 27.3 s time constant (as defined in the standard). The filter has to be designed and implemented in the digital domain.
2. Square multiplier block: the squaring operation resembles the behaviour of an incandescent lamp. The block, together with the high pass filter from the next block, operates as a demodulator. A digital implementation of this algorithm is straightforward.
3. High pass filter and the weighting filter: the blocks consist of three filters. The first high pass filter eliminates the DC component of the voltage, and the standard suggests a first order filter with a –3 dB at 0.05 Hz cut-off frequency. The second filter is a low pass one, and the standard requires a sixth order Butterworth filter with a –3 dB at 35 Hz cut-off

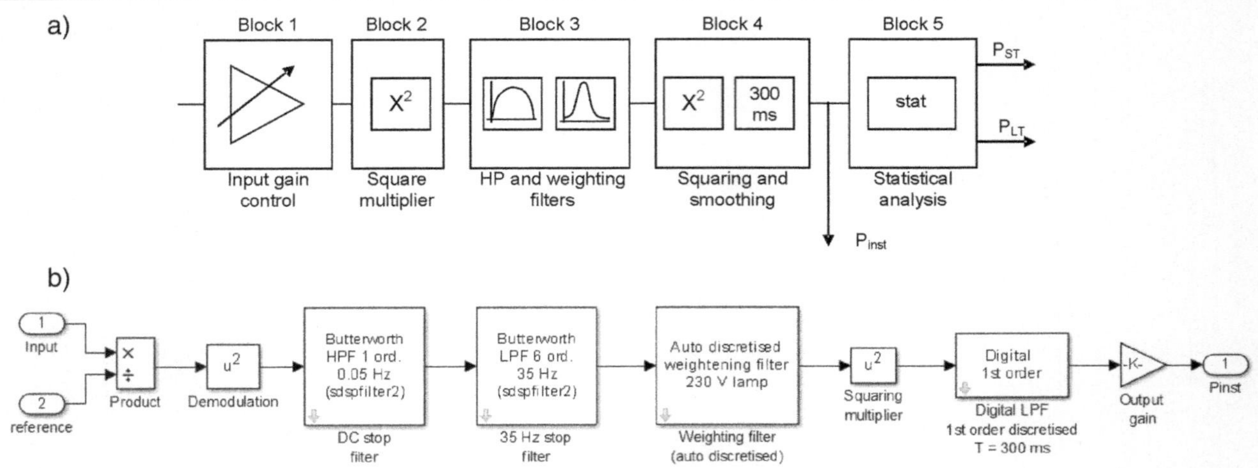

Figure 7 **Block diagram and its Simulink model for instantaneous flicker computation.** Block diagram of the IEC flickermeter according to the standard EN 61000-4-15 **(a)** and its sample model for instantaneous flicker computation in MATLAB/Simulink **(b)**.

frequency (for 230 V systems). A digital implementation of the filter requires utilisation of filter design tools in order to evaluate the parameters of the filter. The last filter is the co-called weighting filter. It simulates the spectral properties of the human visual system. The filter is described as a Laplace transfer function model:

$$G(s) = \frac{k\omega_1 s}{s^2 + 2\lambda s + \omega_1^2} \cdot \frac{1 + s/\omega_2}{(1 + s/\omega_3)(1 + s/\omega_4)}$$

where the parameters k, λ, and ω_i ($i = 1\ldots4$) are described in the standard and are different for 230 V and 110 V systems. Therefore the first operation is to obtain the discrete transfer function so the frequency response of the analogue filter and the discrete implementation are similar in the frequency domain.

4. Squaring and smoothing: this block performs two functions, it simulates the eye-brain perception by means of a squaring operation and emulates the memory effect of the brain. The effects can be implemented as the sliding mean operator, however the standard suggests the utilisation of a first order filter with a 300 ms time constant. The output of this block is instantaneous flicker P_{inst}.
5. Statistical analysis: this is the only block that utilises digital data processing to obtain its output: P_{ST} and P_{LT} flicker levels. The computation is based on the percentiles computation and calculates the flicker according to the formula:

$$P_{\text{ST}} = \sqrt{0.0314P_{0.1} + 0.0525P_{1s} + 0.0657P_{3s} + 0.28P_{10s} + 0.08P_{50s}}$$

where:

$$P_{50s} = (P_{30} + P_{50} + P_{80})/3$$
$$P_{10s} = (P_6 + P_8 + P_{10} + P_{13} + P_{17})/5$$
$$P_{3s} = (P_{2.2} + P_3 + P_4)/3$$
$$P_{1s} = (P_{0.7} + P_1 + P_{1.5})/3$$

and P_x is the x-th percentile of the values of P_{inst} logged during a specified time interval, where x is 0.1, 0.7, 1, 1.5, 2.2, 3, 4, 6, 8, 10, 13, 17, 30, 50, 80. The interval can vary from 1 to 15 m, however, the EN 61000-4-30 standard assumes 10 m for a typical flicker severity evaluation. This conforms with the interval set by EN 61000-4-30 for PQ evaluation. The evaluation of P_{LT} is performed using 12 samples of P_{ST}, hence the long-term flicker describes the flicker severity for the last 2 h. Because the block is actually a discrete implementation in a digital processor, it requires numerical procedures for sorting and mathematical calculation.

The MATLAB/Simulink package provides toolboxes and blocksets for both digital and analogue system design and analysis. Therefore, implementation of the flickermeter in MATLAB/Simulink follows the development stages of the rapid development method. The basic structure of the Simulink diagram that implements the signal processing stages is shown in Figure 7(b). Note that the model closely resembles the description in the standard. Each filter defined by spectral properties is defined in the z domain by means of standard tools available in the MATLAB/Simulink package. The weighting filter requires special procedures. In [11], the transfer functions of the filters were obtained analytically by defining the filter in the s domain and next using a Bilinear Transform to obtain the transfer function in the z domain. This procedure does not match the rapid implementation approach. Filter parameters in the flickermeter blocks are strictly defined in the standard, but their selection depends on the voltage level. On the other hand, there are several discretisation methods and the choice can affect the properties of the flickermeter. The discretisation also depends on sample time, which is related to the hardware platform, and therefore can be unspecified at modelling time. Consequently, the best option is to discretise the filter numerically and let the software perform all necessary computations. The MATLAB/Simulink package supports this operation by providing commands for the immediate discretisation of transfer functions. This functionality is used to obtain a fully automatic discretisation of the weighting filter. Therefore, the filter is defined in the s domain and then numerically discretised using MATLAB commands.

The validation of the model and recalibration can also be done in the simulation environment. It requires additional blocks for test signal generation. The test signals are strictly defined in the EN 61000-4-15 standard. Figure 8 shows a sample flickermeter validation framework.

As in the case of the EPLL implementation in the flickermeter diagram, the signals are reconnected to the driver blocks to access the measurements and store the P_{ST} and P_{LT} values. The C code is automatically generated from the Simulink diagram, compiled, and executed in real-time by the Zynq hardware platform.

1.2.3 Results of the flicker algorithm

The results of the generated flickermeter are shown in Figure 9. The diagram presents 24 hours of measurements. The upper diagram presents two P_{ST} values: one generated by the reference class A PQ analyser (A-eberle PQI-D) and the second generated by the test hardware platform described in Section II B. The Zynq test device ran the code generated automatically from the Simulink diagram. The reference PQ meter is a class A meter (according to IEC 61000-4-30) with the metering uncertainty defined by the standard, e.g. for voltage, the

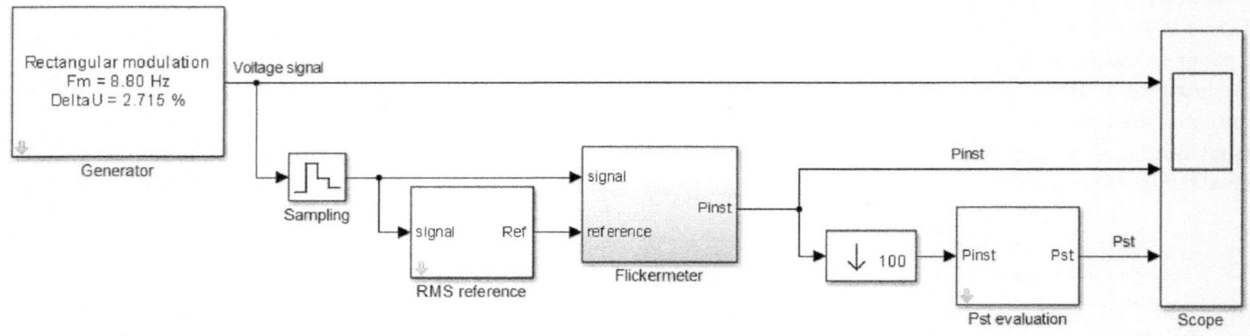

Figure 8 The flickermeter model testing framework.

uncertainty is ± 0.1% of declared voltage (i.e. the grid's nominal). Flicker uncertainty is 5% and is defined by IEC 61000-4-15. For the test device, the metering was done by means of voltage transducers of type LV 25-P [32]. According to the data sheet, the overall accuracy of the transducers is 0.9% of the voltage range, which is 400 V in this case.

The meters were arranged in the PQ metering laboratory in which the building mains was available for metering. The meters were connected to the same voltage signals so the comparison refers to the real metering situation.

The lower diagram presents the difference between the data series. The difference can be explained by the accuracy of the sensors. For the hardware platform, typical off-shelf transducers were used with a conditional board not designed for precise measurements. On the other hand, the typical error allowed in the standard for P_{ST} is 5% when $P_{ST} = 1$.

2 Conclusions

This article presented a method of rapid code generation for implementing and testing of PQ algorithms. The method covers not only the PQ meter development, but

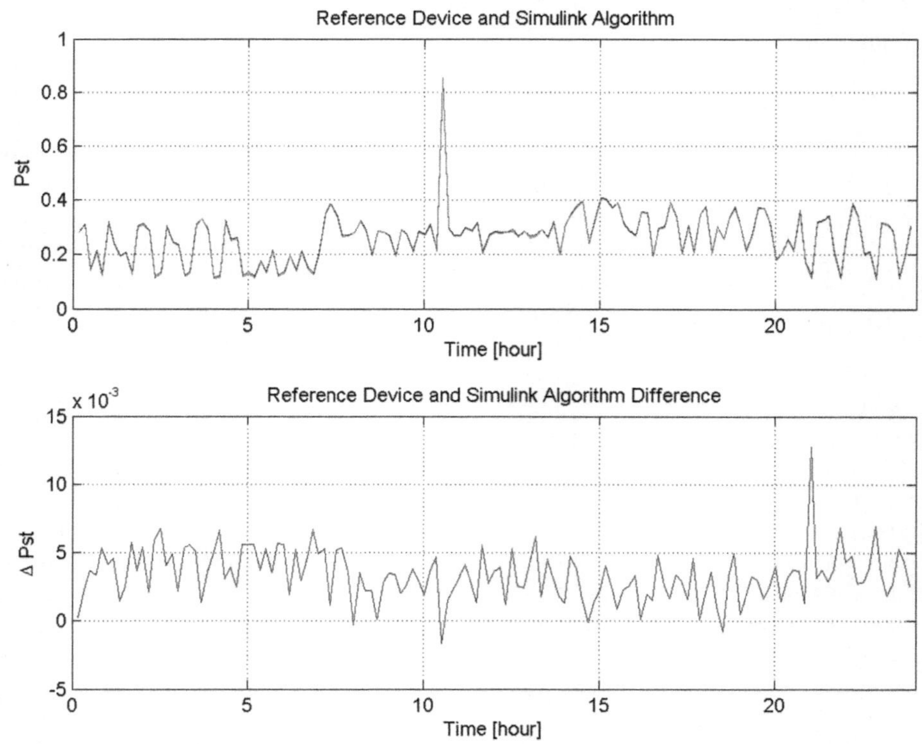

Figure 9 The flickermeter results.

also the preparation of the certification tests and procedures. The MBD approach shortens the time from an idea to an operating device. A user is supported by the functions of a CAE package, can focus on the problem itself, and skip the laborious implementation details of the algorithms under development. In the presented case, Simulink was used as the CAE platform. The efficiency of the presented development path becomes apparent in the implementation of the flicker calculation algorithm. The flickermeter channel consists of a series of filters and data processing blocks. It was first developed and validated in simulation, the C code was then generated and implemented on a real hardware platform. The correctness of the implemented algorithm was confirmed by the match of the generated flicker values with the results of the commercial reference metering device.

PQ measurements have recently become an important topic that has resulted in different regulatory documents. An example is the differences between the EN 50160 standard for supply quality and national regulations. The standard requires e.g. RMS voltage within ± 10% limits in relation to the nominal (or declared) value for low voltage networks. However, in some countries, the limits are different, e.g. in France, the admissible change is from −10% to +6%, and in Spain, it is ± 7%. In addition, the statistical measure differs: the standard requires computation of the 95 and 100 percentiles. The grid codes in Poland for a transmission system require a 99 percentile for flicker and voltage distortion when a wind farm connection is considered. There are also parameters not required in the standard but required in the grid codes, e.g. the THFF (telephone harmonic form factor). The metering interval is also subject to change, e.g. in Norway, a 1 minute interval is used. On the other hand, there is the possibility of change of the limit values, percentile, or even metering interval in the future. This requires additional work in the software of a PQ meter to make the change possible. The real problem arises when non-standard algorithms are required, e.g. the localization of a PQ disturbance source. This functionality is very useful when the introduction of PQ contracts is considered. However, the existing algorithms are not ready for implementation because of lack of efficiency.

The ability to easily add a new algorithm with new functionality to a PQ meter is clearly necessary. Rapid prototyping and implementation makes it possible to test the algorithm and produce the code without manual implementation.

This study highlights the need for a flexible and rapid development method for devices to meet the as yet non-existing standards. The presented approach is a step towards satisfying these demands.

Competing interests
The authors declare that they have no competing interests.

Acknowledgements
This work was supported by the AGH University of Science and Technology under European Regional Development Found, Subsidy no POiG.01.03.01-30-056/12 and by the AGH University of Science and Technology KIC-ASS Grant no 7.7.120.7037.

References
1. MHJ Bollen, IYH Gu, *Signal processing of power quality disturbances* (IEEE Press, New York, NY, USA, 2006)
2. A Zobaa, MM Canteli, R Bansal (eds.), *Application of Signal Processing in Power Quality Monitoring*, in *Power Quality Monitoring, Analysis and Enhancement*) (InTechOpen, 2012). doi:10.5772/21492
3. RV Dell'Aquila, L Balboni, R Morici, *A new approach: modeling, simulation, development and implementation of a commercial grid-connected transformerless PV inverter, SPEEDAM International Symposium on Power Electronics, Electrical Drives, Automation and Motion, 1422–1429*, 2010. doi:10.1109/SPEEDAM.2010.5542040
4. X Yang, B Kirby, Q Zhao, Y Ma, F Xu, *Model-based design process for product development of substation IEDs, 2nd IEEE International Conference & Exhibition (ENERGYCON)*, 2012, pp. 968–974. doi:10.1109/EnergyCon.2012.6348290
5. G Adam, A Zbanţ, G Livint, *New Simulink control block for single phase shunt active power filters, 8th International Symposium on Advanced Topics in Electrical Engineering (ATEE)*, 2013, pp. 1–4. doi:10.1109/ATEE.2013.6563485
6. A Kirubakaran, S Jain, RK Nema, DSP-controlled power electronic interface for fuel-cell-based distributed generation. IEEE Trans. Power Electron. **26**(12), 3853–3864 (2011). doi:10.1109/TPEL.2011.2138162
7. K Vardar, T Sürgevil, E Akpinar, *Rapid prototyping applications on three-phase PWM rectifier and shunt active power filter*. (International Conference on Electrical and Electronics Engineering, 2009), p. 258–262
8. P Meena, KU Rao, D Ravishankar, *Real-time detection and analysis of PQ disturbances with DSP using MATLAB embedded link to code composer studio*. (Third International Conference on Power Systems, 2009), p. 1–5. doi:10.1109/ICPWS.2009.5442725
9. K Kołek, K Piątek, P Włodarczyk, Rapid algorithm prototyping for SCADA applications, (Advances in Electrical and Computer Engineering, in press)
10. MF McGranaghan, S Santoso, Challenges and trends in analyses of electric power quality measurement data. EURASIP J. Adv. Signal Process **57985**, 1–6 (2007). doi:10.1155/2007/57985
11. LW White, S Bhattacharya, A Discrete, MATLAB-Simulink flickermeter model for power quality studies. IEEE Trans. Instrum. Meas. **59**(3), 527–533 (2010). doi:10.1109/TIM.2009.2023121
12. MI Milanes, V Minambres, E Romero, F Barrero, *Quality meter of electric power systems based on IEEE standard 1459-2000*. (Compatibility and Power Electronics, 2009), p. 86–92. doi:10.1109/CPE.2009.5156018
13. M Rogóż, Z Hanzelka, *The design and construction of a power quality parameters recorder*. 9th International Conference. (Electrical Power Quality and Utilisation, 2007), p. 1–6. doi:10.1109/EPQU.2007.4424150
14. J Krizan, L Ertl, M Bradac, M Jasansky, A Andreev, *Automatic code generation from MATLAB/Simulink for critical applications*. (IEEE 27th Canadian Conference on Electrical and Computer Engineering (CCECE), 2014), p. 1–6. doi:10.1109/CCECE.2014.6901058
15. MH Schwarz, H Sheng, A Sheleh, J Boercsoek, *IEEE/ACS International Conference on Computer Systems and Applications, 1058 – 1063 (2008)*, 2008, pp. 1058–1063. doi:10.1109/AICCSA.2008.4493678
16. ET Mekonnen, J Katcha, K Parker, *An FPGA-based digital control development method for power electronics*. (IECON 38th Annual Conference on IEEE Industrial Electronics Society, 2012), p. 222–226. doi:10.1109/IECON.2012.6388804
17. R Selvamuthukumaran, R Gupta, Rapid prototyping of power electronics converters for photovoltaic system application using Xilinx System Generator. IET Power Electron. **7**(9), 2269–2278 (2014). doi:10.1049/iet-pel.2013.0736
18. MATLAB Coder. User's guide, http://www.mathworks.com/help/pdf_doc/coder/coder_ug.pdf. Accessed 5 December 2014
19. Simulink Coder. User's guide, http://www.mathworks.com/help/pdf_doc/rtw/rtw_ug.pdf. Accessed 5 December 2014
20. HDL coder, http://www.mathworks.com/products/hdl-coder/. Accessed 5 December 2014.

21. Fixed-point designer. User's guide, http://www.mathworks.com/help/releases/R2014b/pdf_doc/fixedpoint/FPTUG.pdf. Accessed 5 December 2014.
22. Mars Starter Kits, http://www.enclustra.com/en/products/hardware-kits/mars-starter-kits/. Accessed 18 October 2014.
23. A Gupta, A Porippireddi, VU Srinivasa, A Sharma, M Kadam, Comparative study of single phase PLL algorithms for grid synchronization applications. IJECT **3**(4), 237–245 (2012)
24. S Shinnaka, A Robust, Single-phase PLL system with stable and fast tracking. IEEE Trans. Ind. Appl. **44**(2), 624–633 (2008). doi:10.1109/TIA.2008.916750
25. M Karimi-Ghartemani, M Mojiri, A Safaee, JA Walseth, SA Khajehoddin, P Jain, A Bakhshai, A new phase-locked loop system for three-phase applications. IEEE Trans. Power Electron. **28**(3), 1208–1218 (2013). doi:10.1109/TPEL.2012.2207967
26. XQ Guo, W Wy, HR Gu, Phase locked loop and synchronization methods for grid-interfaced converters: a review. PRZEGLĄD ELEKTROTECHNICZNY (Electric Rev) **87**(4), 182–187 (2011)
27. B Singh, SR Arya, Implementation of single-phase enhanced phase-locked loop-based control algorithm for three-phase DSTATCOM. IEEE Trans. Power. Deliv. **28**(3), 1516–1524 (2013). doi:10.1109/TPWRD.2013.2257876
28. M Karimi-Ghartemani, JA Walseth, *Using the EPLL algorithm as a preprocessor for fault analysis*. (11th International Conference on Information Science, Signal Processing and their Applications (ISSPA), 2012), p. 1377–1382. doi:10.1109/ISSPA.2012.6310508
29. M Karimi-Ghartemani, M Karimi-Ghartemani, Linear and pseudolinear enhanced phased-locked loop (EPLL) structures. IEEE Trans. Ind. Electron. **61**(3), 1464–1474 (2014). doi:10.1109/TIE.2013.2261035
30. Optimization toolbox user's guide, MathWorks, http://www.mathworks.com/help/pdf_doc/optim/optim_tb.pdf. Accessed 18 October 2014
31. Simulink. Developing S-functions, MathWorks, http://www.mathworks.com/help/pdf_doc/simulink/sfunctions.pdf. Accessed 04 December 2014
32. Voltage Transducer LV 25-P, http://www.lem.com/docs/products/lv%2025-p.pdf. Accessed 04 December 2014.

9

Analysis, optimization, and implementation of a hybrid DS/FFH spread-spectrum technique for smart grid communications

Mohammed M Olama[1*], Xiao Ma[2], Stephen M Killough[1], Teja Kuruganti[1], Stephen F Smith[1] and Seddik M Djouadi[2,3]

Abstract

In recent years, there has been great interest in using hybrid spread-spectrum (HSS) techniques for commercial applications, particularly in the Smart Grid, in addition to their inherent uses in military communications. This is because HSS can accommodate high data rates with high link integrity, even in the presence of significant multipath effects and interfering signals. A highly useful form of this transmission technique for many types of command, control, and sensing applications is the specific code-related combination of standard direct sequence modulation with 'fast' frequency hopping, denoted hybrid DS/FFH, wherein multiple frequency hops occur within a single data-bit time. In this paper, error-probability analyses are performed for a hybrid DS/FFH system over standard Gaussian and fading-type channels, progressively including the effects from wide- and partial-band jamming, multi-user interference, and varying degrees of Rayleigh and Rician fading. In addition, an optimization approach is formulated that minimizes the bit-error performance of a hybrid DS/FFH communication system and solves for the resulting system design parameters. The optimization objective function is non-convex and can be solved by applying the Karush-Kuhn-Tucker conditions. We also present our efforts toward exploring the design, implementation, and evaluation of a hybrid DS/FFH radio transceiver using a single field-programmable gate array (FPGA). Numerical and experimental results are presented under widely varying design parameters to demonstrate the adaptability of the waveform for varied harsh smart grid RF signal environments.

Keywords: Hybrid spread-spectrum; Direct sequence; Frequency hopping; Smart grid communications; Non-convex optimization; Receiver sensitivity; FPGA

1 Introduction

Hybrid spread-spectrum (HSS) systems, which combine direct-sequence (DS) and frequency-hopping (FH) spread-spectrum (SS) techniques, are attractive for their strong multiple-access capabilities, resistance to multipath fading and intentional/unintentional jamming, and the security they provide against eavesdroppers [1-6]. In recent years, there has been great interest in using HSS systems for commercial applications, particularly in the Smart Grid.

User requirements for the next generation wireless communication system have been specified for the Smart Grid advanced metering infrastructure (AMI) and distribution automation systems [7]. These requirements demonstrate the need for high capacity and highly secure networks for Smart Grid applications. There is a significant gap between commercially available communications systems and those needed to satisfy the demanding requirements associated with electric utility industry. HSS systems are a promising candidate for Smart Grid applications since they provide high data rates with excellent signal security.

Spreading the signal over a relatively wide bandwidth allows transmission with relatively low power density, leading to low probabilities of detection and interception. HSS systems also provide an inherent security against eavesdroppers because knowledge of the spreading codes is required. The choice of appropriate pseudo-noise (PN) codes and dynamic altering of signal parameters provides the opportunity for a strong security scheme in the

* Correspondence: olamahussemm@ornl.gov
[1]Computational Sciences and Engineering Division, Oak Ridge National Laboratory, P.O. Box 2008, MS 6085, Oak Ridge, TN 37831, USA
Full list of author information is available at the end of the article

physical (PHY) layer of the network [5]; details of these techniques will be addressed in future works. This specific paper will focus on implementation, exploration, and optimization of the parameter space of the HSS system for adapting the technique for application-level requirements in Smart Grid.

Based on the hopping rate, an HSS system is classified into a hybrid direct-sequence/slow frequency hopping (DS/SFH) system or a hybrid direct-sequence/fast frequency hopping (DS/FFH) version. In hybrid DS/FFH systems, multiple frequency hops occur within a single data-bit time. Specifically, each bit is represented by chip transmissions at multiple frequencies. If one or more chips are corrupted by multipath or interference in the RF link, statistically a majority should still be correct. Standard or slow frequency hopping, in contrast, transmits at least one (and usually several) data bits in each hopping interval. DS/FFH systems have not been previously widely implemented in many commercial or industrial applications since fast frequency-hopping rates were limited by the technology of frequency synthesizers. Today's extremely fast hopping speed direct-digital synthesizers (DDSs) [8] are rapidly becoming an alternative to the traditional frequency-agile analog-based phase-locked loop (PLL) synthesizers. Output frequencies with micro-Hertz resolution and sub-degree phase tuning capabilities can thus be readily achieved using a single integrated circuit (IC).

Most of the works related to HSS in the literature have addressed evaluating its performance under different modulation techniques [2], channel conditions [1,3], multi-user interference [2,3], and jamming [4]. However, little research has yet evaluated the performance of a hybrid DS/FFH system under all combinations of the aforementioned cases. Moreover, few efforts have to date attempted to address the design and selection of the HSS system parameters that achieve optimal performance. The work in this paper extends the one in [9] and [10] from a DS system to a hybrid DS/FFH system, in addition to taking jamming impacts into consideration. In [11], the performance of a SFH system was considered. In [2] and [12], the performance of a DS/SFH system over an AWGN channel and with multi-user interference was considered. The performance of an FFH system over fading channels was examined in [13] and extended in [3] to include the effects of partial-band noise jamming. Although [4] and [14] computed the error probability of DS/SFH under jamming tones in both AWGN and Rician fading channels, only a single user was considered. In [15], the optimal spreading sequences for chip-synchronous CDMA are derived by minimizing the average bit error rate under the standard-Gaussian-approximation condition. The work in [16] presents a simulation-based study for evaluating the performance of a hybrid DS/FFH scheme. Some preliminary performance analysis and hardware designs for the hybrid DS/FFH scheme were initially presented in [17-19].

In this paper, error-probability analyses are performed for a hybrid DS/FFH system over standard Gaussian and fading-type channels, progressively including the effects from wide- and partial-band jamming, multi-user interference, and varying degrees of Rayleigh and Rician multipath fading. We present analytical derivations for evaluating the performance in terms of probability of bit error. In addition, an optimization approach is formulated that minimizes the average bit-error probability of a hybrid DS/FFH communication system and solves for the system design parameters that achieve an optimal performance level. The optimization objective function is non-convex and can be solved by applying the Karush-Kuhn-Tucker (KKT) conditions [20]. We also present our efforts toward exploring the design, implementation, and evaluation of a hybrid DS/FFH radio transceiver using a single field-programmable gate array (FPGA). Numerical and experimental results are presented under widely varying design parameters to demonstrate the adaptability of the waveform for varied harsh smart grid RF signal environments.

2 System model

Assume that there are a total of K nodes that represent smart meters or data aggregation points in the Smart Grid wireless network. For the kth node, the transmitted signal is given as

$$s_k(t) = \sqrt{2P}b_k(t)a_k(t)cos\{2\pi(f_c + f_h^k(t))t\} \quad (1)$$

where P is the common transmitted signal power, f_c is the carrier frequency, $\{f_h^k(t)\}$ denotes the hopping frequency of the kth node, the data signal $b_k(t)$ is a sequence of statistically independent, unit-amplitude positive, and negative rectangular pulses of duration T_b, and $a_k(t)$ is the PN-code waveform for the kth node in DSSS and is given as $a_k(t) = \sum_{n=-\infty}^{\infty} a_n^k P_{T_c}(t-nT_c)$, where $\{a_n^k\}$ is the discrete periodic signature sequence assigned to the kth node and $P_{T_c}(t)$ is a rectangular pulse that starts at $t = 0$ and ends at $t = T_c$.

Consider M frequency hopping channels with L (assume L is odd) hops per bit. Let $T = T_b/L$ denote the duration of each hop and $T_c = T_b/NL$ denote the chip duration for the PN-code sequence, where N is the period of the PN-sequence and is also assumed to be odd. The wide-band jamming fully corrupts W hopping channels and another single channel partially (let W_J^P be the part of the channel affected by the partial jamming).

The fading channel considered here is modeled as a general wide-sense-stationary uncorrelated scattering (WSSUS) channel [9]. Following [10,21,22] , the received signal can be described as:

$$y(t) = \sum_{k=1}^{K} \mathrm{Re}\{r_k(t-\tau_k)exp\{2\pi(f_c + f_h^k(t))(t-\tau_k)\}\} + J(t) + n(t) \tag{2}$$

where $J(t)$ and $n(t)$ represent the jamming term and AWGN term that have two-sided spectral densities $N_J/2$ and $N_0/2$,, respectively, and

$$r_k(t) = \gamma_k \int_{-\infty}^{\infty} \beta_k(\tau,t)\sqrt{2P}(t-\tau)a_k(t-\tau)d\tau + \sqrt{2P}(t)a_k(t) \tag{3}$$

where the nonnegative real parameter γ_k is the Rician channel coefficient for the kth node; $\beta_k(\tau,t)$ is a zero-mean complex Gaussian random process that represents the equivalent low-pass time-varying impulse response for the fading channel [10]. The covariance function for the fading process in a WSSUS channel is [22,23].

$$\Lambda_k(\tau,\sigma;t,s) = \frac{1}{2}E\{\beta_k(\tau,t)\beta_k^*(\tau,t)\} = \rho_k(\tau,t-s)\delta(\tau-\sigma) \tag{4}$$

In this paper, we focus on one class of WSSUS channels known as time-selective fading channels [22]; its covariance function is given by $\rho_k(\tau,t-s) = \rho_k(0,t-s)\delta(\tau)$ [23], where the covariance function $\rho_k(0,t-s)$ is defined as

$$\rho_k(0,t) = \begin{cases} 1-\frac{v|t|}{T}, & |t| \le \lambda T_c \\ 0 & \text{, otherwise} \end{cases} \tag{5}$$

where $\lambda = (n+\beta) < N$, n is a positive integer less than N, $0 \le \beta < 1$ and $v = (\lambda T_c)^{-1}T$ [10].

Similar to [24], the time delays and data symbols for the kth node are modeled as mutually independent random variables which are uniformly distributed on [0, T] and {−1, + 1}, respectively. We also assume $\tau_i = 0$ when considering the output of the kth ($k \ne i$) correlation receiver.

3 Error probability analysis

In this section, we first investigate the average error probability for one hop, and then we employ a majority voting scheme to compute the overall error probability for one bit.

For each user k, the other K-1 users are considered as interference. Three different situations may occur in one hop: j out of K-1 users interfere with the same hopping channel of user k and (1) no jamming corrupts the channel, (2) jamming fully corrupts the channel, or (3) jamming partially corrupts the channel. Thus, the total average error probability P_ε^k of one hop for user k can be computed as:

$$\begin{aligned} P_\varepsilon^k &= \sum_{j=0}^{K-1} P_\varepsilon^k(j\,\text{users}) \\ &= \sum_{j=0}^{K-1}\{P_\varepsilon^k(j\,\text{users, no jam}) \\ &\quad + P_\varepsilon^k(j\,\text{users, full jam}) \\ &\quad + P_\varepsilon^k(j\,\text{users, partial jam})\} \end{aligned} \tag{6}$$

where $P_\varepsilon^k(j\,\text{users})$ is the average error probability of one hop due to j interfering users. Expression (6) is equivalent to

$$\begin{aligned} P_\varepsilon^k = \sum_{j=0}^{K-1}\{&P(j\,\text{users, no jam})P^k(\varepsilon|j\,\text{users, no jam}) \\ &+ P(j\,\text{users, full jam})P^k(\varepsilon|j\,\text{users, full jam}) \\ &+ P(j\,\text{users, partial jam})P^k(\varepsilon|j\,\text{users, partial jam})\} \end{aligned} \tag{7}$$

where $P(a,b)$ is the joint probability of events a and b, and $P^k(\varepsilon|a,b)$ is the conditional probability of error, given events a and b have occurred. From the problem formulation, we can obtain:

$$\begin{aligned} P(j\,\text{users, no jam}) &= \binom{K-1}{j}\left(\frac{1}{M}\right)^j\left(1-\frac{1}{M}\right)^{K-1-j}\left(\frac{M-W-1}{M}\right) \\ P(j\,\text{users, full jam}) &= \binom{K-1}{j}\left(\frac{1}{M}\right)^j\left(1-\frac{1}{M}\right)^{K-1-j}\left(\frac{W}{M}\right) \\ P(j\,\text{users, partial jam}) &= \binom{K-1}{j}\left(\frac{1}{M}\right)^j\left(1-\frac{1}{M}\right)^{K-1-j}\left(\frac{1}{M}\right) \end{aligned} \tag{8}$$

The conditional error probabilities for each case of jamming over Rician fading channels are discussed next.

A. *Case 1: No Jamming*

When there is no jamming, the error probability for BPSK modulation is given as [25]:

$$P^k(\varepsilon|j\,\text{users, no jam}) = Q\left(\frac{1}{\sqrt{NSR/2 + I_j^k}}\right) \tag{9}$$

where I_j^k is the interference-to-signal ratio introduced by the other users hopping in user k's channel, $NSR = N_0/2PT$ is the noise-to-signal ratio, and $Q(\cdot)$ is the complementary error function. Following the arguments in [10] and [24], I_j^k is computed as:

$$I_j^k = \frac{2\gamma_k^2\lambda}{N}\left[1-\frac{1}{2}(\nu+1)\frac{\lambda}{N}+\frac{\nu\lambda^2}{3N^2}\right] + \sum_{i=1,i\neq k}^{j+1}\left[\frac{2\gamma_i^2}{N^4}\sum_{l=0}^{n}\{\beta_l R_i(l)R_k(l)\Gamma_1(1/2) + \frac{\beta_l^2}{2}[\zeta_{i,k}(l)+\zeta_{k,i}(l)]\Gamma_1(2/3)+\frac{\beta_l^3}{3}\Delta_i(l)\Delta_k(l)\Gamma_1(3/4)\}\right] + \sum_{i=1,i\neq k}^{j+1}\frac{1}{6N^3}m_{i,k} \tag{10}$$

where $\beta_l = 1$ for $l < n$ and $\beta_n = \beta$, $\Delta_i(l) = R_i(l+1) - R_i(l)$, $\zeta_{i,k}(l) = \Delta_i(l)R_k(l)$, $\Gamma_1(a) = N - \nu(l + a\beta_l)$, $m_{i,k} = 2\sum_{l=1-N}^{N-1} R_i(l)R_k(l) + \sum_{l=1-N}^{N-1} R_i(l)R_k(l+1)$, and $R_i(l)$ is the usual aperiodic autocorrelation function for the PN-sequence.

Different PN-sequences correspond to different aperiodic autocorrelation functions which are functions of the length N of the sequence. In this work, we employ a maximal-length sequence (MLS) as the signature sequence. However, by using an MLS code, there does not exist a closed-form expression of the aperiodic autocorrelation function, $R_i(l)$, for the general MLS code, which prevents us from finding a closed-form expression for I_j^k. However, we can compute a closed-form expression if we know exactly which MLS code is used. Actually, two different MLS codes with the same length will have different aperiodic autocorrelation functions. Therefore, we consider an upper bound on an MLS's aperiodic autocorrelation function derived in [26] to compute an upper bound on the error probability of the HSS system. From [26], we have $R_i(0) = N$ and $R_i(l) < R_u = 1 + \frac{2}{\pi}(N+1)^{\frac{1}{2}}\ln\left(\frac{4N}{\pi}\right)$, $l\neq 0$. Plugging them back into I_j^k in (10) and assuming $\gamma_k = \gamma$ as a constant for simplicity, we get an upper bound on I_j^k as:

$$I_j^k < I_j^u = \frac{2\gamma^2\lambda}{N}\left[1-\frac{1}{2}(\nu+1)\frac{\lambda}{N}+\frac{\nu\lambda^2}{3N^2}\right] + \frac{2\gamma^2 j}{N^4}\Big\{N^2R_u - \frac{2}{3}NR_u\nu - \frac{5}{6}N^2\nu + \frac{4}{3}N^3 + \frac{13}{3}N(n-1)R_u^2 - \frac{17}{6}(n-1)R_u^2\nu + \frac{13}{6}n(n-1)R_u^2\nu + \beta R_u^2(N-\nu n)\left(1+2\beta+\frac{4}{3}\beta^2\right) - \beta^2R_u^2\nu\left(\frac{1}{2}+\frac{4}{3}\beta+\beta^2\right)\Big\} + \frac{j(2N^2+4R_u^2(N-1))}{6N^3} \tag{11}$$

B. *Case 2: Full Jamming*

When jamming fully corrupts the user k's channel, the error probability for BPSK is given as:

$$P^k(\varepsilon|j\text{user, full jam}) = Q\left(\frac{1}{\sqrt{NSR/2 + JSR/2 + I_j^k}}\right) \tag{12}$$

where $JSR = N_J/2PT$ is the jamming-to-signal ratio.

C. *Case 3: Partial Jamming*

When jamming partially corrupts the user k's channel, the error probability includes two portions: one is the part of the channel corrupted and the other is the uncorrupted part. Let $q = W_j^p/(NW_b)$ denote the fraction of the channel jammed, where $W_b = 1/T_b$. Then the error probability for the BPSK case is given as:

$$P^k(\varepsilon|j\text{ users, partial jam}) = qP^k(\varepsilon|j\text{ users, full jam}) + (1-q)P^k(\varepsilon|j\text{ users, no jam}) \tag{13}$$

Based on the arguments above, the error probability per hop, P_ε^k, is obtained. Without loss of generality, we assume the Rician channel coefficients for all users are identical, i.e., $\gamma_k = \gamma$, then, for simplicity, P_ε^k can be represented as P_ε. To compute the error probability for *one bit*, denoted P_E, we employ a majority voting decision scheme given as:

$$P_E = \sum_{d=\frac{L+1}{2}}^{L}\binom{L}{d}(P_\varepsilon)^d(1-P_\varepsilon)^{L-d} \tag{14}$$

Due to the monotonicity of $Q(\bullet)$, using (11) provides an upper bound on P_ε^k and thus an upper bound on P_E. The problem of determining the HSS system parameters for an optimal performance is now discussed in the next section.

4 Optimization problem formulation

In realistic HSS systems, the overall system performance always suffers from practical parameter constraints. Thus, we formulate the problem of minimizing the bit-error performance subject to some representative parameter constraints.

The system design parameters of interest are the number of frequency-hopping channels M, the length of the PN-sequence N, the number of channels fully corrupted by jamming W, and the number of hops per bit L. Assume that these parameters satisfy the following constraints

$$MNW_b - K_1 \leq 0, \quad K_1 > 0 \tag{15}$$

$$W - K_2 M \geq 0, \ 0 \leq K_2 \leq 1 \tag{16}$$

$$K_3 - \frac{1}{LW_b} \leq 0, \quad K_3 > 0 \tag{17}$$

$$M, N, W, L > 0 \tag{18}$$

together with integer constraints on the parameters (i.e., M, N, W, L are positive integers).

The physical meaning of these constraints can be explained as follows: (15) represents that the total bandwidth of the system (MNW_b) is limited by K_1, where $K_1 > 0$; (16) means that the number of frequency channels fully corrupted by the jamming are a portion of the total number of channels, where $0 \leq K_2 \leq 1$; (17) provides a lower bound on the time duration of each hop $\left(\frac{1}{LW_b}\right)$ due to implementation limitations; and (18) restricts all the parameters to be positive.

The optimization problem is to minimize the system's bit- error rate (BER) in (14) with respect to the constraints described in (15) to (18). It can be written as:

$$\min_{M,N,W,L} P_E \quad \text{subject to } (15), (16), (17), (18) \tag{19}$$

The integer constraints are removed in the problem statement and the following analysis because they can be imposed after the solutions of (19) are found. This will be discussed in more detail in the following section.

5 Necessary conditions of the optimization problem

By examining the structure of P_E, we can further relax the constraint (17). Note P_E is a monotonically decreasing function with respect to L, so constraint (17) can be written as:

$$L \leq \frac{1}{K_3 W_b} \tag{20}$$

which means that there is an upper bound on L. Moreover, as P_ε does not depend on L, the error probability P_E reaches its minimum when $L = \frac{1}{K_3 W_b}$ with the other parameters fixed.

Further, it is easy to see that P_E is a monotonically increasing function with respect to P_ε on the interval [0, 1], and M, N, W are all contained only in P_ε; thus, the optimization problem in (19) can be further simplified as:

$$\min_{M,N,W} P_\varepsilon \quad \text{subject to } (15), (16), (18) \tag{21}$$

From Section 3, we have

$$P_\varepsilon = \sum_{j=0}^{K-1} \{P_n P_{nj} + P_f P_{fj} + P_p P_{pj}\} \tag{22}$$

where $P_n = P(j\,users,\ no\ jam)$, $P_f = P(j\,users,\ full\ jam)$, $P_p = P(j\,users,\ partial\ jam)$, $P_{nj} = P^k(\varepsilon | j \text{ users, no jam})$, $P_{fj} = P^k(\varepsilon | j \text{ user, full jam})$, $P_{pj} = P^k(\varepsilon | j \text{ users, partial jam})$. Expression (22) can be further simplified by representing P_{pj} in terms of P_{nj} and P_{fj} as follows:

$$P_\varepsilon = \sum_{j=0}^{K-1} \{(P_n + P_p(1-q))P_{nj} + (P_p q + P_f)P_{fj}\} \tag{23}$$

For convenience, let

$$P_{\varepsilon,j} = (P_n + P_p(1-q))P_{nj} + (P_p q + P_f)P_{fj} \tag{24}$$

and also let $x = (M, N, W)$.

From Section 3, we observe that P_n and P_f are functions of both M and W, while P_p is only a function of M. Moreover, P_{nj} and P_{fj} are functions of N. We can also observe that the error probability to be minimized has a complex structure and is a non-convex function. Thus, to compute the optimal solution, we apply the Karush-Kuhn-Tucker (KKT) [20] conditions to problem (21).

Lemma 1: *(Karush-Kuhn-Tucker Conditions) Let y^* be a local minimum of the following problem*

$$\min f(y) \quad \text{subject to } g_1(y) \leq 0, \ldots, g_m(y) \leq 0 \tag{25}$$

where f and g_i are continuously differentiable functions with appropriate dimensions. Then there exists an unique Lagrange multiplier vector $\mu = (\mu_1, \ldots, \mu_m)$, such that

$$\nabla_y R(y^*, \mu) = 0, \quad \mu_i \geq 0,\ i = 1, \ldots m, \text{ and } \mu_i = 0, \forall i \notin A(y^*) \tag{26}$$

where $R(y,\mu) = f(y) + \sum_{i=1}^{m} \mu_i g_i(y)$ is the Lagrangian function and $A(y^*)$ is the set of active constraints at y^* defined as: For any feasible vector y (the vector that satisfies all constraints), the set of active inequality constraints is given as $A(y) = \{i | g_i(y) = 0\}$ and if $j \notin A(y)$, it is said that the jth constraint is inactive at y.

In addition, if f and g_i are twice continuously differentiable, then there holds $z^T \nabla^2_{yy} R(y^*, \mu) z \geq 0$, for all z in proper dimensions, such that $\nabla g_i (y^*)^T z = 0, \forall i \in A(y^*)$.

Now, the necessary conditions for a local minimum of problem (21) can be derived by applying the KKT conditions as follows:

Theorem 1: *Let $x^* = (M^*, N^*, W^*)$ be a local minimum of the problem (21), then there exists unique $\mu_1 \geq 0$, $\mu_2 \geq 0$, such that*

$$\sum_{j=0}^{K-1} \frac{\partial P_{\varepsilon,j}}{\partial M}\Big|_{x^*} + \mu_1 N^* W_b - \mu_2 K_2 = 0 \tag{27}$$

$$\sum_{j=0}^{K-1} \frac{\partial P_{\varepsilon,j}}{\partial N}\Big|_{x^*} + \mu_1 M^* W_b = 0 \tag{28}$$

$$\sum_{j=0}^{K-1} \frac{\partial P_{\varepsilon,j}}{\partial W}\Big|_{x^*} + \mu_2 = 0 \tag{29}$$

$$M^* N^* W_b - K_1 = 0 \tag{30}$$

$$W^* - K_2 M^* = 0 \tag{31}$$

where

$$\frac{\partial P_{\varepsilon,j}}{\partial M} = \frac{1}{N W_b} \binom{K-1}{j} \left(\frac{1}{M}\right)^{j+1} \left(1 - \frac{1}{M}\right)^{K-1-j} \times \left\{ \frac{\left(N W_b W + W_j^p\right)(jM + M - K)}{M^2 - M} (P_{nj} - P_{fj}) - \frac{N W_b M (jM + 1 - K)}{M^2 - M} P_{nj} \right\} \tag{32}$$

$$\frac{\partial P_{\varepsilon,j}}{\partial W} = \binom{K-1}{j} \left(\frac{1}{M}\right)^{j+1} \left(1 - \frac{1}{M}\right)^{K-1-j} (-P_{nj} + P_{fj}) \tag{33}$$

$$\frac{\partial P_{\varepsilon,j}}{\partial N} = \left(P_n + P_p(1-q)\right) \frac{\partial P_{nj}}{\partial N} + \left(P_p q + P_f\right) \frac{\partial P_{fj}}{\partial N} + \frac{q}{N} P_p P_{nj} - \frac{q}{N} P_p P_{fj} \tag{34}$$

$$\frac{\partial P_{nj}}{\partial N} = \frac{1}{\sqrt{\pi}} e^{-\frac{1}{NSR + 2I_j^k}} \left(NSR + 2I_j^k\right)^{-\frac{3}{2}} \frac{\partial I_j^k}{\partial N} \tag{35}$$

$$\frac{\partial P_{fj}}{\partial N} = \frac{1}{\sqrt{\pi}} e^{-\frac{1}{NSR + JSR + 2I_j^k}} \left(NSR + JSR + 2I_j^k\right)^{-\frac{3}{2}} \frac{\partial I_j^k}{\partial N} \tag{36}$$

In addition, the following inequality holds:

$$\begin{bmatrix} \frac{1}{N^*} \\ \frac{M^*}{K_2} \end{bmatrix}^T \begin{bmatrix} \sum_{j=0}^{K-1} \frac{\partial^2 P_{\varepsilon,j}}{\partial M^2} & \sum_{j=0}^{K-1} \frac{\partial^2 P_{\varepsilon,j}}{\partial M \partial N} + \mu_1 W_b & \sum_{j=0}^{K-1} \frac{\partial^2 P_{\varepsilon,j}}{\partial M \partial W} \\ \sum_{j=0}^{K-1} \frac{\partial^2 P_{\varepsilon,j}}{\partial M \partial N} + \mu_1 & \sum_{j=0}^{K-1} \frac{\partial^2 P_{\varepsilon,j}}{\partial N^2} & \sum_{j=0}^{K-1} \frac{\partial^2 P_{\varepsilon,j}}{\partial N \partial W} \\ \sum_{j=0}^{K-1} \frac{\partial^2 P_{\varepsilon,j}}{\partial M \partial W} & \sum_{j=0}^{K-1} \frac{\partial^2 P_{\varepsilon,j}}{\partial N \partial W} & 0 \end{bmatrix}_{x^*} \begin{bmatrix} \frac{1}{N^*} \\ \frac{M^*}{K_2} \end{bmatrix} \geq 0 \tag{37}$$

Proof:

In order to apply the KKT conditions, we first need to check the types of inequality constraints, to determine whether they are active or inactive inequality constraints.

It is obvious that (18) is inactive at x^*. To check for (15), first assume that (15) is also inactive at x^*, which infers $M^* N^* W_b - K_1 < 0$. However, it should be noted that P_ε is a monotonically decreasing function with respect to both M^* and N^*; thus, $M^* N^* W_b - K_1 < 0$ means there is still an 'increasing space' for either M^* or N^*, such that P_ε can still be reduced by increasing M^* or N^* to $M^* N^* W_b = K_1$, which contradicts that x^* is the local minimum. Thus, (15) is an active constraint at x^*.

To check for (16), first assume that (16) is inactive at x^*, which means $W^* - K_2 M^* > 0$, by applying the KKT necessary conditions (page 316, Proposition 3.3.1 in [20]), we have the unique Lagrange multiplier for (16) $\mu_2 = 0$ and $\sum_{j=0}^{K-1} \frac{\partial P_{\varepsilon,j}}{\partial W}\big|_{x^*} = 0$. However, we should also observe that by (33), $\sum_{j=0}^{K-1} \frac{\partial P_{\varepsilon,j}}{\partial W}\big|_{x^*} \neq 0$ as $P_{nj} \neq P_{fj}$, which leads to a contradiction. Thus, (16) is also an active constraint at x^*.

After specifying the type of each inequality constraint, we can obtain Theorem 1 by applying the KKT conditions in Lemma 1 to problem (21).

Remark 1 *We can similarly obtain second-order sufficiency conditions of the problem by applying the following KKT sufficient conditions (page 320, Proposition 3.3.2 in [20]): If (27) to (37) hold for some x and $\mu_i > 0$, $i = 1, \ldots m$, then x is a strict local minimum.*

Remark 2 *Once the solution is found, the integer constraints need to be imposed. For example, assume M, N, W, L are positive integers; first, round one parameter (e.g., N) to the nearest integer, then plug it back to the problem and re-compute the solution. After that, round the rest of the parameters in a similar fashion.*

Theorem 1 states the necessary conditions for the optimization problem by employing a general PN-sequence. Now, we will employ the MLS code as the PN-sequence in the HSS system and reformulate Theorem 1 explicitly.

Expression (11) describes that I_j^k for an MLS code is upper-bounded by I_j^u. Note that the upper bound of the error probability reserves the same monotonicity with respect to the system parameters (e.g., M, N). Considering the upper bound I_j^u in (11), Theorem 1 remains the same, with the exception that $\frac{\partial I_j^k}{\partial N}$ is replaced with $\frac{\partial I_j^u}{\partial N}$. After performing some derivations, we obtain:

$$\frac{\partial I_j^u}{\partial N} = I_1 + I_2 + I_3 + I_4 \tag{38}$$

where

$$I_1 = \frac{2\gamma^2 \lambda}{N^2} \left(\frac{\lambda}{3N} - \frac{1}{2}\right)$$

$$
\begin{aligned}
I_2 = \frac{2\gamma^2 j}{N^4}\Bigg(& 2NR_\mathrm{u} + N^2R_\mathrm{u}' - \frac{4NR_\mathrm{u}}{3\lambda} - \frac{2N^2R_\mathrm{u}'}{3\lambda} - \frac{5N^2}{2\lambda} + 4N^2 \\
&+\left(\frac{(n-1)R_\mathrm{u}^2}{6} + \frac{N(n-1)R_\mathrm{u}R_\mathrm{u}'}{3}\right)\left(26-\frac{17}{\lambda}\right) \\
&+\frac{13Nn(n-1)R_\mathrm{u}R_\mathrm{u}'}{3\lambda} + \frac{13n(n-1)R_\mathrm{u}^2}{6\lambda} \\
&+\beta\left(1+2\beta+\frac{4}{3}\beta^2\right)\left(1-\frac{n}{\lambda}\right)R_\mathrm{u}\left(2NR_\mathrm{u}'+R_\mathrm{u}\right) \\
&-\beta^2\left(\frac{1}{2}+\frac{4}{3}\beta+\beta^2\right)\frac{R_\mathrm{u}}{\lambda}\left(2NR_\mathrm{u}'+R_\mathrm{u}\right)
\end{aligned}
$$

$$
\begin{aligned}
I_3 = -\frac{8\gamma^2 j}{N^5}\Bigg(& N^2R_\mathrm{u} - \frac{2N^2}{3\lambda}R_\mathrm{u} - \frac{5N^3}{6\lambda} + \frac{4}{3}N^3 \\
&+\frac{13}{3}N(n-1)R_\mathrm{u}^2 - \frac{17N}{6\lambda}(n-1)R_\mathrm{u}^2 + \frac{13N}{6\lambda}n(n-1)R_\mathrm{u}^2 \\
&+\beta R_\mathrm{u}^2 N\left(1-\frac{n}{\lambda}\right) \times \left(1+2\beta+\frac{4}{3}\beta^2\right) \\
&-\beta^2 R_\mathrm{u}^2\frac{N}{\lambda}\left(\frac{1}{2}+\frac{4}{3}\beta+\beta^2\right)\Bigg)
\end{aligned}
$$

$$
I_4 = \frac{j}{6N^3}\left(4N + 8R_\mathrm{u}(N-1)R_\mathrm{u}' + 4R_\mathrm{u}^2\right) - \frac{j}{2N^4}\left(2N^2 + 4R_\mathrm{u}^2(N-1)\right) \tag{39}
$$

Plugging the above equations back into Theorem 1, then we can obtain necessary conditions for the local minimum of the upper bound of the error probability for the MLS code. Sufficient conditions can also be obtained from Remark 1.

Remark 3 *Note that in an MLS code, N is an integer such that $N = 2^n - 1$ where n is a positive integer. Thus, after obtaining solutions of the local minimum of the problem, N in each solution should be rounded to the closest integers in the form of $2^n - 1$ (usually two integers correspond to N in one solution), and the rest of the parameters in the solution should be re-computed and rounded. Then, by comparing the error probabilities resulted from these two sets of parameters, we employ the parameter set with the lower error probability as the local minimum of the problem after re-applying the integer constraints.*

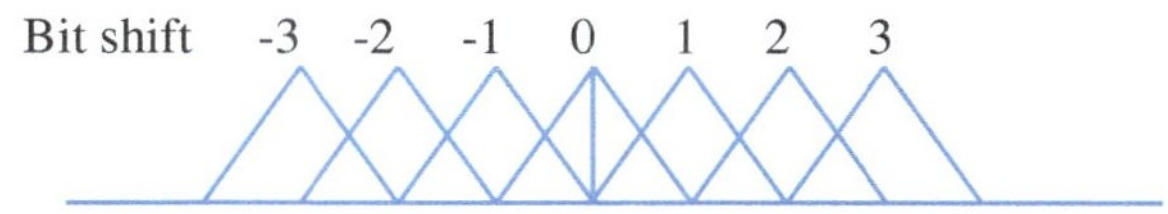

Figure 2 Code-Phase-Shift-Keying modulation for the DS signal.

In the next section, our specific design and implementation of a hybrid DS/FFH radio transceiver using a single FPGA are presented.

6 ORNL specific hybrid DS/FFH design and implementation

The hybrid DS/FFH prototype was designed to demonstrate the fundamental advantages of the HSS system, such as jamming resistance, difficulty of unwanted interception, robust performance, and reasonable cost. The prototype operates in the unlicensed 902 to 928 MHz ISM band, although target applications such as the SG may ultimately use a dedicated frequency band. The system parameters for the prototype are selected based on the available ISM bandwidth and FPGA capabilities and using the analysis presented in the previous section. The selected parameters are considered to be nearly optimal for a typical smart grid environment.

We decided to use the Software Defined Radio (SDR) method for hardware implementation of the hybrid DS/FFH system because of its flexibility in changing the system to evaluate new concepts. This methodology has also proven to be very powerful in that the vast majority of the signal processing components can be placed in a single FPGA. The entire HSS band is down-converted to an intermediate frequency, digitized, and sent to the FPGA. Within the FPGA, look-up-table-based local oscillators down-convert the individual FH channels to baseband. These baseband signals are then decoded using DS correlators and stored in a buffer for subsequent delivery to a host computer.

As shown in Figure 1, the HSS unit splits the 902 to 928 MHz band into ten separate FH channels, each of which sends a DS spread spectrum signal with a 1.25-MHz chipping rate. An analog mixer converts these frequencies up or down for the transmitter or receiver, respectively, for use by the digital-to-analog (D/A) or analog-to-digital (A/D)

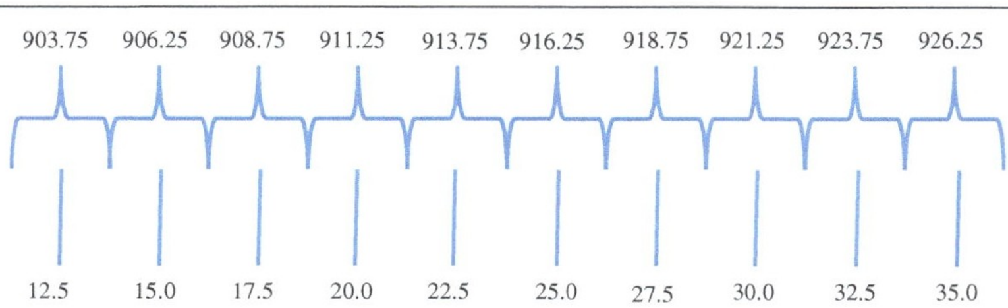

Figure 1 The prototype RF and FPGA frequencies in MHz.

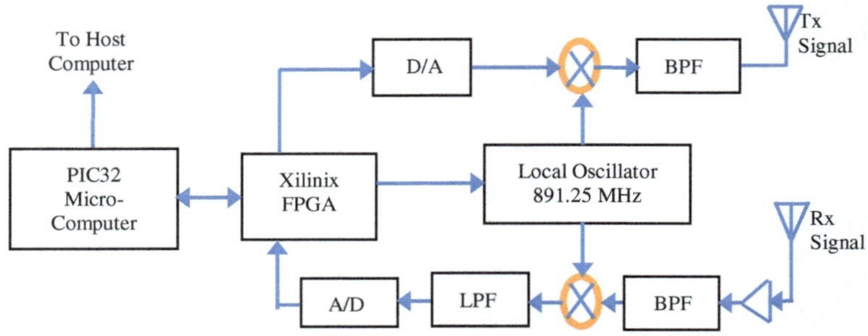

Figure 3 Hardware setup for the hybrid DS/FFH prototype.

converters. The SDR algorithms work over a designated 12.5 to 35.0 MHz frequency range. Each DS signal is a 63-bit length MLS code, although more advanced Gold or Kasami codes could also be used. Three hops per bit are used, and at the receiver a two-out-of-three majority voting decision scheme is employed.

Of particular, interest is the method used for modulating the DS signal. Traditional PSK modulation requires a preamble at the beginning of the packet to determine the reference phase and a Costas Loop [27] or similar mechanism to maintain this phase reference. With HSS in multipath channels, this phase reference is lost after each frequency hop; therefore, we decided to perform the DS modulation by shifting the start time of the code. The incoming signal is correlated with local copies of the shifted code pattern and an early-late voting system determines the amount of shift of the received signal. The correlation algorithm is independent of the carrier phase of the signal. The number of bits that can be encoded by this method is demonstrated by the early-late diagram described in Figure 2.

The bit-shift number refers to the number of bits that the local DS code has been shifted for performing the correlation. To prevent ambiguous results from a correlation being between two bits, only every other bit position is used, which results in 31 positions available for each code word. Four bytes of blank data are sent at the beginning of the packet as a preamble to set the reference DS start time.

A different interpretation of this methodology would be that the DS code is shifted because of a different time-of-flight, similar to GPS or continuous wave radar. Similar to the way GPS can achieve precise time-of-flight resolution, it can be expected that this methodology can be further developed to obtain higher bit capacity. The work in [28] explores this method for multiple users occupying a channel simultaneously.

The HSS channel capacity is calculated by dividing the chip rate, or 1.25 MHz, by the 63-bit code length to get 19,841 DS sequences per second. Since the data is replicated three times for redundancy, the actual throughput is 6,613 DS sequences per second. Since each DS sequence contains 8 bits of data, the data throughput is 52,910 bits per second. The HSS prototype is optimized for reading household utility meters for smart grid applications and thus only requires 32 bytes, although the system has operated successfully with 256-byte packets.

The prototype hybrid DS/FFH system is based on a Xilinx Virtex-4 FPGA for performing the digital signal processing. The hardware setup is described in Figure 3. The

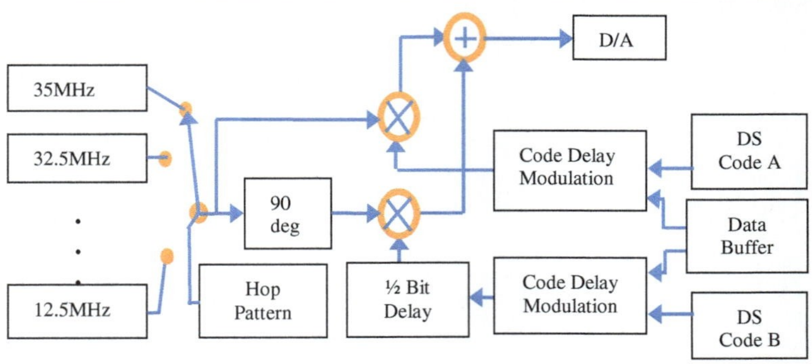

Figure 4 Transmitter portion of the FPGA code.

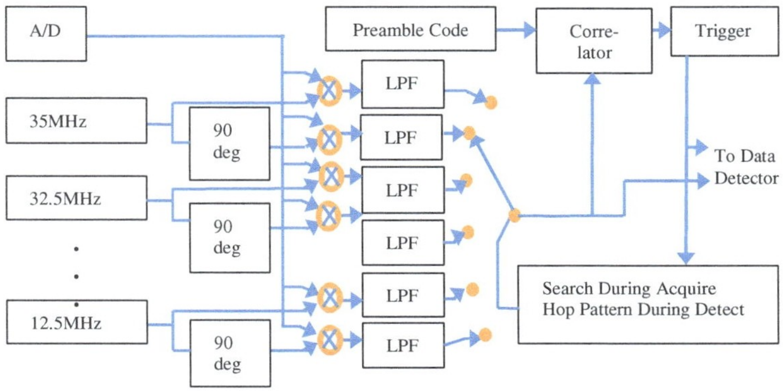

Figure 5 Preamble-detection section of the receiver.

FPGA, A/D, and D/A operate synchronously together at 100 MHz to allow operation on analog signals to a practical limit of 40 MHz. The D/A has 16-bit resolution for a dynamic range of 96 dB, and the corresponding A/D has 14-bit resolution for a dynamic range of 84 dB. The microcomputer loads and unloads data to the FPGA and communicates with sensors and other computers using Ethernet, RS232, or analog signals.

Figure 4 describes the transmitter portion of the FPGA code, which consists of the data buffer, modulator, and ten local oscillators for generating the hopping carriers. Raised-cosine waveshaping is used to reduce the spectral sidebands. The receiver uses the same local oscillators for detecting signals, and all ten channels must be simultaneously received to detect the preamble during jamming situations as illustrated in Figure 5.

To acquire the packet preamble, a spread-spectrum correlator continually looks for the preamble pattern on all channels. Once the preamble is detected, an internal timing sequence compares the signal with shifted copies of the DS code via a simple correlator. The shifted copy of the DS code that provides the strongest correlation then demodulates the actual data. To make the signal detection independent of the carrier phase, both phases of the carrier (I and Q) are correlated with the preamble's code. However, the phase relationship must remain consistent during the duration of the DS sequence.

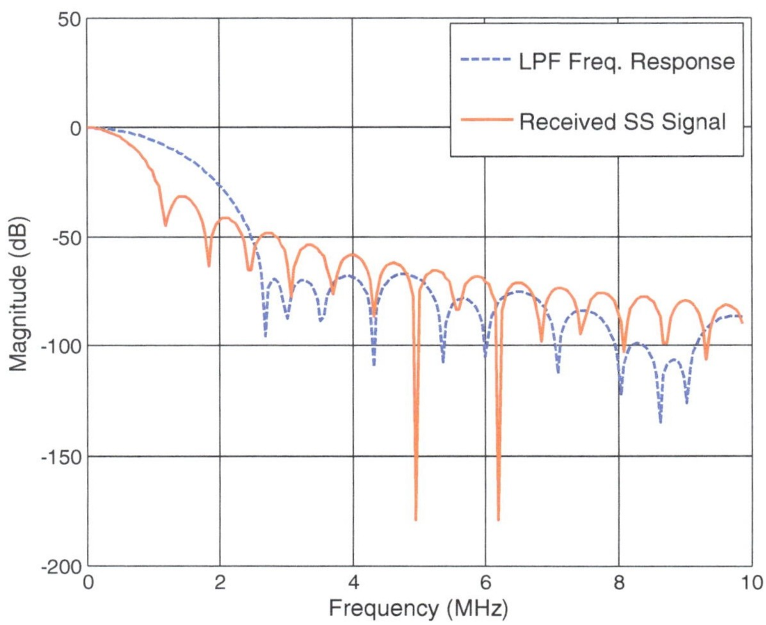

Figure 6 Analytically generated frequency response of the low-pass filter. The frequency response of the digital LPF implemented in the FPGA, superimposed on the frequency spectrum of the received SS signal.

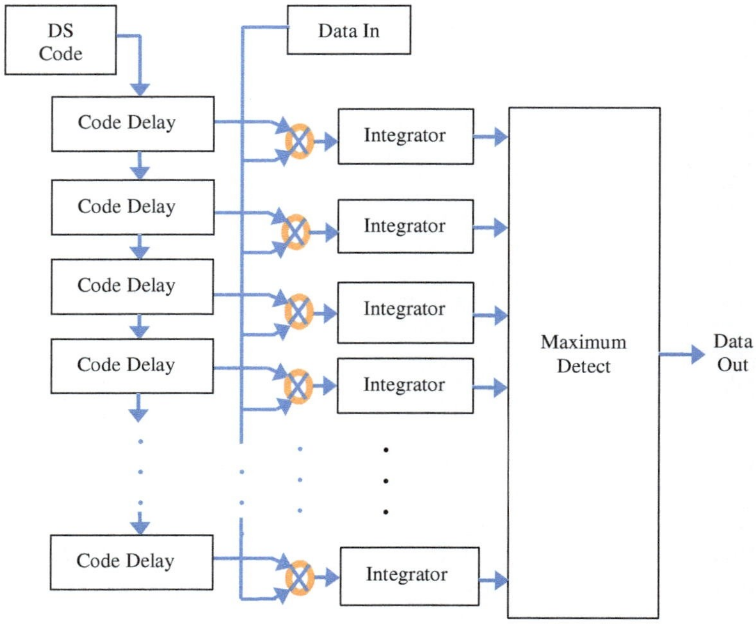

Figure 7 Data-detection section of the receiver.

A key limitation of the radio's selectivity is the digital low-pass filter (LPF) implemented in the FPGA. Because we were limited to integer arithmetic in the FPGA, the filter was implemented as a simple square-window FIR LPF, with four of the filters connected in series. A future implementation of HSS could use a newer generation FPGA with floating-point arithmetic to achieve a filter with better rolloff characteristics and higher ultimate rejection. Figure 6 is an analytically generated plot of the low-pass filter response, superimposed on the frequency spectrum of the spread-spectrum signal. The ultimate rejection level of 70 dB will be apparent in the experimental results presented in the next section.

Once the packet start has been established, the receiver begins listening on specific channels instead of all channels. A simple multiply-and-integrate correlator system is used

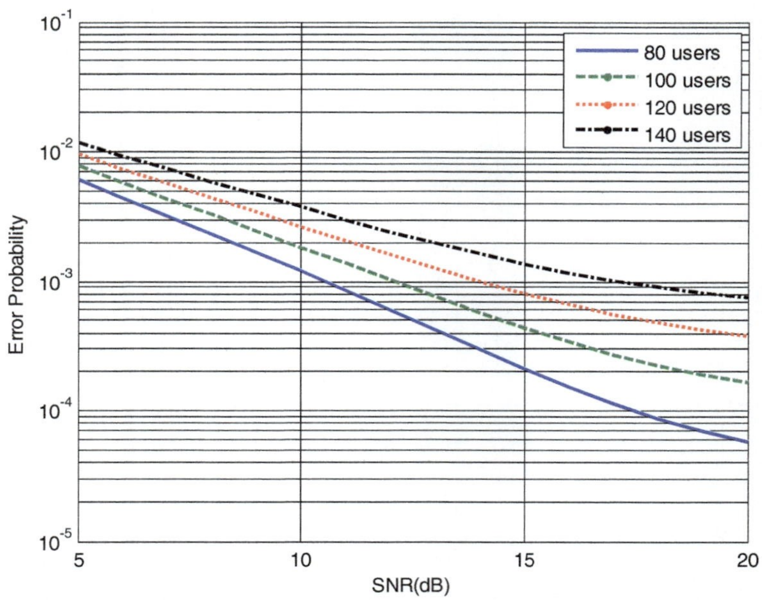

Figure 8 Performance of a hybrid DS/FFH system. Effect of different number of users (multi-user interference).

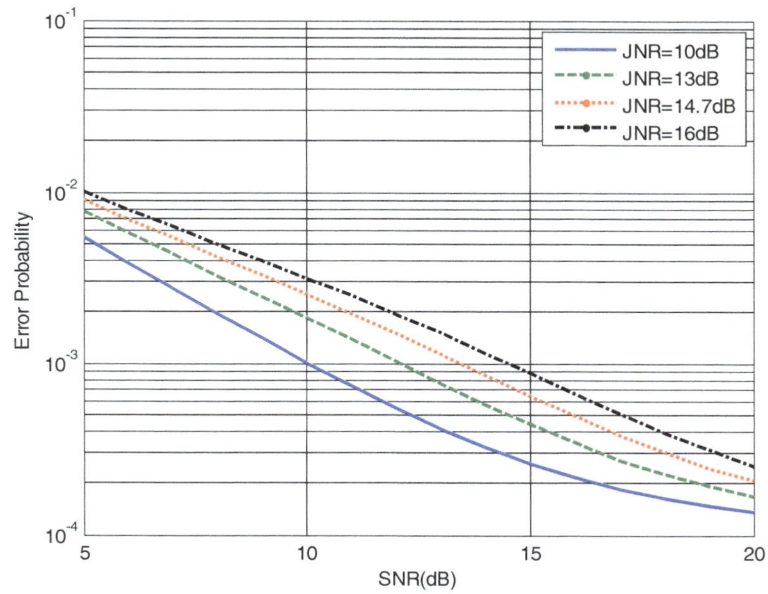

Figure 9 Performance of a hybrid DS/FFH system. Effect of different jamming-to-noise ratios (JNRs).

for signal detection as illustrated in Figure 7. In the next section, we present experimental results to demonstrate the performance of the hybrid DS/FFH prototype.

7 Numerical and experimental results

7.1 Hybrid DS/FFT system performance

We first demonstrate the performance of a hybrid DS/FFH system over Rician time-selective fading channels, progressively including the effects from wide- and partial-band jamming, multi-user interference, and varying degrees of Rician fading. The performance measure is the upper bound of BER described in (14) by employing (11). The parameters of the reference system model considered in this numerical example are total number of users is $K = 100$; number of hops per bit is $L = 5$; number of frequency-hopping channels is $M = 30$; period of PN-sequence in DSSS is $N = 127$; jamming-to-noise ratio (JNR) is 13 dB; number of channels fully jammed is 5; the Rician channel coefficient $\gamma = 0.1$ (represents the channel fading part); channel covariance function scaling factor $\lambda = 10.8$; and the

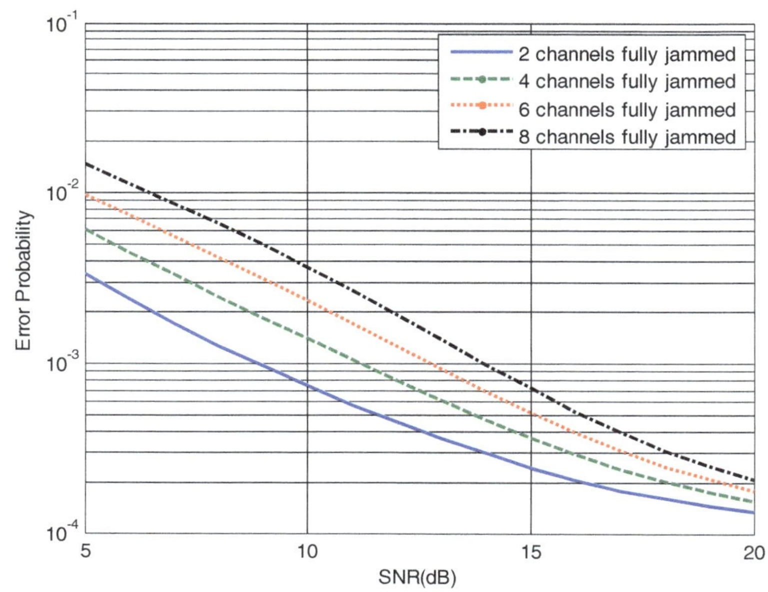

Figure 10 Performance of a hybrid DS/FFH system. Effect of different number of fully jammed channels.

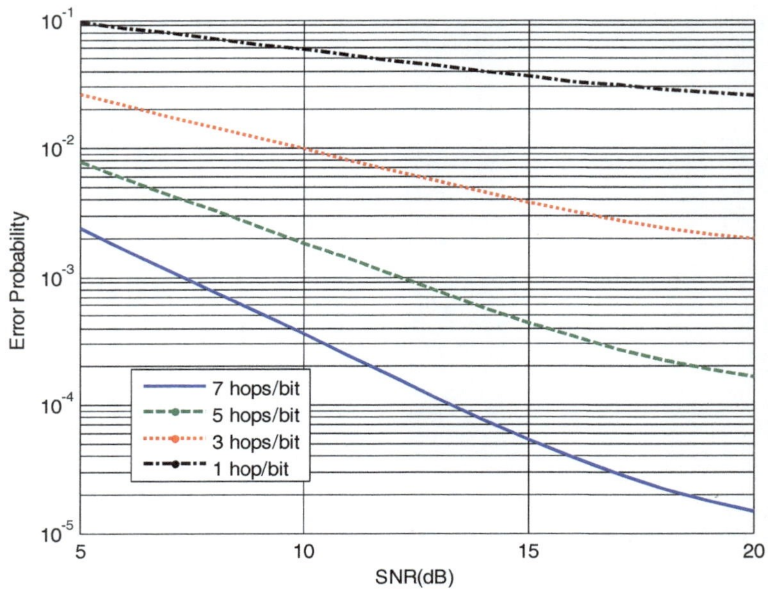

Figure 11 Performance of a hybrid DS/FFH system. Effect of different number of hops per bit.

portion of the channel partially corrupted is 0.4. The parameter space of the HSS system is explored to demonstrate its effectiveness under different conditions and scenarios. In the following analysis, we successively vary one parameter in the reference system model while fixing the other parameters.

Figure 8 shows the effect of different number of continuously transmitting users (multi-user interference) on the performance of a hybrid DS/FFH system. You can observe the high multiple access capability of such a technique, especially at high SNRs. Figure 9 demonstrates the performance for different jamming to noise ratios (JNRs), and Figure 10 demonstrates the performance for varying number of fully jammed channels. You can observe from Figures 9 and 10 the high anti-jamming capability of such a technique, especially at high SNRs. Also, it can be observed that under high SNRs the performance gap reduces for different JNRs and different numbers of fully jammed channels. Figure 11 demonstrates the performance for different numbers of hops per bit. Notice that the

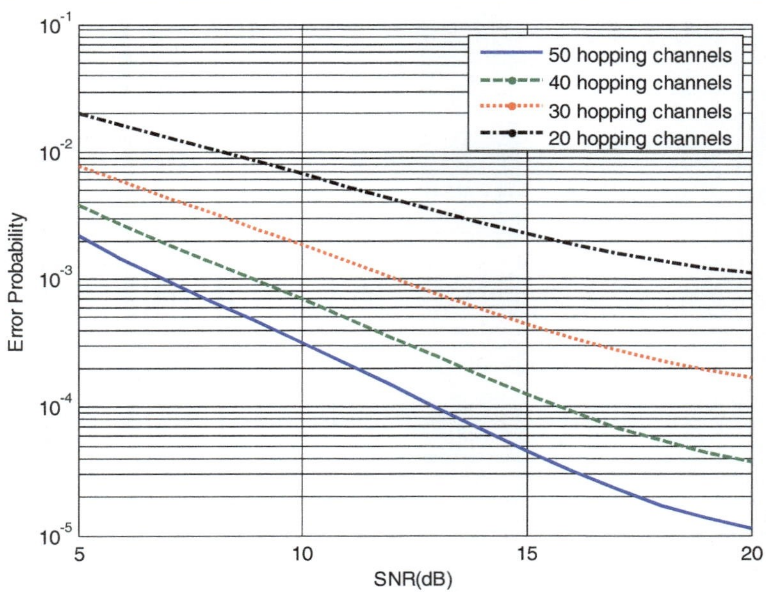

Figure 12 Performance of a hybrid DS/FFH system. Effect of different number of available hopping channels.

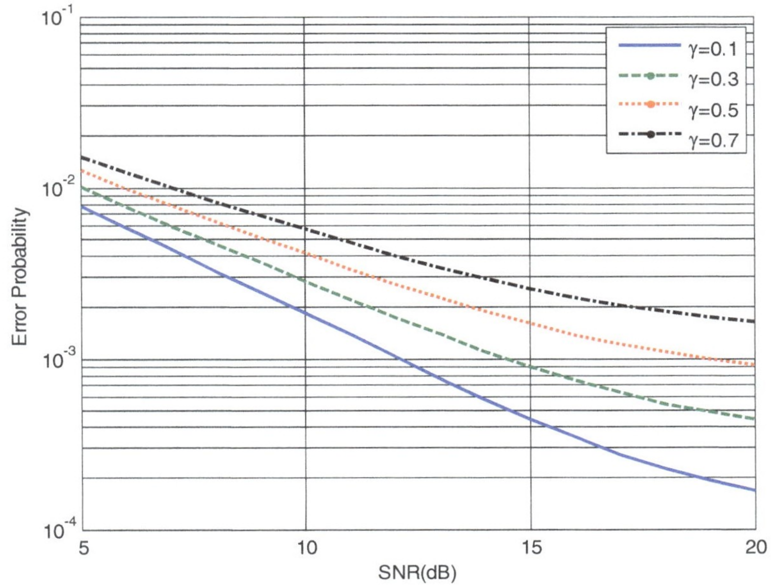

Figure 13 Performance of a hybrid DS/FFH system. Effect of different Rician fading channel parameters.

performance of the DS/FFH system is superior to that of the DS/SFH system (represented by the 1 hop/bit case). Also, notice the high improvement in performance at higher SNRs when increasing the number of hops per bit. This reveals the effectiveness of the proposed technique at high SNRs.

Figure 12 demonstrates the DS/FFH performance for different numbers of available hopping channels. Increasing the number of hopping channels reduces the likelihood of hits from other users using the same spreading PN-code and, therefore, enhances the performance. Figure 13 demonstrates the DS/FFH performance over varying degrees of Rician fading in the channels. You can observe how the performance deteriorates with increasing the fading component in the Rician channel represented by the parameter γ.

Figure 14 demonstrates the DS/FFH performance compared with the other SS systems that include DS, SFH, FFH, DS/SFH, and DS/FFH. It can be observed that the hybrid

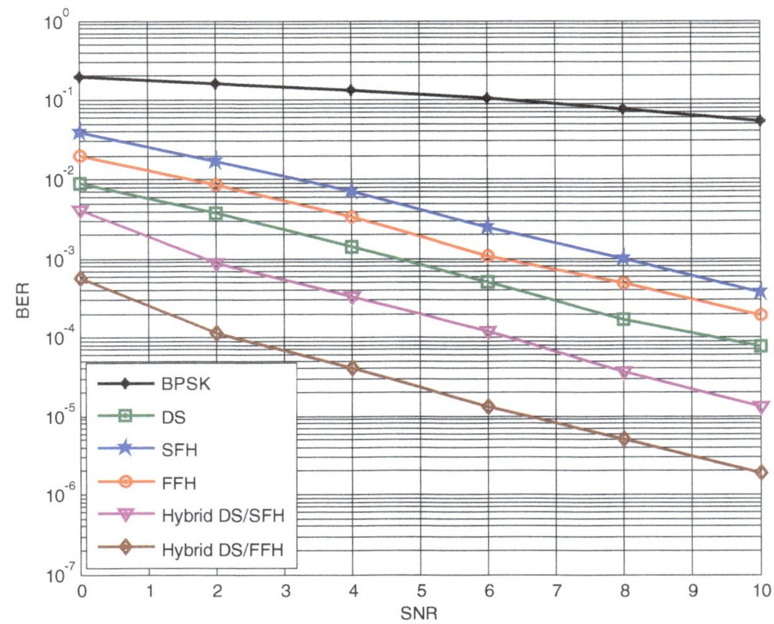

Figure 14 Performance of a two-path Rayleigh hybrid DS/FFH system. Performance comparison.

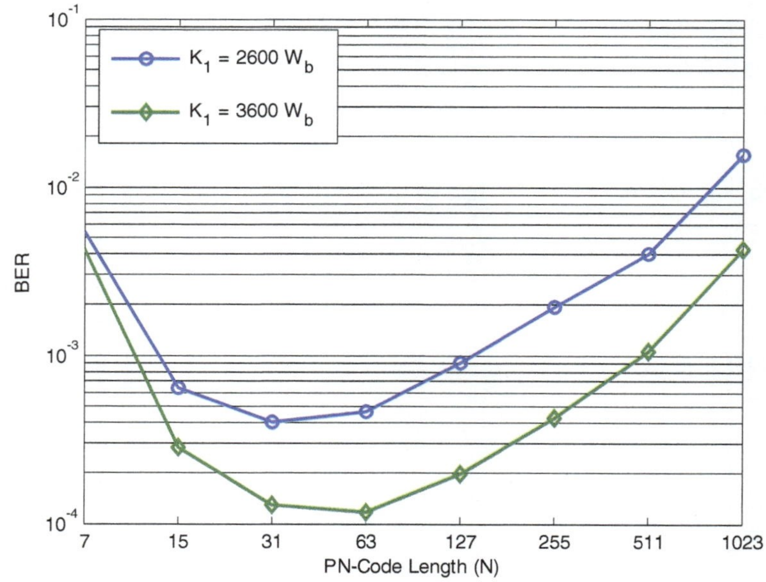

Figure 15 **BER versus DS PN-code length *N*.**

DS/FFH system outperforms the other SS systems. The hybrid DS/FFH system is preferred over the other systems because of its unique advantages, including the better spreading properties gained by frequency hopping and better multipath rejection via the direct-sequence modulation component.

The presented results demonstrate the effectiveness of the proposed hybrid DS/FFH scheme under severe channel conditions and, therefore, indicate that there is a high potential for employing it in complex smart grid communications.

7.2 Optimizing hybrid DS/FFT system performance

We now provide numerical examples to illustrate the results derived in Section 5. For convenience, we only test the necessary conditions that apply to the MLS code. We compute the solutions of the first-order necessary conditions (27) to (31) and impose the integer constraints. Then, the upper bound of the BER, P_E, is plotted for different MLS code lengths N using (14) and (11) to verify the results computed from the derived first-order necessary conditions.

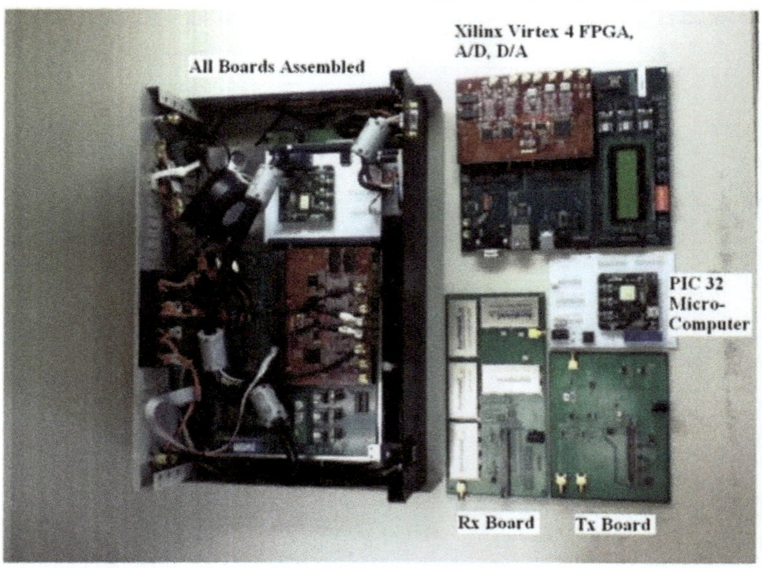

Figure 16 **The implemented hybrid DS/FFH prototype.**

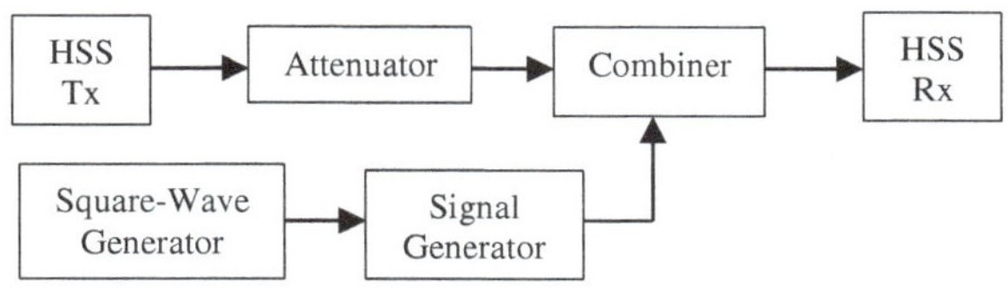

Figure 17 Experimental setup for testing hybrid DS/FFH jamming resistance.

The parameters of the reference hybrid DS/FFH system model considered is the same as described in the previous section ($K = 100$; JNR = 13 dB; $\gamma = 0.1$; *and* $\lambda = 10.8$), in addition to a signal-to-noise ratio (SNR) of 20 dB; finally, the portion of the channel partially corrupted is $q = W_J^p / (NW_b) = 30/N$. Note that the parameters *M, N, W, L* need to be computed for assessing the optimal performance. From the previous analysis, the number of hops per bit is chosen as $L = (1/K_3 W_b) = 5$.

First, we choose $K_1 = 2600\ W_b$ and $K_2 = 0.2$. Then, by applying (27) to (31), we obtain $N = 42$. Because of the integer power-of-two constraint of N ($N = 2^n - 1$), it is rounded to the nearest two integers, 31 and 63. Then by applying (30) and (31) for each integer of N and comparing the corresponding BER of both integers, we see that $N = 31$, $M = 83$, and $W = 17$ results in a smaller BER. The upper bound of the BER in (14) for different PN-code lengths, *N*, is demonstrated in Figure 15, in which we can now observe that at $N = 31$, the BER reaches its minimum. This coincides with the result from the first-order necessary conditions.

Now, we consider $K_1 = 3600\ W_b$, with K_2 unaltered. Through a similar procedure, we obtain $N = 48.3$. After rounding N to 31 and 63, it can found that $N = 63$, $M = 57$, and $W = 11$ results in smaller BER values. Figure 15 demonstrates the upper bound of the BER for different PN-code lengths, *N*, for this scenario. It can now be observed that the BER reaches its minimum at $N = 63$, which also coincides with the result from our analysis.

7.3 Experimental evaluations

Four bi-directional hybrid DS/FFH radio transceivers have been built in our lab and have performed well. The hardware prototype is shown in Figure 16. The sensitivity for the units is −110 dBm to produce an approximately 80% success rate at the packet level. This is 5 dB less sensitive than theoretically possible, but it is expected that the detection algorithms in the SDR could be significantly improved for better overall sensitivity. Also, the radios demonstrated a bit error rate of less than 10^{-6}.

The jamming performance of the system was measured directly with laboratory equipment. The testing method used for the HSS evaluation is shown in Figure 17. The square-wave generator is used at 20 kHz to modulate the signal generator at 100% AM modulation. The test procedure consists of initially transmitting data from the transmitter to the receiver with the signal generator turned off and the attenuator adjusted such that the receiver is operating at an 80% success rate. The attenuator is then reduced 20 dB so the system has a 20-dB margin. Then the signal generator is turned on and ramped up in

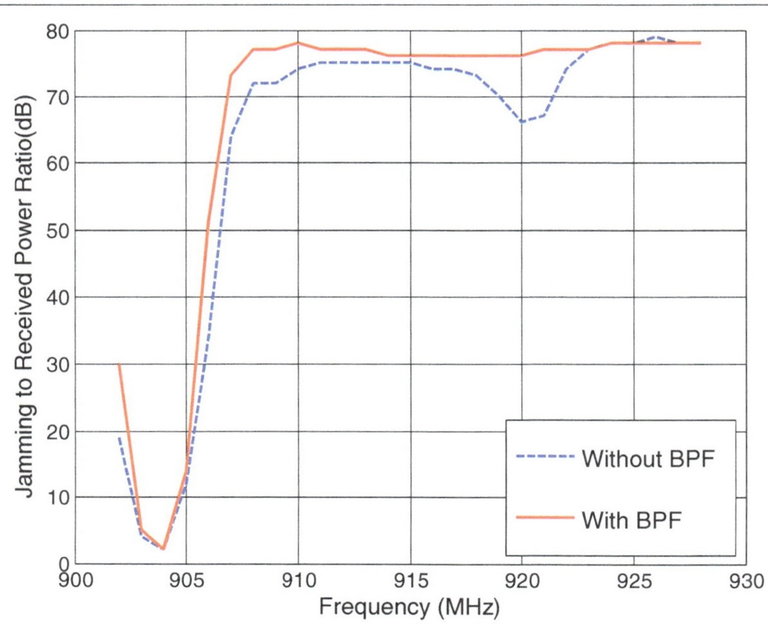

Figure 18 **The hybrid DS/FFH prototype performance while the frequency-hopping feature is disabled and with no jamming.**

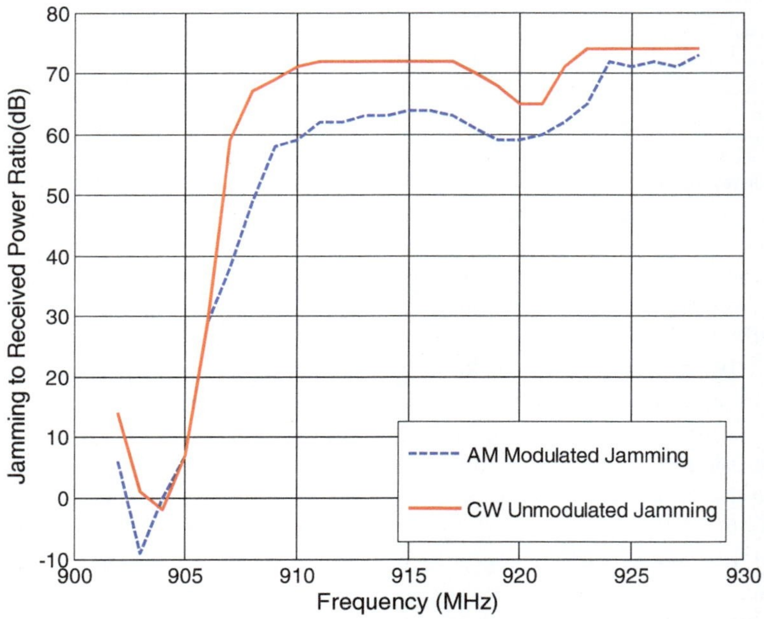

Figure 19 The hybrid DS/FFH prototype performance while the frequency hopping feature is disabled and in the presence of jamming.

power until the receiver has degraded to an 80% success rate. The difference in power between the signal generator (jamming) and the transmitter and attenuator combination (at the 20-dB margin point) is then recorded. This is repeated for signal generator frequencies from 902 to 928 MHz. Versions of the test are performed with and without the AM modulation. This stresses the radio by exposing clipping and other nonlinear effects that are expected in the A/D converter, SDR arithmetic, and analog front-end components.

The first test involved operating the HSS with the hopping feature turned off, so that the filtering capability of the SDR could be measured independently from the hopping benefits. In this test, the intermediate frequency

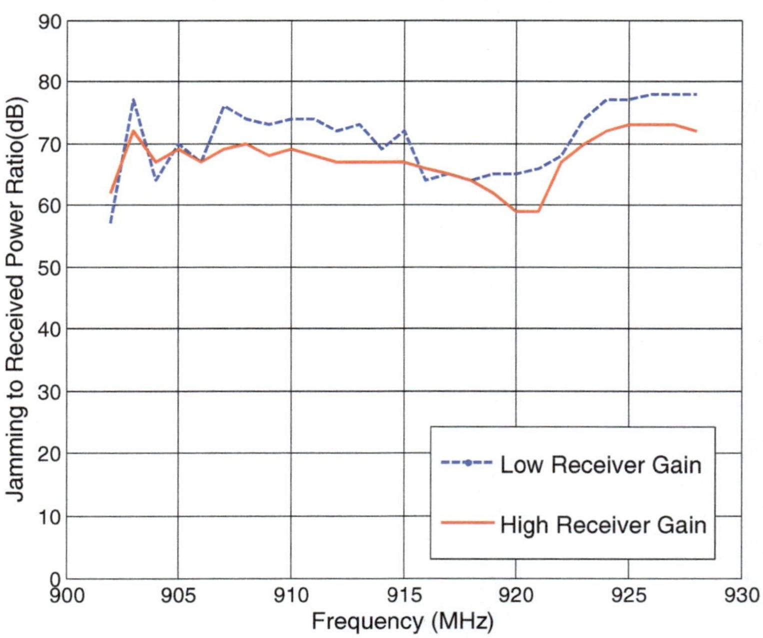

Figure 20 The hybrid DS/FFH prototype performance in the presence of jamming.

was always 12.5 MHz, which also allowed us to insert an analog 12.5 MHz, 3-pole bandpass filter (BPF) in line. This filter lets us operate the radio as a standard analog radio and allows us to do a direct selectivity comparison between the analog and SDR approaches. This comparison was made with the generator AM modulation turned off. The net results are shown in Figure 18. From the filtered version of the results, we still see the dynamic range limitations of the analog components ahead of the filter, which include the front-end amplifiers, surface acoustic wave (SAW) bandpass filters, and first mixer. Figure 19 demonstrates the effect of AM modulation on the jamming signal. Peak values of the jammer signal are used for the comparison. In general, the modulation makes the radio 10 dB more susceptible to jamming.

The main test for HSS is to show that its FH feature will make the system jam-resistant at all jamming frequencies. Experiments showed that the hopping frequencies have to be judiciously chosen such that within a redundant triplet, no two of the three frequencies would be near each other, since this would let a single jammer jam both frequencies. Therefore, the pattern could not be truly random but would need somewhat of a trend. Figure 20 shows the hybrid DS/FFH jamming susceptibility versus frequency. Two receiver gain versions of the HSS were evaluated in this scenario. The difference in gain between the low-gain and high-gain version is 5 dB. Eventually, an automatic adjustment will be developed to choose the best value for a particular environment. It is noticed in Figure 20 that the smaller signal has less distortion and is able to better reject the undesired jamming signal at almost all frequencies.

8 Conclusion

In this paper, the performance of a hybrid DS/FFH system over Rician fading channels was considered. We derived the average BER for a hybrid DS/FFH system that includes the effects from wide- and partial-band jamming, multi-user interference, and/or varying degrees of Rician fading. Numerical results exploring the parameter space of the HSS system have also been presented to demonstrate its effectiveness under different conditions and scenarios. We have also demonstrated a novel non-convex optimization technique that minimizes the bit-error probability of a hybrid DS/FFH communication system under multiple constraints. By employing the Karush-Kuhn-Tucker conditions, the process solves for the optimal system design parameters. In addition, a hardware FPGA-based hybrid DS/FFH prototype was implemented successfully and optimized for a typical smart grid utility application. Experimental results indicate that high resistance of hybrid DS/FFH systems to other jamming and interference signals allows the possibility of intentionally operating several HSS radios in the band simultaneously. For smart grid applications, this would enable a base station to service several clients at the same time, provided the system arranged for different clients to use different hop patterns and DS codes, and possibly even coordinated transmission time windows. The use of hybrid DS/FFH waveform in wireless networks as employed in the smart grid is recommended, as it offers superior resistance to jamming attacks and improves the reliability of transmission compared to existing SS techniques like DS, FH, and hybrid DS/SFH systems.

Competing interests

The authors declare that they have no competing interests.

Authors' contributions

MO carried out the hybrid DS/FFH system performance evaluation studies, conducted the computer simulations that provided numerical results, participated in deriving the analytical expressions and conducting experimental evaluations, and drafted the manuscript. XM derived the analytical expressions for evaluating and optimizing the performance and helped in drafting the manuscript. SK implemented the hybrid DS/FFH radio transceiver and evaluated its performance experimentally. TK shaped the main idea of the study and participated in its design, development, and coordination. SS conceived of the study and participated in investigating and enhancing its security performance. SD guided the analytical derivations and analysis. All authors read and approved the final manuscript.

Acknowledgment

This manuscript has been authored by UT-Battelle, LLC under Contract No. DE-AC05-00OR22725 with the U.S. Department of Energy. The United States Government retains and the publisher, by accepting the article for publication, acknowledges that the United States Government retains a non-exclusive, paid-up, irrevocable, world-wide license to publish or reproduce the published form of this manuscript, or allow others to do so, for United States Government purposes. The Department of Energy will provide public access to these results of federally sponsored research in accordance with the DOE Public Access Plan. In addition, this work has been partially supported by NSF grant CMMI-1334094.

Author details

[1]Computational Sciences and Engineering Division, Oak Ridge National Laboratory, P.O. Box 2008, MS 6085, Oak Ridge, TN 37831, USA. [2]Department of Electrical Engineering and Computer Science, University of Tennessee, 1520 Middle Drive, Knoxville, TN 37996, USA. [3]Department of Electrical Engineering and Computer Science, Masdar Institute of Science and Technology, Masdar City, Abu Dhabi, UAE.

References

1. MP Pursley, Direct sequence spread spectrum communications for multipath channels. IEEE Trans Microwave Theory Tech **50**(3), 653–661 (2002)
2. EA Geraniotis, Noncoherent hybrid DS-SFH spread-spectrum multiple-access communications. IEEE Trans Commun **34**(9), 862–872 (1986)
3. J Zhang, KC Teh, KH Li, Error probability analysis of FFH/MFSK receivers over frequency-selective Rician-fading channels with partial band noise jamming. IEEE Trans Commun **57**(10), 2880–2885 (2009)
4. JH Lee, BS Yu, SC Lee, Probability of error for a hybrid spread spectrum system under tone jamming, in *Proc. of the IEEE Military Communications Conference (MILCOM'90)*, 1990, pp. 410–414
5. MM Olama, X Ma, PT Kuruganti, SF Smith, SM Djouadi, Hybrid DS/FFH spread-spectrum: a robust, secure transmission technique for communication in harsh environments, in *Proc. of the IEEE Military Communications Conference (MILCOM'11)*, 2011, pp. 2136–2141

6. Y Fu, H Leung, Narrow-band interference cancellation in spread-spectrum communication systems using chaos. IEEE Trans Circuit Syst I **48**(7), 847–858 (2001)
7. Security Profile for Advanced Metering Infrastructure (AMI), The Advanced Security Acceleration Project (ASAP-SG), Version 2.0, Jun. 2010. Available at: http://osgug.ucaiug.org/utilisec/amisec/Shared%20Documents/AMI%20Security%20Profile%20(ASAP-SG)/AMI%20Security%20Profile%20-%20v2_0.pdf
8. Analog Devices, A Technical Tutorial on Digital Signal Synthesis, Technical Report, 1999. Available at: http://www.analog.com/media/cn/training-seminars/tutorials/450968421DDS_Tutorial_rev12-2-99.pdf
9. PA Bello, Characterization of randomly time-variant linear channels. IEEE Trans Commun Syst **11**, 360–393 (1963)
10. DE Borth, MB Pursley, Analysis of direct-sequence spread-spectrum multiple access communication over Rician fading channels. IEEE Trans Commun **27** (10), 1566–1577 (1979)
11. EA Geraniotis, MB Pursley, Error probabilities for slow-frequency-hopped spread-spectrum multiple-access communications over fading channels. IEEE Trans Commun **30**(5), 996–1010 (1982)
12. EA Geraniotis, Coherent hybrid DS-SFH spread-spectrum multiple-access communications. IEEE Trans Commun **3**(5), 695–705 (1985)
13. B Solaiman, A Glavieux, A Hillion, Error probability of fast frequency hopping spread spectrum with BFSK modulation in selective Rayleigh and selective Rician fading channels. IEEE Trans Commun **38**(2), 233–240 (1990)
14. C Park, JH Lee, Probability of error for a hybrid DS/SFH spread spectrum system over a Rician fading channel in the presence of multiple-tone jamming, in *Proc. of the IEEE Second International Symposium on Spread Spectrum Techniques and Applications (ISSSTA'92)*, 1992, pp. 123–126
15. CC Chen, K Yao, K Umeno, E Biglieri, Design of spread-spectrum sequences using chaotic dynamical systems and ergodic theory. IEEE Trans Circuit Syst I **48**(9), 1110–1114 (2001)
16. MM Olama, SF Smith, PT Kuruganti, X Ma, Performance study of hybrid DS/FFH spread-spectrum systems in the presence of frequency-selective fading and multiple-access interference, in *Proc. of the IEEE International Communications Quality and Reliability (CQR) Conference*, 2012
17. X Ma, MM Olama, T Kuruganti, SF Smith, SM Djouadi, Determining system parameters for optimal performance of hybrid DS/FFH spread-spectrum, in *Proc. of the IEEE Military Communication Conference (MILCOM'12)*, 2012, pp. 1888–1893
18. M Killough, MM Olama, T Kuruganti, SF Smith, FPGA-based implementation of a hybrid DS/FFH spread-spectrum transceiver, in *Proc. of the World Congress in Computer Science, Computer Engineering, and Applied Computing (WORLDCOMP'13)*, 2013
19. MM Olama, SM Killough, T Kuruganti, TE Carroll, Design, implementation, and evaluation of a hybrid DS/FFH spread-spectrum radio transceiver, in *Proc. of the IEEE Military Communication Conference (MILCOM'14)*, 2014
20. DP Bertsekas, *Nonlinear Programming*, 3rd edn. (Athena Scientific, Belmont, Massachusetts, 1999)
21. M Schwartz, WR Bennett, S Stein, *Communication Systems and Techniques* (McGraw-Hill, New York, 1966)
22. HL Van Trees, *Detection, Estimation, and Modulation Theory, Part 111* (Wiley, New York, 1971)
23. RS Kennedy, *Fading Dispersive Communication Channels* (Wiley, New York, 1969)
24. MB Pursley, Performance evaluation for phase coded spread spectrum multiple-access communication-Part I: system analysis. IEEE Trans Comm **25**, 759–799 (1977)
25. TS Rappaport, *Wireless Communications: Principles and Practice* (Prentice Hall, New Jersey, 1996)
26. DV Sarwate, An upper bound on the aperiodic autocorrelation function for a maximal-length sequence. IEEE Trans Inf Theory **30**(4), 685–687 (1984)
27. D Taylor, Introduction to synchronous communications, a classic paper by. J Costas Proc IEEE **90**(8), 1459–1460 (2002)
28. Y-R Tsai, M-ary Spreading-Code-Phase-Shift-Keying modulation for DSSS multiple access systems. IEEE Trans Comm **57**(11), 3220–3224 (2009)

10

Adaptive reconfigurable V-BLAST type equalizer for cognitive MIMO-OFDM radios

Mehmet Tahir Ozden

Abstract

An adaptive channel shortening equalizer design for multiple input multiple output-orthogonal frequency division multiplexing (MIMO-OFDM) radio receivers is considered in this presentation. The proposed receiver has desirable features for cognitive and software defined radio implementations. It consists of two sections: MIMO decision feedback equalizer (MIMO-DFE) and adaptive multiple Viterbi detection. In MIMO-DFE section, a complete modified Gram-Schmidt orthogonalization of multichannel input data is accomplished using sequential processing multichannel Givens lattice stages, so that a Vertical Bell Laboratories Layered Space Time (V-BLAST) type MIMO-DFE is realized at the front-end section of the channel shortening equalizer. Matrix operations, a major bottleneck for receiver operations, are accordingly avoided, and only scalar operations are used. A highly modular and regular radio receiver architecture that has a suitable structure for digital signal processing (DSP) chip and field programable gate array (FPGA) implementations, which are important for software defined radio realizations, is achieved. The MIMO-DFE section of the proposed receiver can also be reconfigured for spectrum sensing and positioning functions, which are important tasks for cognitive radio applications. In connection with adaptive multiple Viterbi detection section, a systolic array implementation for each channel is performed so that a receiver architecture with high computational concurrency is attained. The total computational complexity is given in terms of equalizer and desired response filter lengths, alphabet size, and number of antennas. The performance of the proposed receiver is presented for two-channel case by means of mean squared error (MSE) and probability of error evaluations, which are conducted for time-invariant and time-variant channel conditions, orthogonal and nonorthogonal transmissions, and two different modulation schemes.

Keywords: V-BLAST; MIMO-DFE; MIMO-OFDM; Cognitive radio; Software defined radio

1 Introduction

The fundamental problem in the design of future wireless communication systems is to reliably and efficiently transmit and receive information signals over imperfect channels using substantially high data rates. One successful approach adopted in several wireless standards such as digital audio broadcasting (DAB), digital video broadcasting (DVB-T), local area networking (LAN), and metropolitan area networking (MAN) is orthogonal frequency division multiplexing (OFDM) in which the entire bandwidth is divided into several narrow subbands so that the frequency response over each individual subband is relatively flat, and each subband channel occupies only a small fraction of the original bandwidth. Nevertheless, OFDM-based wireless communication systems can support peak rates of 54 Mbps at most, which is not enough to cover services the providers offer nowadays. Throughputs far beyond 54 Mbps can be provided when multiple input multiple output (MIMO) system approach is applied especially in a rich scattering environment [1,2]. Hence, the combination of OFDM and MIMO technologies [3,4] constitutes the basis for next-generation wireless communication systems such as IEEE 802.11n for wireless local area networks (WLAN), IEEE 802.16e for MAN, and evaluation of higher generational cellular systems.

The performance achieved through MIMO technology also entails a considerable increase in signal processing complexity in receiver, and there exists a major challenge in designing low-complexity receivers for multichannel wireless systems. A significant development for MIMO communications is the proposition of V-BLAST architecture [5], and two important areas of research on V-BLAST

Correspondence: mtozden@pirireis.edu.tr
Piri Reis University, Tuzla, 34940 Istanbul, Turkey

receivers are reduction of complexity, i.e., avoidance of matrix inversions, and extension to broadband implementation. Consequently, the recent research such as [6-8] focused on reducing the complexity of V-BLAST receiver architectures for frequency selective channels by developing efficient matrix inversion operations.

An important problem in realizing OFDM system designs, however, is the appendage of a cyclic prefix (CP) with a length at least equal to the channel length to each block of N IFFT coefficients, and this application may not be adequate in case the length of CP, ξ, is large relative to the data length, N, so that the channel throughput is reduced by a factor $N/(N+\xi)$. In addition, information about the channel length may not even be available in some cases. Accordingly, it is desired to design systems that guarantee a certain amount of throughput, i.e., $N/(N+\xi) \cong 1$, in all possible channel conditions. An elegant solution to this problem is to implement a time domain equalizer to shorten the channel memory and hence reduce the CP overhead [9].

A recent development in the design of next-generation wireless communication systems is the cognitive radio, built on a software radio, which functions as an intelligent system that is aware of its environment and uses the methodology of understanding-by-building to learn from the environment and adapt to statistical variations in the input stimuli in order to establish reliable communication by efficient utilization of the radio spectrum [10]. The concept of software radio on the other hand relies on the development of DSP technology, that is flexible, reconfigurable, and reprogrammable by software to adapt to an environment where there are multiple services, standards, and frequency bands [11]. Correspondingly, the infrastructure in a software radio system is generally required to use reconfigurable VLSI hardware components such as DSP chip sets [12], FPGAs [13], embedded processors [14], and even general purpose processors [15].

A typical cognitive radio cycle includes spectrum sensing, analysis, reasoning, and adaptation to new operating parameter steps [16]. The cognitive radio can detect the availability of a portion of frequency band through spectrum sensing and analysis steps [17]. During the reasoning step, it determines the optimum operating parameters, so that no harmful interference to other users of the spectrum is generated due to its transmission. In the adaptation step, the radio switches to transmission and reception mode using its reconfigurability and reprogrammability property and tunes its operating parameters according to its best response strategy. Another emerging requirement for cognitive radios is location and environment awareness that involves modeling the capabilities of human beings and bats for realization of advanced and autonomous location and environment awareness features [18]. Adaptive positioning, determining the coordinates of a cognitive radio in space, is a step towards realization of accurate location awareness in cognitive radios. The author has recently proposed a spectrum estimation method in [19] and a range estimation method in [20] that are suitable for spectrum sensing and positioning functions of cognitive radios, respectively. In this paper, we focus on the reception mode of operation of cognitive MIMO-OFDM radios and propose a new minimum MSE channel shortening equalizer design, which consists of adaptive fron-tend MIMO-DFE and multiple Viterbi detection sections.

The optimum solution for the channel shortening equalization problem can be found using one of these constraints: (1) unit tap constraint and (2) unit energy constraint; and the performances under these two criteria were compared for single input single output (SISO) and MIMO channel shortening equalization in [21,22], respectively. Since the findings in these papers show that the unit energy constrained channel shortener equalization resulted in better performance, we have used unit energy constraint for the MIMO channel shortening optimization problem under consideration in this paper. Accordingly, the contributions of the paper can be stated as follows: (1) the proposed equalizer has a front-end MIMO-DFE as opposed to the MIMO feed forward equalizer (MIMO-FFE) in [22], (2) a modified version of sequential processing multichannel lattice stages (SPMLSs) [23] is utilized in the design of front-end MIMO-DFE and a complete modified Gram-Schmidt orthogonalization of multichannel input data, which avoids matrix inversions, enables scalar only operations and contributes to the flexibility, reconfigurability, and reprogrammability of the receiver, is attained, (3) the proposed equalizer can be viewed as a V-BLAST receiver for frequency selective channels, (4) spectrum sensing or range estimation can be accomplished at no cost by simply reconfiguring the front-end MIMO-DFE as multichannel spectral analysis or positioning filter as shown in [19,20], respectively, and (5) a detailed computational complexity and performance analysis is presented. The first contribution is important from the perspective of interference removal and, by means of that, error performance, whereas the second one is considered the key since matrix inversion is a major bottleneck in the design of embedded receiver architectures that increases computational complexity [24]. The third one is relevant since the receiver operations of a V-BLAST system can be considered as performing Gram-Schmidt orthogonalization [25], whereby inter-symbol interference (ISI) as well as interchannel interference (ICI) effects are suppressed. The fourth contribution is crucial from the point of view of cognitive radio operation cycle, so that the filter structure of MIMO-DFE can be reused for spectrum sensing and range estimation functions of cognitive radio, and finally, a comparative computational complexity and performance analysis of the proposed equalizer with respect to the other benchmark equalizers such as MIMO-DFE, MIMO-FFE, and multichannel Viterbi equalizer (VE), has been provided, which to the best of the author's knowledge, does not exist in the literature.

Various MIMO-DFEs for MIMO ISI channels have been proposed in the literature [26-28] after the introduction of the finite length MIMO-DFE in [29], and it has been delineated by Ginis and Cioffi in [30] that DFE is the basic principle behind the BLAST detection algorithm. Very recently, a QR decomposition-based MIMO-DFE has been presented by Wang et al. in [31]. In QR decomposition approaches, the Q matrix is implicitly formed and then used to compute the R matrix; whereas in the Gram-Schmidt approach, the inverse of the R is implicitly formed and then used to compute the Q matrix. As a consequence of this fact, Regalia and Bellanger showed in [32] that there exists a duality between QR and lattice methods and the possibility of combining elements of both approaches to obtain new hybrid algorithms. With respect to developing these hybrid algorithms, Ling showed in [33] that an orthogonal Givens rotation-based algorithm algebraically coincides with the recursive modified Gram-Schmidt-based lattice algorithm in [34]. In accordance with this perspective, we modify the SPMLS using Givens rotation-based lattice algorithms of [33] on the structure of the SPMLS, so that a sequential processing multichannel Givens lattice stage (SPMGLS) is obtained. SPMLSs are known for their modular, order-recursive, and regular structure as we previously used them in the decision feedback equalization of nonlinear communication channels in [35]. Additionally, good numerical properties are incurred by the use of Givens rotation-based lattice algorithms.

Subsequent to Givens lattice realization of the front-end MIMO-DFE, we perform a systolic array implementation of multiple adaptive Viterbi detectors [36], thereby a highly concurrent receiver structure is obtained. A two-channel (2×2) problem is considered in this presentation due to the ease of explanation and space limitations in developing the method. However, it is considered straightforward to apply the method to any number of channels, i.e., to massive MIMO implementations for next-generation wireless systems [37], at the expense of increased complexity. Even though the complete orthogonality and thereby the suppression of ISI and ICI is accomplished in the minimum mean square error sense for any number of transmit and receive antennas, the performance achieved in terms of MSE as well as in the probability of error will depend on how much the channel is ill-conditioned.

The organization of this paper is as follows. In Section 2, the adaptive multichannel channel shortening equalization optimization problem is introduced. In Section 3, we describe the adaptive multiple Viterbi detection section of the proposed equalizer. The computational complexity computations are treated in Section 4. In Section 5, we present the experimental results, and finally, Section 6 is concerned with the discussion of results and conclusions. $(\bullet)^*$ represents the complex conjugate of $(\bullet)$. $(\bullet)^T$, and $(\bullet)^H$ stands for the transpose and the Hermitian transpose of $(\bullet)$, respectively. The variables m, i, and n are global while all other variables are local. The variable m represents the stage number while i and n are the time indexes related to data and coefficients, respectively, till we equate them in Section 3 to have a single time index.

2 Optimization problem statement

We consider the discrete-time baseband equivalent 2×2 channel shortening equalization problem depicted in Figure 1, where the number of transmitters (M_T) and receivers (M_R) is assumed equal, so that the number of antennas (M) is $M = M_T = M_R = 2$. Note that the kth baseband channel in Figure 1 models the effects of serial-to-parallel (S/P) and parallel-to-serial (P/S) conversions, the addition and removal of CP, the IDFT and DFT operations, and the physical baseband channel itself as delineated in Figure 2. Accordingly, the input signal to the kth receiver can be expressed as the sum of transmitted signals corrupted by ISI, ICI, and the noise:

$$y_k(n) = \sum_{\ell=1}^{M} \sum_{j=0}^{\mu_{\ell,k}} x_\ell(j) h_{\ell,k}(n-j) + u_k(n), \quad k = 1, 2, \tag{1}$$

in which $h_{\ell,k}(n)$ is the impulse response of the channel between the ℓth transmitter and the kth receiver, whereas $x_\ell(n)$ is the transmitted sequence, and $\mu_{\ell,k}$ is the memory of channel $h_{\ell,k}(n)$. Accordingly, $h_{\ell,k}(n)$ for $\ell \neq k$ constitute ICI, and the elements of $h_{\ell,k}(n)$ for $\ell = k$ and $\mu_{\ell,k} \neq 0$ amount to ISI. $u_k(n)$ denotes the kth channel noise.

In the adaptive two-channel channel shortening equalizer design problem, the objective is to find an exponentially windowed, least squares (LS) solution for the coefficients of the kth adaptive DFE and the corresponding kth adaptive desired impulse response (ADIR) filter that minimizes the kth cost function:

$$J_k(n) = \sum_{i=0}^{n} \beta^{n-i} \left| e_n^k(i) \right|^2 \tag{2}$$

at each time instant n, where $k = 1, 2$ and β is the exponential weighting factor. Herein, the error signal, $e_n^k(i)$, is given by:

$$e_n^k(i) = d_n^k(i) - \hat{d}_n^k(i). \tag{3}$$

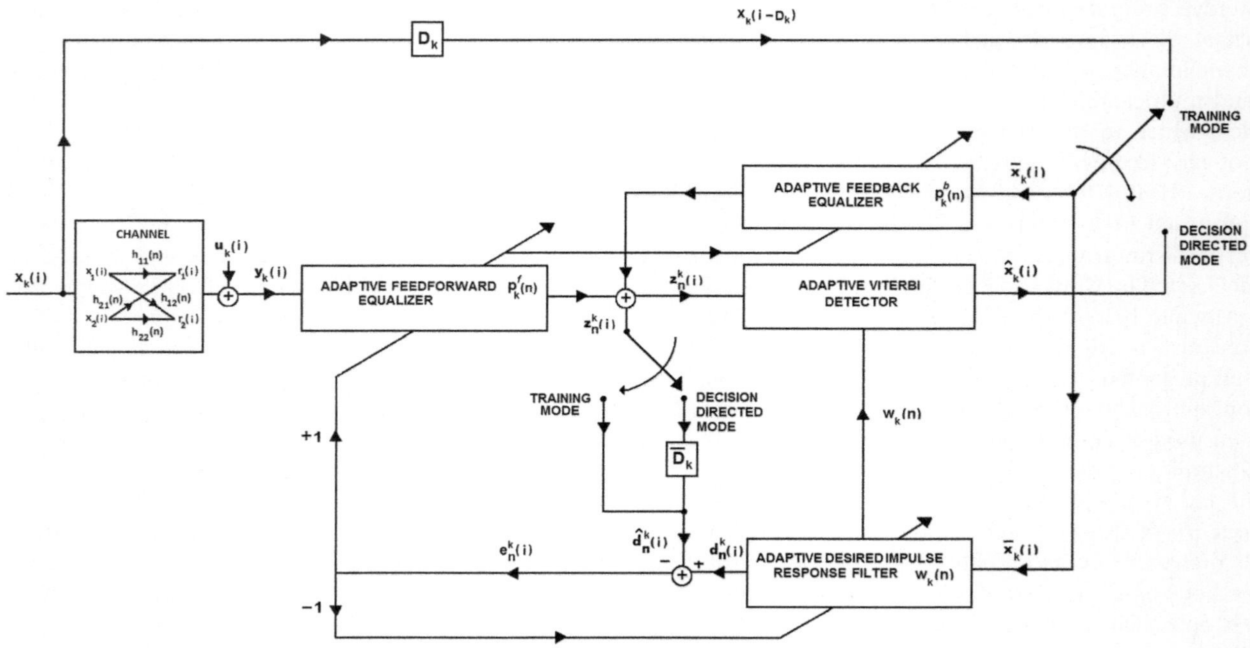

Figure 1 A block diagram of the baseband 2 × 2 channel shortening equalization problem.

The kth desired signal, $d_n^k(i)$, at the output of the kth ADIR is expressed as:

$$d_n^k(i) = \sum_{n=0}^{N_k^d-1} w_k(n)\bar{x}_k(i-n), \tag{4}$$

whereas the estimate of kth desired signal, $\hat{d}_n^k(i)$, in training mode of operation, is equal to the output signal of the kth DFE:

$$\hat{d}_n^k(i) = z_n^k(i) = \sum_{n=0}^{N_k^f-1} p_k^f(n)y_k(i-n) + \sum_{n=0}^{N_k^b-1} p_k^b(n)\bar{x}_k(i-n), \tag{5}$$

and it is identical to the delayed output signal of the kth DFE:

$$\hat{d}_n^k(i) = z_n^k\left(i-\overline{D}_k\right) = \sum_{n=0}^{N_k^f-1} p_k^f(n)y_k\left(i-\overline{D}_k-n\right) + \sum_{n=0}^{N_k^b-1} p_k^b(n)\bar{x}_k\left(i-\overline{D}_k-n\right) \tag{6}$$

in decision directed mode. Note that N_k^d in (4) and N_k^f and N_k^b in Equations (5) and (6) are the length of the kth ADIR filter, and the lengths of feed forward and feedback sections of the kth DFE, respectively, whereas $\left(N_k^d-1\right)$, $\left(N_k^f-1\right)$, and $\left(N_k^b-1\right)$ are the corresponding filter and equalizer memories. Accordingly, $\overline{D}_k$ is the delay experienced by the output signal of kth DFE, $z_n^k(i)$, during the Viterbi processing. The input signal ($\bar{x}_k(i)$) to the feedback section and to the kth ADIR filter in Equations (4),(5), and

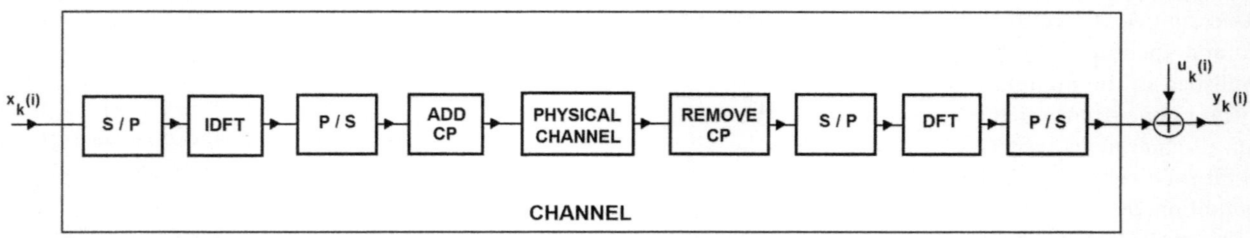

Figure 2 Block diagram of the kth baseband channel model.

(6) equals to the delayed kth channel input signal in training mode ($\bar{x}_k(i) = x_k(i - D_k)$), while it is equal to the kth detected signal in decision directed mode of operation of the receiver ($\bar{x}_k(i) = \hat{x}_k(i)$). Herein, D_k represents the delay experienced by the kth input signal, $x_k(i)$, through the corresponding channel during the training mode of operation.

Subsequently, we define the input vector to the kth DFE, $\mathbf{y}_k(i)$, at time instant i, as:

$$\mathbf{y}_k(i) = \left[y_k(i), y_k(i-1), \ldots, y_k\left(i - N_k^f + 1\right), \bar{x}_k(i), \bar{x}_k(i-1), \ldots, \bar{x}_k\left(i - N_k^b + 1\right)\right]^T, \tag{7}$$

and the corresponding coefficient vector, $\mathbf{p}_k(n)$, at time instant n, as:

$$\mathbf{p}_k(n) = \left[p_k^f(n), p_k^f(n-1), \ldots, p_k^f\left(n - N_k^f + 1\right), p_k^b(n), p_k^b(n-1), \ldots, p_k^b\left(n - N_k^b + 1\right)\right]^T. \tag{8}$$

The input vector to the kth ADIR filter at time instant i and its coefficient vector at the time instant n are also defined as:

$$\bar{\mathbf{x}}_k(i) = \left[\bar{x}_k(i), \bar{x}_k(i-1), \ldots, \bar{x}_k\left(i - N_k^d + 1\right)\right]^T \tag{9}$$

and

$$\mathbf{w}_k(n) = \left[w_k(n), w_k(n-1), \ldots, w_k\left(n - N_k^d + 1\right)\right]^T. \tag{10}$$

Note that we assume, without loss of generality, that $N_k^b \leq N_k^f$ for the kth DFE.

The main concern of the exponentially weighted LS problem under consideration is thus to find, at each time n, the kth optimal coefficient vectors, $\mathbf{p}_k(n)$ and $\mathbf{w}_k(n)$, that would minimize the cost function:

$$J_k(n) = \sum_{i=0}^{n} \beta^{n-i} \left|\mathbf{w}_k^H(n)\bar{\mathbf{x}}_k(i) - \mathbf{p}_k^H(n)\mathbf{y}_k(i)\right|^2, \tag{11}$$

which can be expressed in matrix form as follows:

$$J_k(n) = \mathbf{w}_k^H(n)\, \mathbf{R}_{\bar{x}_k\bar{x}_k}(n)\, \mathbf{w}_k(n) - \mathbf{w}_k^H(n)\, \mathbf{R}_{\bar{x}_k y_k}(n)\, \mathbf{p}_k(n) - \mathbf{p}_k^H(n)\, \mathbf{R}_{y_k\bar{x}_k}(n)\, \mathbf{w}_k(n) + \mathbf{p}_k^H(n)\, \mathbf{R}_{y_k y_k}(n)\, \mathbf{w}_k(n). \tag{12}$$

Herein, $\mathbf{R}_{y_k y_k}(n)$ is the $\left(N_k^f + N_k^b\right) \times \left(N_k^f + N_k^b\right)$ correlation matrix of $\mathbf{y}_k(i)$, which is given by:

$$\mathbf{R}_{y_k y_k}(n) = \sum_{i=0}^{n} \beta^{n-i}\mathbf{y}_k(i)\mathbf{y}_k^H(i), \tag{13}$$

$\mathbf{R}_{\bar{x}_k y_k}(n)$ is the $N_k^d \times \left(N_k^f + N_k^b\right)$ cross-correlation matrix of $\bar{\mathbf{x}}_k(n)$ and $\mathbf{y}_k(n)$, which can be expressed as:

$$\mathbf{R}_{\bar{x}_k y_k}(n) = \sum_{i=0}^{n} \beta^{n-i}\bar{\mathbf{x}}_k(i)\mathbf{y}_k^H(i), \tag{14}$$

and $\mathbf{R}_{\bar{x}_k\bar{x}_k}(n)$ is the $N_k^d \times N_k^d$ autocorrelation matrix of the kth ADIR filter input data vector $\bar{\mathbf{x}}_k(i)$, and is found by:

$$\mathbf{R}_{\bar{x}_k\bar{x}_k}(n) = \sum_{i=0}^{n} \beta^{n-i}\bar{\mathbf{x}}_k(i)\bar{\mathbf{x}}_k^H(i). \tag{15}$$

Note that $\mathbf{R}_{\bar{x}_k y_k}(n) \triangleq \mathbf{R}_{y_k\bar{x}_k}^H(n)$. Subsequently, the kth optimal coefficient vector for the equalizer is determined by differentiating $J_k(n)$ with respect to $\mathbf{p}_k(n)$, setting the derivative to zero, and solving for $\mathbf{p}_k(n)$:

$$\mathbf{p}_k^{opt}(n) = \mathbf{R}_{y_k y_k}^{-1}(n)\mathbf{R}_{y_k\bar{x}_k}(n)\mathbf{w}_k(n). \tag{16}$$

In order to find a solution for the optimal coefficient vector of the kth ADIR filter, we substitute Equation (16) back into the cost function in Equation (12) and attain the following quadratic form in $\mathbf{w}_k(n)$:

$$J_k(n) = \mathbf{w}_k^H(n)\left[\mathbf{R}_{\bar{x}_k\bar{x}_k}(n) - \mathbf{R}_{\bar{x}_k y_k}(n)\mathbf{R}_{y_k y_k}^{-1}(n)\mathbf{R}_{y_k\bar{x}_k}(n)\right]\mathbf{w}_k(n). \tag{17}$$

Then, the expression enclosed by square brackets in Equation (17), a symmetric $N_k^d \times N_k^d$ matrix, is defined as:

$$\widetilde{\mathbf{R}}_{\bar{x}_k y_k}(n) = \mathbf{R}_{\bar{x}_k\bar{x}_k}(n) - \mathbf{R}_{\bar{x}_k y_k}(n)\mathbf{R}_{y_k y_k}^{-1}(n)\mathbf{R}_{y_k\bar{x}_k}(n), \tag{18}$$

so that the cost function in Equation (17) can be restated as:

$$J_k(n) = \mathbf{w}_k^H(n)\widetilde{\mathbf{R}}_{\bar{x}_k y_k}(n)\mathbf{w}_k(n). \tag{19}$$

In minimizing the expression in Equation (19), a unit energy constraint, $\mathbf{w}_k^H(n)\mathbf{w}_k(n) = 1$, is applied to the kth ADIR filter coefficients to avoid the trivial null equalizer solution [21], and the following Lagrangian expression is formed:

$$L(\mathbf{w}_k(n), \lambda) = \mathbf{w}_k^H(n)\widetilde{\mathbf{R}}_{\bar{x}_k y_k}(n)\mathbf{w}_k(n) + \lambda\left(\mathbf{w}_k^H(n)\mathbf{w}_k(n) - 1\right). \tag{20}$$

After taking the derivative of the expression in (20) and equating to zero, we get:

$$\widetilde{\mathbf{R}}_{\bar{x}_k y_k}(n)\, \mathbf{w}_k^{\mathrm{opt}}(n) = \lambda\, \mathbf{w}_k^{\mathrm{opt}}(n), \tag{21}$$

which shows that the optimal kth ADIR coefficient vector $\mathbf{w}_k^{\mathrm{opt}}(n)$ and λ are the unit magnitude eigenvector and eigenvalue of the matrix $\widetilde{\mathbf{R}}_{\bar{x}_k y_k}(n)$, respectively. If the expression on the righthand side of Equation (21) is substituted for $\widetilde{\mathbf{R}}_{\bar{x}_k y_k}(n)\mathbf{w}_k(n)$ in Equation (19) and also

$\mathbf{w}_k^H(n)$ in Equation (19) is replaced with $\mathbf{w}_k^{Hopt}(n)$, then the minimum cost can be stated as follows:

$$J_k(n) = \lambda\, \mathbf{w}_k^{Hopt}(n)\mathbf{w}_k^{opt}(n) = \lambda_{\min} \tag{22}$$

which demonstrates that the cost function is minimized by choosing $\mathbf{w}_k^{opt}(n)$ to be equal to the eigenvector of the matrix $\widetilde{\mathbf{R}}_{\tilde{x}_k y_k}$ and that the corresponding eigenvalue λ is the minimum eigenvalue of the matrix $\widetilde{\mathbf{R}}_{\tilde{x}_k y_k}$ and is represented with $\lambda_{\min}$. Consequently, the optimal coefficient vectors for the kth equalizer and the kth ADIR filter, $\mathbf{p}_k^{opt}(n)$ and $\mathbf{w}_k^{opt}(n)$, are given by Equations (16) and (21), respectively.

2.1 V-BLAST type MIMO-DFE

We would like to use a V-BLAST type design approach for the front-end filter of the proposed equalizer, and we also require the design of a single, multichannel, and compact equalizer structure, so that two separate equalizers and direct evaluations as in (16) are avoided, and the same filter can be reconfigured as spectral analysis or positioning filter. These objectives can be accomplished by considering the equivalence of V-BLAST and modified Gram-Schmidt orthogonalization operations, and therefore by completely orthogonalizing the two-channel input data of DFE using SPMGLSs, which provide scalar only operations, good numerical properties as well as modularity, regularity, order recursiveness, and reconfigurability to the solution of equalization problem under consideration. Hereupon, we present the modifications we propose to make in SPMLSs so as to obtain SPMGLSs and then the design of front-end multichannel DFE using SPMGLSs.

2.2 SPMGLS

A SPMGLS has a block structure as shown in Figure 3, and the input signal vectors to a SPMGLS are defined as follows: the input forward prediction error vector:

$$\mathbf{f}_{\ell-1}(i) = \left[f_{\ell-1}^{0}(i), f_{\ell-1}^{1}(i), \ldots, \ldots, f_{\ell-1}^{p-1}(i), f_{\ell-1}^{p}(i)\right]^T, \tag{23}$$

the backward prediction error vector:

$$\mathbf{b}_{\ell-1}(i) = \left[b_{\ell-1}^{0}(i), b_{\ell-1}^{1}(i), \ldots, \ldots, b_{\ell-1}^{p-1}(i), b_{\ell-1}^{p}(i)\right]^T, \tag{24}$$

and the estimation error vector:

$$\mathbf{e}_{\ell-1}(i) = \left[e_{\ell-1}^{0}(i), e_{\ell-1}^{1}(i), \ldots, \ldots, e_{\ell-1}^{p-1}(i), e_{\ell-1}^{p}(i)\right]^T. \tag{25}$$

The elements of input forward and backward prediction error vectors in Equations (23) and (24) are orthogonalized by using self-orthogonalization processors (SOPs), which are triangular-shaped processors in Figure 3. The outputs of SOPs are the orthogonalized forward prediction error vector:

$$\hat{\mathbf{f}}_{\ell-1}(i) = \left[\hat{f}_{\ell-1}^{0}(i), \hat{f}_{\ell-1}^{1}(i), \ldots, \ldots, \hat{f}_{\ell-1}^{p-1}(i), \hat{f}_{\ell-1}^{p}(i)\right]^T \tag{26}$$

and the orthogonalized backward prediction error vector:

$$\hat{\mathbf{b}}_{\ell-1}(i) = \left[\hat{b}_{\ell-1}^{0}(i), \hat{b}_{\ell-1}^{1}(i), \ldots, \ldots, \hat{b}_{\ell-1}^{p-1}(i), \hat{b}_{\ell-1}^{p}(i)\right]^T. \tag{27}$$

The elements of $\hat{\mathbf{f}}_{\ell-1}(i)$ are fed into a forward prediction reference-orthogonalization processor (ROP) in order to predict the elements of $\mathbf{b}_{\ell-1}(i-1)$ and to produce the stage output back prediction error vector $\mathbf{b}_\ell(i)$. The elements of $\hat{\mathbf{b}}_{\ell-1}(i)$ are fed into a ROP to perform p-channel joint process estimation and to produce the stage output estimation error vector $\mathbf{e}_\ell(i)$. Subsequently, the elements of $\hat{\mathbf{b}}_{\ell-1}(i)$ are delayed and are also fed into another ROP to obtain the stage output forward prediction error vector $\mathbf{f}_\ell(i)$.

There are two types of processing cells, single and double circular processors, in a SPMGLS as in the original SPMLS in [23]. Nevertheless, we change the processing equations implemented in these processing cells with the equations of the square root version of the Givens algorithm in [33]. The interconnections and signals propagating through these processing cells are shown in Figure 4. The processing cells symbolized with a double circle, which are also called boundary (angle computer) cells, perform the following equations:

$$d(i) = \sqrt{\beta d^2(i-1) + |x_{ref}(i)|^2} \tag{28}$$

$$c(i) = \frac{\sqrt{\beta}\, d(i-1)}{d(i)} \tag{29}$$

$$s(i) = x_{ref}^*(i)/d(i). \tag{30}$$

From Equations (28),(29), and (30), it can be shown that the parameters $c(i)$ and $s(i)$ satisfy the equation:

$$\begin{bmatrix} c(i) & s(i) \\ -s^*(i) & c(i) \end{bmatrix} \begin{bmatrix} \sqrt{\beta}\, d(i-1) \\ x_{ref}(i) \end{bmatrix} = \begin{bmatrix} d(i) \\ 0 \end{bmatrix}. \tag{31}$$

The connection between the input, $\gamma_{\text{in}}(i)$, and output, $\gamma_{\text{out}}(i)$, likelihood variables is defined as:

$$\gamma_{\text{out}}(i) = c(i)\gamma_{\text{in}}(i). \tag{32}$$

On the other hand, the processing cells symbolized with a single circle, which are called internal (rotator) cells, perform the following equations:

$$x_{\text{out}}(i) = c(i)x_{\text{in}}(i) - s^*(i)\sqrt{\beta}\,\kappa(i-1) \tag{33}$$

and

$$\kappa(i) = c(i)\sqrt{\beta}\,\kappa(i-1) + s(i)x_{\text{in}}(i). \tag{34}$$

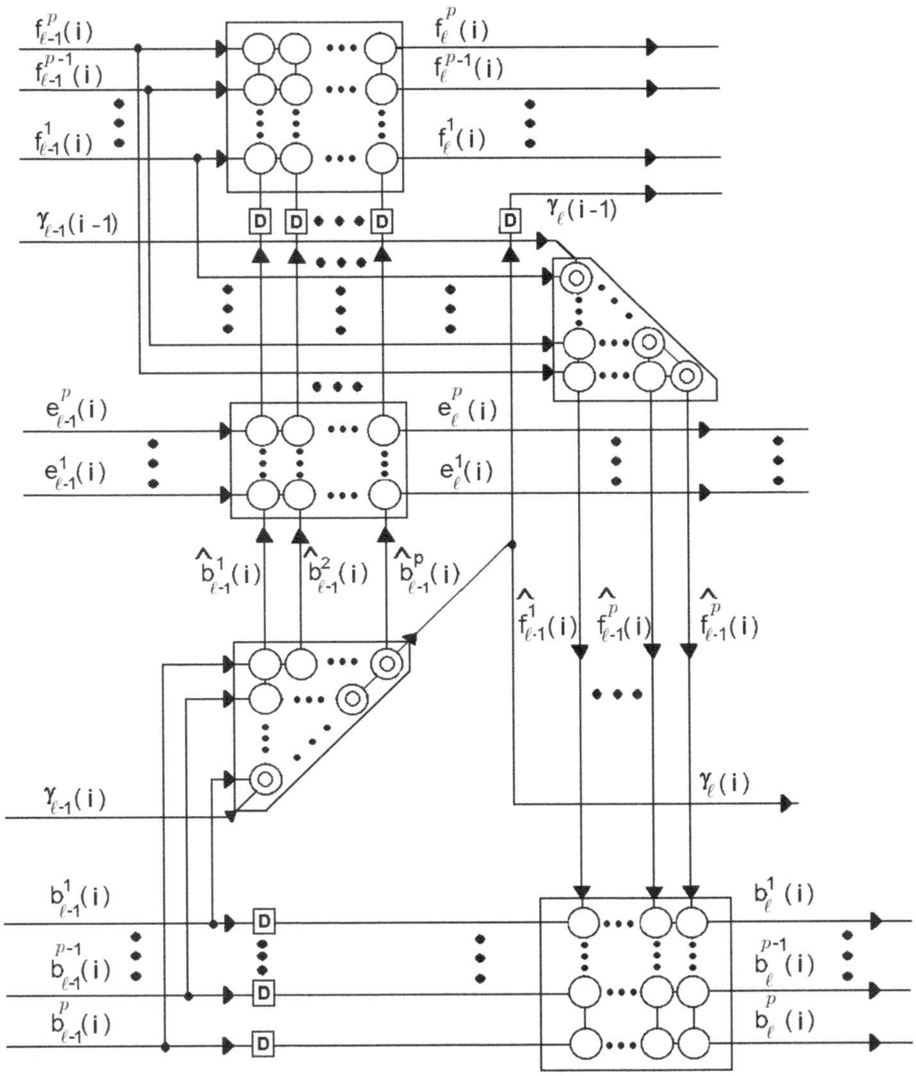

Figure 3 A block diagram of SPMGLS.

Accordingly, Equations (31), (33), and (34) can be combined into:

$$\begin{bmatrix} c(i) & s(i) \\ -s^*(i) & c(i) \end{bmatrix} \begin{bmatrix} \sqrt{\beta} d(i-1) & \sqrt{\beta}\kappa(i-1) \\ x_{\text{ref}}(i) & x_{\text{in}}(i) \end{bmatrix} = \begin{bmatrix} d(i) & \kappa(i) \\ 0 & x_{\text{out}}(i) \end{bmatrix}, \tag{35}$$

where

$$\mathbf{Q}(i) = \begin{bmatrix} c(i) & s(i) \\ -s^*(i) & c(i) \end{bmatrix} \tag{36}$$

with $|\mathbf{Q}(i)| = 1$, so that it performs Givens plane rotation in a complex plane, and thereby the stability of Givens algorithm is guaranteed.

2.3 Sequential givens lattice orthogonalization

The V-BLAST processing is made possible by utilizing SPMGLSs in the design of MIMO-DFE, so that the number of channels at different sections of the proposed multichannel lattice DFE is different due to the sequential processing nature of SPMGLSs. Therefore, we carry out the exponentially weighted LS optimization problem by taking into consideration each of these sections separately and assume that the proposed equalizer is comprised of three cascaded equalizers, which are two-channel, three-channel, and four-channel lattice sections; and we use a different index for each section while using m to indicate a stage in the whole equalizer. Henceforth, we focus on the case $N_1^f = N_2^f$ as it brings out the essential ideas

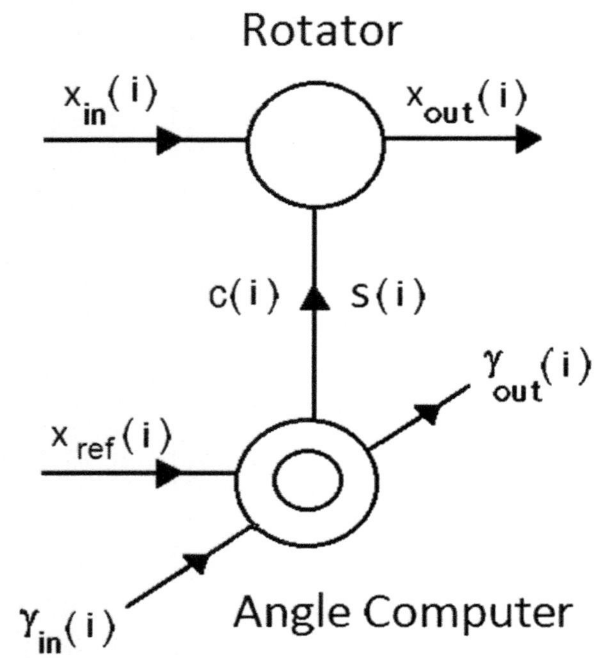

Figure 4 Processing cells in a SPMGLS.

without unduly complicating the development and loss of generality.

In order to sequentially solve the exponentially weighted LS optimization problem under consideration, we first organize the elements of input signal vectors $\mathbf{y}_1(i) = [y_1(i), \ldots, y_1(i-\ell+1)]^T$ and $\mathbf{y}_2(i) = [y_2(i), \ldots, y_2(i-\ell+1)]^T$, according to the natural ordering of SPMGLSs as:

$$\bar{\mathbf{y}}_\ell(i) = \begin{bmatrix} y_1(i) \\ y_2(i) \\ y_1(i-1) \\ y_2(i-1) \\ ---- \\ y_1(i-\ell+1) \\ y_2(i-\ell+1) \end{bmatrix} \tag{37}$$

and input to two-channel stages for which the stage number m has a range of values given by $0 < m \leq \left(N_1^f - N_1^b\right)$.

Accordingly, we redefine Equations (13) and (14) using this new data vector as follows:

$$\Re_{\bar{y}_\ell \bar{y}_\ell}(n) = \sum_{i=0}^{n} \beta^{n-i} \bar{\mathbf{y}}_\ell(i) \bar{\mathbf{y}}_\ell{}^H(i), \tag{38}$$

and

$$\Re_{\bar{x}_k \bar{y}_\ell}(n) = \sum_{i=0}^{n} \beta^{n-i} \bar{\mathbf{x}}_k(i) \bar{\mathbf{y}}_\ell^H(i), \tag{39}$$

where $k = 1, 2$. The orthogonalization of input data using SPMGLSs corresponds to the transformation of (38) and (39) into:

$$\Xi_\ell(n) = \sum_{i=0}^{n} \beta^{n-i} \Omega_\ell(n) \bar{\mathbf{y}}_\ell(i) \bar{\mathbf{y}}_\ell^H(i) \Omega_\ell^H(n) \tag{40}$$

and

$$\Gamma_{\ell,k}(n) = \sum_{i=0}^{n} \beta^{n-i} \Omega_\ell(n) \bar{\mathbf{y}}_\ell(i) \bar{\mathbf{x}}_k^H(i) \tag{41}$$

respectively. Here, $\Omega_\ell(n)$ is the $2\ell \times 2\ell$ lower triangular transformation matrix and is realized stage-by-stage using 2×2 lower triangular transformation matrices:

$$\mathbf{L}_\ell(n) = \begin{bmatrix} 1 & 0 \\ \hat{\kappa}_\ell(n-1) & 1 \end{bmatrix} \tag{42}$$

whose diagonal elements are all equal to unity at time instant n, and $\hat{\kappa}_\ell(n)$ is the reflection coefficient computed at the single circular cell in the triangular shaped self-orthogonalization processor of the ℓth two-channel SPMGLS. Then, the lattice joint process estimation coefficients are computed by means of:

$$\Theta_{\ell_k}(n) = \Xi_\ell^{-1}(n) \Gamma_{\ell,k}(n), \tag{43}$$

where $\Theta_{\ell_k}(n)$ represents the kth row of the $2 \times 2\ell$ lattice joint process estimation reflection coefficient matrix $\Theta_\ell(n)$ which is also sequentially computed stage-by-stage using 2×2 joint process estimation coefficient matrices:

$$\Delta_\ell(n) = \begin{bmatrix} \bar{\kappa}_{\ell,1,1}(n) & \bar{\kappa}_{\ell,1,2}(n) \\ \bar{\kappa}_{\ell,2,1}(n) & \bar{\kappa}_{\ell,2,2}(n) \end{bmatrix}, \tag{44}$$

in which $\bar{\kappa}_{\ell_{k,j}}(n)$ is the jth reflection coefficient related to the estimation of the kth desired signal, and it is computed at the (k, j)th single circular cell of the square shaped reference-orthogonalization processor related to joint process estimation at the ℓth two-channel SPMGLS. Note that the matrix inversion operation in Equation (16) is transformed into a simple scalar inversion operation in (43) due to the diagonal nature of $\Xi_\ell(n)$.

After the processing of input signals by two-channel lattice stages, the first estimation error signal, $\bar{x}_1(i) = e_1(i)$, which corresponds to the detected and fed back signal of the first channel, is incorporated at the $\left(N_1^f - N_1^b + 1\right)th$ stage as the third channel. Accordingly, we expand the optimization problem by organizing the elements of the input data vectors $\mathbf{y}_1(i) = [y_1(i), \ldots, y_1(i-\alpha+1)]^T$, $\mathbf{y}_2(i) =$

$\left[y_2(i), \ldots, y_2(i-\alpha+1)\right]^T$, and $\mathbf{x}_1(i) = [\bar{x}_1(i-1), \ldots, \bar{x}_1(i-\alpha+1)]^T$ as follows:

$$\bar{\mathbf{y}}_\alpha(i) = \begin{bmatrix} y_1(i) \\ y_2(i) \\ \bar{x}_1(i-1) \\ ---- \\ y_1(i-\alpha+1) \\ y_2(i-\alpha+1) \\ \bar{x}_1(i-\alpha+1) \end{bmatrix}, \qquad (45)$$

and input to the three-channel lattice section, where the stage number (m) takes values in the range given by $\left(N_1^f - N_1^b\right) < m \leq \left(N_2^f - N_2^b\right)$. Subsequently, we solve the optimization problem in (43) once again with the new input vector, in which case $\Omega_\alpha(n)$ and $\Theta_\alpha(n)$ are the $3\alpha \times 3\alpha$ lower triangular transformation and the $3 \times 3\alpha$ lattice joint process estimation coefficient matrices, respectively. $\Omega_\alpha(n)$ is computed sequentially by means of 3×3 lower triangular transformation matrices, $\mathbf{L}_\alpha(n)$ and $\Theta_\alpha(n)$, and is similarly realized stage-by-stage making use of 3×3 joint process estimation coefficient matrices, $\Delta_\alpha(n)$, at time instant n.

Finally, the optimization problem is expanded one more time with the inclusion of the second estimation error signal, $\bar{x}_2(i) = e_2(i)$, which is related to the detected and fed back signal of the second channel, and this time, the elements of input data vectors $\mathbf{y}_1(i) = \left[y_1(i), \ldots, y_1(i-\vartheta+1)\right]^T$, $\mathbf{y}_2(i) = \left[y_2(i), \ldots, y_2(i-\vartheta+1)\right]^T$, $\bar{\mathbf{x}}_1(i) = [\bar{x}_1(i-1), \ldots, \bar{x}_1(i-\vartheta+1)]^T$, and $\bar{\mathbf{x}}_2(i) = [\bar{x}_2(i-1), \ldots, \bar{x}_2(i-\vartheta+1)]^T$ are organized as:

$$\bar{\mathbf{y}}_\vartheta(i) = \begin{bmatrix} y_1(i) \\ y_2(i) \\ \bar{x}_1(i-1) \\ \bar{x}_2(i-1) \\ ---- \\ y_1(i-\vartheta+1) \\ y_2(i-\vartheta+1) \\ \bar{x}_1(i-\vartheta+1) \\ \bar{x}_2(i-\vartheta+1) \end{bmatrix}, \qquad (46)$$

where the stage number (m) is in the range given by $\left(N_2^f - N_2^b\right) < m \leq N_2^f$ due to four-channel processing. Similar to two-channel and three-channel cases, we solve the optimization problem in (43) using the new data vector in Equation (46), in which case $\Omega_\vartheta(n)$ and $\Theta_\vartheta(n)$ are $4\vartheta \times 4\vartheta$ lower triangular transformation, and $4 \times 4\vartheta$ joint process estimation error coefficient matrices at the time instant n, respectively. Similar to previous cases, these matrices are computed stage-by-stage by the use of 4×4 lower triangular transformation matrices, $\mathbf{L}_\vartheta(n)$, and 4×4 joint process estimation error coefficient matrices, $\Delta_\vartheta(n)$, at time instant n, respectively.

2.4 Computation of error order updates

Due to the sequential nature of the proposed lattice structure, we carry out the multichannel error order update task by taking into consideration two-channel, three-channel, and four-channel sections separately, and therefore we assume that the filter is comprised of three cascaded filters as described in the previous subsection. The prediction and joint state estimation errors for the end of the observation interval $n = i$ at the output of the ℓth order two-channel equalizer section, where $0 < m \leq \left(N_1^f - N_1^b\right)$, can be stated in terms of lattice parameters and the $(\ell-1)$th order prediction errors as follows:

$$\begin{bmatrix} f_\ell^1(n) \\ f_\ell^2(n) \end{bmatrix} = \begin{bmatrix} f_{\ell-1}^1(n) \\ f_{\ell-1}^2(n) \end{bmatrix} + \begin{bmatrix} \bar{\kappa}_{\ell,1,1}^{f*}(n-1) & \bar{\kappa}_{\ell,1,2}^{f*}(n-1) \\ \bar{\kappa}_{\ell,2,1}^{f*}(n-1) & \bar{\kappa}_{\ell,2,2}^{f*}(n-1) \end{bmatrix} \times \begin{bmatrix} 1 & 0 \\ \hat{\kappa}_\ell^{f*}(n-2) & 1 \end{bmatrix} \begin{bmatrix} b_{\ell-1}^1(n-1) \\ b_{\ell-1}^2(n-1) \end{bmatrix} \qquad (47)$$

and

$$\begin{bmatrix} b_\ell^1(n) \\ b_\ell^2(n) \end{bmatrix} = \begin{bmatrix} b_{\ell-1}^1(n-1) \\ b_{\ell-1}^2(n-1) \end{bmatrix} + \begin{bmatrix} \bar{\kappa}_{\ell,1,1}^{b*}(n-1) & \bar{\kappa}_{\ell,1,2}^{b*}(n-1) \\ \bar{\kappa}_{\ell,2,1}^{b*}(n-1) & \bar{\kappa}_{\ell,2,2}^{b*}(n-1) \end{bmatrix} \times \begin{bmatrix} 1 & 0 \\ \hat{\kappa}_\ell^{b*}(n-1) & 1 \end{bmatrix} \begin{bmatrix} f_{\ell-1}^1(n) \\ f_{\ell-1}^2(n) \end{bmatrix} \qquad (48)$$

where the lower triangular and square coefficient matrices are generated in triangular-shaped self-orthogonalization and square-shaped reference-orthogonalization processors in a two-channel SPMGLS as previously defined in Equations (42) and (44). The joint process estimation error updates are accordingly given as:

$$\begin{bmatrix} e_\ell^1(n) \\ e_\ell^2(n) \end{bmatrix} = \begin{bmatrix} e_{\ell-1}^1(n) \\ e_{\ell-1}^2(n) \end{bmatrix} + \begin{bmatrix} \bar{\kappa}_{\ell,1,1}^{e*}(n-1) & \bar{\kappa}_{\ell,1,2}^{e*}(n-1) \\ \bar{\kappa}_{\ell,2,1}^{e*}(n-1) & \bar{\kappa}_{\ell,2,2}^{e*}(n-1) \end{bmatrix} \times \begin{bmatrix} 1 & 0 \\ \hat{\kappa}_\ell^{e*}(n-1) & 1 \end{bmatrix} \begin{bmatrix} b_{\ell-1}^1(n) \\ b_{\ell-1}^2(n) \end{bmatrix}. \qquad (49)$$

We then multiply the lower triangular and square coefficient matrices in Equations (47), (48), and (49), and make the following definitions:

$$\mathbf{\Gamma}_\ell^f(n) = \begin{bmatrix} \Gamma_{\ell,1,1}^f(n) & \Gamma_{\ell,1,2}^f(n) \\ \Gamma_{\ell,2,1}^f(n) & \Gamma_{\ell,2,2}^f(n) \end{bmatrix} = \begin{bmatrix} \bar{\kappa}_{\ell,1,1}^f(n) + \bar{\kappa}_{\ell,1,2}^f(n)\hat{\kappa}_\ell^f(n-1) & \bar{\kappa}_{\ell,1,2}^f(n) \\ \bar{\kappa}_{\ell,2,1}^f(n) + \bar{\kappa}_{\ell,2,2}^f(n)\hat{\kappa}_\ell^f(n-1) & \bar{\kappa}_{\ell,2,2}^f(n) \end{bmatrix}, \qquad (50)$$

$$\mathbf{\Gamma}^{b}_{\ell}(n) = \begin{bmatrix} \Gamma^{b}_{\ell,1,1}(n) & \Gamma^{b}_{\ell,1,2}(n) \\ \Gamma^{b}_{\ell,2,1}(n) & \Gamma^{b}_{\ell,2,2}(n) \end{bmatrix} = \begin{bmatrix} \bar{\kappa}^{b}_{\ell,1,1}(n) + \bar{\kappa}^{b}_{\ell,1,2}(n)\hat{\kappa}^{b}_{\ell}(n) & \bar{\kappa}^{b}_{\ell,1,2}(n) \\ \bar{\kappa}^{b}_{\ell,2,1}(n) + \bar{\kappa}^{b}_{\ell,2,2}(n)\hat{\kappa}^{b}_{\ell}(n) & \bar{\kappa}^{b}_{\ell,2,2}(n) \end{bmatrix}, \tag{51}$$

and

$$\mathbf{\Gamma}^{e}_{\ell}(n) = \begin{bmatrix} \Gamma^{e}_{\ell,1,1}(n) & \Gamma^{e}_{\ell,1,2}(n) \\ \Gamma^{e}_{\ell,2,1}(n) & \Gamma^{e}_{\ell,2,2}(n) \end{bmatrix} = \begin{bmatrix} \bar{\kappa}^{e}_{\ell,1,1}(n) + \bar{\kappa}^{e}_{\ell,1,2}(n)\hat{\kappa}^{e}_{\ell}(n) & \bar{\kappa}^{e}_{\ell,1,2}(n) \\ \bar{\kappa}^{e}_{\ell,2,1}(n) + \bar{\kappa}^{e}_{\ell,2,2}(n)\hat{\kappa}^{e}_{\ell}(n) & \bar{\kappa}^{e}_{\ell,2,2}(n) \end{bmatrix}, \tag{52}$$

in order to obtain compact versions of the Equations (47), (48), and (49) as follows:

$$\begin{bmatrix} f^{1}_{\ell}(n) \\ f^{2}_{\ell}(n) \end{bmatrix} = \begin{bmatrix} f^{1}_{\ell-1}(n) \\ f^{2}_{\ell-1}(n) \end{bmatrix} + \mathbf{\Gamma}^{f*}_{\ell}(n-1) \begin{bmatrix} b^{1}_{\ell-1}(n-1) \\ b^{2}_{\ell-1}(n-1) \end{bmatrix}, \tag{53}$$

$$\begin{bmatrix} b^{1}_{\ell}(n) \\ b^{2}_{\ell}(n) \end{bmatrix} = \begin{bmatrix} b^{1}_{\ell-1}(n-1) \\ b^{2}_{\ell-1}(n-1) \end{bmatrix} + \mathbf{\Gamma}^{b*}_{\ell}(n-1) \begin{bmatrix} f^{1}_{\ell-1}(n) \\ f^{2}_{\ell-1}(n) \end{bmatrix}, \tag{54}$$

and

$$\begin{bmatrix} e^{1}_{\ell}(n) \\ e^{2}_{\ell}(n) \end{bmatrix} = \begin{bmatrix} e^{1}_{\ell-1}(n) \\ e^{2}_{\ell-1}(n) \end{bmatrix} + \mathbf{\Gamma}^{e*}_{\ell}(n-1) \begin{bmatrix} b^{1}_{\ell-1}(n) \\ b^{2}_{\ell-1}(n) \end{bmatrix}. \tag{55}$$

The development of prediction and joint process estimation error order updates from $(\alpha-1)$th order to αth for the three-channel section, where the stage number (m) takes values in the range given by $\left(N^{f}_{1} - N^{b}_{1}\right) < m \leq \left(N^{f}_{2} - N^{b}_{2}\right)$, is carried out in a similar fashion to the two-channel section, and they can be expressed in compact form with the following equations:

$$\begin{bmatrix} f^{1}_{\alpha}(n) \\ f^{2}_{\alpha}(n) \\ f^{3}_{\alpha}(n) \end{bmatrix} = \begin{bmatrix} f^{1}_{\alpha-1}(n) \\ f^{2}_{\alpha-1}(n) \\ f^{3}_{\alpha-1}(n) \end{bmatrix} + \mathbf{\Gamma}^{f*}_{\alpha}(n-1) \begin{bmatrix} b^{1}_{\alpha-1}(n-1) \\ b^{2}_{\alpha-1}(n-1) \\ b^{3}_{\alpha-1}(n-1) \end{bmatrix}, \tag{56}$$

$$\begin{bmatrix} b^{1}_{\alpha}(n) \\ b^{2}_{\alpha}(n) \\ b^{2}_{\alpha}(n) \end{bmatrix} = \begin{bmatrix} b^{1}_{\alpha-1}(n-1) \\ b^{2}_{\alpha-1}(n-1) \\ b^{3}_{\alpha-1}(n-1) \end{bmatrix} + \mathbf{\Gamma}^{b*}_{\alpha}(n-1) \begin{bmatrix} f^{1}_{\alpha-1}(n) \\ f^{2}_{\alpha-1}(n) \\ f^{3}_{\alpha-1}(n) \end{bmatrix}, \tag{57}$$

and

$$\begin{bmatrix} e^{1}_{\alpha}(n) \\ e^{2}_{\alpha}(n) \\ e^{2}_{\alpha}(n) \end{bmatrix} = \begin{bmatrix} e^{1}_{\alpha-1}(n) \\ e^{2}_{\alpha-1}(n) \\ e^{3}_{\alpha-1}(n) \end{bmatrix} + \mathbf{\Gamma}^{e*}_{\alpha}(n-1) \begin{bmatrix} b^{1}_{\alpha-1}(n) \\ b^{2}_{\alpha-1}(n) \\ b^{3}_{\alpha-1}(n) \end{bmatrix}, \tag{58}$$

where

$$\mathbf{\Gamma}^{f}_{\alpha}(n) = \begin{bmatrix} \bar{\kappa}^{f}_{\alpha,1,1}(n) & \bar{\kappa}^{f}_{\alpha,1,2}(n) & \bar{\kappa}^{f}_{\alpha,1,3}(n) \\ \bar{\kappa}^{f}_{\alpha,2,1}(n) & \bar{\kappa}^{f}_{\alpha,2,2}(n) & \bar{\kappa}^{f}_{\alpha,2,3}(n) \\ \bar{\kappa}^{f}_{\alpha,3,1}(n) & \bar{\kappa}^{f}_{\alpha,3,2}(n) & \bar{\kappa}^{f}_{\alpha,3,3}(n) \end{bmatrix} \times \begin{bmatrix} 1 & 0 & 0 \\ \hat{\kappa}^{f}_{\alpha,2,1}(n-1) & 1 & 0 \\ \hat{\kappa}^{f}_{\alpha,3,1}(n-1) & \hat{\kappa}^{f}_{\alpha,3,2}(n-1) & 1 \end{bmatrix},$$

$$\Gamma^{b}_{\alpha}(n) = \begin{bmatrix} \bar{\kappa}^{b}_{\alpha,1,1}(n) & \bar{\kappa}^{b}_{\alpha,1,2}(n) & \bar{\kappa}^{b}_{\alpha,1,3}(n) \\ \bar{\kappa}^{b}_{\alpha,2,1}(n) & \bar{\kappa}^{b}_{\alpha,2,2}(n) & \bar{\kappa}^{b}_{\alpha,2,3}(n) \\ \bar{\kappa}^{b}_{\alpha,3,1}(n) & \bar{\kappa}^{b}_{\alpha,3,2}(n) & \bar{\kappa}^{b}_{\alpha,3,3}(n) \end{bmatrix} \begin{bmatrix} 1 & 0 & 0 \\ \hat{\kappa}^{b}_{\alpha,2,1}(n) & 1 & 0 \\ \hat{\kappa}^{b}_{\alpha,3,1}(n) & \hat{\kappa}^{b}_{\alpha,3,2}(n) & 1 \end{bmatrix},$$

and

$$\mathbf{\Gamma}^{e}_{\alpha}(n) = \begin{bmatrix} \bar{\kappa}^{e}_{\alpha,1,1}(n) & \bar{\kappa}^{e}_{\alpha,1,2}(n) & \bar{\kappa}^{e}_{\alpha,1,3}(n) \\ \bar{\kappa}^{e}_{\alpha,2,1}(n) & \bar{\kappa}^{e}_{\alpha,2,2}(n) & \bar{\kappa}^{e}_{\alpha,2,3}(n) \\ \bar{\kappa}^{e}_{\alpha,3,1}(n) & \bar{\kappa}^{e}_{\alpha,3,2}(n) & \bar{\kappa}^{e}_{\alpha,3,3}(n) \end{bmatrix} \begin{bmatrix} 1 & 0 & 0 \\ \hat{\kappa}^{b}_{\alpha,2,1}(n) & 1 & 0 \\ \hat{\kappa}^{b}_{\alpha,3,1}(n) & \hat{\kappa}^{b}_{\alpha,3,2}(n) & 1 \end{bmatrix}.$$

The prediction and joint process estimation error order update equations from $(\vartheta-1)$th order to ϑth order for the four-channel section, where the stage number (m) is in the range given by $\left(N^{f}_{2} - N^{b}_{2}\right) < m \leq N^{f}_{2}$, can be derived by following a similar procedure to two- and three-channel sections, and 4×1 error order update matrices can be subsequently obtained with 4×4 lower triangular and square coefficient matrices.

2.5 Matrix visualization

In order to visualize the cascading and functioning of two-channel, three-channel, and four-channel sections as a single equalizer, we provide a matrix representation of sequential Givens lattice orthogonalization by considering $\left(N^{f}_{1}, N^{b}_{1}\right) = (13,7)$ and $\left(N^{f}_{2}, N^{b}_{2}\right) = (13,2)$ DFEs for the first and second channels, and also organizing the elements of input data vectors $\mathbf{y}_1(n) = [y_1(n), y_1(n-1), \ldots, y_1(n-11), y_1(n-12)]^T$, $\mathbf{y}_2(n) = [y_2(n), y_2(n-1), \ldots, y_2(n-11), y_2(n-12)]^T$, $\mathbf{x}_1(n) = [\bar{x}_1(n-1), \ldots, \bar{x}_1(n-6)]^T$, and $\mathbf{x}_2(n) = [\bar{x}_2(n-1)]^T$ as columns of a matrix:

$$\begin{bmatrix} y_1(n) & y_1(n-1) & \ldots & y_1(n-6) & y_1(n-7) & \ldots & y_1(n-11) & y_1(n-12) \\ y_2(n) & y_2(n-1) & \ldots & y_1(n-6) & y_2(n-7) & \ldots & y_2(n-11) & y_2(n-12) \\ & & & & \bar{x}_1(n-1) & \ldots & \bar{x}_1(n-5) & \bar{x}_1(n-6) \\ & & & & & & & \bar{x}_2(n-1) \end{bmatrix} \tag{59}$$

by taking into consideration different numbers of parameters in the feed forward and feedback channels and shifting properties of input data. This matrix helps us to visualize the orthogonalization process, and thus to draw a diagram of the four channel DFE structure under consideration as in Figure 5. Note that the elements of the first and second rows are related to the input signals of the

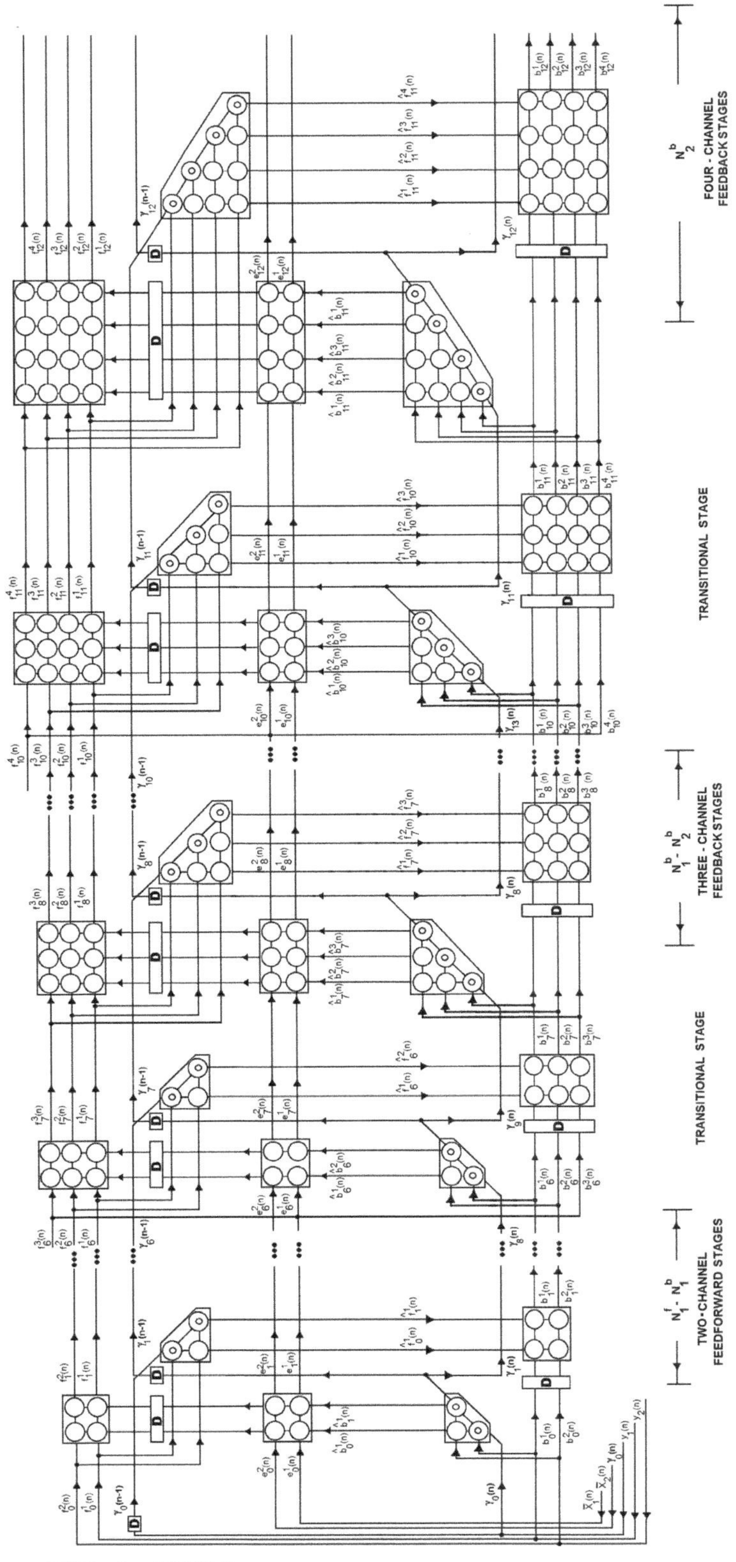

Figure 5 Diagram of four-channel DFE using SPMGLSs.

first and the second channels of the DFE under consideration, while the third and fourth rows are associated with the detected and fed back signals. Lattice orthogonalization begins with the elements of the first two rows using two-channel sequential lattice processing stages until the first fed back channel is incorporated as the new channel at a transitional stage, which is the $\left(N_1^f - N_1^b + 1\right)$th stage. Then, the orthogonalization continues with three-channel lattice stages until the fourth channel, which is related to the detected and fed back signal of the second channel, is taken into the process at another transitional stage, which is the $\left(N_2^f - N_2^b + 1\right)$th stage, and so the orthogonalization of input data finalizes with four-channel stages when the mean squared estimation error performance requirements are met, and thereby the kth desired signal, $d^k(n)$, is sequentially estimated using self-orthogonalized backward prediction error signals as follows:

$$\begin{aligned}\hat{d}_n^k(i) = & \sum_{m=1}^{\left(N_1^f - N_1^b\right)} \sum_{j=1}^{2} \bar{\kappa}_{m,k,j}^*(n-1)\hat{b}_{m-1}^j(i) \\ & + \sum_{m=\left(N_1^f - N_1^b+1\right)}^{\left(N_2^f - N_2^b\right)} \sum_{j=1}^{3} \bar{\kappa}_{m,k,j}^*(n-1)\hat{b}_{m-1}^j(i) \\ & + \sum_{m=\left(N_2^f - N_2^b+1\right)}^{N_2^f} \sum_{j=1}^{4} \bar{\kappa}_{m,k,j}^*(n-1)\hat{b}_{m-1}^j(i). \end{aligned} \quad (60)$$

Here, the first and second summations represent the estimation accomplished by the two-channel and three-channel sections, respectively, and the third summation is connected with the four-channel estimation section. In each section, $\bar{\kappa}_{m,k,j}(n)$ represents the jth estimation reflection coefficient at the mth stage related to the kth channel as defined in the previous subsection, and $\hat{b}_{m-1}^j(i)$ represents the jth element of the self-orthogonalized backward prediction error signal vector, $\hat{\mathbf{b}}_{m-1}(i)$, at the input of the mth stage. The self-orthogonalized backward prediction error vector, $\hat{\mathbf{b}}_{m-1}(i)$, is produced by the lower triangular transformation of the input backward prediction error vector, $\mathbf{b}_{m-1}(i)$, using $\mathbf{L}_m(n)$, and this operation is accomplished at the triangular shaped self-orthogonalization processor (related to forward prediction) of the mth SPMGLS. Note that the sizes of vectors, $\hat{\mathbf{b}}_{m-1}(i)$ and $\mathbf{b}_{m-1}(i)$, and matrix, $\mathbf{L}_m(n)$, at different sections of the proposed lattice equalizer are as follows: 2×1 and 2×2 in two-channel section, 3×1 and 3×3 in three-channel section, and 4×1 and 4×4 in four-channel section, respectively.

3 Adaptive multiple systolic Viterbi detection

In order to achieve an all systolic equalizer architecture, we propose to use the systolic array processor approach in [36] for the design and implementation of Viterbi detection, so that a high degree of computational concurrency is obtained by operating simultaneously and in synchronization with the rest of equalizer circuitry. Accordingly, the most computationally intensive operation in the Viterbi detection of sent data is related to the comparator metric, and a systolic computation of this metric is accomplished by multiply and accumulate operations:

$$\mathrm{STMO}_k^j = \mathrm{OSMI}_k^j + z_n^k(i) + \sum_{n=0}^{3N_k^d-3} c_k(n)x_k(i-n), \; j=1,2,\ldots,\upsilon, \quad (61)$$

for υ branches leading from states at time instant $i-1$ to each state at time instant i as illustrated in Figure 6. Herein, υ stands for the alphabet size, OSMI_k^j and STMO_k^j represent old survivor state metric input (OSMI) and state metric output (STMO) for the jth branch leading to the state constituted by the vector $\left[x_k(i-1)x_k(i-2)\ldots x_k(i-N_k^d+1)\right]$ related to the kth channel of equalizer, and $z_n^k(i)$ is the kth equalizer output as given in (5). Note that the elements of coefficient vector $\mathbf{c}_k(i)$:

$$\mathbf{c}_k(n) = \left[c_k(n), c_k(n-1), \ldots, c_k\left(n - 2N_k^d + 2\right)\right]^T \quad (62)$$

are computed as:

$$c_k(n) = \sum_{i=0}^{N_k^d-1} w_k(i)w_k(n-i) \quad (63)$$

where $\mathbf{w}_k$ represents the coefficient vector for the kth ADIR filter as defined in (10). Note that a processing element (PE) of the systolic array in Figure 6 is symbolized with a circled PE, and the memory of computation cycle is designated as $L = 3N_k^d - 3$ for the ease of illustration.

4 Computational complexity

The computational complexity can be calculated by considering two main sections of the proposed channel shortening equalizer. The first section implements the MIMO-DFE while the second one is related to the Viterbi processing. The number of operations required for the MIMO-DFE can be calculated by thinking about the number of operations per stage and the number of stages. The number of operations for a single SPMGLS with two, three, and four channels have been computed by making use of complexity calculations in [23] and [33] as 84,171, and 288, respectively. There are $\left(N_1^f - N_1^b\right)$ stages in two-channel section, and thereby the number of required

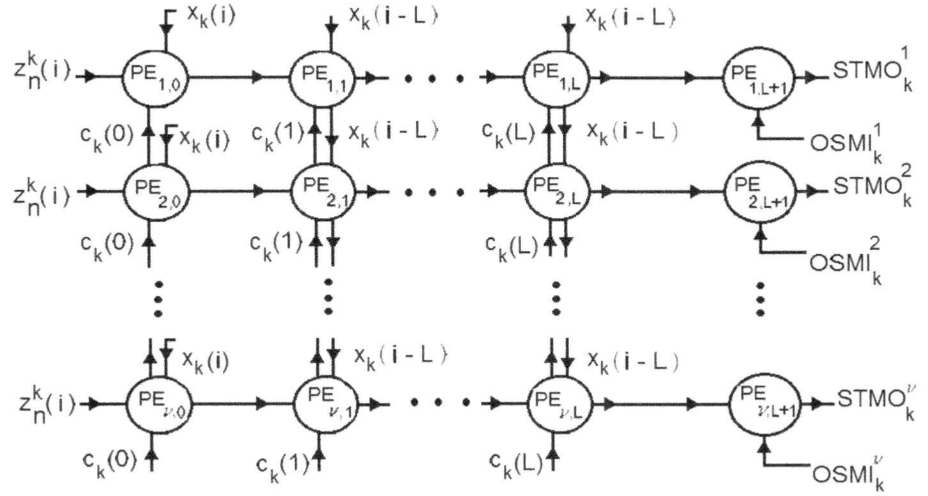

Figure 6 υ **branches from states at time instant** $i-1$ **to each state at time instant** i.

operations for this section sums up to $84\left(N_1^f - N_1^b\right)$. Similarly, the number operations for three and four channel sections are calculated as $171\left(N_1^b - N_2^b\right)$ and $288\,N_2^b$, respectively. Therefore, the total number of operations for the MIMO-DFE in Figure 5 becomes $84N_1^f + 87N_1^b + 117N_2^b$.

The complexity calculation for systolic Viterbi section can be accomplished as follows. The total number of processing elements in the systolic array that implements Viterbi processing per channel is given as $\upsilon \times N_k^d$ in [36]. Each element in this array performs one addition and one multiplication, which are counted together as one operation. Then, the total number of operations for the systolic implementation of M Viterbi detectors are calculated as $M \times \upsilon \times N_k^d$. Accordingly, the total computational complexity for the proposed equalizer taking into account both MIMO-DFE and multiple Viterbi detector sections becomes $84N_1^f + 87N_1^b + 117N_2^b + M\upsilon N_k^d$.

We compare the computational complexity of the proposed channel shortening equalizer using a front-end feed forward equalizer (CSFFE) and a front-end decision feedback equalizer (CSDFE) with those of VE in Figures 7,

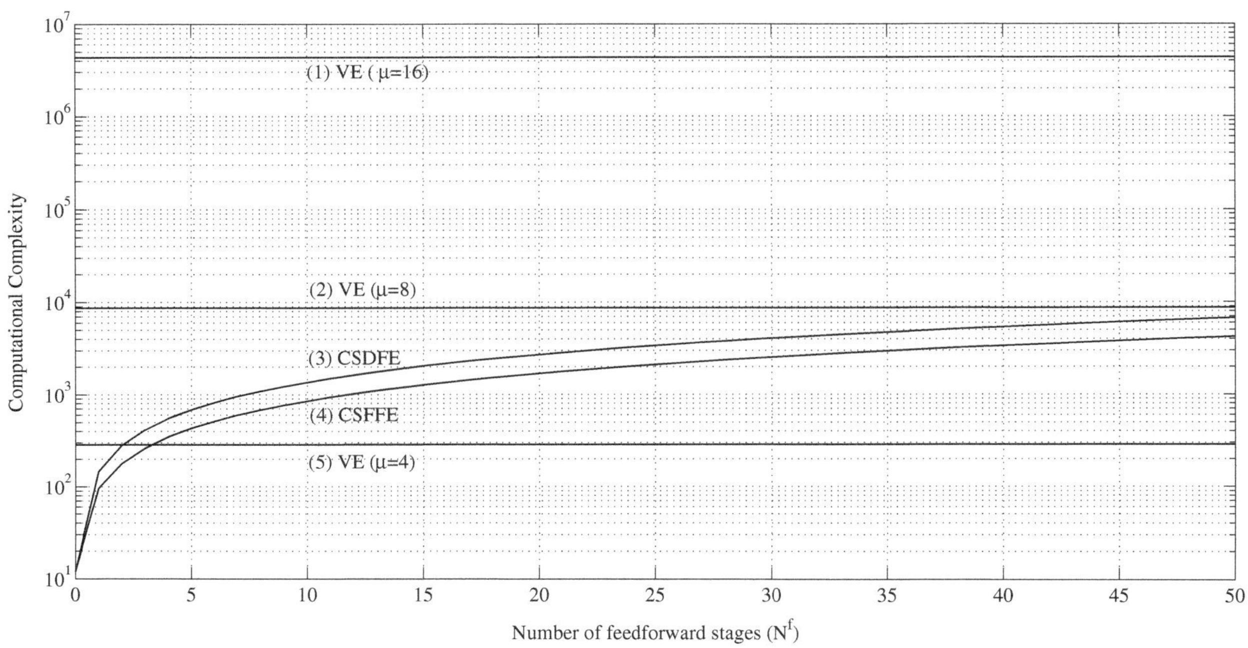

Figure 7 Comparative computational complexity of the proposed equalizer vs. VE when $M = 2$, $\upsilon = 2$, **and** $N^d - 1 = 2$.

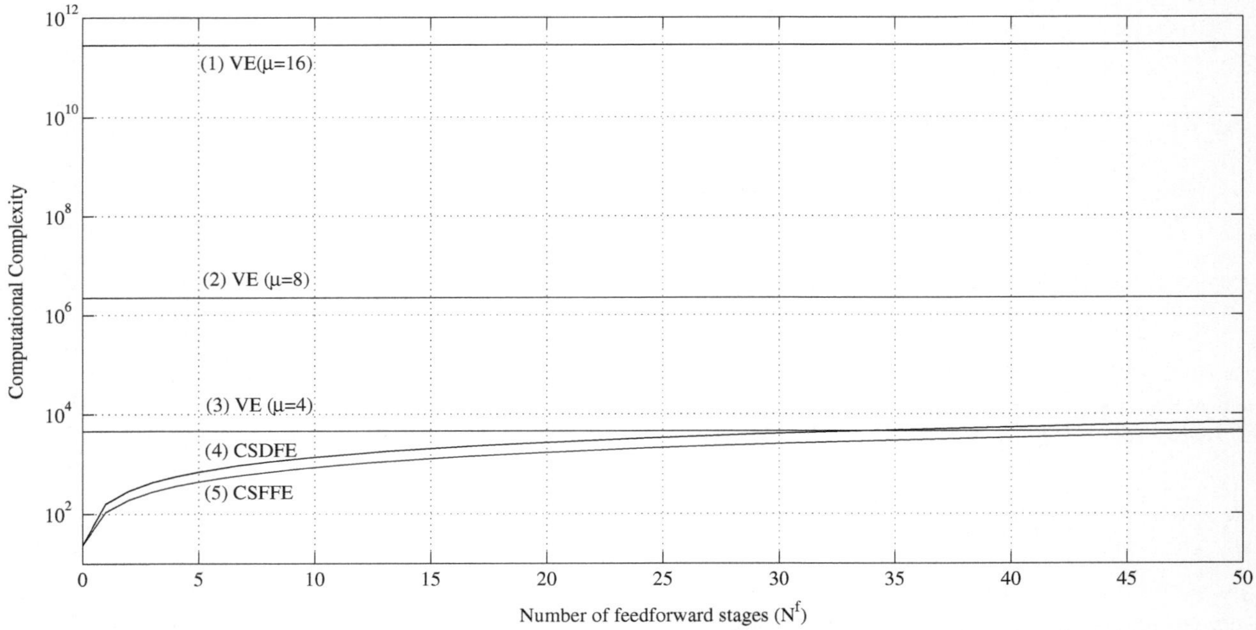

Figure 8 Comparative computational complexity of the proposed equalizer vs. VE when $M = 2$, $\upsilon = 4$, and $N^d - 1 = 2$.

8, and 9. When generating the complexity curves, we assumed that $N^f = N_1^f = N_2^f$ and $N^d = N_k^d = 3$ for both CSFFE and CSDFE and additionally $N_b = N_1^b = N_2^b = 0.25 \times N^f$ for CSDFE and $N^b = 1$ for CSFFE. The computational complexity for VE is calculated by taking into account the most computationally demanding operation, i.e., comparative metric calculation, and is given as $M \times (4\mu + 3) \times \upsilon^{\mu-1}$ [38], where each channel from transmitter to receiver is assumed to have a memory of μ. In Figure 7, we plotted the complexity curves when $M = 2$, $\upsilon = 2$, and $\mu = 4, 8, 16$. It shows that the proposed method is not computationally advantageous when the channel memory ($\mu = 4$) is two times the ADIR filter memory ($N^d - 1 = 2$). However, the complexity of

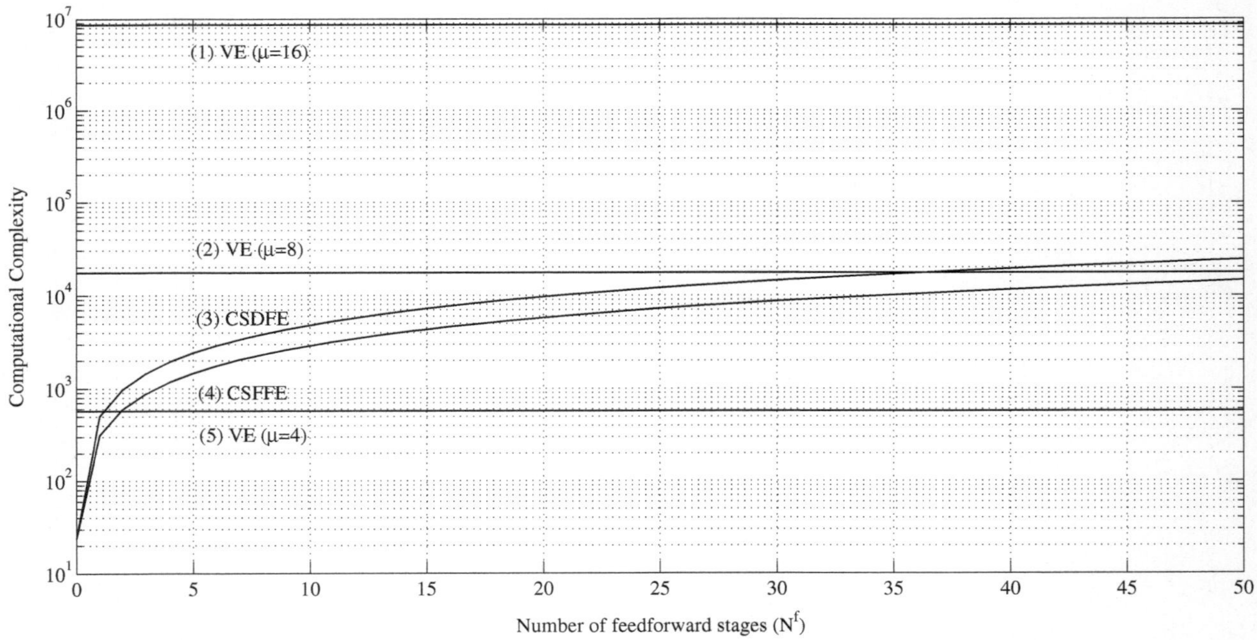

Figure 9 Comparative computational complexity of the proposed equalizer vs. VE when $M = 4$, $\upsilon = 2$, and $N^d - 1 = 2$.

the proposed method becomes advantageous regardless of feed forward filter length (N^f) used when the channel memory is four times the ADIR filter memory and far more beneficial when the channel memory is eight times the ADIR filter memory. Figure 8 displays that the computational advantage of the proposed method comparing to VE becomes more attractive when the alphabet size is increased from $\upsilon = 2$ to $\upsilon = 4$ even for the channel memory values of $\mu = 4, 8$; and Figure 9 demonstrates that the computational advantage becomes less pronounced when the number of antennas increases from $M = 2$ to 4.

The computational complexity vs. the number of antennas analysis has also been carried out due to the recent interest in massive MIMO for next-generation wireless systems [37]. We would like to point out that the computational complexities of CSDFE and CSFFE are larger than DFE and FFE, respectively, by an amount of $M \times \upsilon \times N^d$, assuming $N^d = N^d_k, \forall k$. In this analysis, we assumed that $N^f = N^f_k = 12, \forall k$, $N^b = N^b_k = 0.25 \times N^f$, and $N^b = N^b_k = 1, \forall k$. In Figure 10, the computational complexity vs. number antennas curves for CSDFE, CSFFE, DFE, and FFE when $\upsilon = 256$ and $N_d - 1 = 8$, and M is increasing from 0 and to 150, are presented. For smaller values of υ and $N_d - 1$, the difference between the computational complexities of CSDFE and DFE or CSFFE and FFE, respectively, is minor, and the curves of CSDFE and DFE or CSFFE and FFE can not be discriminated, which implies that the performance increase by the implementation of CSDFE instead of DFE or CSFFE instead of FFE, respectively, is achieved at the expense of almost negligible computational complexity cost, as will be clearer in the next section. In Figure 11, we compare the computational complexity of the proposed method with those of VE when the channel memory values of $\mu = 4, 8$, and 16 are used, the alphabet size is $\upsilon = 2$, the ADIR memory is assumed as $N_d - 1 = 2$, and M is increasing from 0 to 150. Finally, we repeat the same comparison in Figure 12 for $\upsilon = 4$ in order to demonstrate the effect of using larger alphabet size. Figure 11 shows that the proposed method is computationally advantageous comparing to VE when channel memory is larger than eight ($\mu > 8$), and in Figure 12, it can be seen that the computational complexity advantage of the proposed method improves when the alphabet size is increased from $\upsilon = 2$ to $\upsilon = 4$, and the proposed method becomes less complex comparing to VE even when $\mu = 8$.

5 Experimental results

The performance of the proposed receiver was investigated by means of MSE and probability of error simulations. In these evaluations, we considered linear time-invariant channels with spectral nulls as well as time-varying channels. The following channel impulse response matrices were defined in order to be used in simulations that demonstrate performance of the proposed equalizer:

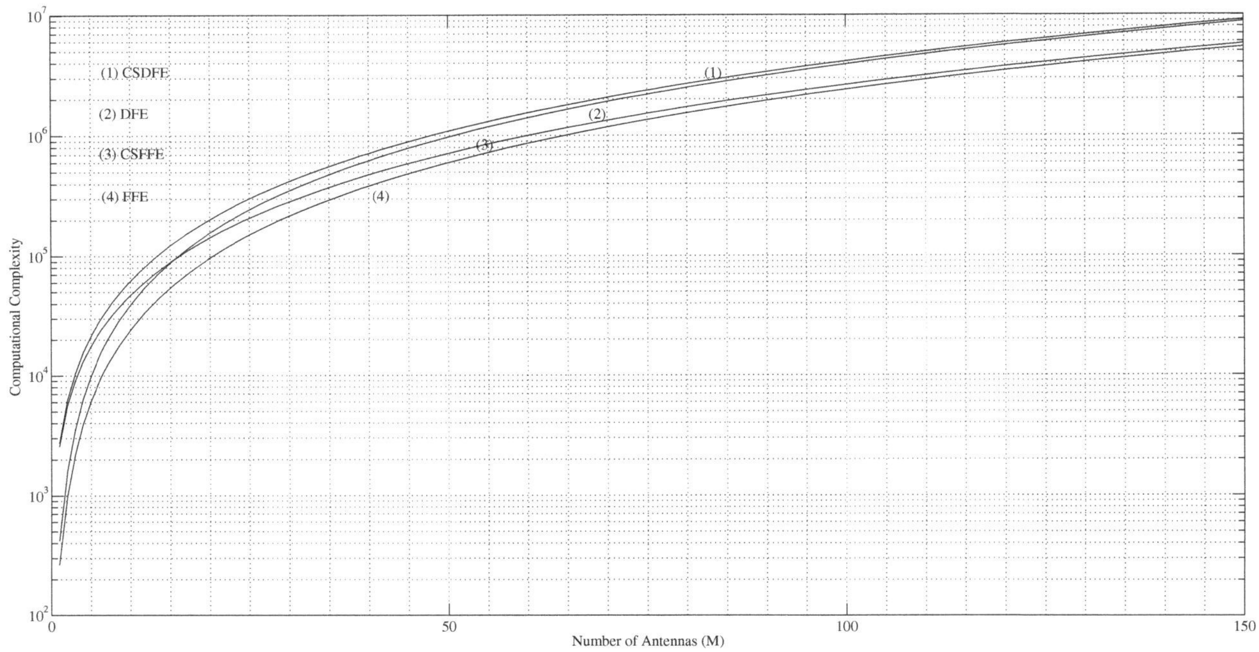

Figure 10 Comparative computational complexities of CSDFE, DFE, CSFFE, and FFE when $\upsilon = 256$ and $N^d - 1 = 8$.

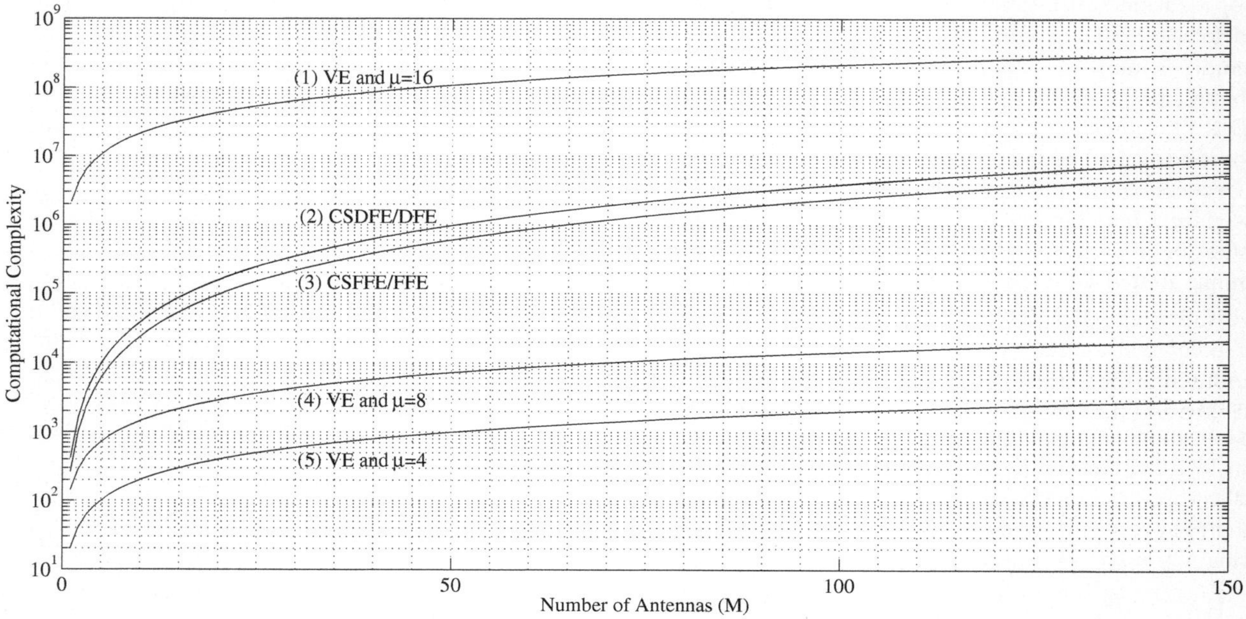

Figure 11 Comparative computational complexity of the proposed equalizer vs. VE when $\upsilon = 2$ and $N^d - 1 = 2$.

$$\mathbf{h}_1(n) = \begin{bmatrix} h_a(n) & \rho.h_a(n-1) \\ \rho.h_a(n-1) & h_a(n) \end{bmatrix}$$
$$\mathbf{h}_2(n) = \begin{bmatrix} \delta(n) & \rho.\delta(n-1) \\ \rho.\delta(n-1) & \delta(n) \end{bmatrix}$$
$$\mathbf{h}_3(n) = \begin{bmatrix} h_b(n) & \rho.h_b(n-1) \\ \rho.h_b(n-1) & h_b(n) \end{bmatrix} \quad (64)$$

where $\delta(n)$ represents the dirac delta function, and the channel impulse response $h_a(n)$ is defined as $h_a(n) = \sum_{j=0}^{4} a_j\delta(n-j)$, which has spectral nulls and a large eigenvalue spread ($\chi = 1317.65$). Herein, the channel coefficients have been taken from [39], and are given by:

$$\mathbf{a}_j = [0.227045\ 0.460091\ 0.688136\ 0.460091\ 0.227045] \quad (65)$$

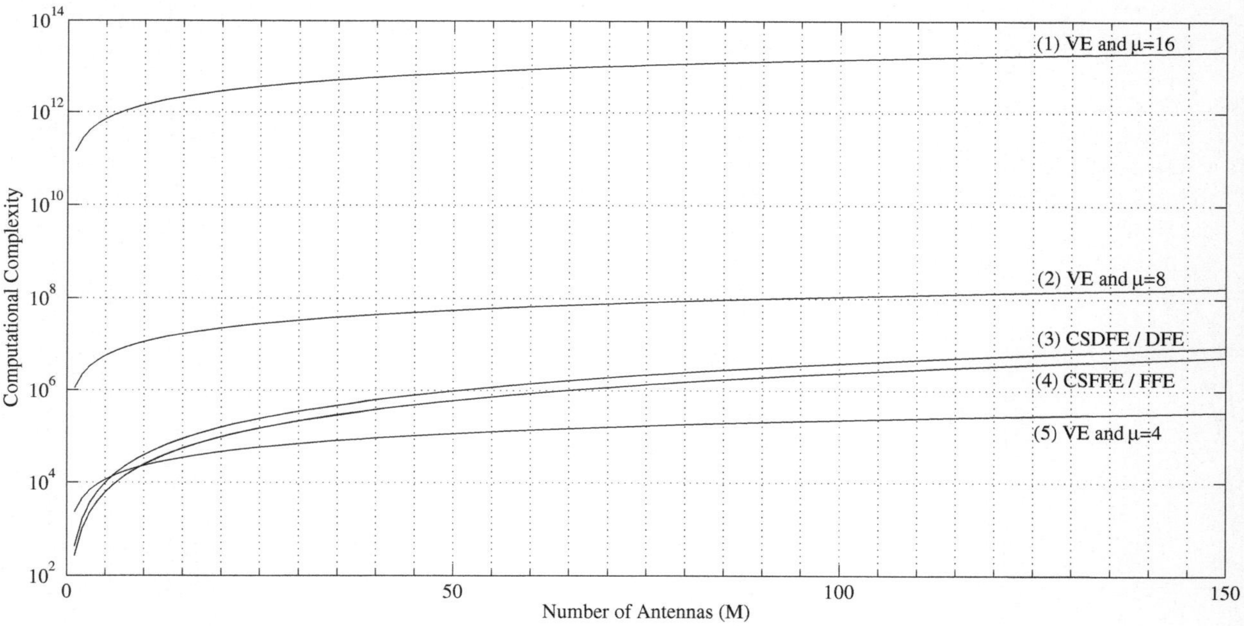

Figure 12 Comparative computational complexity of the proposed equalizer vs. VE when $\upsilon = 4$ and $N^d - 1 = 2$.

with $\sum_{j=0}^{4} a_j^2 = 1$. Note that eigenvalue spread (χ) was determined by using the method described in [40], assuming a feed forward equalizer with a memory of $N^f - 1 = 18$, and a noise variance of 0.001. The time-varying channel impulse response is defined as $h_b(n) = \sum_{j=0}^{4} b_j(n)\delta(n-j)$, where $b_j(n)$ represents the jth time-variant attenuation factor, generated independently using the improved version of Jakes' Rayleigh fading model in [41] by assuming a data rate of 100 kbytes/s and doppler shifts of $f_D = 50$ Hz and $f_D = 10$ Hz ; $b_j(n)$ is also normalized such that $\sum_{j=0}^{4} b_j^2(n) = 1$ for all n. By choosing the same channel for both direct and indirect channels, we demonstrate the performance under such a severe distortion situation that both ISI and ICI are significant, and therefore, the use of DFE is justified. On the other hand, we could have generated a longer channel with the same spectral characteristics (or eigenvalue spread), in which case we would not be able to benchmark the performance with those of VE, as the simulation of VE becomes computationally cumbersome for longer channels. Note that ρ represents the gain factor for the effect of ICI and takes values between 0 and 1. In this presentation, we consider two values of the gain factor, $\rho = 0$ and $\rho = 1$, which correspond to completely orthogonal and nonorthogonal transmissions, respectively [42].

In the simulations for the performance evaluation of the proposed method (CSDFE/CSFFE) with respect to FFE, DFE, and VE, and also in the simulations for the performance comparison of the proposed method when different ADIR filter memories are used, the input signal $x(n)$ applied to the channel was made of uniformly distributed bipolar $(+/-)1/\sqrt{2}$ random numbers because of relative simplicity it provided in simulations. Note that the uniformly distributed bipolar random numbers represent BPSK modulation supported by the IEEE 802.11n WLAN standard [43]. In order to account for the modulations that are both in the IEEE 802.11n WLAN and IEEE 802.16e MAN standards and to demonstrate the effect of higher modulation on probability of error performance, we also performed CSFFE/CSDFE simulations by using the input signal made of uniformly distributed random numbers taking values from $(1/2+i/2, -1/2+i/2, -1/2-i/2, 1/2-i/2)$, which represents QPSK modulation ($i = \sqrt{-1}$), and compared against the performance results of the proposed method for BPSK modulation in both time-invariant and time-variant channel cases. Moreover, we assume the complete knowledge of time-invariant channel and the knowledge of time-variant channel memory in VE and adaptive Viterbi equalizer (AVE) performance evaluations, respectively. On the other hand, the proposed method does not need any information about channel; however, if information about channel is already available and an evaluation on the badness of channel can be carried out, the result of this evaluation can used to determine the ADIR filter memories of the proposed equalizer so as to improve the performance.

The channel noise signal was additive white Gaussian noise (AWGN) with zero mean and is uncorrelated with the input signal. The received signal-to-noise ratio (SNR) per each channel of the receiver is defined as:

$$\mathrm{SNR}_k \triangleq \frac{1}{2\sigma_{n_k}^2} \tag{66}$$

where $\sigma_{n_k}^2$ is the variance of AWGN for the kth channel. Accordingly, SNRs for all channels of the receiver are equal, and the system SNR is defined as $\mathrm{SNR} = \mathrm{SNR}_k$ for $k = 1, 2$. The exponential weighting factors were 0.99 and 1.0 for the front-end equalizer and ADIR filter, respectively, when the channel was time-invariant. In time-varying channel case, they were assumed as 0.975 and 1 in order to better track the signal. The probability of error evaluations were conducted using 4×10^5 samples so that 2×10^5 samples per channel were used, and the simulations were carried out in training mode of receiver operation. The delays (D_k) for the desired signals in this mode of operation were assumed equal ($D = D_k$) for $k = 1, 2$, and D was chosen so as to minimize MSE; that is, $D = (N^f - 1 + \aleph - 1)/2$ when only front-end FFE is utilized, $D = \left(N^f - 1 + N^b - 1 + \aleph - 1\right)/2$ when only front-end DFE is implemented, $D = \left(N^f - 1 + \aleph - 1\right)/2 - N^d + 1$ for CSFFE, and $D = \left(N^f - 1 + N^b - 1 + \aleph - 1\right)/2 - N^d + 1$ for CSDFE. Note that the memories of feed forward and feedback sections of DFEs, ADIR filters, and channel impulse responses for $k = 1, 2$ were assumed equal, i.e., $N^f - 1 = N_k^f - 1$, $N^b - 1 = N_k^b - 1$, $N^d - 1 = N_k^d - 1$, and $\aleph - 1 = \aleph_k - 1$. The noise variance per channel during MSE simulations was 0.001.

5.1 Time-invariant channel

The objective of simulations with the channel matrix $\mathbf{h}_1(n)$ is to display the performance of the proposed CSDFE with respect to CSFFE, FFE, DFE, and VE. In the simulations, we have taken into account the modularity and regularity properties of SPMGLSs and started simulations with FFE; after observing the performance, we altered the equalizer to CSFFE, then we added new SPMGLSs to alter the equalizer to DFE, subsequently to CSDFE. The memory of FFE ($N^f - 1$) was 18, while the memory of feedback channels ($N^b - 1$) for DFE was 4, and the memory of ADIR ($N^d - 1$) filter was 2.

In Figure 13, we present the MSE performance of the proposed equalizer when orthogonal transmission is used, i.e., $\rho = 0$, and also compare its performance to those of CSFFE, FFE, and DFE together with the performance for the channel matrix $\mathbf{h}_2(n)$. In Figure 14, we provide the corresponding probability of error performance for orthogonal transmission and compare the performance

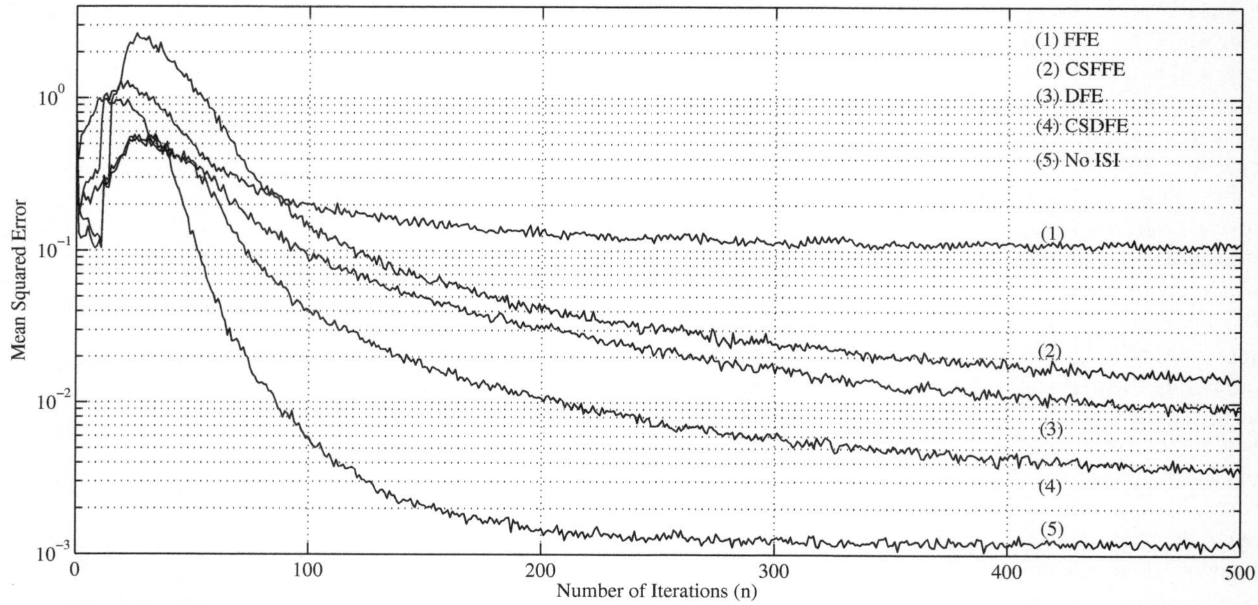

Figure 13 MSE performance of the proposed equalizer for the channel matrix $\mathbf{h}_1(n)$ when $\rho = 0$.

of CSDFE with those of CSFFE, FFE, DFE, VE, and with the performance for the channel matrix $\mathbf{h}_2(n)$. It can be seen in these figures that the performance of CSDFE is better than those of CSFFE, FFE, and DFE, respectively. It also has the closest probability of error performance to that of VE. Note that neither MSE nor probability of error performances for the channel matrix $\mathbf{h}_2(n)$ with $\rho = 0$ change with the use of different equalizers, CSDFE, CSFFE, DFE, or FFE, since this channel matrix with $\rho = 0$ does not include ISI and ICI components. Figures 15 and 16, on the other hand, display MSE and probability of error comparisons when $\rho = 1$, and also provide comparison with respect to the performance of DFE when the channel matrix $\mathbf{h}_2(n)$ is used, since the channel matrix $\mathbf{h}_2(n)$ with $\rho = 1$ does not have ISI components.

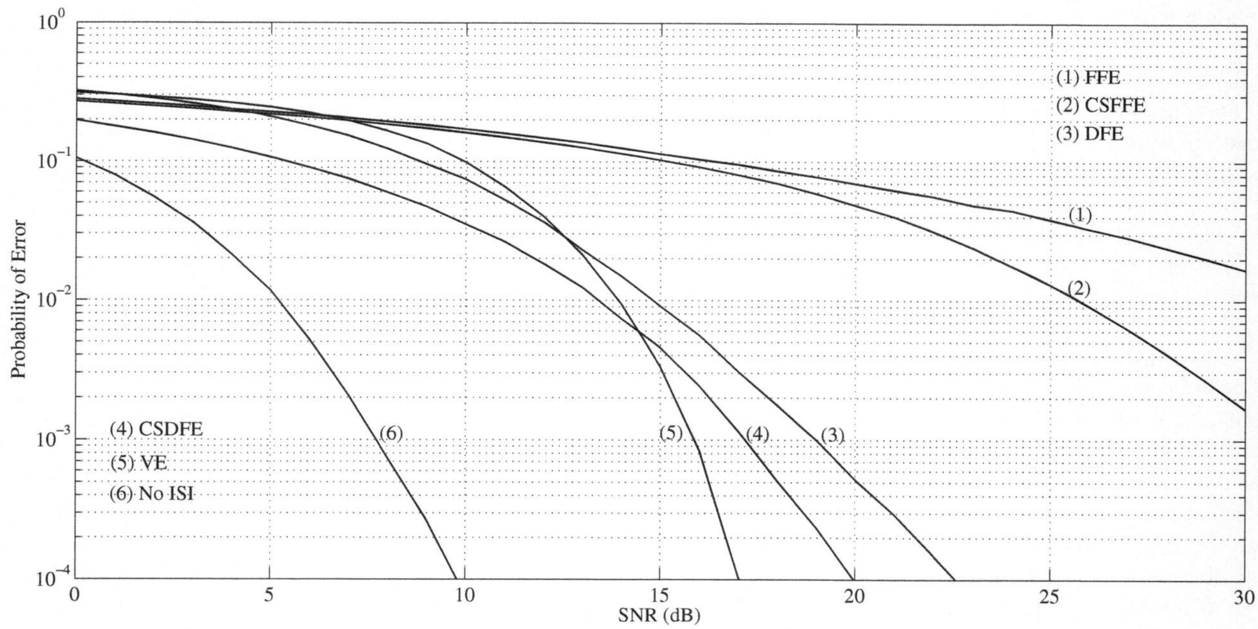

Figure 14 Probability of error performance of the proposed equalizer for the channel matrix $\mathbf{h}_1(n)$ when $\rho = 0$.

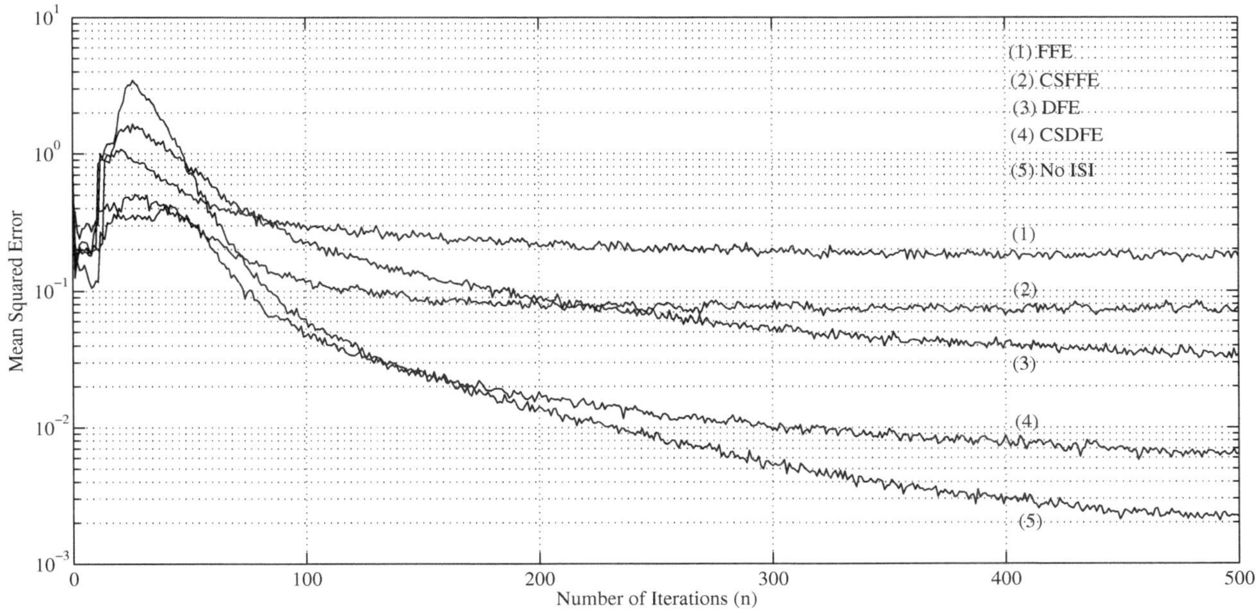

Figure 15 MSE performance of the proposed equalizer for the channel matrix $\mathbf{h}_1(n)$ when $\rho = 1$.

The effect of ICI on the performance of CSDFE and CSFFE can seen by comparing MSE values between Figures 13 and 15 and by collating probability of error values between Figures 14 and 16. It can be deduced from these comparisons that the performance improvement that can achieved by the combination of CSDFE and orthogonal transmission is far more beneficial than using CSFFE with orthogonal transmission. Furthermore, Figures 17 and 18 show the performance improvement that can be achieved by using an ADIR filter memory $(N^d - 1)$ of 3 instead of 2 for both CSFFE and CSDFE in $\rho = 0$ and $\rho = 1$ cases, respectively.

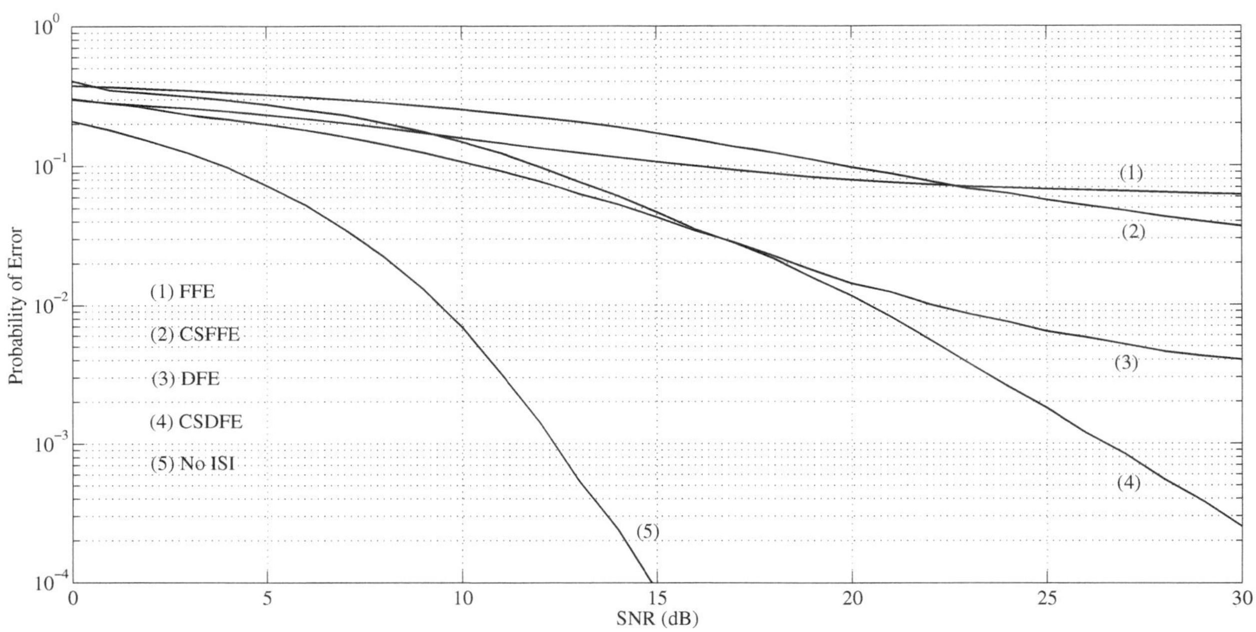

Figure 16 Probability of error performance of the proposed equalizer for the channel matrix $\mathbf{h}_1(n)$ when $\rho = 1$.

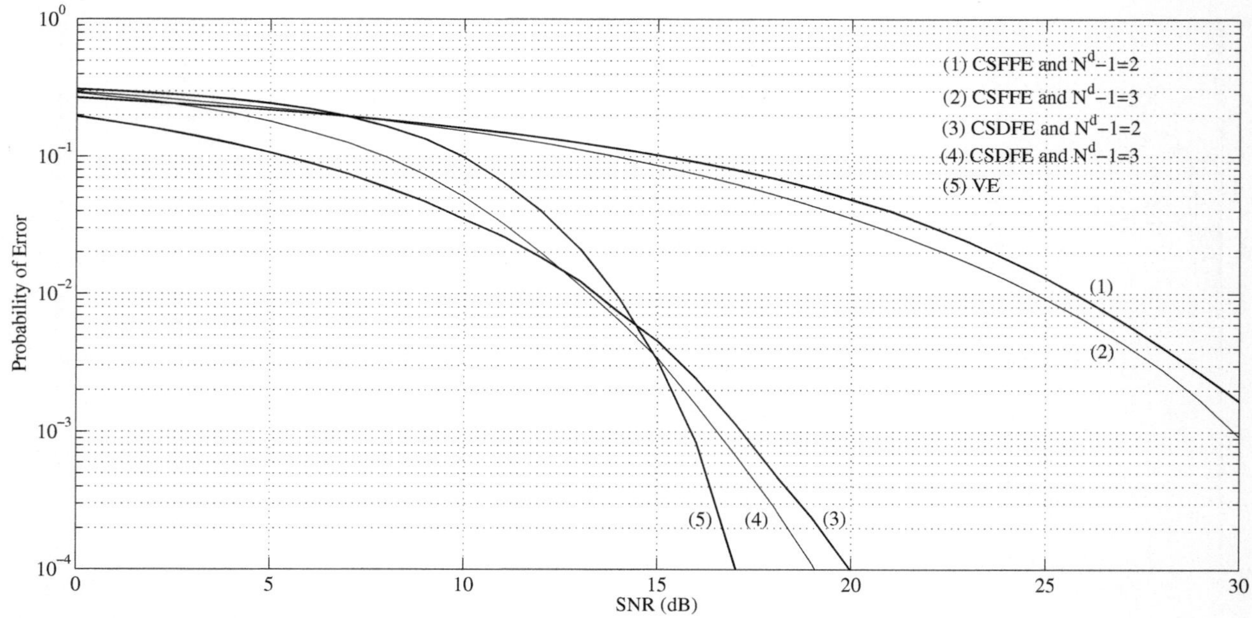

Figure 17 Probability of error comparison for different ADIR filter memories when $\mathbf{h}_1(n)$ is used and $\rho = 0$.

In order to investigate the effect of channel memory for a given ADIR filter memory on the performance of the proposed method, we generated channels with longer impulse responses than that of $h_a(n)$, nevertheless, we made sure that these channels have exactly the same spectral characteristics or eigenvalue spread with $h_a(n)$. We then produced the corresponding channel matrices using the same channel impulse responses for direct and indirect paths as in $\mathbf{h}_1(n)$ and repeated the aforementioned experiments and found out that the performance was not different from the ones displayed in Figures 13, 14, 15, 16, 17, and 18. Consequently, it can be said that, the eigenvalue spread of channel, not the memory as the channel shortener name implies, determines the performance of the proposed equalizer.

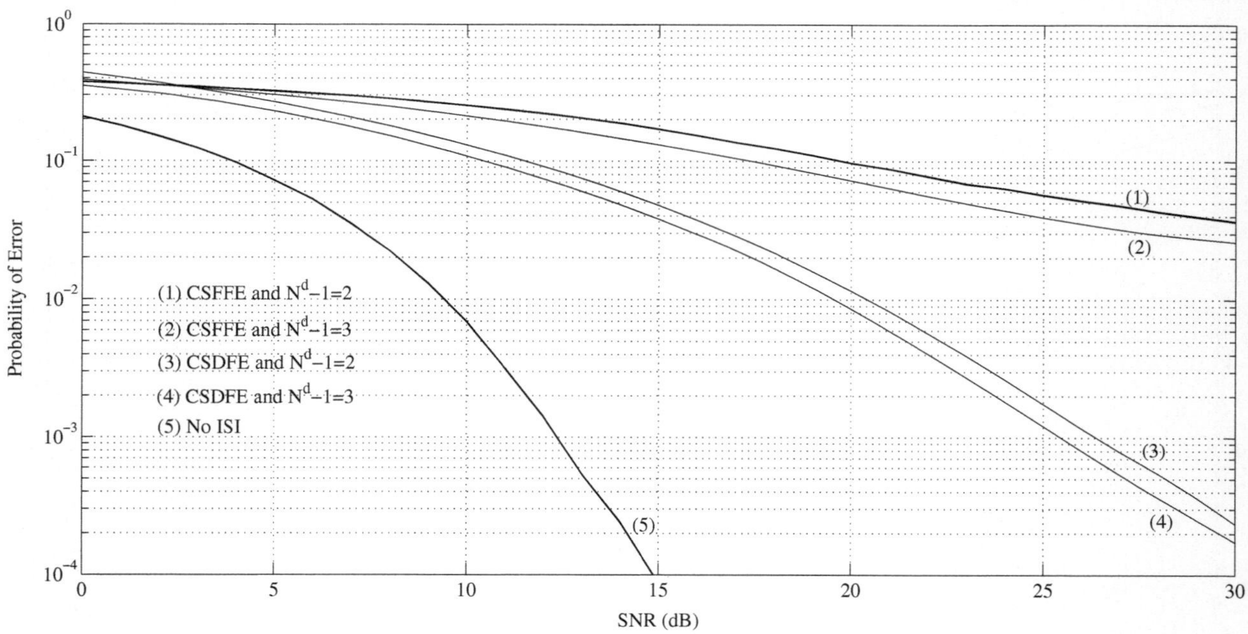

Figure 18 Probability of error comparison for different ADIR memories when $\mathbf{h}_1(n)$ is used and $\rho = 1$.

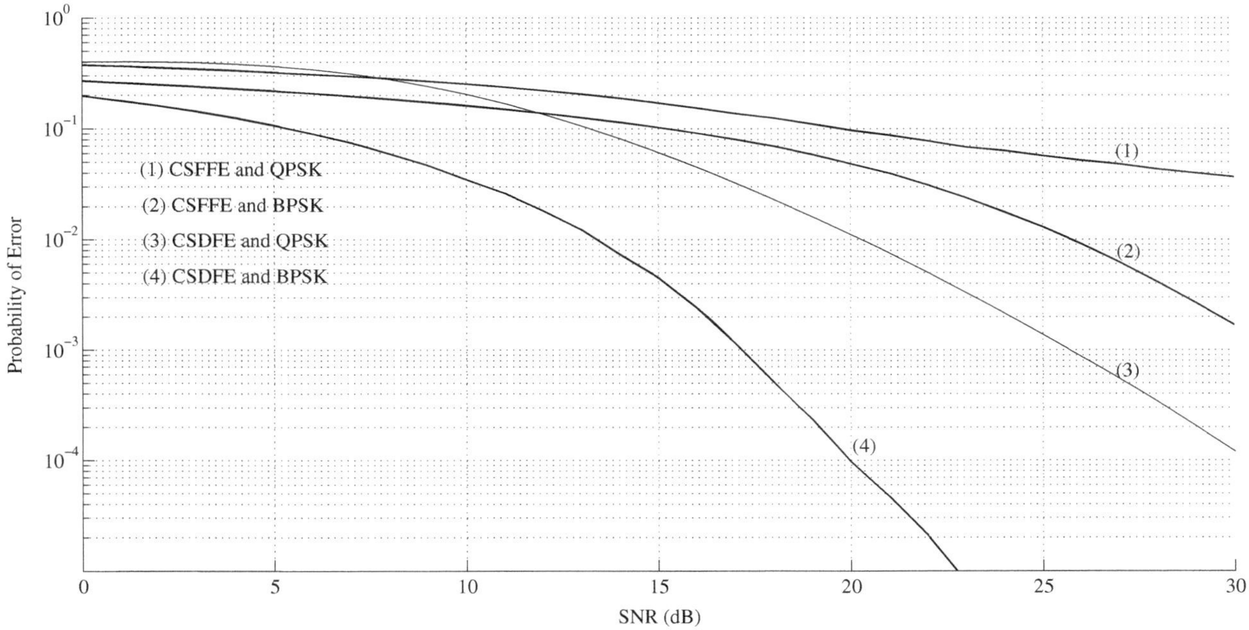

Figure 19 Probability of error comparison for different modulation schemes when $\mathbf{h}_1(n)$ is used and $\rho = 0$.

Subsequently, we examine the effect of using a higher modulation scheme on the probability of error performance of the proposed method, and in Figures 19 and 20, we present the performance degradation caused by switching modulation from BPSK to QPSK in orthogonal and nonorthogonal transmission cases, respectively.

5.2 Time-variant channel

Our objective in the simulations using the channel matrix $\mathbf{h}_3(n)$ is to present the performance of the proposed equalizer under two different time-varying channel conditions. Figure 21 presents the MSE performance results when $f_D = 50$ Hz and provides comparisons between CSFFE

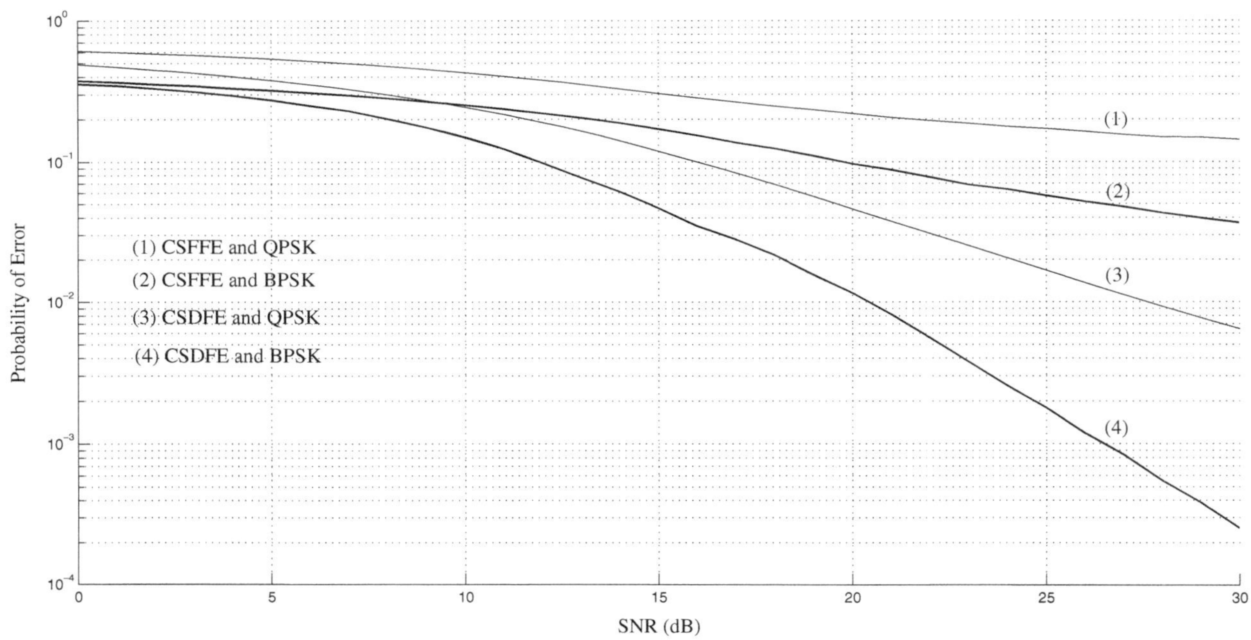

Figure 20 Probability of error comparison for different modulation schemes when $\mathbf{h}_1(n)$ is used and $\rho = 1$.

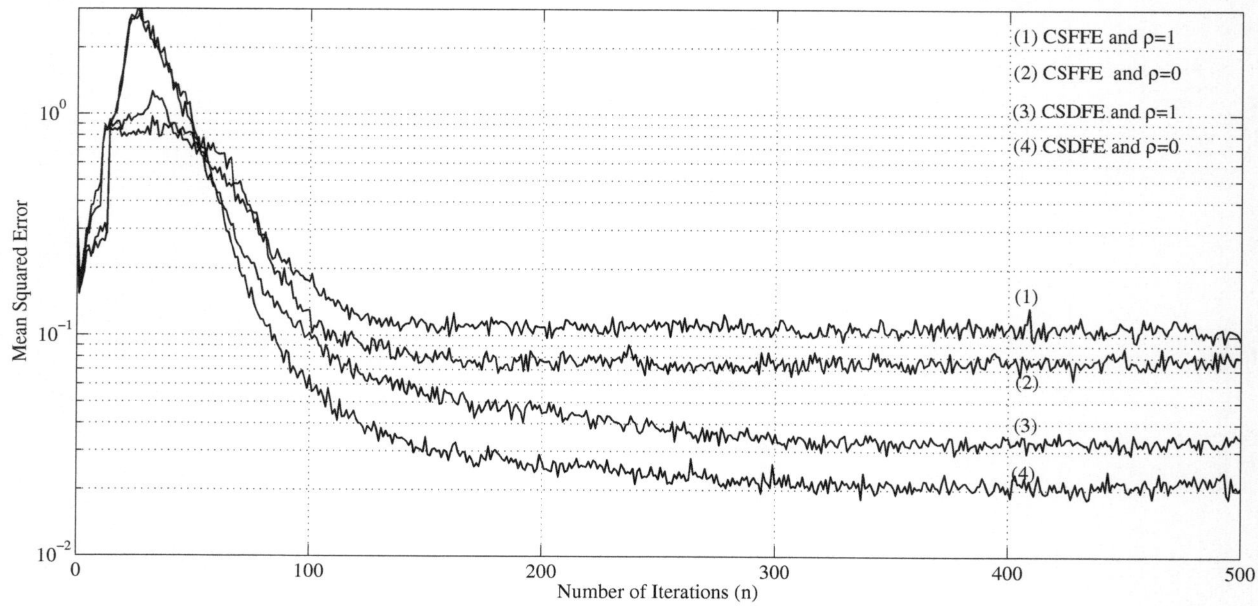

Figure 21 MSE performance of the proposed equalizer for the channel matrix $\mathbf{h}_3(n)$ when $f_D = 50$ Hz.

and CSDFE when $\rho = 1$ and $\rho = 0$. In Figure 22, we show the corresponding probability of error performances. Note that the probability of error values of CSFFE for $\rho = 1$ and $\rho = 0$ saturate approximately at SNR= 25 dB to 2×10^{-2} and 1.5×10^{-2}, respectively, the probability of error values of CSDFE, on the other hand, saturate approximately at SNR= 27 dB to 10^{-3} and 6.3×10^{-4}, which are closer to the probability of error values of AVE when $\rho = 0$, that converges to approximately 23.4×10^{-5} at SNR= 23 dB.

The second experiment under time-variant channel conditions was carried out using a lower doppler frequency so that the effect of doppler frequency on the equalizer performance can be demonstrated. When we

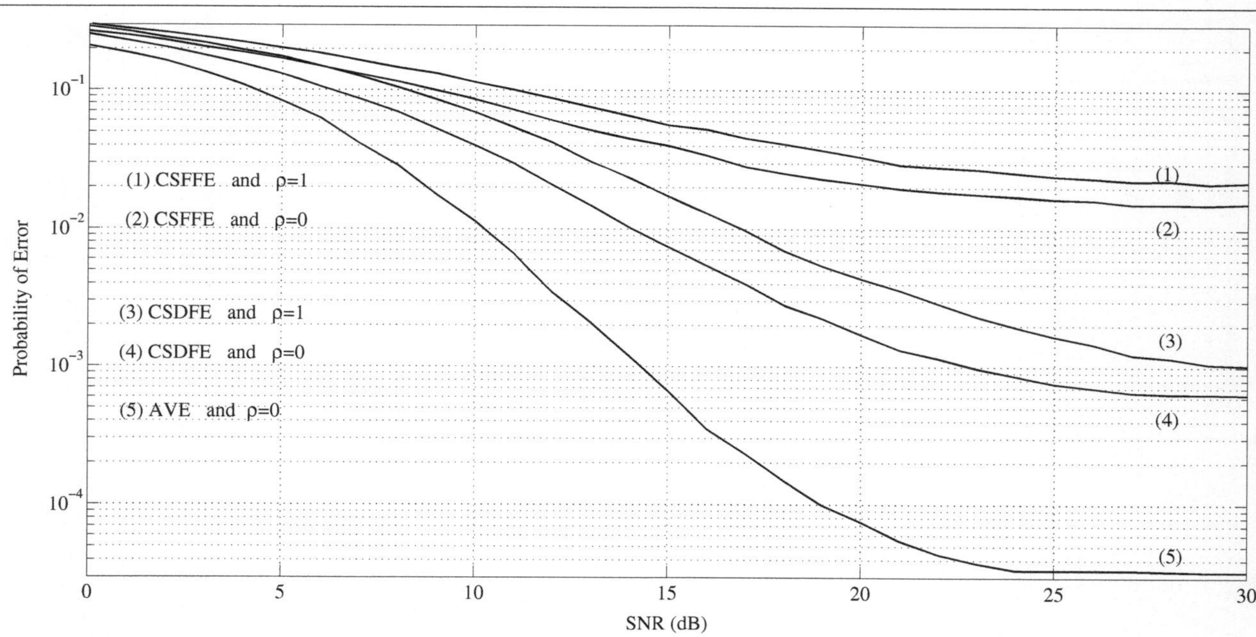

Figure 22 Probability of error performance of the proposed equalizer for the channel matrix $\mathbf{h}_3(n)$ when $f_D = 50$ Hz.

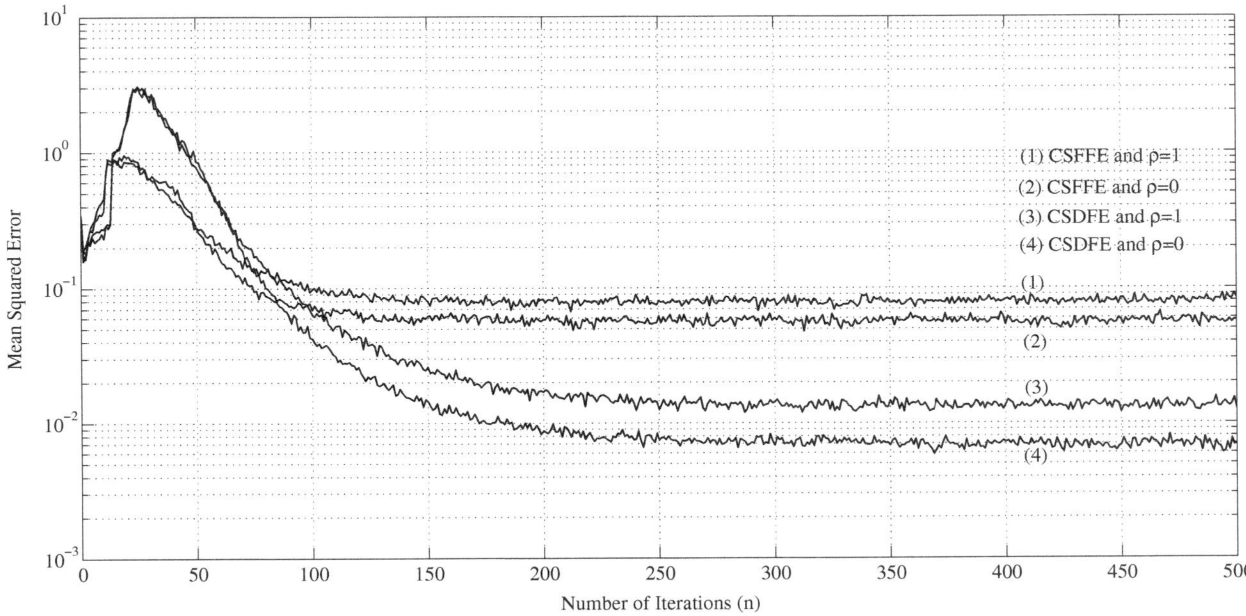

Figure 23 **MSE performance of the proposed equalizer for the channel matrix $\mathbf{h}_3(n)$ when $f_D = 10$ Hz.**

compare the MSE performance results in Figure 23 with those of Figure 21, which is related to the MSE performance for $f_D = 50$ Hz, we see that the MSE performances of CSFFE for $\rho = 1$ and $\rho = 0$ when $f_D = 50$ Hz are 1.33 and 1.22 times, respectively, higher than those of CSFFE when $f_D = 10$ Hz. The same comparison for CSDFE yields that the MSE performances when $f_D = 50$ Hz are 2 and 2.85 times higher for $\rho = 0$ and $\rho = 1$, respectively, than the MSE performances when $f_D = 10$ Hz. Similar evaluations can be done for the probability of error performances, and it can be seen in Figure 24 that the probability of error curves of CSFFE for $\rho = 0$ and $\rho = 1$ cases saturate to lower values than those of CSFFE in Figure 22. On the other hand, whereas probability of error curves of

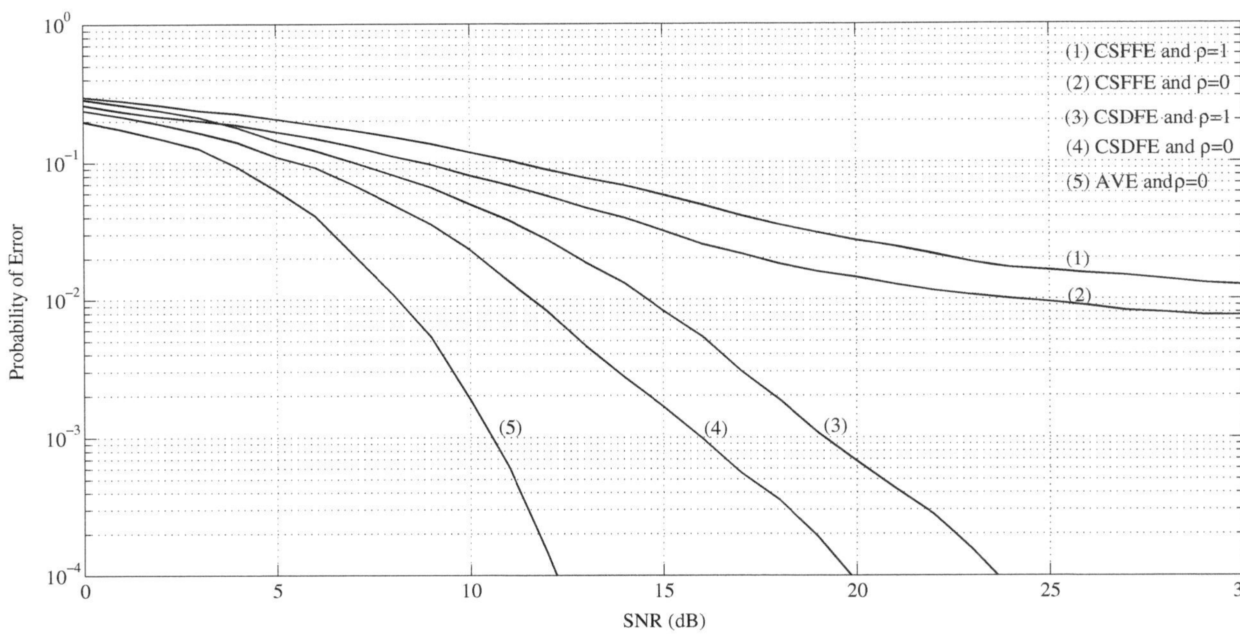

Figure 24 **Probability of error performance of the proposed equalizer for the channel matrix $\mathbf{h}_3(n)$ when $f_D = 10$ Hz.**

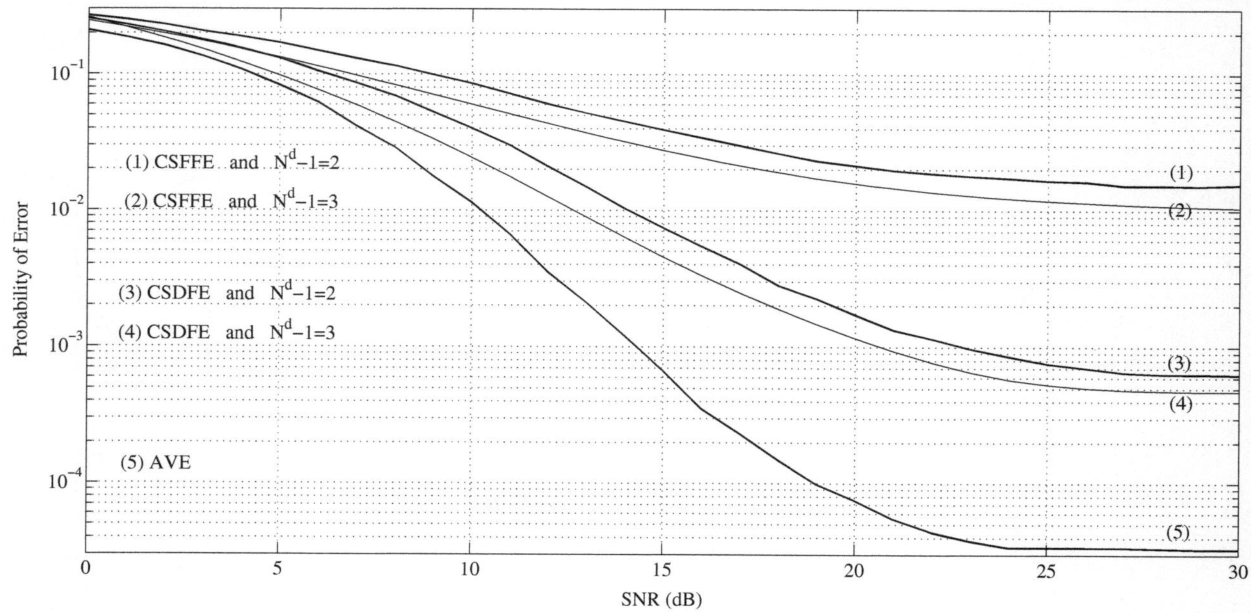

Figure 25 **Probability of error comparison for different ADIR filter memories when $f_D = 50$ Hz and $\rho = 0$.**

CSDFE for $\rho = 0$ and $\rho = 1$ in Figure 22 reach to the probability of error value of 10^{-2} at SNRs of 14 and 17 dB, respectively, they converge to the same probability of error value at approximately 3.5 dB lower SNRs, i.e., at 10.5 and 13.5 dB in Figure 24.

We have performed two more probability of error simulations for the time-variant channel case using $f_D = 50$ Hz, the first of which was related to the performance improvement that can be gained by using an ADIR filter memory $(N^d - 1)$ of 3 instead of 2. Accordingly, Figures 25 and 26 depict the achieved performance improvement for both CSFFE and CSDFE in $\rho = 0$ and $\rho = 1$ cases. The second simulation was for the effect of using QPSK modulation instead of BPSK, and Figures 27 and 28 therefore render

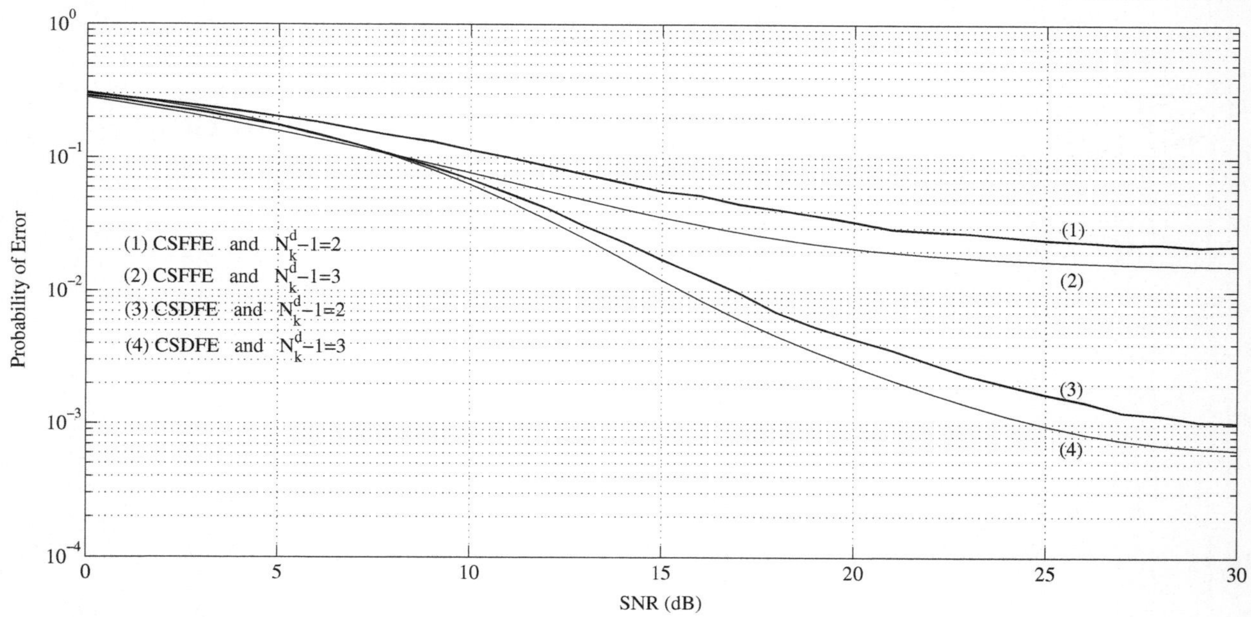

Figure 26 **Probability of error comparison for different ADIR filter memories when $f_D = 50$ Hz and $\rho = 1$.**

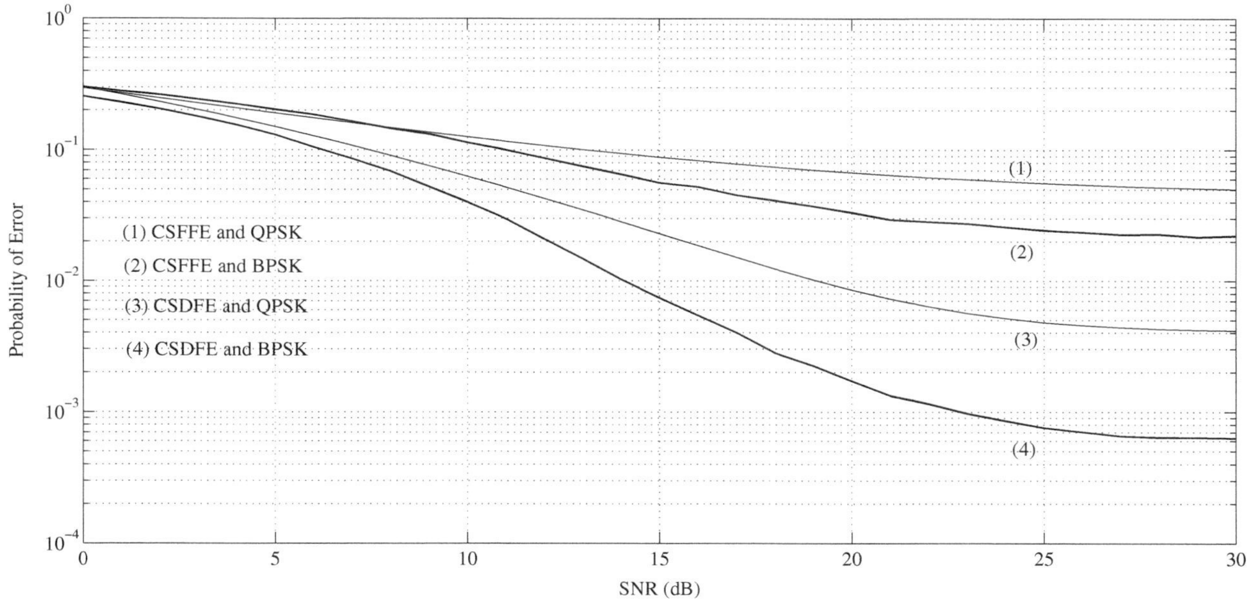

Figure 27 Probability of error comparison for different modulation schemes when $f_D = 50$ Hz and $\rho = 0$.

the probability of error performance comparison when using QPSK modulation instead of BPSK in $\rho = 0$ and $\rho = 1$ cases, respectively.

It can be seen in these figures that the performance improvement attainable through orthogonal transmission under time-variant channel conditions is not as significant as time-invariant channel conditions.

6 Conclusions

A V-BLAST type channel shortening equalizer design for cognitive MIMO-OFDM radios has been presented. The V-BLAST property for frequency-selective channels is realized by completely orthogonalizing the input data using SPMGLSs, so that a systolic MIMO-DFE is accomplished at the front-end of the proposed channel

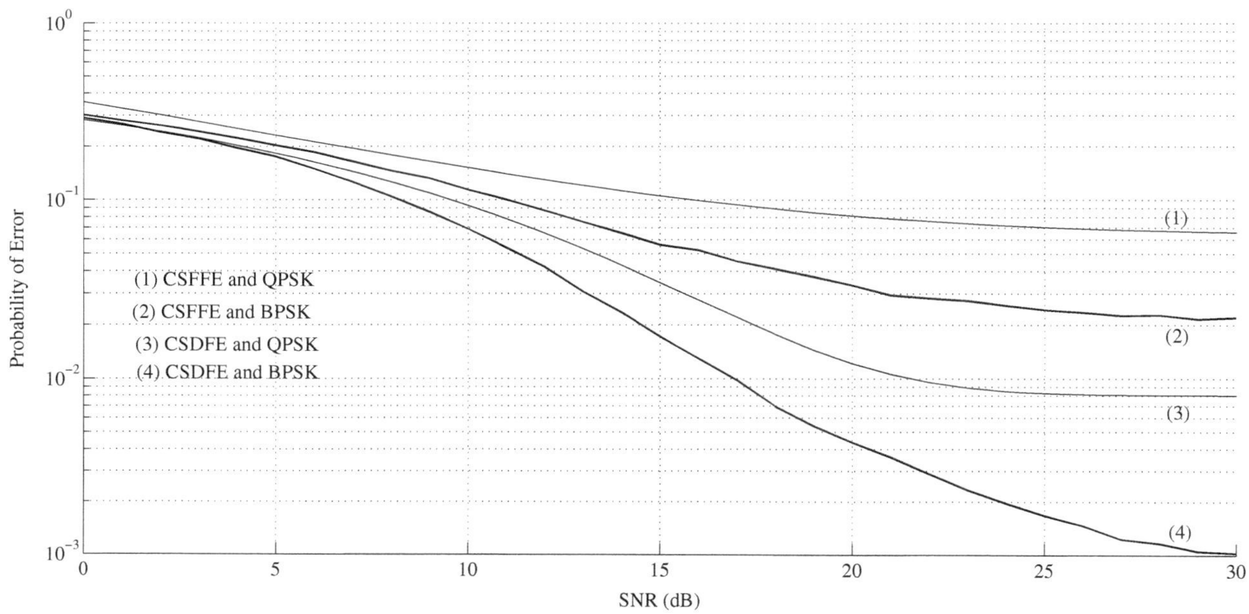

Figure 28 Probability of error comparison for different modulation schemes when $f_D = 50$ Hz and $\rho = 1$.

shortening equalizer. Accordingly, ISI and ICI effects are suppressed due to completely orthogonalizing nature of the receiver structure. A systolic array implementation of multiple adaptive Viterbi detectors is also utilized in order to realize a channel shortening equalizer with high degree of computational concurrency. The matrix inversions, which are significant bottlenecks in receiver design, are avoided, scalar only operations are enabled. A highly modular, regular, order-recursive, and simple receiver architecture, which is suitable for the DSP chip- and FGPA-based signal processing implementations of MIMO-OFDM wireless communication systems, is obtained.

Spectrum sensing and positioning functions, important tasks for cognitive radios, can be accomplished at no cost by simply reconfiguring the front-end MIMO-DFE filter as spectral analysis and positioning filters, respectively. These properties make the proposed equalizer a good candidate for software defined cognitive radio receiver realizations of MIMO-OFDM systems.

The computational complexity of the proposed equalizer has been provided by separately taking into account the MIMO-DFE and multiple adaptive Viterbi detector sections. Then, the total complexity was compared to the complexity of VE for different channel and ADIR filter memories and different alphabet and antenna sizes.

The performance has been supplied in terms of MSE and probability of error analysis for orthogonal and nonorthogonal transmissions under time-invariant and time-variant channel conditions using two different modulation schemes, and it has been demonstrated that desirable performance results can be attained particularly under time-invariant channel conditions when orthogonal transmission and BPSK modulation is used together with CSDFE implementation. It has been also shown that the performance of a CSDFE is between those of Viterbi and DFEs under time-invariant as well as time-variant channel conditions.

It has been revealed that the channel shortener equalizer is indeed a reduced complexity Viterbi equalizer with its ADIR filter memory functioning as a trade-off parameter between performance and complexity for a given channel matrix. Another important property of the proposed equalizer is that it does not need channel information.

Competing interests

The author declares that he has no competing interests.

Acknowledgements

The author is grateful to the Editor Prof. Costas Berberitis and anonymous reviewers for their useful comments.

References

1. GJ Foschini, MJ Gans, On limits of wireless communications in a fading environment when using multiple antennas.Wireless Pers. Commun. **6**(3), 311–335 (1998). doi:10.1023/A:1008889222784
2. GG Raleigh, JM Cioffi, Spatio-temporal coding for wireless communication. IEEE Trans. Commun. **46**(3), 357–366 (1998). doi:10.1109/26.662641
3. JR GL Stüber, SW Barry, YG Mclaughlin, MA Li, TG Ingram, Pratt, Broadband MIMO-OFDM wireless communications. Proc. IEEE. **92**(2), 271–294 (2004). doi:10.1109/JPROC.2003.821912
4. H Bölcskei, Advances in smart antennas - MIMO-OFDM wireless systems : basics, perspectives, and challenges. IEEE Wireless Commun. **13**(4), 31–37 (2006). doi:10.1109/MWC.2006.1678163
5. PW Wolniansky, GJ Foschini, GD Golden, RA Valenzuela, in *Proceedings of URSI Int. Symp. Signals, Systems and Electronics*. V-BLAST: an architecture for realizing very high data rates over the rich-scattering wireless channel (IEEE, Pisa, 1998), pp. 295–300. doi:10.1109/ISSSE.1998.738086
6. Q Liu, L Yang, in *Proceedings of the 2004 IEEE Joint Conf. 10th Asia-Pacific Conf. Commun., and 5th Int. Symp. Multi-Dimensional Mobile Commun.* A simplified method for V-BLAST detection in MIMO OFDM communication, vol. 1 (IEEE, 2004), pp. 30–33. doi:10.1109/APCC.2004.1391645
7. W Bei, Z Qi, in *Proceedings of the IEEE 4th Int.Conf. Wireless Commun. Netw. and Mobile Comput. (WiCOM'08)*. A low complex V-BLAST detection algorithm for MIMO-OFDM system (IEEE, Dalian, 2008), pp. 1–4. doi:10.1109/WiCom.2008.386
8. L Ma, K Dickson, J McAllister, J McCanny, QR decomposition-based matrix inversion for high performance embedded MIMO receivers. IEEE Trans. Signal Process. **59**(4), 1858–1867 (2011). doi:10.1109/TSP.2011.2105485
9. RK Martin, CR Johnson, Adaptive equalization: transitioning from single-carrier to multicarrier systems. IEEE Signal Process. Mag. **22**(6), 108–122 (2005). doi:10.1109/MSP.2005.1550193
10. S Haykin, Cognitive radio: brain-empowered wireless communications. IEEE J. Sel. Areas Commun. **23**(2), 201–220 (2005). doi:10.1109/JSAC.2004.839380
11. M Dillinger, K Madani, N Alonistioti, *Software Defined Radio: Architectures, Systems and Functions*. (John Wiley and Sons, England, 2003)
12. O Anjum, T Ahonen, F Garzia, J Nurmi, C Brunelli, H Berg, State of the art baseband DSP platforms for software defined radio: a survey.EURASIP. J. Wirel. Commun. Netw. **2011**, 5 (2011). doi:10.1186/1687-1499-2011-5
13. K He, L Crockett, R Stewart, Dynamic reconfiguration technologies based on FPGA in software defined radio system. J. Signal Process. Syst. **69**(1), 75–85 (2012). doi:10.1007/s11265-011-0646-2
14. J Im, M Cho, Y Jung, Y Jung, J Kim, A low-power and low-complexity baseband processor for MIMO-OFDM WLAN systems. J. Signal Process. Syst. **68**(1), 19–30 (2012). doi:10.1007/s11265-010-0570-x
15. AP Vinod, E M-K Lai, A Omondi, Special issue on signal processing for software defined radio handsets. J. Signal Process. Syst. **62**(2), 113–115 (2011). doi:10.1007/s11265-009-0428-2
16. B Wang, RJ Ray Liu, Advances in cognitive radio networks: a survey. IEEE J. Sel. Topics Signal Process. **5**(1), 5–23 (2011). doi:10.1109/JSTSP.2010.2093210
17. E Axell, G Leus, EG Larsson, HV Poor, Spectrum sensing for cognitive radio: state-of-the-art and recent advances. IEEE Signal Process. Mag. **29**(3), 101–116 (2012). doi:10.1109/MSP.2012.2183771
18. H Celebi, H Arslan, Enabling location and environment awareness in cognitive radios. Computer Commun. **31**, 1114–1125 (2008). 10.1016/j.comcom.2008.01.006
19. MT Ozden, Adaptive multichannel sequential lattice prediction filtering method for ARMA spectrum estimation in subbands. EURASIP J. Adv. Signal Process. **2013**, 9 (2013). doi:10.1186/1687-6180-2013-9
20. MT Ozden, in *Proceedings of the 2014 IEEE/ION Position, Location, and Navigation Symposium (PLANS 2014)*. Adaptive multichannel sequential lattice prediction filtering method for range estimation in cognitive radios, (2014), pp. 426–433. doi:10.1109/PLANS.2014.6851400
21. N Al-Dhahir, JM Cioffi, Efficiently computed reduced-parameter input-aided MMSE equalizers for ML detection: a unified approach. IEEE Trans. Inf. Theory. **42**(3), 903–915 (1996). doi:10.1109/18.490553
22. N Al-Dhahir, FIR channel-shortening equalizers for MIMO ISI channels. IEEE Trans. Commun. **49**(2), 213–218 (2001). doi:10.1109/26.905867
23. F Ling, JG Proakis, A generalized multichannel least squares lattice algorithm based on sequential processing stages. IEEE Trans. Acoust. Speech Signal Process. **32**(2), 381–389 (1984). doi:10.1109/TASSP.1984.1164325

24. J Ma, GY Li, BH Juang, Signal processing in cognitive radio. Proc. IEEE. **97**(5), 805–823 (2009). doi:10.1109/JPROC.2009.2015707
25. GJ Foschini, Layered space-time architecture for wireless communications in a fading environment when using multielement antennas. Bell. Labs. Tech. J. **1**(2), 41–59 (1996). doi:10.1002/bltj.2015
26. A Lozano, C Papadias, Layered space-time receivers for frequency-selective wireless channels. IEEE Trans. Commun. **50**(1), 65–73 (2002). doi:10.1109/26.975751
27. Y Jiang, J Li, WW Hager, Uniform channel decomposition for MIMO communications. IEEE Trans. Signal Process. **53**(11), 4283–4294 (2005). doi:10.1109/TSP.2005.857052
28. R Merched, NR Yousef, Fast techniques for computing finite-length MIMO MMSE decision feedback equalizers. IEEE Trans. Signal Process. **54**(2), 701–711 (2006). doi:10.1109/TSP.2005.861900
29. N Al-Dhahir, AH Sayed, The finite length multi-input multi-output MMSE-DFE. IEEE Trans. Signal Process. **48**(10), 2921–2936 (2000). doi:10.1109/78.869048
30. G Ginis, JM Cioffi, On the relation between V-BLAST and the GDFE. IEEE Commun. Lett. **5**(9), 364–366 (2001). doi:10.1109/4234.951378
31. J Wang, Y Jiang, GE Sobelman, Iterative computation of FIR MIMO MMSE-DFE with flexible complexity-performance tradeoff. IEEE Trans. Signal Process. **61**(9), 2394–2404 (2013). doi:10.1109/TSP.2013.2245325
32. PA Regalia, MG Bellanger, On the duality between fast QR methods and lattice methods in least squares adaptive filtering. IEEE Trans. Signal Process. **39**(4), 879–891 (1991). doi:10.1109/78.80910
33. F Ling, Givens rotation based least squares lattice and related algorithms. IEEE Trans. Signal Process. **39**(7), 1541–1551 (1991). doi:10.1109/78.134393
34. F Ling, D Manolakis, JG Proakis, A recursive modified Gram-Schmidt algorithm for least-squares estimation. IEEE Trans Acoust. Speech Signal Process. **34**(4), 829–835 (1986). doi:10.1109/TASSP.1986.1164877
35. MT Ozden, AH Kayran, Adaptive multichannel decision feedback equalization for Volterra type nonlinear communication channels. AEU - Int. J. Electron. Commun. **62**(6), 430–437 (2008). doi:10.1016/j.aeue.2007.06.005
36. JD Provence, SC Gupta, Systolic arrays for Viterbi processing in communication systems with a time-dispersive channel. IEEE Trans. Commun. **36**(10), 1148–1156 (1988). doi:10.1109/26.7532
37. E Larsson, O Edfors, F Tufvesson, ML Marzetta, Massive MIMO for next generation wireless systems. IEEE Commun. Mag. **52**(2), 186–195 (2014). 10.1109/MCOM.2014.6736761
38. HC Myburgh, JC Olivier, Low complexity MLSE equalization in highly dispersive Rayleigh fading channels.EURASIP. J. Adv. Signal Process. **2010**(1), 874874 (2010). doi:10.1155/2010/874874
39. JG Proakis, *Digital Communications* (McGraw-Hill, New York, 2000)
40. S Haykin, *Adaptive Filter Theory*, 4th edn. (Prentice-Hall, New Jersey, 2001)
41. Y Rosa Zheng, C Xiao, Simulation models with correct statistical properties for Rayleigh fading channels. IEEE Trans. Commun. **51**(6), 920–928 (2003). doi:10.1109/TCOMM.2003.813259
42. W Zhang, X-G Xia, KB Letaief, Space-time/frequency coding for MIMO-OFDM in next generation broadband wireless systems. IEEE Wireless Commun. **14**(3), 32–43 (2007). doi:10.1109/MWC.2007.386610
43. AF Molisch, *Wireless Communications*, 2nd edn. (John Wiley and Sons, UK, 2011)

11

Nanoscale molecular communication networks: a game-theoretic perspective

Chunxiao Jiang[1*], Yan Chen[2] and K J Ray Liu[2]

Abstract

Currently, communication between nanomachines is an important topic for the development of novel devices. To implement a nanocommunication system, diffusion-based molecular communication is considered as a promising bio-inspired approach. Various technical issues about molecular communications, including channel capacity, noise and interference, and modulation and coding, have been studied in the literature, while the resource allocation problem among multiple nanomachines has not been well investigated, which is a very important issue since all the nanomachines share the same propagation medium. Considering the limited computation capability of nanomachines and the expensive information exchange cost among them, in this paper, we propose a game-theoretic framework for distributed resource allocation in nanoscale molecular communication systems. We first analyze the inter-symbol and inter-user interference, as well as bit error rate performance, in the molecular communication system. Based on the interference analysis, we formulate the resource allocation problem as a non-cooperative molecule emission control game, where the Nash equilibrium is found and proved to be unique. In order to improve the system efficiency while guaranteeing fairness, we further model the resource allocation problem using a cooperative game based on the Nash bargaining solution, which is proved to be proportionally fair. Simulation results show that the Nash bargaining solution can effectively ensure fairness among multiple nanomachines while achieving comparable social welfare performance with the centralized scheme.

Keywords: Nanocommunication; Molecular communication; Game theory; Non-cooperative game; Cooperative game; Nash bargaining

1 Introduction

Nanotechnology, a manipulation of matter on an atomic and molecular scale, makes the design and fabrication of nanoscale components become a reality. Such nanoscale components can be used to assemble basic structural and functional devices, called nanomachines, which are able to perform basic and simple tasks at the nanolevel, such as computing, data storing, sensing, and actuation [1]. Recently, this promising technology catalyzes a new communication paradigm - 'nanocommunications,' arousing the researchers' great interests in both industrial and academic fields. Nanocommunications refer to the information exchange between nanomachines, which can be realized through nanomechanical, acoustic, electromagnetic, and chemical or molecular communication means.

*Correspondence: jchx@tsinghua.edu.cn
[1] Department of Electronic Engineering, Tsinghua University, Beijing 100084, China
Full list of author information is available at the end of the article

It is expected that nanocommunication networks can be applied in many different scenarios including human health monitoring, food and water quality control, air pollution control, as well as aggressive chemical agent detection [2].

Since molecule migration naturally occurs within both living organisms and abiotic components, molecular communication is considered as the most promising approach for nanocommunication networks [3,4], where the transmission and reception of information are realized through molecules. In the molecular communication model, the nanotransmitter releases molecules, which are modulated and coded to carry information, into the fluid medium. The molecules propagate to the receiver through the medium, which are demodulated and decoded to restore the information that the transmitter intends to convey. There are mainly three molecule prorogation models: walkway-based model, flow-based model, and diffusion-based model [5]. In the walkway-based model, the molecules propagate through physical pathways connecting the transmitter to the receiver, such as molecular motors [6]. In the flow-based model, the propagation of the released molecules is controlled by the predefined flow and turbulence in the medium, such as pheromonal communication [2]. In the diffusion-based model, the molecules propagate through their spontaneous diffusion in the fluid medium, such as calcium signaling among cells [7]. In this paper, we focus on the diffusion-based model since it represents the most general and widespread molecular communication architecture found in nature.

In the literature, various topics about the diffusion-based molecular communications have been studied, including channel model [8-13], modulation and coding [14-18], and receiver design [19,20]. The earlier works regarding diffusion-based molecular communications were mainly focused on channel capacity analysis [8-13], where a commonly accepted channel model is based on Brownian motion [21]. In [8], Eckford analyzed the achievable bound on information rate for the diffusion-based channel with one-dimensional Brownian motion. As an extension, Brownian motion of molecules in a fluid medium with drift velocity instead of static environment was analyzed in [13]. In addition to the channel analysis, a comprehensive physical end-to-end model, including molecule emission, diffusion, and reception, was proposed in [5]. Meanwhile, simulation-based approaches for exploring the diffusive molecules were conducted in [22] and [23]. As for channel noise and interference analysis, Pierobon and Akyildiz studied molecule sampling and counting noise in [16], as well as the inter-symbol and co-channel interference in [17]; Kadloor et al. presented an additive inverse Gaussian noise channel model in [14]. To enhance system performance, various coding schemes were also introduced, including a forward error correction coding scheme in [15] and a rate-delay trade-off network coding scheme in [18]. In terms of receiver design, an optimal receiver design based on weighted sum detectors was proposed in [19], and a ligand-binding reception model was studied in [20]. Moreover, the consensus problem and relaying role under diffusion-based molecular communication were recently studied in [24] and [25], respectively.

However, the resource allocation issue in molecular communication networks has not yet been investigated. In traditional electromagnetic wave wireless communications, when multiple users share the same resource, e.g., power and spectrum, how to allocate the resource among different users is very important to guarantee the high system efficiency while maintaining fairness [26,27]. Similarly, in the diffusion-based molecular communications, there can be multiple transmitters sharing the same fluid medium. In such a case, interference will occur since the molecules from different transmitters are identical and indistinguishable [13]. Therefore, how to perform resource allocation among multiple transmitters, molecule emission control, is an essential problem in practical molecular communication systems. Generally, the computational capability of nanomachines is low, and the information exchange cost among them is expensive. Thus, distributed resource allocation algorithms are strongly favorable in molecular communications. Moreover, different nanomachines may be fabricated by different manufacturers and may have different objectives, e.g., in human body monitoring nanosensor networks [3], different biomedical sensors may have different functionalities and may be deployed by different doctors. Considering these problems, in this paper, we propose a game-theoretic framework for distributed resource allocation in diffusion-based molecular communications. The game-theoretical model provides distributed molecule emission control solutions to achieve high system efficiency, while guaranteing fairness among different nanomachines. Game theory has been corroborated as an effective tool for modeling different problems in traditional distributed wireless communication networks [28]. To the best of our knowledge, this is the first work that uses game theory to study the interactions among nanomachines in molecular communication networks, which is expected to exemplify the application of game theory in the nanocommunication and networking field.

The contributions of this paper can be summarized as follows.

1. We analyze the inter-symbol and inter-user interference in the molecular communication system with amplitude modulation. Based on the interference analysis, we find the optimal decision threshold using the maximum-a-posterior (MAP) detection method and derive the corresponding bit error rate (BER) performances of different transmitters.
2. We propose a game-theoretic framework to model the resource allocation problem in nanoscale molecular communication systems. Specifically, we focus on the molecule emission control issue by using a two-transmitter and one-receiver system as an example. In the proposed game-theoretic framework, the players are the transmitters whose objectives are to maximize their own utilities, the strategy of each player is the number of emitted molecules, and the utility function is related with the BER performance.
3. Based on the game-theoretic framework, we formulate the molecule emission control problem as a non-cooperative emission control game, where the Nash equilibrium (NE) is derived and proved theoretically as the unique NE. In order to improve the system efficiency while guaranteeing fairness, we

further model the problem using a cooperative game based on the Nash bargaining solution, where we prove that the Nash bargaining solution of the proposed emission control game is a proportionally fair solution.

The rest of this paper is organized as follows. We first introduce the system model of molecular communication in Section 2. Then, we analyze the inter-symbol and inter-user interference and BER performance in Section 3. The non-cooperative and cooperative molecule emission control games with simulation results are discussed in Section 4. Finally, conclusions are drawn in Section 5.

2 System model

2.1 Network entity

As shown in Figure 1, in this paper, we consider a molecular communication system with two transmitters and one receiver. However, we would like to point out that since the interference from different transmitters is additive due to the independence among transmitters, all the analyses and results in this paper can be easily extended to the multi-transmitter and multi-receiver case. Moreover, the malicious nanonodes are not taken into account in this paper. As we can see from Figure 1, there are four components in a diffusion-based molecular communication system: transmitter, molecule, propagation medium, and receiver, whose functionalities are described as follows.

- *Transmitter*: Each transmitter can independently generate and emit molecules into the propagation medium. It is responsible for the modulation of a series of input symbols $s(t) \in \{0, 1\}$ by controlling the speed, number, or time of the emitted molecules. We assume that the transmitters can perfectly control the release time and number of the molecules, while having no control on the motion of the molecules once the molecules escape from the transmitter [5,9].
- *Molecule*: A molecule is an indivisible object that can be released by the transmitter and absorbed by the receiver by means of chemical reactions. The molecules carry information of the transmitters and randomly diffuse in the propagation medium. They are considered as identical and undistinguishable between each other. Moreover, the interactions among molecules are not taken into account in general [14]. Therefore, the trajectories of all molecules in the medium are independent of each other.
- *Propagation medium*: The propagation medium is made of some kind of fluid, where the molecules can freely diffuse inside. Relative to a single molecule, the space of the medium is considered as infinite in any dimension. The Brownian motion is a generally accepted model for the propagation of molecules in the medium, which can be characterized by two parameters: drift velocity and diffusion constant determined by the physical properties of the fluid medium [29]. In this paper, we only consider the diffusion effect of molecules. Note that [13] considered both diffusion and drift effects, and all our analyses and results can be easily extended to that case.
- *Receiver*: When the molecules arrive at the receiver, they are absorbed by the receiver and disappear from the medium. It is assumed that the receiver can perfectly measure the time when it absorbs a single molecule and use such information to determine the

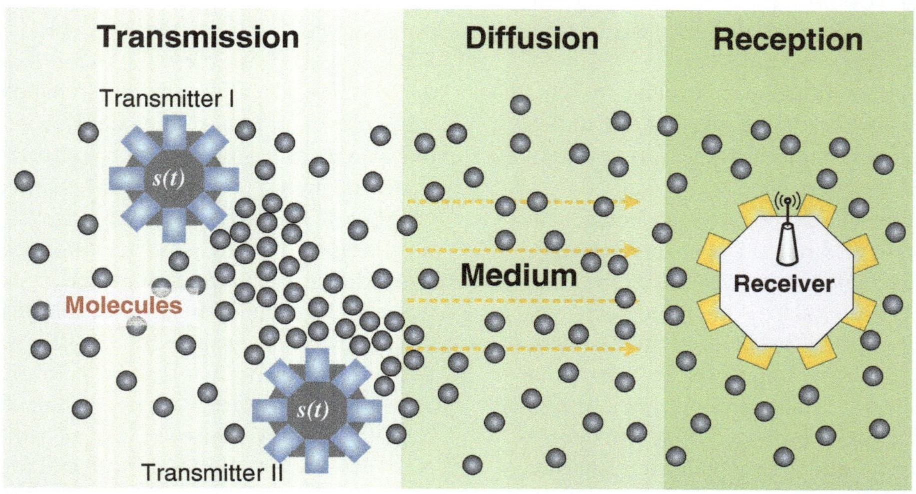

Figure 1 System model with two transmitters and one receiver.

information sent by the transmitter. Moreover, the synchronization between the transmitter and receiver is also assumed to be perfect [30-32]. Note that the focus of this paper is the interference and resource allocation in molecular communication networks, while the synchronization issue is out of the scope. In molecular communication networks, the receiver is usually more powerful than the transmitter, having a larger size, which is usually designed to be able to communicate with microscale machines using electromagnetic waves as well.

2.2 Modulation

At the transmitter, the input symbols $s(t)$ can be modulated onto the molecules with different approaches:

- *Frequency modulation*: controlling the density/speed of the emitted molecules [5]
- *Timing modulation*: controlling the emission time of the molecules [14]
- *Amplitude modulation*: controlling the number of the emitted molecules [15]

In this paper, we consider the amplitude modulation. Specifically, at the beginning of each time slot, the transmitter releases n molecules to represent transmitting bit '1' and releases 0 molecule to represent transmitting bit '0', i.e., only one bit transmission per time slot, as shown in Figure 2. The length of each time slot t_s is considered as identical for all transmitters. For the reception side, the receiver counts and accumulates the number of absorbed molecules during each time slot, comparing the number with some predefined threshold to make a decision on receiving bit '1' or '0'. In such a case, the number of molecules emitted at time slot t, $e(t)$, is

$$e(t) = s(t) \cdot n, \tag{1}$$

where $s(t)$ is a series of '0-1' symbols. Note that n can be regarded as the molecule emission power.

2.3 Channel model

In molecular communication systems, the communication channel is the fluid medium and the propagation model is based on the Brownian motion [21]. Due to the random behavior of the molecules with Brownian motion, after escaping from the transmitter, the time they arrive at the receiver is probabilistic. Suppose the transmitter is located at the origin and releases a molecule at time $t = 0$. Let $X(t)$ denote the position of the released molecule at some specific time t. According to [17], under the Brownian motion in three-dimensional space, the probability density of $X(t)$ is

$$P_X(x,t) = \frac{1}{\sqrt{(4\pi Dt)^3}} \exp\left(-\frac{x^2}{4Dt}\right), \tag{2}$$

where x represents the distance from the original emission point and D is the aforementioned diffusion constant. Equation (2) illustrates the prorogation model of the molecules in the medium, which can be regarded as the impulse response of the Brownian motion-based channel. Note that Brownian motion has been corroborated to be practical and extensively adopted in the literature, as in [5,10,17]. According to the aforementioned amplitude modulation model, only the molecules absorbed by the receiver within time t_s are considered as useful ones. Suppose the distance between the transmitter and receiver is d and the radius of the receiver is R. Then, according to [15], the probability that a molecule is absorbed by the receiver within t_s can be calculated by

$$P_a(d,t_s) = \frac{d}{R}\text{erfc}\left(\frac{d-R}{\sqrt{4Dt_s}}\right), \tag{3}$$

where the error function $\text{erfc}(x) = \frac{2}{\sqrt{\pi}}\int_x^\infty e^{-u^2}du$. Figure 3 shows the relationship between the absorbing probability P_a and the distance d under different lengths of time slot t_s. We can see that $P_a(d,t_s)$ is a decreasing function in terms of distance d. This is consistent with the intuition that the farther from the transmitter, the more difficult the receiver can absorb the molecule. On the other hand, the absorbing probability $P_a(d,t_s)$ is an increasing function in terms of the length of time slot t_s. It is also intuitive that the longer the reception time, the higher probability that the molecule can reach the receiver.

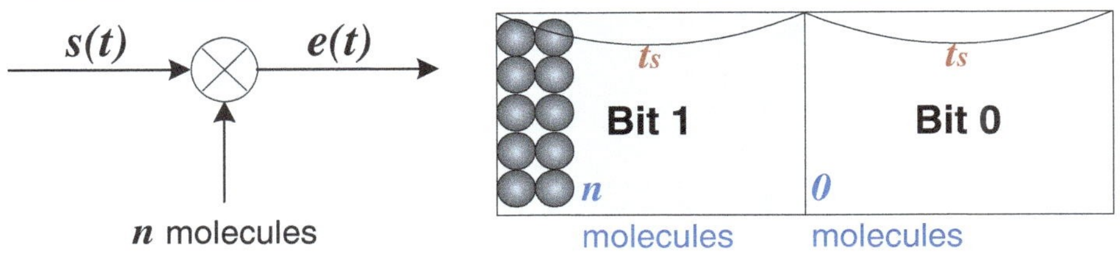

Figure 2 Amplitude modulation.

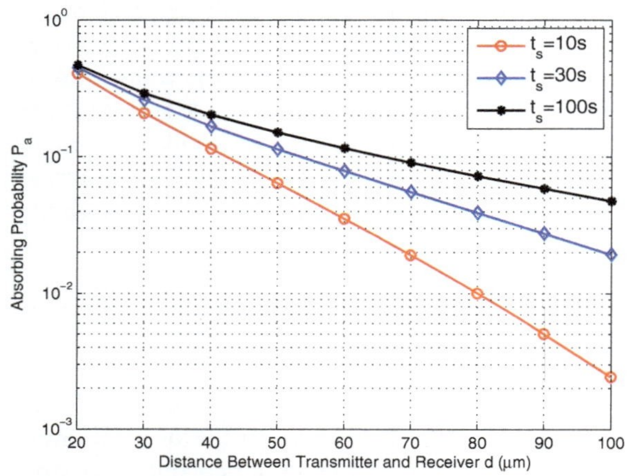

Figure 3 Absorbing probability versus the distance between the transmitter and receiver. The radius of the receiver is $R = 10\ \mu$m, and the diffusion coefficient is $D = 79.4\ \mu m^2/s$ [29].

the interference analysis, we can further derive the optimal decision threshold based on MAP method and the corresponding BER performance.

According to the system model, if the transmitter intends to send bit '1', it would release n molecules to the medium. Let N_a denote the number of molecules absorbed by the receiver located d away from the transmitter within time period t_s. Since all the molecules are independent of each other and each reaches the receiver with probability $P_a(d, t_s)$, N_a follows a binomial distribution as follows [15,33,34]:

$$N_a \sim \mathcal{B}\left(n, P_a(d, t_s)\right). \tag{4}$$

When n is sufficiently large (e.g., $n \geq 20$), which is generally satisfied by the molecular communication systems, the binomial distribution can be approximated by a normal distribution [15,33]. In such a case, the distribution of N_a can approximated by

$$N_a \sim \mathcal{N}\left(nP_a(d, t_s), nP_a(d, t_s)(1 - P_a(d, t_s))\right). \tag{5}$$

3 Interference analysis

In this section, we analyze the interference issues in the molecular communication system. Similar to the traditional wireless communication systems, there are inter-symbol interference and inter-user interference in molecular communication systems. For each transmitter, the molecules released in the previous time slots may leak into the current time slot, i.e., the current symbol, which causes the inter-symbol interference (ISI), as shown in the left part of Figure 4. Moreover, the molecules released by one transmitter may also leak into the time slot of other transmitters, which causes the inter-user interference (IUI), as shown in the right part of Figure 4. Based on

3.1 Inter-symbol interference

The ISI defined in traditional wireless communication systems represents the power leakage of one transmitted symbol into the following symbols, which is usually caused by multi-path propagation. Similarly, in the molecular communication system, due to the random nature of the Brownian motion-based diffusion process, molecules emitted in previous time slots may arrive at the receiver in the current time slot, which leads to the interference to the current bit reception. Such interference is defined as ISI in molecular communications, as shown in the left part of Figure 4. Suppose that the current reception time slot is m; let us denote N_{a,it_s} as the number of molecules

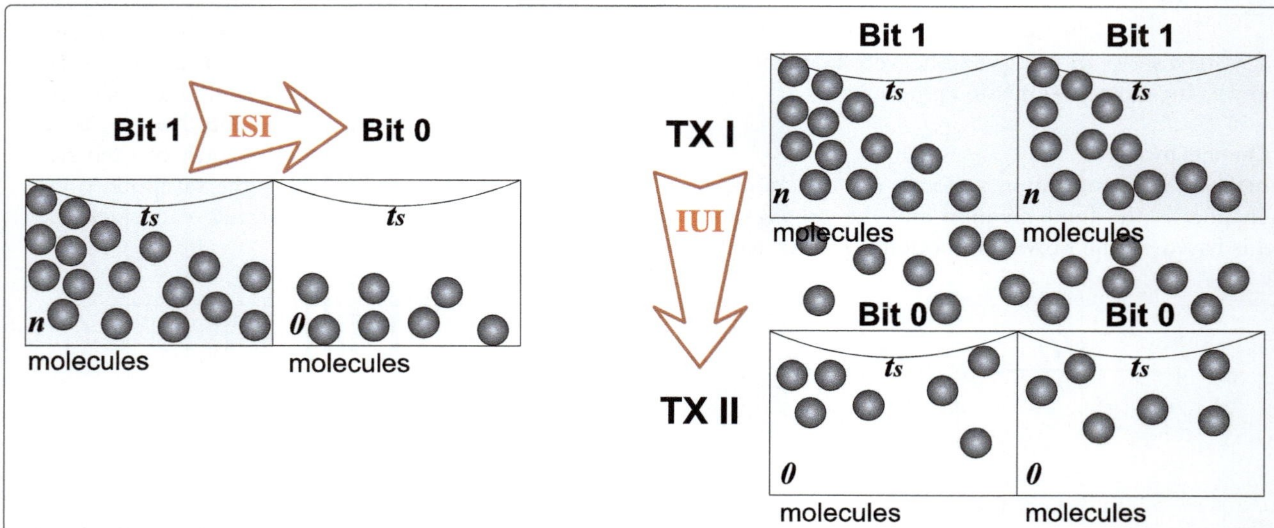

Figure 4 Inter-symbol interference and inter-user interference illustration.

that were sent at the beginning of i time slots before, i.e., at $(m - i)t_s$, and leak into the current time slot m. In such a case, N_{a,it_s} should follow the distribution as follows:

$$N_{a,it_s} \sim \frac{1}{2}\mathcal{N}\left(nP_a(d,(i+1)t_s),\ nP_a(d,(i+1)t_s) \times (1 - P_a(d,(i+1)t_s))\right) - \frac{1}{2}\mathcal{N}\left(nP_a(d,it_s),\ nP_a(d,it_s)(1 - P_a(d,it_s))\right). \quad (6)$$

where $\frac{1}{2}$ is based on the assumption that bits '1' and '0' are equally generated; the first term represents the total number of molecules that were sent at $mt_s - it_s$ and absorbed by the receiver within all subsequent $i+1$ time slots, while the second term represents that within all subsequent i time slots, both of which follow the normal distribution as defined in (5). Thus, we can have the distribution of total interference caused by all previous time slots, N_{ISI}, as follows:

$$N_{\mathrm{ISI}} = \sum_{i=1}^{\infty} N_{a,it_s} \sim \sum_{i=1}^{\infty}\frac{1}{2}\mathcal{N}\left(nP_a(d,(i+1)t_s),\ nP_a(d,(i+1)t_s)(1 - P_a(d,(i+1)t_s))\right) - \sum_{i=1}^{\infty}\frac{1}{2}\mathcal{N}\left(nP_a(d,it_s),\ nP_a(d,it_s)(1 - P_a(d,it_s))\right). \quad (7)$$

According to [33], the experiment results have shown that only the interference from one previous time slot, i.e., the ISI from the last symbol, needs to be considered. In such a case, the distribution of ISI N_{ISI}, can be approximated by

$$N_{\mathrm{ISI}} \sim \frac{1}{2}\mathcal{N}\left(nP_a(d,2t_s),\ nP_a(d,2t_s)(1 - P_a(d,2t_s))\right) - \frac{1}{2}\mathcal{N}\left(nP_a(d,t_s),\ nP_a(d,t_s)(1 - P_a(d,t_s))\right). \quad (8)$$

3.2 Inter-user interference

The IUI defined in traditional wireless communication systems represents the interference power from one transmitter to the other. Similarly, in molecular communications, when there are two or more transmitters releasing molecules into the medium, they would interfere each other at the receiver side, as shown in the right part of Figure 4. Let us consider the two-transmitter case as shown in Figure 1, where the receiver is located d_1 away from transmitter (TX) I and d_2 away from TX II. The emission power of TX I is n_1 and that of TX II is n_2, i.e., releasing n_1 or n_2 molecules to represent bit '1'. Similar to the aforementioned ISI analysis, the IUI for TX I is caused by the molecules which leaked from TX II, and vice versa. For the symbol of TX II transmitted at time slot m, the molecules leaked to it should be considered from both the previous symbol at time slot $m-1$ and the current symbol of TX I. There are four cases: '00', '01', '10', and '11', where '10', for example, means TX I transmitted '1' in the previous time slot and transmits '0' in the current time slot.

For case '00', there is no IUI since no molecule is released. For case '01', only the current symbol '1' of TX I causes interference to TX II. In this case, the number of molecules that leaks into the current symbol of TX II from TX I, denoted by N_{01}^{I}, follows normal distribution as

$$N_{01}^{\mathrm{I}} \sim \mathcal{N}\left(n_1P_a(d_1,t_s),\ n_1P_a(d_1,t_s)\left(1 - P_a(d_1,t_s)\right)\right) = \mathcal{N}\left(n_1P_{1c},\ n_1P_{1c}\left(1 - P_{1c}\right)\right), \quad (9)$$

where $P_{1c} = P_a(d_1,t_s)$ defined in (3) represents the absorbing probability of molecules of TX I's current symbol at the current time slot, which is interference to TX II. For case '10', only the previous symbol '1' of TX I causes interference to TX II. In this case, the number of molecules that leaks into the current symbol of TX II from TX I, denoted by N_{10}^{I}, also follows normal distribution as

$$N_{10}^{\mathrm{I}} \sim \mathcal{N}\left(n_1P_a(d_1,2t_s),\ n_1P_a(d_1,2t_s)(1 - P_a(d_1,2t_s))\right) - \mathcal{N}\left(n_1P_a(d_1,t_s),\ n_1P_a(d_1,t_s)(1 - P_a(d_1,t_s))\right) = \mathcal{N}\left(n_1P_{1p},\ n_1P_{1p}\left(1 - P_{1p}\right)\right) - \mathcal{N}\left(n_1P_{1c},\ n_1P_{1c}\left(1 - P_{1c}\right)\right), \quad (10)$$

where $P_{1p} = P_a(d_1,2t_s)$ represents the absorbing probability of molecules of TX I's previous symbol at both the previous and current time slots. Note that the IUI in case '10' is quite similar to the ISI in (8), while the difference is that ISI means the interference from the transmitter's own previous symbol, while IUI in case '10' means the interference from the other transmitter's previous symbol. For case '11', both the previous and current symbols of TX I can cause interference to TX II. Thus, the number of molecules that leaks into the current symbol of TX II from TX I, denoted by N_{11}^{I}, follows another normal distribution as

$$N_{11}^{\mathrm{I}} \sim \mathcal{N}\left(n_1P_{1c}, n_1P_{1c}(1-P_{1c})\right) + \mathcal{N}\left(n_1P_{1p}, n_1P_{1p}\left(1-P_{1p}\right)\right) - \mathcal{N}\left(n_1P_{1c},\ n_1P_{1c}\left(1 - P_{1c}\right)\right). \quad (11)$$

Overall, based on the assumption that the probabilities of all the four cases are identical, i.e., $P_{00} = P_{01} = P_{10} =$

$P_{11} = \frac{1}{4}$, the IUI from TX I to TX II, denoted by $N_{\mathrm{IUI}}^{\mathrm{I}}$, follows the distribution as follows:

$$N_{\mathrm{IUI}}^{\mathrm{I}} = \frac{1}{4}\left(N_{00}^{\mathrm{I}} + N_{01}^{\mathrm{I}} + N_{10}^{\mathrm{I}} + N_{11}^{\mathrm{I}}\right) \sim \frac{1}{4}\mathcal{N}\left(2n_1P_{1p}, 4n_1P_{1c}\left(1-P_{1c}\right)+2n_1P_{1p}\left(1-P_{1p}\right)\right). \quad (12)$$

Similarly, the IUI from TX II to TX I, denoted by $N_{\mathrm{IUI}}^{\mathrm{II}}$, follows the distribution as follows:

$$N_{\mathrm{IUI}}^{\mathrm{II}} \sim \frac{1}{4}\mathcal{N}\left(2n_2P_{2p}, 4n_2P_{2c}\left(1-P_{2c}\right)+2n_2P_{2p}\left(1-P_{2p}\right)\right), \quad (13)$$

where $P_{2c} = P_a(d_2, t_s)$ and $P_{2p} = P_a(d_2, 2t_s)$ can be calculated by (3). Note that multiplexing techniques can be adopted to alleviate the IUI, which is not the focus of this paper. The techniques from traditional wireless communications may be used to improve the multi-access performance. For examples, the time-division multiplexing techniques can be easily applied in molecular communications, where the molecular transmissions are scheduled by time slots and by utilizing the Neural Delay Boxes (NDBs) connecting different molecular transmitters to the shared medium [35]. Moreover, the space multiplexing techniques can also be applied in molecular multiple access, e.g., MIMO communication based on molecular diffusion [36]. The IUI analysis in this paper can be easily extended to the multi-transmitter scenario. When there are K transmitters, one transmitter may be interfered by other $K-1$ transmitters. Since each transmitter is with four cases: '00', '01', '10', and '11', there are totally $4(K-1)$ cases. Since the interference is additive, there are totally $4(K-1)$ summations in (11). Therefore, the complexity would be linear with the number of transmitters, i.e., $\mathcal{O}(K)$.

3.3 Bit error rate performance

BER is defined as the number of bit errors divided by the total number of transmitted bits during an evaluated time interval, which is usually approximated by the bit error probability, i.e., the probability that bit '0' or '1' is wrongly decoded by the receiver. In the amplitude modulation-based system, the receiver compares the number of molecules absorbed in one time slot with some predefined threshold τ to decode the bit. In such a case, the BER performance is heavily related with the threshold τ. Therefore, we need to first derive the optimal τ that minimizes the BER of each transmitter.

Let us consider the two-transmitter case in Figure 1. When TX I transmits bit '0', i.e., no molecule is released, the number of molecules absorbed by the receiver within current time slot, denoted by N_0^{I}, includes both ISI from its previous symbol and IUI from TX II, which follows normal distribution as follows:

$$N_0^{\mathrm{I}} = N_{\mathrm{ISI}}^{\mathrm{I}} + N_{\mathrm{IUI}}^{\mathrm{II}} \sim \mathcal{N}\left(\mu_{0,\mathrm{I}}, \sigma_{0,\mathrm{I}}^2\right), \quad (14)$$

where $N_{\mathrm{ISI}}^{\mathrm{I}}$ can be calculated by (8) and $\mu_{0,\mathrm{I}}, \sigma_{0,\mathrm{I}}^2$ are

$$\mu_{0,\mathrm{I}} = \frac{1}{2}\left(n_1P_{1p} - n_1P_{1c} + n_2P_{2p}\right), \quad (15)$$

$$\sigma_{0,\mathrm{I}}^2 = \frac{1}{4}n_1P_{1c}\left(1-P_{1c}\right) + \frac{1}{4}n_1P_{1p}\left(1-P_{1p}\right) + \frac{1}{4}n_2P_{2c}\left(1-P_{2c}\right) + \frac{1}{8}n_2P_{2p}\left(1-P_{2p}\right). \quad (16)$$

On the other hand, when TX I transmits bit '1', i.e., n_1 molecules are released, the number of molecules absorbed by the receiver within current time slot, denoted by N_1^{I}, follows another normal distribution as follows:

$$N_1^{\mathrm{I}} = \mathcal{N}\left(n_1P_{1c}, n_1P_{1c}(1-P_{1c})\right) + N_0^{\mathrm{I}} \sim \mathcal{N}\left(\mu_{1,\mathrm{I}}, \sigma_{1,\mathrm{I}}^2\right), \quad (17)$$

where the first term represents the number of molecules absorbed by the receiver among the n_1 molecules released by TX I, and $\mu_{1,\mathrm{I}}, \sigma_{1,\mathrm{I}}^2$ are

$$\mu_{1,\mathrm{I}} = \frac{1}{2}(n_1P_{1p} + n_1P_{1c} + n_2P_{2p}), \quad (18)$$

$$\sigma_{1,\mathrm{I}}^2 = \frac{1}{4}n_1P_{1c}\left(1-P_{1c}\right) + \frac{5}{4}n_1P_{1p}\left(1-P_{1p}\right) + \frac{1}{4}n_2P_{2c}\left(1-P_{2c}\right) + \frac{1}{8}n_2P_{2p}\left(1-P_{2p}\right). \quad (19)$$

At the receiver side, to demodulate TX I's information, it is confronted with the following hypothesis test based on the absorbed molecules Y.

$$\mathcal{H}_0: \quad Y = N_0^{\mathrm{I}} \sim \mathcal{N}\left(\mu_{0,\mathrm{I}}, \sigma_{0,\mathrm{I}}^2\right) \quad (20)$$

$$\mathcal{H}_1: \quad Y = N_1^{\mathrm{I}} \sim \mathcal{N}\left(\mu_{1,\mathrm{I}}, \sigma_{1,\mathrm{I}}^2\right). \quad (21)$$

With the hypothesis test model above, we can derive the optimal decision that minimizes the BER using the MAP detection method. Since we assume that bits '0' and '1' are uniformly distributed, i.e., $p(\mathcal{H}_0) = p(\mathcal{H}_1)$, the MAP test is equivalent to the the likelihood-ratio test (LRT) as follows:

$$\frac{p(Y|\mathcal{H}_1)}{p(Y|\mathcal{H}_0)} = \frac{\sigma_{0,\mathrm{I}}}{\sigma_{1,\mathrm{I}}}\exp\left\{-\frac{(Y-\mu_{1,\mathrm{I}})^2}{2\sigma_{1,\mathrm{I}}^2} + \frac{(Y-\mu_{0,\mathrm{I}})^2}{2\sigma_{0,\mathrm{I}}^2}\right\} \quad (22)$$

$$= \frac{\sigma_{0,\mathrm{I}}}{\sigma_{1,\mathrm{I}}}\exp\left\{-\frac{1}{2}\left(\frac{1}{\sigma_{1,\mathrm{I}}^2} - \frac{1}{\sigma_{0,\mathrm{I}}^2}\right)Y^2 + \left(\frac{\mu_{1,\mathrm{I}}}{\sigma_{1,\mathrm{I}}^2} - \frac{\mu_{0,\mathrm{I}}}{\sigma_{0,\mathrm{I}}^2}\right)Y - \frac{1}{2}\left(\frac{\mu_{1,\mathrm{I}}^2}{\sigma_{1,\mathrm{I}}^2} - \frac{\mu_{0,\mathrm{I}}^2}{\sigma_{0,\mathrm{I}}^2}\right)\right\}. \quad (23)$$

By setting the LRT test $\frac{p(Y|\mathcal{H}_1)}{p(Y|\mathcal{H}_0)}$ as 1, we can find the MAP decision threshold for demodulating information from TX I as

$$\tau_1 = \text{round}\left(\frac{\sqrt{2A\ln C + B^2} - B}{A}\right), \tag{24}$$

where *round* represents the rounding operation since the decision threshold should be an integer; the parameters A, B, and C are as follows:

$$A = -\left(\frac{1}{\sigma_{1,\mathrm{I}}^2} - \frac{1}{\sigma_{0,\mathrm{I}}^2}\right), \quad B = \frac{\mu_{1,\mathrm{I}}}{\sigma_{1,\mathrm{I}}^2} - \frac{\mu_{0,\mathrm{I}}}{\sigma_{0,\mathrm{I}}^2},$$
$$C = \frac{\sigma_{1,\mathrm{I}}^2}{\sigma_{0,\mathrm{I}}^2}\exp\left\{\frac{1}{2}\left(\frac{\mu_{1,\mathrm{I}}^2}{\sigma_{1,\mathrm{I}}^2} - \frac{\mu_{0,\mathrm{I}}^2}{\sigma_{0,\mathrm{I}}^2}\right)\right\}. \tag{25}$$

Based on the optimal decision threshold, we can calculate the BER of information from TX I as follows:

$$\begin{aligned} P_e^{\mathrm{I}} &= \frac{1}{2}\left(P\left(N_0^{\mathrm{I}} \ge \tau_1\right) + P\left(N_1^{\mathrm{I}} \le \tau_1\right)\right) \\ &= \frac{1}{2}\left(1 - Q\left(\frac{\tau_1 - \mu_{1,\mathrm{I}}}{\sigma_{1,\mathrm{I}}}\right) + Q\left(\frac{\tau_1 - \mu_{0,\mathrm{I}}}{\sigma_{0,\mathrm{I}}}\right)\right), \end{aligned} \tag{26}$$

where $Q(x) = \frac{1}{\sqrt{2\pi}}\int_x^{\infty} e^{-\frac{u^2}{2}}du$. Similarly, we can derive the BER of information from TX II as

$$P_e^{\mathrm{II}} = \frac{1}{2}\left(1 - Q\left(\frac{\tau_2 - \mu_{1,\mathrm{II}}}{\sigma_{1,\mathrm{II}}}\right) + Q\left(\frac{\tau_2 - \mu_{0,\mathrm{II}}}{\sigma_{0,\mathrm{II}}}\right)\right), \tag{27}$$

where τ_2, $\mu_{0,\mathrm{II}}$, $\mu_{1,\mathrm{II}}$, $\sigma_{0,\mathrm{II}}$, and $\sigma_{1,\mathrm{II}}$ are dual with τ_1, $\mu_{0,\mathrm{I}}$, $\mu_{1,\mathrm{I}}$, $\sigma_{0,\mathrm{I}}$, and $\sigma_{1,\mathrm{I}}$; hence, the detailed expressions are omitted here. In Figure 5, we plot the relationships between P_e^{I}, P_e^{II} and n_1 and n_2, respectively. From Figure 5a, we can see that TX I's BER P_e^{I} is a decreasing function in terms of n_1 and an increasing function in terms of n_2. This is consistent with traditional wireless communication systems that the higher emission power TX I adopts, i.e., the more molecules, n_1, are transmitted to represent bit '1', the lower BER performance can be achieved. On the other hand, when TX II enhances its emission power n_2, the interference to TX I increases, due to which the BER performance of TX I is degraded. For TX II's BER performance shown in Figure 5b, an opposite phenomenon can be found, where P_e^{II} is a decreasing function in terms of its own emission power n_2, while an increasing function in terms of TX I's emission power n_1. Thus, an important problem in molecular communication systems is how to control the transmitters' emission power to achieve not only efficiency but also fairness. In the next section, we will solve this technical problem using game theory.

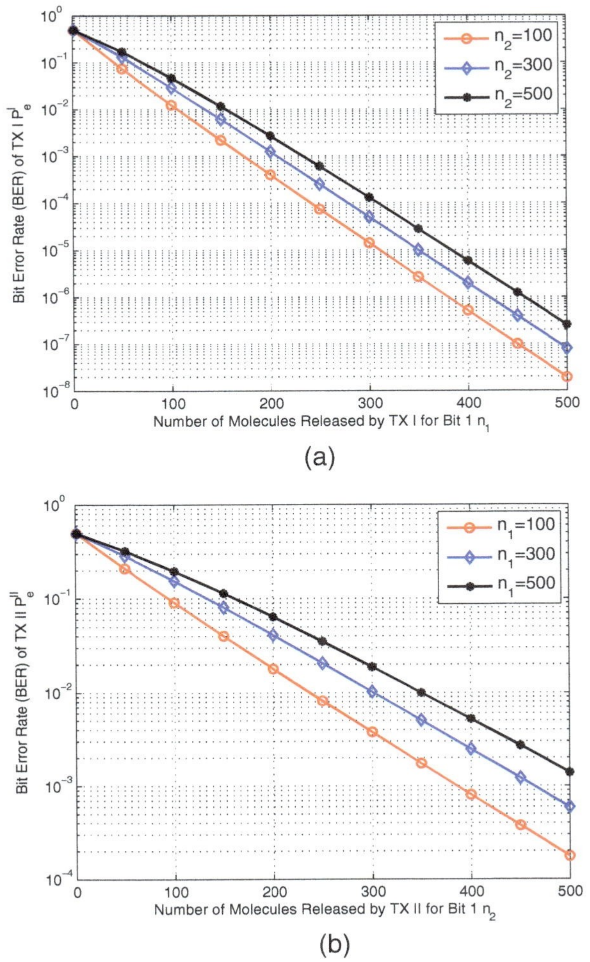

Figure 5 Bit error rate performance. The radius of the receiver is $R = 10\ \mu\text{m}$. The distances between TX I, TX II and the receiver are $d_1 = 30\ \mu\text{m}$ and $d_2 = 40\ \mu\text{m}$. The diffusion coefficient is $D = 79.4\ \mu\text{m}^2/\text{s}$ [29]. **(a)** Bit error rate of TX I. **(b)** Bit error rate of TX II.

4 Molecule emission control game

In this section, we discuss how to control the molecule emission for both transmitters in Figure 1, i.e., how to control n_1 for TX I and how to control n_2 for TX II. As discussed in the introduction, the centralized method can globally optimize all the transmitters' molecule emission by minimizing the total BER $P_e^{\mathrm{I}} + P_e^{\mathrm{II}}$ derived in (26) and (27). However, the centralized method needs a nanoscale server/coordinator to collect each transmitter's location information, and each coordinator responds with the optimal emission control scheme, which inevitably incurs huge communication and energy cost for the nanoscale molecular machines. Moreover, minimizing the total BER $P_e^{\mathrm{I}} + P_e^{\mathrm{II}}$ may lead to unfairness phenomenon, e.g., one transmitter is with extremely low BER and the other is with extremely high BER. In the following, we first discuss a non-cooperative game formulation of the emission control problem and then consider a cooperative game formulation based on Nash bargaining to improve the system efficiency and ensure the fairness between TX I and TX II. The Nash bargaining solution shows its merits of low computational cost, low information exchange cost, and proportional fairness.

4.1 Non-cooperative game formulation

The game theory models strategic interactions among players using formalized incentive structures. It not only provides game models for an efficient self-enforcing distributed design, but also derives well-defined equilibrium criteria to study the optimality of game outcomes for various game scenarios (static or dynamic, complete or incomplete information, non-cooperative or cooperative) [37]. Generally speaking, a game in the strategic form has three elements: $\mathcal{G} = \langle \mathcal{P}, \mathbf{S}, \mathbf{U} \rangle$, where $\mathcal{P} = \{1, 2, \dots\}$ represents a set of players, $\mathbf{S} = \{\mathcal{S}_1, \mathcal{S}_2, \dots\}$ represents a tuple containing the strategy spaces of all players, and $\mathbf{U} = \{U_1, U_2, \dots\}$ represents a tuple containing payoff functions of all players, which measures the outcome of each player. In a fully distributed molecular communication system, each transmitter has the incentive to minimize the BER of its released information at the receiver. In such a case, we can model the non-cooperative molecule emission control game between TX I and TX II, $\mathcal{G}_c = \langle \mathcal{P}_c, \mathbf{S}_c, \mathbf{U}_c \rangle$, as follows:

- *Player*: The players of $\mathcal{G}_c$ are TX I and TX II in the system, i.e.,

$$\mathcal{P}_c = \{\text{TX I, TX II}\}. \tag{28}$$

- *Strategy*: The strategy for each transmitter is defined as its molecules emission power: n_1 for TX I and n_2 for TX II, i.e., how many molecules are released to represent bit '1'. We can see that the strategy space of each transmitter is finite and discrete. Suppose the molecule emission constraints for TX I and TX II are $N^{\mathrm{I}}_{\max}$ and $N^{\mathrm{II}}_{\max}$. Thus, the strategy spaces of TX I and TX II, denoted by $\mathcal{S}_1$ and $\mathcal{S}_2$, respectively, are as follows:

$$\mathbf{S}_c = \{\mathcal{S}_1, \mathcal{S}_2\}, \quad \text{where} \begin{cases} \mathcal{S}_1 = \left\{0, 1, \dots, N^{\mathrm{I}}_{\max}\right\}, \\ \mathcal{S}_2 = \left\{0, 1, \dots, N^{\mathrm{II}}_{\max}\right\}. \end{cases} \tag{29}$$

  Moreover, the pure strategy profile of game $\mathcal{G}_c$, which is defined as a set of strategies including each player's specified action, can be written as (n_1, n_2) [38].
- *Utility function*: The utility function should be defined proportional to the player's benefit. Since each transmitter expects a lower BER, the utility should be defined inversely proportional to the BER performance. Considering that the worst case of BER is $\frac{1}{2}$ (when each transmitter adopts strategy $n_1 = n_2 = 0$ and the receiver cannot distinguish between bit '1' and '0'), we can define the utility function as the difference between $\frac{1}{2}$ and BER to ensure a positive utility. Let U_1 and U_2 denote the utility functions of TX I and TX II, respectively, which can be written as follows:

$$\mathbf{U}_c = \{U_1, U_2\},$$
$$\text{where} \begin{cases} U_1 = \frac{1}{2} - P_e^{\mathrm{I}} = \frac{1}{2}\left(Q\left(\frac{\tau_1 - \mu_{1,\mathrm{I}}}{\sigma_{1,\mathrm{I}}}\right) - Q\left(\frac{\tau_1 - \mu_{0,\mathrm{I}}}{\sigma_{0,\mathrm{I}}}\right)\right), \\ U_2 = \frac{1}{2} - P_e^{\mathrm{II}} = \frac{1}{2}\left(Q\left(\frac{\tau_2 - \mu_{1,\mathrm{II}}}{\sigma_{1,\mathrm{II}}}\right) - Q\left(\frac{\tau_2 - \mu_{0,\mathrm{II}}}{\sigma_{0,\mathrm{II}}}\right)\right). \end{cases} \tag{30}$$

  According to the utility function above, we can see that U_1 is an increasing function in terms of n_1 and a decreasing function in terms of n_2. On the other hand, U_2 is a decreasing function in terms of n_1 but an increasing function in terms of n_2.

Based on the game model of $\mathcal{G}_c = \langle \mathcal{P}_c, \mathbf{S}_c, \mathbf{U}_c \rangle$ above, we can further find its NE. Since the utility functions of $\mathcal{G}_c$ are monotonic in terms of both transmitters' strategies, the NE can be guaranteed to exist and is unique. Apparently, the NE is that each transmitter uses the highest molecule emission power, which would lead to low system efficiency compared with the centralized scheme. Therefore, to improve the system efficiency, we formulate the emission control problem as a cooperative game using the Nash bargaining solution in next subsection.

4.2 Cooperative game formulation based on Nash bargaining

In game theory, a cooperative game studies how a group of players should cooperate with each other when non-cooperation leads to an unfavorable outcome for each player, e.g., in the non-cooperative molecule emission control game discussed above, each transmitter adopts the highest molecule emission power and heavily interferes with each other. Under such circumstances, the players have the incentive to cooperatively enhance the system efficiency in conjunction with fairness. The bargaining game is one branch of the cooperative game, where two players try to reach an agreement on trading/sharing a limited amount of resources. These two individual players have a choice to bargain with each other so that both of them can gain benefit higher than that without cooperation. In a bargaining game, since there might be an infinite number of social optimal agreement points (i.e., bargaining solutions), three kinds of bargaining solutions - egalitarian solution, Kalai-Smorodinsky solution, and Nash bargaining solution [38] - were proposed to refine the multiple bargaining solutions, among which the Nash bargaining solution is mostly widely adopted with the emphasis of fairness and social optimality [39]. In the following, we formulate the emission control game between two transmitters as a bargaining game and find the Nash bargaining solution for them.

The emission control bargaining game for molecular communications can be described as follows. Let $\mathcal{P}_c = \{\text{TX I}, \text{TX II}\}$ denote the set of two players, i.e., two nanoscale transmitters. Let $\mathbf{W} = \{(U_1, U_2)\} \subset \mathbb{R}^2$ represent the set of feasible payoff allocations that each transmitter can get if they can reach an agreement to cooperate. Let $\mathbf{U}_{\min} = \left(U_{\min}^{\mathrm{I}}, U_{\min}^{\mathrm{II}}\right)$, where $U_{\min}^{\mathrm{I}}$ and $U_{\min}^{\mathrm{II}}$ represent the minimal payoffs that TX I and TX II can be satisfied if both of them cooperate with each other. One way is to set the $U_{\min}^{\mathrm{I}}$ and $U_{\min}^{\mathrm{II}}$ to be the NE of the non-cooperative molecule emission control game analyzed in the previous subsection. This means that both of the transmitters agree to cooperate only if their respective utility cannot be less than that of the non-cooperative scenario. However, in such a case, fairness cannot be guaranteed since the TX, which can obtain a much higher utility in the non-cooperative scenario, can still obtain a much higher utility than the other in this cooperative scenario, which may lead to the failure of cooperation. Therefore, as in most cases, to ensure fairness among transmitters, we set $U_{\min}^{\mathrm{I}} = U_{\min}^{\mathrm{II}} = 0$, which means that as long as the utility is larger than 0, both transmitters would be cooperative. In such a case, all the possible solutions in $\mathbf{W}$ satisfy the individual rationality, i.e., $\left\{(U_1, U_2) | U_1 \geq U_{\min}^{\mathrm{I}}, U_2 \geq U_{\min}^{\mathrm{II}}\right\} \subset \mathbf{W}$ is the set of all the possible bargaining solutions. This is also practical since the NE of the non-cooperative game usually corresponds to the scenario that each transmitter has little utility, i.e., $U_{\min}^{\mathrm{I}} \to 0$, $U_{\min}^{\mathrm{II}} \to 0$. Thus, the pair $(\mathbf{W}, \mathbf{U}_{\min})$ can be defined as the emission control bargaining game. We can see that there are numerous possible operating points for TX I and TX II in $\mathbf{W}$. The Nash bargaining solution (NBS), defined as follows, provides a unique, fair and efficient operating point in $\mathbf{W}$ with the idea that after the minimal requirements are satisfied for both transmitters, the rest of the resources are allocated proportionally to each of them according to their conditions.

Definition 1: In the molecule emission control bargaining game $(\mathbf{W}, \mathbf{U}_{\min})$, a solution $\mathbf{U}^* = \left(U_1^*, U_2^*\right)$ is said to be an NBS in $\mathbf{W}$ if the following axioms are satisfied.

1. *Pareto optimality*: There does not exist a point $\mathbf{U}' = \left(U_1', U_2'\right)$ other than $\mathbf{U}^*$ in $\mathbf{W}$, satisfying that $U_1' \geq U_1^*$ and $U_2' \geq U_2^*$.
2. *Symmetry*: If the feasible set $\mathbf{W}$ is symmetric in terms of two transmitters, for example, when TX I and TX II are homogenous, then $\mathbf{U}^*$ is also symmetric, i.e., $U_1^* = U_2^*$.
3. *Independence of irrelevance alternatives*: If $\mathbf{U}^* \in \mathbf{W}' \subset \mathbf{W}$ is an NBS for $(\mathbf{W}, \mathbf{U}_{\min})$, then it is also an NBS for $\left(\mathbf{W}', \mathbf{U}_{\min}\right)$.
4. *Independence of linear transformations*: For any linear transformation ψ, if $\mathbf{U}^*$ is an NBS for $(\mathbf{W}, \mathbf{U}_{\min})$, then $\psi(\mathbf{U}^*)$ is an NBS for $(\psi(\mathbf{W}), \psi(\mathbf{U}_{\min}))$.

Among the axioms, the physical meaning of Pareto optimality is that there exists no other operating point that can lead to superior performance for one transmitter without degrading the performance of the other. In other words, Pareto optimality ensures that all the system resources have been fully utilized. Axioms 2 to 4 are related to fairness, where the symmetry axiom requires that if the feasible set is completely symmetric, then TX I and TX II have exactly the same utility, the irrelevance alternatives axiom can help to refine the feasible solution set by eliminating the irrelevant ones, and the linear property asserts that the NBS is scale-invariant. We can see that an NBS can not only achieve a high system efficiency, but also guarantee the fairness issue. It has been shown that there is a unique NBS that satisfies all the axioms in *Definition 1* [38], wherein the solution maximizes the product of TX I's and TX II's utilities, i.e.,

$$\mathbf{U}^* = \left(U_1^*, U_2^*\right) = \arg \max_{n_1 \leq N_{\max}^{\mathrm{I}}, n_2 \leq N_{\max}^{\mathrm{II}}} U_1(n_1, n_2) \cdot U_2(n_1, n_2). \tag{31}$$

In such a case, the Nash bargaining equilibrium for both transmitters can be calculated by solving the optimization problem as follows:

$$\begin{aligned} \left(n_1^*, n_2^*\right) = \arg\max_{n_1, n_2} \frac{1}{4} & \left(Q\left(\frac{\tau_1 - \mu_{1,\mathrm{I}}}{\sigma_{1,\mathrm{I}}}\right) - Q\left(\frac{\tau_1 - \mu_{0,\mathrm{I}}}{\sigma_{0,\mathrm{I}}}\right)\right) \\ & \times \left(Q\left(\frac{\tau_2 - \mu_{1,\mathrm{II}}}{\sigma_{1,\mathrm{II}}}\right) - Q\left(\frac{\tau_2 - \mu_{0,\mathrm{II}}}{\sigma_{0,\mathrm{II}}}\right)\right) \\ \text{s.t.} \quad n_1 \leq N_{\max}^{\mathrm{I}}, & \quad n_2 \leq N_{\max}^{\mathrm{II}} \end{aligned} \tag{32}$$

Since it is difficult to derive closed-form expressions for $U_1(n_1, n_2)$ and $U_2(n_1, n_2)$, the optimization problem above has to be numerically solved. Suppose $U(n_1, n_2) = U_1(n_1, n_2) \cdot U_2(n_1, n_2)$, the NBS can be found by using the gradient descent method as follows:

$$n_1(t+T) = n_1(t) + \lambda \left[U(n_1(t)+1, n_2(t)) - U(n_1(t), n_2(t))\right], \tag{33}$$

$$n_2(t+T) = n_2(t) + \lambda \left[U(n_1(t), n_2(t)+1) - U(n_1(t), n_2(t))\right], \tag{34}$$

where T is the updating period, λ is an integer step size which can be set as 1, and $[U(n_1(t)+1, n_2(t)) - U(n_1(t), n_2(t))]$ and $[U(n_1(t), n_2(t)+1) - U(n_1(t), n_2(t))]$ are the approximated partial derivatives of $U(n_1, n_2)$ over n_1 and n_2, respectively. Note that in the implementation, $U(n_1, n_2)$ is not calculated using (30), which, instead, is evaluated by the transmitters. Therefore, they need to exchange their utility $U_1(t)$ and $U_2(t)$ with each other to find the NBS. In Algorithm 1, we summarize the distributed NBS algorithm. Note that the structure of the algorithm can also be applied into multiple TX and RX scenarios, where an extra time slot is required for multiple transmitters exchanging their utility information, i.e., the transmitters sequentially broadcast their utility in step 3 of Algorithm 1.

Algorithm 1 Distributed NBS algorithm

1: • Initialize $n_1(0) = n_2(0) = 0$.
2: **while** $n_1(t+T) \neq n_1(t)$ or $n_2(t+T) \neq n_2(t)$ **do**
3: • Both transmitters evaluate the average utility $U_1(n_1(t), n_2(t))$ and $U_2(n_1(t), n_2(t))$ within $\frac{T}{3}$ slots and exchange the utility.
4: • TX I evaluates the average utility $U_1(n_1(t) + 1, n_2(t))$ within $\frac{T}{3}$ slots.
5: • TX II evaluates the average utility $U_2(n_1(t), n_2(t))$ within $\frac{T}{3}$ slots.
6: • TX I updates $n_1(t+T)$ using (33).
7: • TX I updates $n_2(t+T)$ using (34).
8: **end while**

From the definition of NBS, the Pareto optimality quantitatively shows the high efficiency of NBS, while fairness is only illustrated in a qualitative manner. To further quantitatively verify the fairness property of NBS, we introduce the concept of 'proportional fairness', which is a widely used metric to evaluate fairness performance in wired networks [40]. Proportional fairness is a compromise between fairness and performance and based upon maintaining a balance between two competing interests, i.e., trying to maximize the social welfare while at the same time satisfying each user's minimal level of requirement. According to the general definition of proportional fairness [40], the specific definition of proportional fairness in the emission control bargaining game for molecular communications can be described as follows.

Definition 2: In the molecule emission control bargaining game $(\mathbf{W}, \mathbf{U}_{\min})$, a solution $\mathbf{U}^{\mathrm{F}} = (U_1^{\mathrm{F}}, U_2^{\mathrm{F}})$ is said to be proportionally fair when any change in $\mathbf{U}^{\mathrm{F}} = (U_1^{\mathrm{F}}, U_2^{\mathrm{F}})$ would result in the sum of proportional changes of the utilities being non-positive, i.e.,

$$\frac{U_1' - U_1^{\mathrm{F}}}{U_1^{\mathrm{F}}} + \frac{U_2' - U_2^{\mathrm{F}}}{U_2^{\mathrm{F}}} \le 0, \quad \forall (U_1', U_2') \in \mathbf{W}. \tag{35}$$

From the definition, we can see that the physical meaning of proportional fairness is that at this point, if one transmitter intends to improve its performance with some increasing percentage, then the other transmitter would suffer from some degrading performance with some decreasing percentage, and the increasing percentage is no larger than the decreasing percentage. Therefore, the Pareto optimality is a special case of proportional fairness. In the following theorem, we will show that the NBS of the emission control bargaining game derived by (31) is a proportionally fair solution.

Since the ln function is concave and monotonic, the optimization problem in (31) is equivalent to the following problem:

$$\begin{aligned} \mathbf{U}^* &= (U_1^*, U_2^*) \\ &= \arg \max_{n_1 \le N_{\max}^{\mathrm{I}}, n_2 \le N_{\max}^{\mathrm{II}}} \ln (U_1(n_1, n_2) \cdot U_2(n_1, n_2)). \end{aligned} \tag{36}$$

Let us define $\hat{U}_1(n_1, n_2) = \ln (U_1(n_1, n_2))$ and $\hat{U}_2(n_1, n_2) = \ln (U_2(n_1, n_2))$. Thus, the gradients of $\hat{U}_1(n_1, n_2)$ and $\hat{U}_2(n_1, n_2)$ at the NBS point $\mathbf{U}^* = (U_1^*, U_2^*)$ are $\partial \hat{U}_1(n_1, n_2)/\partial U_1 \big|_{U_1^*}$ and $\partial \hat{U}_2(n_1, n_2)/\partial U_2 \big|_{U_2^*}$. Since the NBS is the unique point that can optimize both (31) and (36), for any point deviating from the NBS point, denoted by $\mathbf{U}' = (U_1', U_2')$, the following inequality should hold:

$$\left.\frac{\partial \hat{U}_1(n_1, n_2)}{\partial U_1}\right|_{U_1^*} (U_1' - U_1^*) + \left.\frac{\partial \hat{U}_2(n_1, n_2)}{\partial U_2}\right|_{U_2^*} (U_2' - U_2^*) \le 0 \tag{37}$$

$$\Rightarrow \frac{U_1' - U_1^*}{U_1^*} + \frac{U_2' - U_2^*}{U_2^*} \le 0. \tag{38}$$

The first inequality means for all feasible point $\mathbf{U}' = (U_1', U_2') \in \mathbf{W}$ that is different from NBS point $\mathbf{U}^*$, the overall change of two transmitters' utilities is negative according to the gradient.We can see that the second inequality is same as the definition of proportional fairness in (35). Therefore, we can conclude that the NBS of the emission control bargaining game is equivalent to the proportional fairness.

4.3 Simulation results

In this subsection, we conduct a simulation to evaluate the performance of the non-cooperative and cooperative game theoretic analyses. In the simulation, the length of each time slot is set as $t_s = 10$ s, the radius of the receiver is configured as $R = 10$ μm, and the diffusion coefficient is an empirical parameter set as $D = 79.4$ μm^2/s

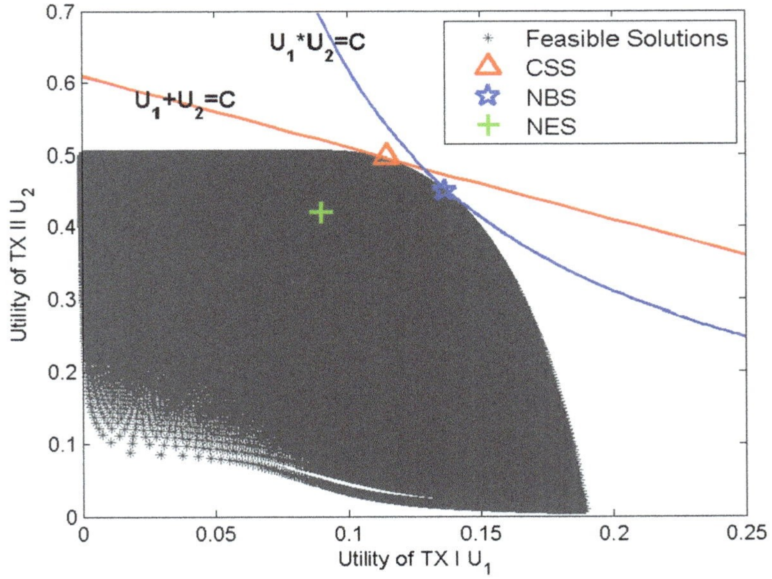

Figure 6 Physical meanings of three solutions.

[29]. As for the location settings, the distance between TX I and the receiver is fixed at $d_1 = 100$ μm, while the distance between TX II and the receiver d_2 varies from 50 to 150 μm. The molecule emission constraint for each transmitter is 1,000, i.e., $N^{\mathrm{I}}_{\max} = N^{\mathrm{II}}_{\max} = 1,000$. In the simulation, we compare the performances of three kinds of schemes as follows:

- *Centralized scheme solution (CSS)*: The centralized scheme is to maximize the social welfare of the two transmitters, i.e.,

$$\max_{n_1 \leq N^{\mathrm{I}}_{\max}, n_2 \leq N^{\mathrm{II}}_{\max}} U_1(n_1, n_2) + U_2(n_1, n_2). \tag{39}$$

- *Nash equilibrium solution (NES)*: Nash equilibrium means the solution of the non-cooperative emission control game, i.e., $n_1 = N^{\mathrm{I}}_{\max}$, $n_2 = N^{\mathrm{II}}_{\max}$.
- *Nash bargaining solution (NBS)*: Nash bargaining means the solution of the cooperative emission control game, i.e.,

$$\max_{n_1 \leq N^{\mathrm{I}}_{\max}, n_2 \leq N^{\mathrm{II}}_{\max}} U_1(n_1, n_2) \cdot U_2(n_1, n_2). \tag{40}$$

We first conduct a simulation to illustrate the physical meaning of the three schemes as shown in Figure 6. The dotted area represents all the feasible solutions for U_1 and

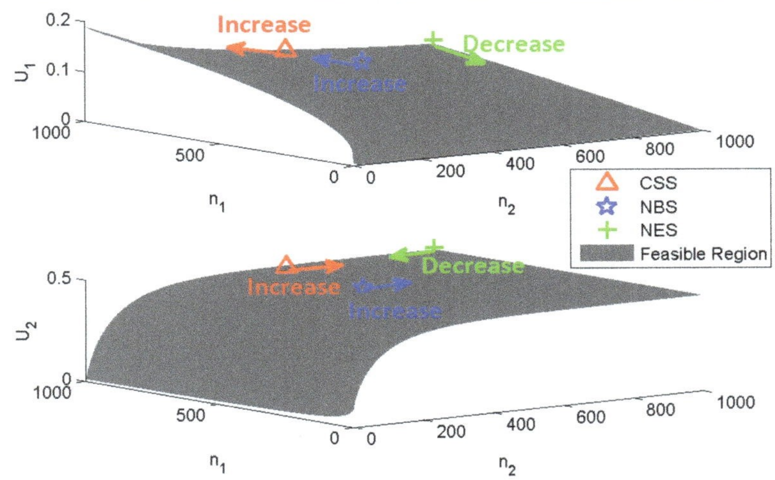

Figure 7 Individual utility comparison.

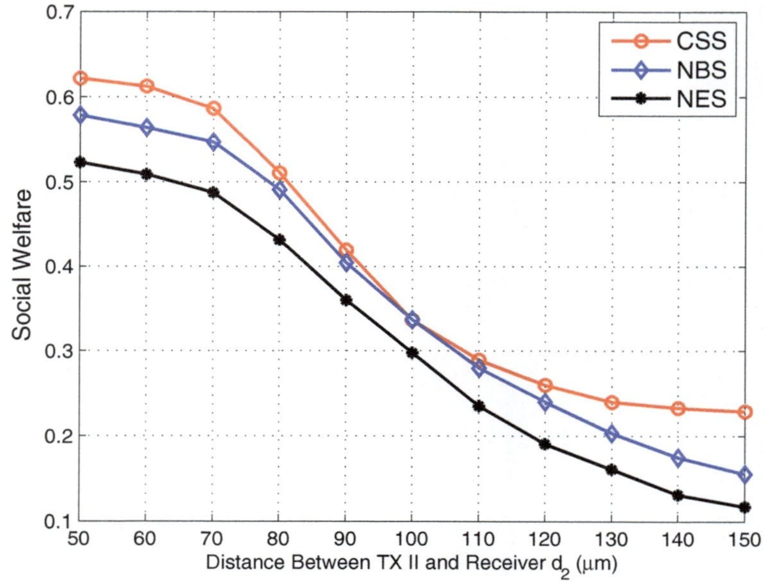

Figure 8 Social welfare comparison.

U_2 when the location of TX I is $d_1 = 100\ \mu\text{m}$ and the location of TX II is $d_2 = 50\ \mu\text{m}$. The feasible set is obtained by simulating all possible strategy profiles (n_1, n_2), where $1 \le n_1 \le 1,000$ and $1 \le n_2 \le 1,000$ (the cases of $n_1 = 0$ and $n_1 = 0$ are omitted). We can see that for the centralized scheme which maximizes the total utilities, the corresponding solution, i.e., CSS denoted by the red triangle, is just the tangency point between the line $U_1 + U_2 = C$ and the feasible set. For the Nash bargaining scheme that maximizes the product of two transmitters' utilities, the corresponding solution, i.e., NBS denoted by the blue pentagon, is just the tangency point between the line $U_1 \cdot U_2 = C$ and the feasible set. Since both transmitters use the highest molecule emission power at the Nash equilibrium solution of the non-cooperative game, their utilities are impaired by the interference, which can

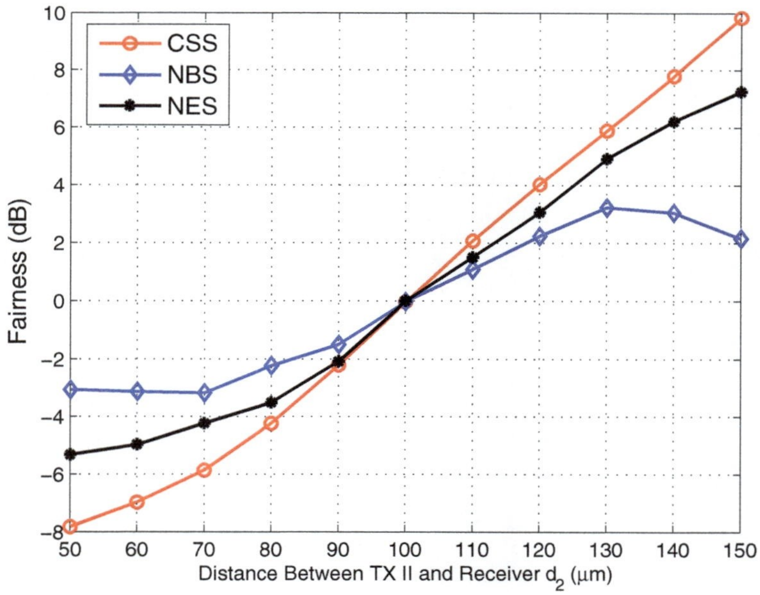

Figure 9 Fairness comparison.

be seen by the NES denoted by the green plus. However, with NES, no transmitter has the incentive to deviate. As shown in Figure 7, which illustrates the utilities of both transmitters versus their strategies, n_1 and n_2, we can see that at the NES denoted by the green plus, there is no deviation direction that can increase either transmitter's utility. While for the CSS and NBS, each transmitter can increase its utility by unilaterally deviating from the current strategy. Therefore, although the NES provides a low system efficiency, it is an equilibrium that neither transmitter will deviate since any deviation will lead to a utility loss.

In Figure 8, we show the total utilities of both transmitters for three schemes versus TX II's location d_2, i.e., social welfare comparison. We can see that the centralized scheme always has the best performance since its objective function is to maximize the social welfare, while the Nash equilibrium scheme has the worst performance since both transmitters adopt the highest molecule emission power and thus severely interfere with each other. The performance of the Nash bargaining scheme is in the middle, and the gap from the centralized scheme is larger when the two transmitters have very different locations with regard to the receiver, i.e., when 50 μm $\leq d_2 \leq$ 70 μm and 130 μm $\leq d_2 \leq$ 150 μm (note that $d_1 =$ 100 μm). That is because, in order to balance the utilities between TX I and TX II, the NBS has to sacrifice some system performance to ensure fairness. Moreover, the difference between the two transmitters' utilities is proportional to the difference between their locations relative to the receiver $(d_1 - d_2)$, which leads to the fact that the larger difference $(d_1 - d_2)$, the more system performance is required to be compromised in order to guarantee fairness. We can also see that when TX I and TX II are in similar locations, i.e., 90 μm $\leq d_2 \leq$ 110 μm, the performances of CSS and NBS are also similar with each other due to the nearly symmetric settings. In addition, to evaluate the fairness performances of those three schemes, we plot the utility ratio of two transmitters in Figure 9, i.e., $10 \log (U_1(n_1, n_2)/U_2(n_1, n_2))$. Note that the vertical axis is in dB, which means that the closer to 0 dB, the better the fairness performance. Different from the social welfare performance, we can see that the NBS always has the best fairness performance, which further verifies that the NBS is a trade-off between performance and fairness.

5 Conclusions

In this paper, we provided a game-theoretic framework for nanoscale molecular communication systems. We analyzed the inter-symbol and inter-user interference, optimal decision threshold, and BER performance in the molecular communication system with amplitude modulation. Based on the interference analysis, we formulated the molecule emission control problem as a non-cooperative emission control game, where the Nash equilibrium was found and proved to be unique. In order to improve the system efficiency while guaranteeing fairness, we further modeled the proposed molecule emission control problem using a cooperative game based on Nash bargaining solution, which was proved to be proportionally fair. Finally, a simulation was conducted to compare the performance among the centralized scheme, Nash equilibrium scheme, and Nash bargaining scheme. The results verified that the NBS can achieve a trade-off between social welfare performance and user fairness.

Competing interests

The authors declare that they have no competing interests.

Acknowledgements

This work was funded by projects 61371079, 61273214, 61271267, and 91338203 supported by NSFC China, and the Postdoctoral Science Foundation funded project. The work was done during Chunxiao Jiang's visit to the University of Maryland.

Author details

[1]Department of Electronic Engineering, Tsinghua University, Beijing 100084, China. [2]Department of Electrical anc Computer Engineering, University of Maryland, College Park, MD 20742, USA.

References

1. IF Akyildiz, J Jornet, The internet of nano-things. IEEE Wireless Commun. **17**(6), 58–63 (2010)
2. IF Akyildiz, F Brunetti, C Blazquez, Nanonetworks: a new communication paradigm. Comput. Networks. **52**(12), 2260–2279 (2008)
3. S Hiyama, Y Moritani, Molecular communication: harnessing biochemical materials to engineer biomimetic communication systems. Networks. **1**(1), 20–30 (2010)
4. T Nakano, MJ Moore, F Wei, AV Vasilakos, J Shuai, Molecular communication and networking: opportunities and challenges. IEEE Trans. Nanobiosci. **11**(2), 135–148 (2012)
5. M Pierobon, IF Akyildiz, A physical end-to-end model for molecular communication in nanonetworks. IEEE J. Sel. Areas Commun. **28**(4), 602–611 (2010)
6. S F Walsh, D Balasubramaniam, T Botvich, T Suda, SF Nakano, MO Bush, in *Third International Conference on NanoNetworks and Workshops (NanoNet)*. Foghlu, Hybrid DNA and enzymatic based computation for address encoding, link switching and error correction in molecular communication (USA Boston, MA, September 2008), pp. 28–38
7. T T Nakano, M Suda, R Moore, A Egashira, K Enomoto, in *Genetic and Evolutionary Computation Conference (GECCO)*. Arima, Molecular communication for nanomachines using intercellular calcium signaling (USA Washington D.C., June 2005), pp. 478–481
8. A Eckford, in *Information Sciences and Systems (CISS)*. Nano communication with Brownian motion (USA Baltimore, MD, March 2007), pp. 14–16
9. B Atakan, OB Akan, On channel capacity and error compensation in molecular communication. Trans. Comput. Syst. Biol. X. **5410**, 59–80 (2008)
10. B Atakan, OB Akan, Deterministic capacity of information flow in molecular nanonetworks.Nano Commun. Networks. **1**(1), 31–42 (2010)
11. D Arifler, Capacity analysis of a diffusion-based short-range molecular nano-communication channel. Comput. Networks. **55**(6), 1426–1434 (2011)
12. M A Einolghozati, A Sardari, F Beirami, in *IEEE ISIT*. Fekri, Capacity of discrete molecular diffusion channels (Saint Petersburg Russia, August 2011), pp. 723–727

13. S Kadloor, RS Adve, AW Eckford, Molecular communication using Brownian motion with drift. IEEE Trans. NanoBiosci. **11**(2), 89–99 (2012)
14. KV Srinivas, AW Eckford, RS Adve, Molecular communication in fluid media: the additive inverse Gaussian noise channel. IEEE Tran. Inform. Theory. **58**(7), 4678–4692 (2012)
15. MS Leeson, MD Higgins, Forward error correction for molecular communications.Nano Commun. Networks. **3**(1), 161–167 (2012)
16. M Pierobon, IF Akyildiz, Diffusion-based noise analysis for molecular communication in nanonetworks. IEEE Trans. Signal Process. **59**(6), 2530–2547 (2011)
17. M Pierobon, IF Akyildiz, in *Proc. IEEE ICC*. Intersymbol and co-channel interference in diffusion-based molecular communication (Ottawa Canada, June 2012), pp. 6126–6131
18. BD Unluturk, D Malak, OB Akan, Rate-delay tradeoff with network coding in molecular nanonetworks. IEEE Trans. Nanotechnol. **12**(2), 120–128 (2013)
19. A Noel, KC Cheung, R Schober, Optimal receiver design for diffusive molecular communication with flow and additive noise. IEEE Trans. NanoBiosci. **13**(3), 350–362 (2014)
20. M Pierobon, IF Akyildiz, Noise analysis in ligand-binding reception for molecular communication in nanonetworks. IEEE Trans. Signal Process. **59**(9), 4168–4182 (2011)
21. I Karatzas, SE Shreve, *Brownian Motion and Stochastic Calculus*. (Springer, New York, 1991)
22. MJ Moore, T Suda, K Oiwa, Molecular communication: modeling noise effects on information rate. IEEE Trans. Nanobiosci. **8**(2), 169–180 (2009)
23. I I Llatser, N Pascual, A Garralda, M Cabellos-Aparicio, E Pierobon, in *Proc. IEEE GLOBECOM*. Alarcon, Exploring the physical channel of diffusion-based molecular communication by simulation (USA Houston, TX, December 2011), pp. 1–5
24. M A Einolghozati, A Sardari, F Beirami, in *Inf. Sci. Syst. (CISS)*. Fekri, Consensus problem under diffusion-based molecular communication (USA Baltimore, MD, March 2011), pp. 1–6
25. A Einolghozati, FF Mohsen Sardari, in *IEEE ISIT*. Relaying in diffusion-based molecular communication (Istanbul Turkey, July 2013), pp. 1844–1848
26. Z Han, KJR Liu, Non-cooperative power control game and throughput game over wireless networks. IEEE Trans. Commun. **53**(10), 1625–1629 (2005)
27. Z Han, ZJ Ji, KJR Liu, Non-cooperative resource competition game by virtual referee in multi-cell OFDMA. IEEE J. Sel. Areas Commun. **25**(6), 1079–1090 (2007)
28. KJR Liu, B Wang, *Cognitive Radio Networking and Security: A Game Theoretical View*. (Cambridge University Press, Cambridge, 2010)
29. JJ Kim, K Park, Modulated insulin delivery from glucose-sensitive hydrogel dosage forms. J. Controlled Release. **77**(1-2), 39–47 (2001)
30. T Zhou, L Chen, K Aihara, Molecular communication through stochastic synchronization induced by extracellular fluctuations. Phys. Rev. Lett. **95**(17), 178103 (2005)
31. S Abadal, IF Akyildiz, Automata modeling of quorum sensing for nanocommunication networks. Nano Commun. Networks. **2**(1), 74–83 (2011)
32. H ShahMohammadian, GG Messier, S Magierowski, Blind synchronization in diffusion-based molecular communication channels. IEEE Commun. Lett. **17**(11), 2156–2159 (2013)
33. MS Kuran, HB Yilmaz, T Tugcu, B Ozerman, Energy model for communication via diffusion in nanonetworks. Nano Commun. Networks. **1**(2), 86–95 (2010)
34. MS Kuran, HB Yilmaz, T Tugcu, IF Akyildiz, Interference effects on modulation techniques in diffusion based nanonetworks. Nano Commun. Networks. **3**(1), 65–73 (2012)
35. S H Tezcan, FN Oktug, in *IEEE ICC Workshop on Molecular and Nanoscale Communications*. Kok, Neural delay lines for TDMA based molecular communication in neural networks (Ottawa Canada, June 2012), pp. 6209–6213
36. P-C L-S Meng, K-C Yeh, IF Chen, *Akyildiz, MIMO communications based on molecular diffusion*. (CA, Anaheim, December 2012), pp. 5380–5385
37. Z Ji, KJR Liu, Cognitive radios for dynamic spectrum access-dynamic spectrum sharing: a game theoretical overview. IEEE Commun. Mag. **45**(5), 88–94 (2007)
38. G Owen, *Game Theory*. (Academic Press, New York, 2001)
39. Z Han, Z Ji, KJR Liu, Fair multiuser channel allocation for OFDMA networks using Nash bargaining solutions and coalitions. IEEE Trans. Commun. **53**(8), 1366–1376 (2005)
40. F Kelly, Charging and rate control for elastic traffic. Eur. Trans. Telecommun. **28**(1), 33–37 (1997)

12

Energy-efficient downlink resource management in self-organized OFDMA-based two-tier femtocell networks

Adnan Shahid, Saleem Aslam, Hyung Seok Kim and Kyung-Geun Lee*

Abstract

Femtocell is a novel technology that is used for escalating indoor coverage as well as the capacity of traditional cellular networks. However, interference is the limiting factor for performance improvement due to co-channel deployment between macrocells and femtocells. The traditional network planning is not feasible because of the random deployment of femtocells. Therefore, self-organization approaches are the key to having successful deployment of femtocells. This study presents the joint resource block (RB) and power allocation task for the two-tier femtocell network in a self-organizing manner, with the concern to minimizing the impact of interference and maximizing the energy efficiency. In this study, we analyze the performance of the system in terms of the energy efficiency, which is composed of both the transmission and circuit power. Most of the previous studies investigate the performance regarding the throughput requirement of the two-tier femtocell network while the energy efficiency aspect is largely ignored. Here, the joint allocation task is modeled as a non-cooperative game which is demonstrated to exhibit pure and unique Nash equilibrium. In order to reduce the complexity of the proposed non-cooperative game, the joint RB and power allocation task is divided into two subproblems: an RB allocation and a particle swarm optimization-based power allocation. The analysis of the proposed game is carried out in terms of not only energy efficiency but also throughput. With practical 3rd Generation Partnership Project (3GPP) Long-Term Evolution (LTE) parameters, the simulation results illustrate the superior performance of the proposed game as compared to the traditional methods. Also, the comparison is carried out with the joint allocation scheme which only considers the throughput as the objective function. The results illustrate that significant performance improvement is achieved in terms of energy efficiency with slight loss in the throughput. The analysis in regard to energy efficiency and throughput of the two-tier femtocell network is carried out in terms of the performance metrics, which include convergence, impact of varying RBs, impact of femtocell density, and the fairness index.

Keywords: RB allocation; Power allocation; Non-cooperative game; Particle swarm optimization; Nash equilibrium

1 Introduction

Femtocell is a promising technology for expanding indoor coverage as well as the capacity of traditional cellular networks [1]. Therefore, it has attracted attention from academia, industry, and standardization forums. Although femtocell can escalate the performance of indoor users, interference is the critical factor in this regard because of the co-channel deployment of macrocell and femtocell. Femtocell is serviced by a small cellular base station (SBS), which is specifically designed for indoor users or small business. An important characteristic of femtocell is that they are installed randomly and connected to the core macrocell network by using an existing backhaul link, which could be digital subscriber line (DSL), optical fiber, etc. This backhaul link makes it possible for communication with the macrocells and femtocells. The benefits obtained by exploiting femtocell can be divided into two categories: operator and customer perspectives. From the operator's perspective, femtocell alleviates the burden on the macrocell network by offloading the traffic from the macrocell to the femtocell network. As far as the customer is concerned, high throughput, reduced power, and reliable communication are achievable by the reduced

* Correspondence: kglee@sejong.ac.kr
Department of Information and Communication Engineering, Sejong University, 98 Gunja-dong, Gwangjin-gu, Seoul 143-747, Republic of Korea

distance between the femto users (FUEs) and the connected SBS [2].

A hierarchical two-tier femtocell network is usually implemented in a co-channel environment owing to the efficient spectrum utilization. The interference in this two-tier network is bottle neck in the overall performance. Principally, there are two types of interferences that exist in the two-tier femtocell network: cross-tier and co-tier [3]. Within the context of femtocells, the former is the interference between macrocells and femtocells while the latter is among femtocells. Thereby, it is imperative and challenging to deal with the interferences in such an unplanned femtocell network.

Traditional network planning and interference management strategies are not feasible for the two-tier femtocell networks. This is in accordance to the fact that prior information regarding the positions and number of femtocells is generally unknown to the operators. Therefore, self-organization provides a vital solution for the management of this unplanned femtocell network. On the other hand, the interference can also be adequately managed by employing the self-organization approaches and consequently improves the system performance [4]. The self-organization concept originates from cognitive radio technology in which there is a little or no involvement of centralized entity. Most of the existing literature on the two-tier network only considers the throughput as the performance measure of the network, while the energy efficiency aspect is largely ignored [5]. For the dense deployment of femtocells, energy efficiency becomes a prime concern that needs significant attention. Besides this, the energy efficiency has recently attracted the attention from academia, industry, and standardization forums because of the rapid growth in the demand of users.

This motivates us to investigate the energy efficiency aspect for the self-organized two-tier femtocell networks.

1.1 Contributions

In this study, the joint allocation task for the downlink two-tier femtocell networks is modeled as a non-cooperative game. Our main contributions in this study are categorically defined as under:

1. An energy-efficient resource block (RB) and power allocation task is modeled as a non-cooperative game for the two-tier femtocell network. Precisely, the players (SBSs) in the game interact with the environment autonomously and learn the action profile (RB and power levels) with the concern to maximizing the energy efficiency. Under the assumption of the availability of channel gains, the proposed game demonstrated to exhibit pure and unique Nash equilibrium.
2. The utility function of the proposed game is designed in a manner that maximizes the energy efficiency and minimizes the impact of interference in the two-tier network.
3. The complexity of the joint RB and power allocation task of the proposed game is decomposed into two subproblems: an RB allocation and a particle swarm optimization (PSO)-based power allocation. This decomposition significantly minimizes the complexity in each iteration of the proposed game.
4. The analysis of the proposed game is carried out regarding energy efficiency and throughput. Additionally, two different scenarios are taken into consideration for having an in-depth analysis of the proposed game. The comparison of the proposed game is carried out regarding joint allocation game with throughput maximization [6] and traditional methods.
5. Simulation results are computed regarding energy efficiency and throughput in terms of the performance measures which include convergence, impact of varying RBs, impact of femtocell density, and the fairness index.

1.2 Related work

The related work is divided into three parts: different approaches for the self-organized resource management [6-10], the game-theoretic approaches for resource management [11-13], and the energy-efficient resource management for two-tier femtocell networks [14,15].

The authors in [6] propose a joint resource and power allocation in self-organized femtocell networks by exploiting a potential game. However, the whole analysis is carried out in terms of the throughput of the macrocell and femtocell networks. The authors in [7] propose a utility-based signal to interference and noise ratio (SINR) that reduces the cross-tier interference in a femtocell network. However, they do not cater the co-tier interference component, which is also termed as the bottle neck in performance enhancement for the shared channel environment. The authors in [8] propose a heuretic approach for resource allocation and power control for femtocell networks. However, the assumption of their study is that the information is thoroughly exchanged among the SBSs for performance improvement which is not practical in the real environment. The authors in [9] employ a novel docitive Q-learning for self-organized resource allocation in femtocell networks. However, it takes time to learn the learning mechanism for the optimal strategies, which makes it unsuitable for the real cases. The authors in [10] propose a joint subchannel and power allocation for the downlink of femtocell networks. Specifically, they have exploited the convex optimization and iterative approach for performance improvement of the network. However, they have only considered the throughput of the system and ignored the energy efficiency aspect that we are presenting in this study.

Game-theoretic approaches, which are appropriate for distributed resource management, have been extensively applied for the two-tier femtocell networks. The authors in [11] present a joint channel allocation and power control by using game learning mechanisms for cognitive radio networks. They utilize the regret learning method for convergence to the Nash equilibrium. Moreover, the comparison with genetic algorithm in centralized framework justifies their mechanism. The results illustrate that the inculcation of no-regret learning within the context of no-cooperation game theory has similar performance to that of the centralized one at the cost of increased complexity. The authors in [12] propose a distributed power allocation scheme using a game theoretic framework for the overlay scenario in cognitive radio networks. More precisely, the game theoretic proposition aims for maximizing the secondary users' throughput while keeping the interference inculcated on other secondary users and primary networks below the specified threshold. However, the uplink transmission case consideration imposes a huge burden to the FUEs in the network. The authors in [13] present a novel Stackelberg game for the resource allocation problem within the context of femtocell networks. In the game, the macro base station (MBS) which is the leader selects the resources for meeting the demand of its users followed by the allocation of followers with the concern to maximizing the throughput. They take the spectrum sharing into account while assuming the fixed power. However, the joint allocating task significantly improves the performance. All the investigation [11-13] carried out only considers the throughput enhancement of the network without taking into consideration the energy efficiency.

Most of the literature focuses on the throughput enhancement of the network; however, the limited battery capacity leads to the attention on the energy efficiency aspect of the two-tier femtocell networks. Although the energy efficiency aspect is largely ignored, a few efforts have been made in this domain, which includes [14] and [15]. The authors in [14] propose a non-cooperative game for subcarrier allocation and power control that aims at maximizing the energy efficiency. The authors in [15] investigate the non-cooperative power optimization game for enhancing energy efficiency. Both works [14] and [15] investigate the resource management for uplink case, which is indeed important for the limited power batteries. On the other hand, the growing demands of users and dense deployment of femtocells attract attention on the energy efficiency in the downlink of the femtocell networks. This will contribute to minimization of the energy consumption of the voice and data networks. Therefore, in this study, we are presenting the energy-efficient downlink resource management for the two-tier femtocell networks.

According to the best of the authors' knowledge, the proposed game-based joint RB and power allocation of the two-tier femtocell networks is unique and has not been investigated so far. The exploitation of both RB and power allocation not only maximize the energy efficiency but also maximize the throughput which is shown in the results. This is in accordance to the fact that interference components are also included in the energy efficiency maximization expression which accordingly enhances the throughput also. On the other hand, the decomposition of the joint allocation task into RB allocation and PSO-based power allocation makes this study a unique one. The concern of the decomposition of the joint allocation task is to minimize the complexity which makes it more appropriate for real systems. On the other hand, two different scenarios are taken into account for having an in-depth analysis of the proposed game.

The rest of this paper is structured as follows: the system model and proposed framework are presented in Section 2. Energy-efficient non-cooperative game is presented in Section 3, which includes detailed information of the proposed game. The simulation results regarding the energy efficiency and throughput in terms of the performance measures are presented in Section 4. Finally, Section 5 concludes the article.

2 System model

2.1 Proposed framework

The considered framework for the proposed non-cooperative game for RB and power allocation in two-tier femtocell networks is shown in Figure 1, where macrocells are underlaid with multiple femtocells. Under the violation of the macro user (MUE) performance, MBS alters the concerning SBS to change the action profile accordingly. This information is conveyed by the existed backhaul link that connects MBS and SBS. The benefits gained from the proposed game are: First, energy efficiency is achieved for the two-tier femtocell network without jeopardizing the macrocell performance. Second, a less complex game is developed by decomposing the joint allocation task into two subproblems. Third, a self-organizing downlink resource management is achieved with no involvement of centralized entity.

The interference scenarios that we considered in this study are explicitly shown in Figure 2. Specifically, we are considering the downlink case of the two-tier femtocell networks. The cross-tier interference component which is the interference between the macrocell and femtocell is represented by $i1$. On the other hand, the co-tier interference which is the interference among femtocells is

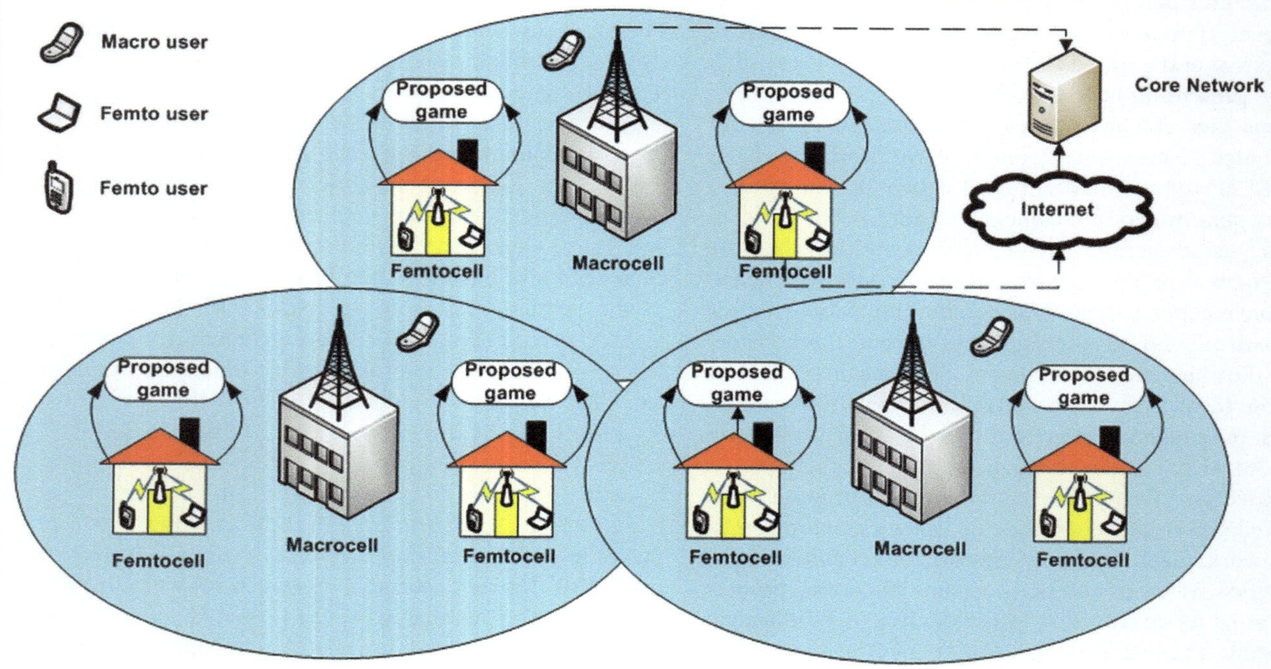

Figure 1 Proposed framework of the non-cooperative game for two-tier femtocell networks.

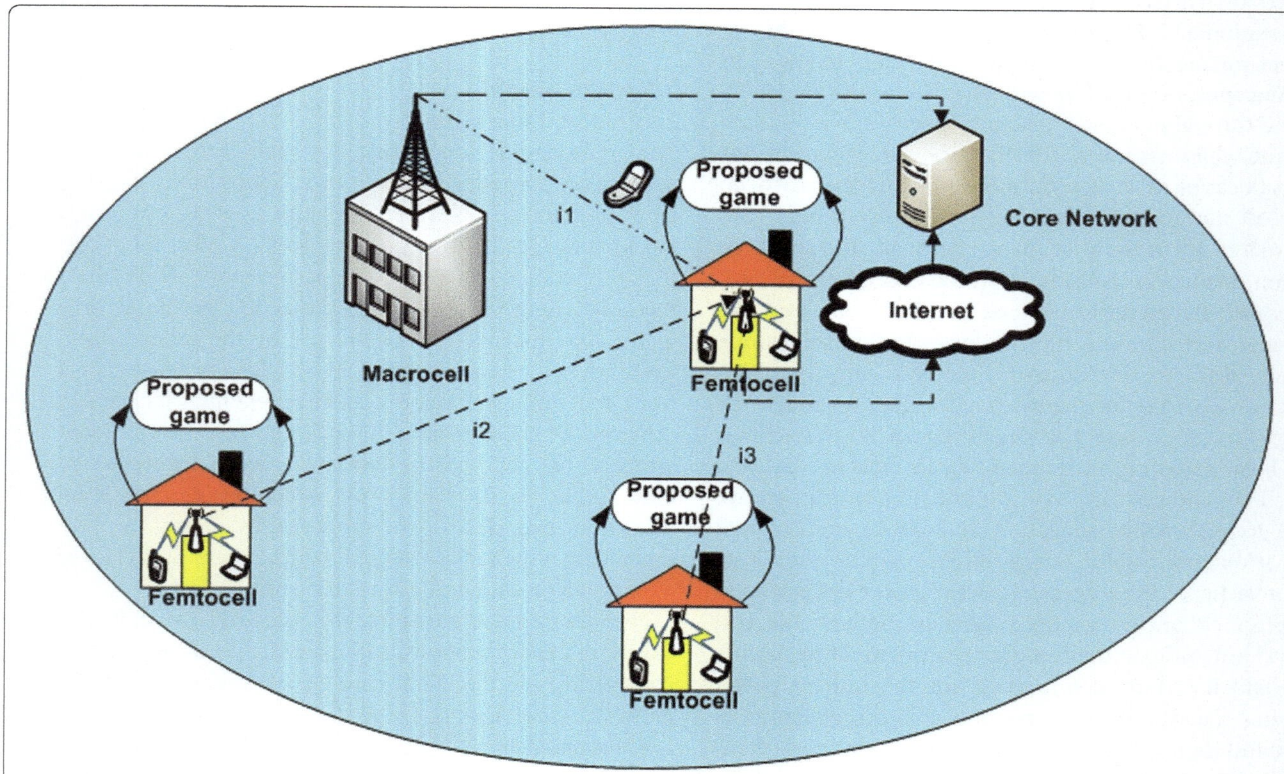

Figure 2 Cross-tier and co-tier interferences.

depicted by $i2$ and $i3$. Owing to the co-channel environment, the aforementioned interference creates significant impact on the performance of the network. The proposed non-cooperative game-based joint resource management mitigates both the interference components and enhances the energy efficiency accordingly.

2.2 Energy efficiency performance criterion

The repeatedly growing demands of users attract the attention from academia, industry, and standardization forums towards the energy-efficient design of wireless networks [16-19]. Generally, there are two methods for computing the energy efficiency of the system. The first one considers the amount of information in bits transmitted per Joule (bits/J) [16,17], while the other takes into account the transmission power [18,19].

In this study, we are using the first energy-efficient performance criterion which is the ratio of the transmission capacity to the transmission power [16,17]. Moreover, the transmission power also contains the circuit power. The energy efficiency (EE) in bits/J for the OFDMA-based system can be written as:

$$\mathrm{EE} = \frac{W \log_2\left(1 + \frac{h^2 p}{\sigma^2}\right)}{p_c + p} \tag{1}$$

where W is the bandwidth, p_c is the circuit power, p is the transmission power, h is the channel gain, and σ^2 is the noise power. The reason behind the exploitation of the energy efficiency performance metric is to transmit more information using less power.

2.3 Problem formulation

In this study, we consider a two-tier OFDMA-based femtocell network comprising of L macrocells, where each macrocell is serviced by an MBS at its center. In addition, each macrocell is overlaid with N femtocells, where each femtocell is serviced by an SBS. Specifically, we are considering the downlink case for the resource management in the two-tier femtocell network along with the universal frequency reuse-1 for each cell. The concern of employing the frequency reuse-1 is due to the fact that each cell has access of all the pool of RBs. The total Y MUEs are randomly deployed within the coverage area of a macrocell, and X FUEs are assumed to be in the indoor environment. The closed group formation (CSG) within the femtocell is assumed in which a certain number of FUEs can only be a part of the femtocell network. Concerning the co-channel deployment, the total pool of Q RBs is taken into account here. On the other hand, the number of RBs acquired by each SBS at any time is assumed to be G such that $G \le Q$. In addition, the total R power levels are considered which can be utilized by each SBS on the acquired G RBs under the constraint of maximum power transmission. The tight synchronization amount the OFDMA subcarriers is taken into account, which correspond to the fact that interference is only inculcated when there is transmission on the same RBs.

The transmission power of lth MBS and nth FBS on acquired G RBs is given by $p_l^{g,M} = \left\{p_l^{1,M}, \ldots, p_l^{G,M}\right\}$ and $p_n^{g,F} = \left\{p_n^{1,F}, \ldots, p_n^{G,F}\right\}$. The maximum power constraint on each base station is represented as P_{MAX}^{M} and P_{MAX}^{F} such that $\sum_{g=1}^{G} p_n^{g,M} \le \mathrm{P}_{\mathrm{MAX}}^{M}$ and $\sum_{g=1}^{G} p_n^{g,F} \le \mathrm{P}_{\mathrm{MAX}}^{F}$, respectively.

The performance of the proposed non-cooperative game is then analyzed in terms of the energy efficiency and throughput. In the computation of SINR ratio for femtocells, we evaluate the average SINR with respect to users on each RB. To this end, the SINR at the xth FUE of nth SBS operating on th gth RB is given by,

$$\mathrm{SINR}_n^{g,F} = \frac{1}{X}\sum_{x=1}^{X} \frac{p_n^{g,F} h_{nn,x}^{g,F-F}}{\sigma^2 + \underbrace{\sum_{i=1, i\neq n}^{N} p_i^{g,F} h_{in,x}^{g,F-F} \delta_{k_g k_i}}_{\text{Femtocomponents}} + \underbrace{\sum_{l=1}^{L} p_l^{g,M} h_{\mathrm{ln},x}^{g,M-F} \delta_{k_g k_l}}_{\text{Macrocomponents}}},$$

$$\forall n = 1, \ldots, N, g = 1, \ldots, G, x = 1, \ldots, X \tag{2}$$

where $h_{nn,x}^{g,F-F}$ is the channel gain between xth FUE and nth SBS operating on th gth RB; $h_{in,x}^{g,F-F}$ is the channel gain between xth FUE of nth SBS and ith SBS operating on th gth RB; $h_{\mathrm{ln},x}^{g,M-F}$ is the link gain between xth FUE of nth SBS and lth MBS operating on th gth RB; $\delta_{k_g k_l}$ is the interference function that corresponds to the inculcation of the interference for the same set of RBs, i.e., if $k_g = k_l$, then $\delta_{k_g k_l} = 1$, otherwise, $\delta_{k_g k_l} = 0$, and finally, σ^2 is the noise power. In addition, the femto components are the co-tier interference components, whereas the macro components are the cross-tier one.

The SINR at yth MUE of lth MBS that have allocated th gth RB is given by,

$$\mathrm{SINR}_l^{g,M} = \frac{1}{Y}\sum_{y=1}^{Y} \frac{p_l^{g,M} h_{ll,y}^{g,M-M}}{\sigma^2 + \underbrace{\sum_{i=1, i\neq l}^{L} p_i^{g,M} h_{il,y}^{g,M-M} \delta_{k_g k_i}}_{\text{Macrocomponents}} + \underbrace{\sum_{n=1}^{N} p_n^{g,F} h_{nl,y}^{g,F-M} \delta_{k_g k_l}}_{\text{Femtocomponents}}},$$

$$\forall l = 1, \ldots, L, g = 1, \ldots, G, y = 1, \ldots, Y \tag{3}$$

The femtocell capacity of the nth SBS that has acquired G RBs among the Q pool of RBs is given by,

$$\Delta_n^F = \frac{W}{G}\sum_{g=1}^{G} \log_2\left(1 + \mathrm{SINR}_n^{g,F}\right), \quad \forall n = 1, \ldots, N \tag{4}$$

The energy efficiency of the nth SBS is represented as:

$$\mathrm{EE}_n^F = \frac{\frac{W}{G}\sum_{g=1}^{G}\log_2\left(1+\mathrm{SINR}_n^{g,F}\right)}{\sum_{g=1}^{G}\left(p_n^{g,F}+p_c\right)}, \forall n=1,\ldots,N \quad (5)$$

where p_c is the circuit power and W is the total system bandwidth.

The total energy efficiency of the two-tier femtocell networks is given by,

$$\mathrm{EE}^F = \sum_{n=1}^{N}\mathrm{EE}_n^F = \sum_{n=1}^{N}\frac{\sum_{g=1}^{G}\frac{W}{G}\log_2\left(1+\mathrm{SINR}_n^{g,F}\right)}{\sum_{g=1}^{G}\left(p_n^{g,F}+p_c\right)} \quad (6)$$

Concerning the joint RB and power allocation, the optimization problem is formulated as:

$$\max \frac{\frac{W}{G}\sum_{g=1}^{G}\log_2\left(1+\mathrm{SINR}_n^{g,F}\right)}{\sum_{g=1}^{G}\left(p_n^{g,F}+p_c\right)}, \forall n=1,\ldots,N, G\in\{1,2,\ldots,Q\} \quad (7)$$

s.t.

$$\mathrm{C1}: \sum_{g=1}^{G}\frac{W}{G}\log_2\left(1+\mathrm{SINR}_n^{g,F}\right)\geq\Gamma_F, \forall n$$

$$\mathrm{C2}: \sum_{g=1}^{G}\frac{W}{G}\log_2\left(1+\mathrm{SINR}_l^{g,M}\right)\geq\Gamma_M, \forall l$$

$$\mathrm{C3}: \sum_{g=1}^{G}p_n^{g,F}\leq P_{\mathrm{MAX}}^F, \forall n$$

$$\mathrm{C4}: p_n^{g,F}\in\left[0,\frac{P_{\mathrm{MAX}}^F}{\mathrm{G}}\right], \forall n, g$$

$$\mathrm{C5}: \frac{1}{X}\sum_{x=1}^{X}\left(\sum_{g=1}^{G}\sum_{l=1}^{L}p_l^{g,M}h_{\mathrm{ln},x}^{g,M-F}\right)\leq\zeta_n^{M-F}, \forall n, x$$

$$\mathrm{C6}: \frac{1}{X}\sum_{x=1}^{X}\left(\sum_{g=1}^{G}\sum_{i=1}^{N-1}p_i^{g,F}h_{in,x}^{g,F-F}\right)\leq\zeta_n^{F-F}, \forall n, x$$

The optimization problem in (6) aims at maximizing the energy efficiency subject to the minimum throughput requirement of the femtocells C1 and the maximum power constraint C3. Additionally, per RB power constraint is also taken into account as in C4. The performance of the macrocell users is protected by the incorporation of the constraint C2 in the problem formulation. The accumulated cross-tier and co-tier interference components are represented by the constraints C5 and C6. The concern of inculcating the aforementioned interference limits is that both the energy efficiency and throughput are enhanced accordingly.

2.4 Notations and assumptions

The notations presented in Table 1 are used in the rest of the paper.

3 Energy-efficient non-cooperative game-based resource block and power allocation

In this section, the energy-efficient non-cooperative game is presented, including modeling of the game as RB and power allocation, existence to the pure and unique Nash equilibrium, proposed self-organizing framework, and algorithmic details. The game theoretic approaches are the most suitable for the resource management of the self-organized femtocell network. Specifically, non-cooperative game has been extensively applied for the resource allocation problems for the wireless networks [20].

3.1 Non-cooperative game

We model the joint RB and power allocation problem as a non-cooperative game. Generally, the game is represented by a tuple $\mathcal{G} = \{N, \{A_n\}, \{U_n(.)\}_{n\epsilon N}\}$, where N is the finite set of players, A_n is the set of action or strategies (action profile) that each player can opt, and U_n is the utility function that is associated with each player. In other words, the utility function corresponds to the level of satisfaction of each nth player. Moreover, the utility function depends not only on the action profile of nth player A_n but also on the other ones $A_{-n} = \{A_1, A_2,\ldots, A_{n-1}, A_{n+1},\ldots, A_N\}$. Precisely, the utility function can be thought of as a function that maps the action profile A_n into a real number $\mathbb{R}$ such that U_n: $A_n \rightarrow \mathbb{R}$. In a game, the general goal of each nth is to maximize its utility function while keeping the interest of other players into account.

In the proposed game, the players are the set of SBSs which interacts with the environment in a self-organizing manner for deciding the best action profile. The action profile associated with each nth player is composed of two parts: RB and power levels. Mathematically, the action profile can be written as $A_n = \left(S_n^F, P_n^F\right)$, where $S_n^F = \left\{S_n^1, S_n^2, \ldots, S_n^G\right\}$ and $P_n^F : \sum_{g=1}^{G}p_n^{g,F} \leq \mathrm{P}_{\mathrm{MAX}}^F$. In other words, S_n^F is the selection of G RBs by each nth SBS, and P_n^F is the power value on the selected RBs. Each player involved in a game makes the decision in a self-organizing manner, and their decision has strong impact on others. The concern of the study is the maximization of the energy efficiency of the two-tier femtocell network without compromising the macrocell performance. Therefore, the utility function U_n is the same as the energy efficiency and is written as:

Table 1 Notation and assumptions

Parameters	Meaning
MBS	Macro base station
SBS	Femto base station
MUE	Macro user
FUE	Femto user
RB	Resource block
L	Number of macrocells
N	Number of femtocells or number of players
Y	Number of macro users
X	Number of femto users
Q	Total pool RBs
G	Number of RBs acquired by each SBS
W	System bandwidth
R	Number of power levels utilized by SBS
$p_l^{g,M}$	Transmission power of lth MBS on th gth RB
$p_n^{g,F}$	Transmission power of nth SBS on th gth RB
P_{MAX}^{M}	Maximum transmission power of MBS
P_{MAX}^{F}	Maximum transmission power of SBS on G RBs
p_c	Circuit power
$h_{nn,x}^{g,F-F}$	Channel gain between nth SBS and xth FUE operating on th gth RB
$h_{ln,x}^{g,M-F}$	Channel gain between lth MBS and xth FUE of nth SBS operating on th gth RB
Δ_n^F	Achieved throughput by nth SBS
EE_n^F	Achieved energy efficiency by nth SBS
Γ_F	Minimum capacity of SBS
ζ_n^{M-F}	Accumulated cross-tier interference limit on nth SBS
ζ_n^{F-F}	Accumulated co-tier interference limit on nth SBS
A_n	Action profile associated with nth player
U_n	Utility function associated with nth player
S_n^F	Selection of G RBs by nth SBS
P_n^F	Power allocation of G RBs by nth SBS
A_n^*	Optimal action profile
S_n^{*F}	Optimal RB allocation
P_n^{*F}	Optimal power allocation
T	Number of particles in PSO-based power allocation
O_j	jth particle position of PSO-based power allocation

$$U_n(A_n, A_{-n}) = \frac{\frac{W}{G}\sum_{g=1}^{G}\log_2\left(1+\mathrm{SINR}_n^{g,F}\right)}{\sum_{g=1}^{G}\left(p_n^{g,F}+p_c\right)}, \forall n = 1, \ldots, N \tag{8}$$

Definition 1: Given the fixed S_{-n}^F and P_{-n}^F, the best response dynamics of the RB and power allocation are given by,

$$\left(S_n^F, P_n^F\right) = \arg\max_{S_n^F, P_n^F} U_n\left(S_n^F, P_n^F \mid S_{-n}^F, P_{-n}^F\right) \tag{9}$$

where S_{-n}^F and P_{-n}^F are the action profile of all the players except the nth one.

3.2 Existence and uniqueness of Nash equilibrium

A set of action profile is termed as Nash equilibrium if the players involve in the game cannot deviate profitably given the action profile of other players [21]. For the proposed game, the Nash equilibrium is defined as:

Definition 2: A set of resource blocks and power allocation action profile $A^* = \left\{A_1^*, A_2^*, \ldots, A_N^*\right\}$ is termed as the Nash equilibrium if and only the following condition is satisfied,

$$U_n\left(A_n^*, A_{-n}^*\right) \geq U_n\left(A_n, A_{-n}^*\right) \forall n = \{1, 2, \ldots N\} \tag{10}$$

where $A_{-n}^* = \left\{A_1^*, A_2^*, \ldots, A_{n-1}^*, A_{n+1}^*, \ldots, A_N^*\right\}$ is the composite action profile of all the players in the Nash equilibrium except the nth player. The composite action profile corresponds to the RB and power allocation.

Primarily, the Nash equilibrium of the proposed game exists if the following two conditions are satisfied [22]:

1) The action profile A_n is a non-empty, convex, and compact subset of some Euclidean space $\mathbb{R}$.
2) The utility function $U_n(A_n, A_{-n})$ is a continuous and quasi-concave.

Since the transmission power and RBs are finite non-empty sets, therefore, condition 1 is satisfied for each femtocell. As far as condition 2 is concerned, the utility function in (8) is differentiable and strictly concave with respect to the composite action profile (RB and power levels) [23]. Thus, condition 2 is also satisfied, and the existence of the Nash equilibrium is guaranteed in accordance to the aforementioned conditions. We have the following theorems.

Theorem 1: The Nash equilibrium point exists in the proposed game $\mathcal{G}$.
Theorem 2: The proposed game has a unique Nash equilibrium [15].

3.3 Resource block and power allocation

In this subsection, we describe the RB and power allocation of the proposed game. Generally, the optimal joint RB and power allocation is an NP-hard problem [22]. In this manner, to reduce the complexity of the proposed game, the joint allocation task is decomposed into two subproblems: an RB allocation and a PSO-based power allocation. Primarily, in each iteration of the proposed game, two steps need to be carried out: RB allocation and power allocation. Firstly, the RB allocation is done given the power allocation of all the players. Secondly, the power allocation on the selected RBs is executed out which have been selected in the previous step. In both steps, the goal is to maximize the energy efficiency under the network constraints C1 to C6. The detail of the individual step is elaborated as below.

3.3.1 Resource block allocation

In this subsection, we describe the RB allocation algorithm which is based on the algorithm in [23]. However, the difference here is that we analyze the energy efficiency rather than the throughput. On the other hand, the multichannel environment is considered in our proposed game. Specifically, each player (SBS) in our proposed game acquires G RBs at any time among the total Q RBs. Under the assumption of the availability of link gains of all the other players, the individual RB problem can be written as:

$$S_n^{*F} = \max_{S_n^F} U_n\left(S_n^F | P_{-n}^F\right) \tag{11}$$

It can be noticed from the above expression that the concern of the RB allocation step is to maximize the utility function, which indirectly maximizes the energy efficiency. The assumption here is that power levels of the other players are acquired on each player, and this information is repeatedly acquired via control channels from the existed backhaul link.

Substituting Equations (2) and (5) in (11), we get,

$$S_n^{*F} = \max_{S_n^F} \frac{\sum_{g=1}^{G} \frac{W}{G}\left(1 + \frac{p_n^{g,F} h_{nn}^{g,F-F}}{I_n^{g,F}}\right)}{\sum_{g=1}^{G}\left(p_n^{g,F} + p_c\right)} \tag{12}$$

where $I_n^{g,F} = \sigma^2 + \sum_{i=1,i\neq n}^{N} p_i^{g,F} h_{in}^{g,F-F} \delta_{k_g k_i} + \sum_{i=1}^{L} p_l^{g,F} h_{ln}^{g,M-F} \delta_{k_g k_i}$ is the interference comprising of cross-tier and co-tier components. Assuming that ρ is the SINR value and substituting the power values of the SBSs $p_n^{g,F} = \frac{\rho I_n^{g,F}}{h_{nn}^{g,F-F}}$ in (12), we get

$$S_n^{*F} = \max_{S_n^F} \frac{\sum_{g=1}^{G} \frac{W}{G}(1+\rho)}{\sum_{g=1}^{G}\left(\frac{\rho I_n^{g,F}}{h_{nn}^{g,F-F}} + p_c\right)} \tag{13}$$

It can be intuitively seen that the above expression be written as:

$$S_n^{*F} = \min_{S_n^F} \sum_{g=1}^{G} \frac{I_n^{g,F}}{h_{nn}^{g,F-F}} \tag{14}$$

The above expression reveals that the contiguous group of G RBs is allocated by each SBS whose accumulated $\sum_{g=1}^{G} \frac{I_g^{g,F}}{h_{nn}^{g,F-F}}$ value is minimum.

3.3.2 Power allocation

In the power allocation step, the PSO is exploited for obtaining the optimum power values on the selected RBs in the previous step. The reason for incorporating the PSO is that it is less complex and gives better results in less time. Owing to the various considered power levels, the PSO-based power allocation evaluates the best power values and minimizes the complexity of the game. On the other hand, the main concern of breaking the joint allocation task into two subproblems is to minimize the complexity of the proposed game.

3.3.3 Particle swarm optimization

PSO is a population-based biologically inspired algorithm, inspired by the bird flocking and fish schooling mechanisms. These types of algorithms are specifically useful where the sample space is very large, the parameters of interest are dynamic, and there is little information exchanged between the users (particles) [24].

Generally speaking, the PSO algorithm starts with populating the swarm of particles, where each particle represents a potential solution. The swarm is similar to the population, while a particle is equivalent to the individual. Each particle is associated with a position and velocity in search space. In each iteration of the algorithm, the fitness is computed using (8) and both the velocity and position of each particle are updated according to (15) and (16), respectively. For the PSO-based power allocation, the fitness function is the same as the utility function that we have designed for the proposed game. The velocity of each particle is updated according to the finest two known positions, the personal best position (pbest) and the neighborhood best position (nbest), where pbest is the best position that the individual particle has visited and nbest is the best position that the particle and its neighborhood have visited. When the whole swarm is considered as the neighbor, then the nbest is termed as global best gbest and for small neighborhood nbest is equivalent to local best lbest.

$$V_j^{\text{new}} = V_j + a_1 r_1\left(\text{pbest}_j - O_j\right) + a_2 r_2\left(\text{nbest}_j - O_j\right) \quad j = 1, 2, \ldots, T \tag{15}$$

$$O_j^{\text{new}} = O_j + V_j^{\text{new}} \tag{16}$$

where:

- a_1 and a_2 are termed as acceleration coefficients whose job is to control the influence in the search process.
- T is the number of particles in a swarm.

- r_1 and r_2 are two random numbers uniformly distributed in the interval from (0, 1).
- O_j, O_j^{new}, V_j, and V_j^{new} represent the current and updated position and velocity of the jth particle.

3.3.4 Particle swarm optimization-based power allocation

In a PSO-based power allocation step, the power levels on the selected RBs are evaluated with the concern to maximizing the energy efficiency. Under the assumptions of the availability of the power values of the other $N - 1$ players, the power allocation problem can be written as:

$$P_n^{*F} = \max_{P_n^F} U_n\left(P_n^F | P_{-n}^F\right) \tag{17}$$

The abovementioned power allocation is achieved by exploiting the PSO-based optimization. Principally, there are three main steps in the PSO-based optimization: particle encoding, fitness computation, and velocity and position upgradation. In our considered PSO-based power allocation, each particle is represented by a G-dimensional vector, where each element represents the power level utilized by the SBS. In addition, R power levels can be exploited here with the concern to maximizing the utility function. As far as the fitness of the particle is considered, we use the utility function in (8) as the fitness function for the PSO-based power allocation. More precisely, the PSO iteratively maximizes the utility function and yields best results in terms of power allocation on the selected group of G RBs by each SBS. The velocity and position of particles is upgraded by (15) and (16).

3.4 Proposed self-organizing framework and algorithm

The proposed self-organizing framework is shown in Figure 3. Specifically, the framework comprises of three main phases: *sensing*, *learning*, and *tuning*. These three phases actually resemble the cognitive radio cycle [25]. This self-organizing procedure is executed by each SBS involved in a game.

Initially, the random assignment is done in terms of RBs and power levels by each SBS. In the *sensing* phase, each SBS interacts with the environment and acquires link gains. Here, we assume that the availability of link gains is strictly known on the SBSs, and this leads to the optimal performance of the proposed game in terms of the Nash equilibrium. In the *learning* phase, the proposed game is executed by each SBS based on the availability of link gains which are acquired in *sensing* phase. Concerning the high complexity of the joint RB and power allocation task, the joint problem is decomposed into two main tasks: an RB allocation and a PSO-based power allocation. In the RB allocation step, each SBS acquires the required group of G RBs according to (14) under the available of power levels of other players. In the second step, the power allocation is carried out on the selected RBs by exploiting PSO. Because of the various choices of power levels, the PSO-based optimization rules out the power levels in less time with the concern to maximizing the energy efficiency. Lastly, in the *tuning* phase, the decision is to be made by each SBS, i.e., whether to change the action profile or not. In addition, the changed action profile is also broadcasted to the neighbors so that they can also tune themselves accordingly.

Algorithm: Proposed self-organized non-cooperative game based RB and power allocation

Input:

- All link gains that connect macrocells and femtocells on the pool of Q RBs.
- Initialization of the parameters related to PSO: *pbest*, *nbest*, O and V : (T , G)-dimensional.
- Randomly initialize the RBs and power allocation by each SBS.
- Set the proposed game maximum iteration count: ITER

Output:

- Resource block and power allocation, $A^* = \{S_n^F, P_n^F\}$

```
while (t < ITER)
    for each SBS (player) n do
        Select G RBs according to S_n^{*F} = min_{S_n^F} Σ_{g=1}^{G} I_n^{g,F} / h_{nn}^{g,F-F}
        do
            for each particle j do
                if Utility function(O_j) > Utility function(pbest_j)
                    Pbest_j ← O_j
                end if
                        if Utility function (pbest_j) > Utility function (nbest_j)
                            nbest_j ← pbest_j
                    end if
            end for
            for each particle j do
                        Compute new velocity V_j^{new} using (15)
                        Compute new position O_j^{new} using (16)
                end for
            end dowhile (required fitness achieved or interaction count passed)
        P_n^{*F} ← best(O_j^{new})
        Update S_n^F and P_n^F based upon the newly completed ones: S_n^{*F} and P_n^{*F}
        end for
    end while
    return A^* = {S_n^{*F}, P_n^{*F}}
```

As far as the complexity of the proposed non-cooperative game is concerned, the joint allocation task that we have considered in the game is decomposed into two subproblems: an RB allocation and a PSO-based power allocation. In each iteration of the proposed game, RB allocation and power allocation are carried out separately, and this significantly reduces the complexity of the original joint allocation task. Firstly, RB allocation is carried out, and then the PSO-based power allocation executes on the selected RBs which decides the optimal power levels on them. The concern of exploiting PSO for power allocation is that it gives the optimal power values without parsing the whole set of power values. Under the assumption of the availability of channel gains on the SBS, the proposed non-cooperative game converges to pure and unique Nash equilibrium. This information about the channel gains is repeatedly acquired on the control channels and helps in achieving the best performance. The proposed

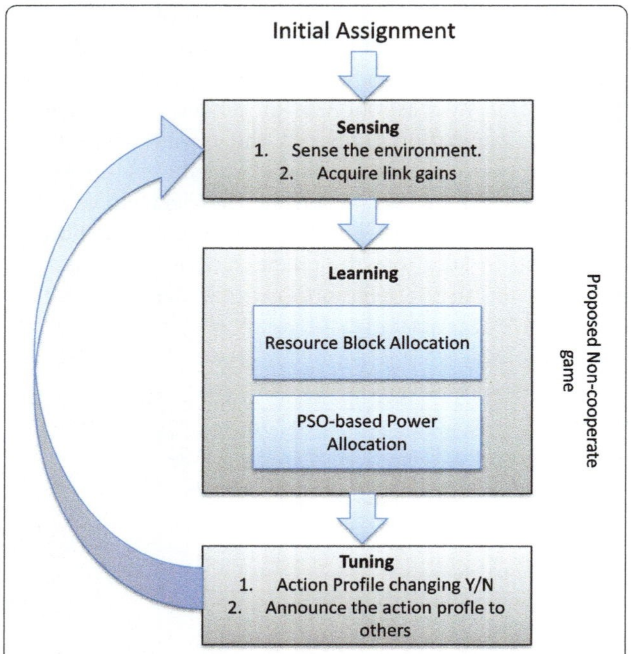

Figure 3 Proposed self-organizing framework.

game is also scalable in the sense that any SBS can be in the part of the game any time. However, this may lead to a slightly slower convergence to the Nash equilibrium, where the convergence corresponds to the energy efficiency maximization of the femtocell networks without compromising the macrocell performance. As far as the comparison with the joint scheme is concerned, the joint optimization task is computationally intense.

4 Simulation results and analysis

In this section, the simulation setup of the considered femtocell environment is presented and then the results are presented thereafter. The analysis of the proposed game is carried out regarding energy efficiency and throughput of the femtocell networks. In addition, the performance is evaluated in terms of the performance measures such as convergence, impact of varying RBs, impact of femtocell density, and the fairness index. For the sake of comparison, the following methods have been taken into account: joint resource and power allocation game with throughput maximization (JRPAG-TM) [6], proportional fair-non-cooperative power optimization game (PF-NPOG), and proportional fair-fixed power (PF-FP). In JRPAG-TM, a potential game is utilized for enhancing the throughput without considering the energy efficiency. The concern of comparing the proposed game with JRPAG-TM is to classify the gains and losses in energy efficiency and throughput. Furthermore, the two power levels for PF-FP are incorporated: 10 and 17 dBm. In order to have an in-depth analysis, two scenarios are taken into consideration: scenario 1 and scenario 2. In scenario 1, the number of SBS is taken to be $N = 20$, while in scenario 2 $N = 40$. All the simulations of the proposed game are done in MATLAB.

4.1 Simulation setup

We consider an OFDMA-based urban setting environment, operating at 1,850 MHz. For the sake of simplicity, we consider a singular macrocell environment $L = 1$ with a radius of 1,000 m. However, for multicell environment, the cross-tier interference components also become a critical factor along with the co-tier for single cell. The performance of the system degrades somehow because of the addition of this interference component. The achieved convergence for multicell environment is also reduced due to cross-tier interference component. In addition, there are N femtocells underlaid in a macrocell, where each femtocell is of radius 40 m. The Y MUEs and X FUEs are randomly deployed within the coverage area of macrocell and femtocell. An important clarification is to be made here is that we are considering the average SINR computation among the X FUEs and is in accordance to (2).

In our proposed game, each SBS operates in a self-organizing manner and rules out the action profile (RB and power levels) with the concern to maximizing the energy efficiency without compromising the macrocell performance. According to the 3rd Generation Partnership Project (3GPP) Long-Term Evolution (LTE) specifications, the total system bandwidth $W = 1,850$ MHz is taken into consideration, which is composed of several RBs. These RBs are utilized by both macrocells and femtocells because of the co-channel deployment. Quantitatively, the total pool of RBs $Q = 50$ is taken into account. Furthermore, the number of RBs that SBSs can acquire any time is assumed to be $G = 5$. In order to have a thorough analysis, various values of G are employed such as $G = \{2, 4, 6, 8, \text{and } 10\}$. As far as the power levels are concerned, we are incorporating $R = 100$ power levels uniformly distributed in the range from −80 to 23 dBm. The maximum power constraint that each SBS can utilize is $P^F_{\text{MAX}} = 23$ dBm. Furthermore, CSG mode of operation of femtocell is concerned; in which only particular users can be part of the femtocell network. The static circuit power of $p_c = 100$ mW is taken into account here for the computation of energy efficiency for the femtocell network.

The pathloss (PL) models that we have utilized in this study for the two-tier femtocell networks are in accordance to the 3GPP [26]. The PL models are:

a) From MBS to MUE or FUE (indoor): PL(dB) = 15.3 + 37.6 $\log_{10}d$.
b) From MBS to MUE or FUE (outdoor): PL(dB) = 15.3 + 37.6 $\log_{10}d$.
c) From SBS to MUE or FUE (different): PL(dB) = 7 + 56 $\log_{10}d + \gamma$
d) From SBS to FUE (own): PL(dB) = 37 + 20 $\log_{10}d + \gamma$

where d is the distance between MBS/SBS and MUE/FUE, γ is the penetration loss and its value is taken to be 15 dB. The shadowing factor of 8 and 4 dB is also contemplated here for indoor and outdoor, respectively.

The proposed game is executed for 100 iterations, and within each iteration, each SBS interacts in a self-organizing manner for evaluating the resources until the Nash equilibrium is achieved. Additionally, the results are captured for 1,000 different independent instances of the game. In the PSO-based power allocation, the PSO is executed for 200 iterations for obtaining the optimum results in terms of maximize energy efficiency (fitness function). The parameters for the PSO utilized for the power allocation are given in Table 2.

4.2 Simulations results

4.2.1 Convergence in terms of energy efficiency and throughput

The convergence characteristics of the proposed game in terms of energy efficiency and throughput for both the scenarios are depicted in Figures 4 and 5. It can be seen that energy efficiency of the proposed game for both the scenarios increases with the increase of game iterations as shown in Figure 4a,b. This is in accordance to the fact that each SBS competes for resources (RB and power levels) in a competitive manner until the optimal performance, i.e., the Nash equilibrium point is achieved. However, the proposed game in scenario 2 takes longer time to converge to the Nash equilibrium point. The reason is that the large number of SBSs (players) in scenario 2 leads to slower convergence to the Nash equilibrium because of increased competition. On the other hand, the addition of players in scenario 2 corresponds to the reduced energy efficiency. The comparison of the proposed non-cooperative game with JRPAG-TM, PF-NPOG, PF-FP (10 dBm), and PF-FP (17 dBm) illustrates the superior performance in both the scenarios. On the other hand, the comparison with PF-NPOG illustrates the significance of employing the joint allocation task within the game. In PF-NPOG, power is optimized in accordance with the utility function, while RBs are allocated randomly. Another important point to be noticed is that the performance gap between the proposed game and PF-NPOG becomes wider by increasing from N = 20 to 40 (scenario 1 to scenario 2). This is due to the random allocation of RBs, which degrades the performance with the increase of SBSs. The comparison with PF-FP with two different fixed power levels (10 and 17 dBm) also illustrates the superior performance of the proposed game. The concern of employing the energy efficiency within this framework is elaborated here, i.e., each scenario, to which a large power corresponds, reduces energy efficiency, and this is in accordance with the designed utility function for the proposed game.

Table 2 Particle swarm optimization parameters

Parameters	Values
Population size, T	16
Acceleration coefficients, a_1 and a_2	2.05
$[V_{min}, V_{max}]$	$[-G, G]$

Figure 5 illustrates the convergence in terms of throughput versus the game iterations for both the scenarios. The convergence in terms of throughput of the proposed game also increases with the increase of game iterations. Although the utility function of the game is designed by taking into account the energy efficiency, the throughput will also be adequately improved by taking into account various RBs and power levels. Similar reason for slow convergence holds for throughput in Figure 5 as in energy efficiency in Figure 4. The comparison with the PF-NPOG, PF-FP (10 dBm), and PF-FP (17 dBm) illustrates the superior performance of the proposed game in terms of throughput. However, the comparison with JRPAG-TM illustrates a slight loss in the throughput for each scenario. The reason being that in JRPAG-TM, the objective is solely to enhance the throughput without considering energy efficiency. In PF-NPOG, for each scenario, only power allocation is considered in the utility function of the game whereas the RBs are randomly allocated, and this corresponds to a significant inferior performance as compared to the proposed game. A similar performance trend with PF-FP exists as in energy efficiency. It can be seen that a larger fixed power of 17 dBm corresponds to reduced throughput, and this is due to the fact that utility function is designed in a manner that is aligned with the energy efficiency. Therefore, a better performance is achieved for reduced power in both the scenarios.

4.2.2 Impact of varying RBs on energy efficiency and throughput

The impact of varying RBs on energy efficiency and throughput of our proposed game is illustrated in Figures 6 and 7. The energy efficiency is plotted versus the number of RBs acquired by each SBS G in Figure 6. In addition, two different pools of resources are taken into consideration: Q = 25 and 50. It is illustrated in Figure 6a,b that energy efficiency of the proposed game for both the scenarios increases with the increase of G. The reason is that

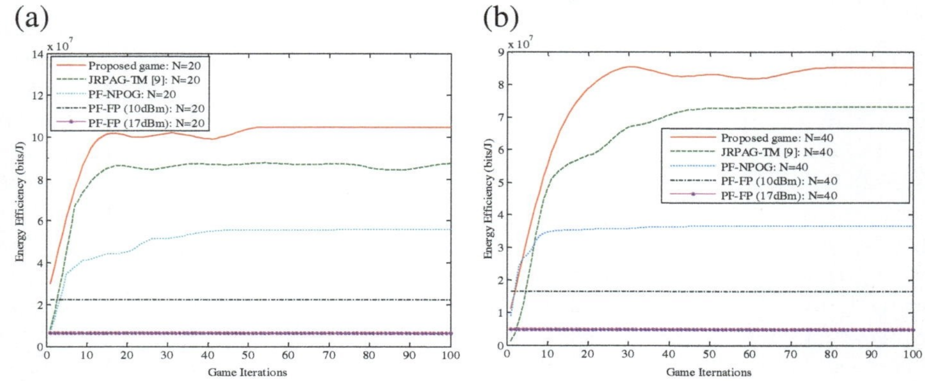

Figure 4 Convergence in terms of energy efficiency: **(a)** scenario 1 and **(b)** scenario 2.

the increased number of RBs acquired per SBS results in the accumulation of energy efficiency values, and this is in accordance with (4). On the other hand, increase in the pool of RBs Q 25 to 50 also contributes to the escalation of energy efficiency for both the scenarios. However, the reduced energy efficiency value for scenario 2 is observed, and this is due to the increased number of SBSs. As far as the comparison of the proposed game with JRPAG-TM, PF-NPOG, PF-FP (10 dBm), and PF-FP (17 dBm) is concerned, the proposed game outperforms the considered methods. A performance gap of around 15% is observed while comparison with JRPAG-TM for each scenario. The comparison with the PF-NPOG illustrates that the performance gap is around 40% for scenario 1 while 50% for scenario 2. A high performance gap exists because in the proposed game, both the RB and power allocation is taken into considerations while in PF-NPOG RB allocation is done randomly. On the other hand, 10% increased performance improvement is observed for scenario 2, and this is due to the increased competition among players. The comparison with PF-FP for both the scenarios also illustrates a significant performance improvement. Precisely, PF-FP (10 dBm) exhibits higher energy efficiency value as compared to PF-FP (17 dBm). The reason is that the utility function is aligned with the energy efficiency, which contributes to the better performance with reduced power.

Figure 7 illustrates the impact of varying the RBs on throughput of the femtocell network for both the scenarios. Firstly, the increasing trend of the proposed game in terms of throughput is observed by the increase of acquired RBs per SBS, i.e., G. Second, 20% increase in the throughput is observed by doubling the total pool of RBs from Q = 25 to 50. This is in accordance to the fact that accumulated throughput is achieved by the increase of G. On the other hand, the increase in the pool of RBs Q also contributes to the escalation of throughput for each scenario. Therefore, the throughput is enhanced with the increase of either G or Q for each scenario. Nevertheless, a reduced throughput for scenario 2 is also observed as compared to scenario 1, and this is due to the increase competition. The comparison of the proposed game with PF-NPOG, PF-FP (10 dBm), and PF-FP (17 dBm) in terms of throughput illustrates the superior performance of the proposed game.

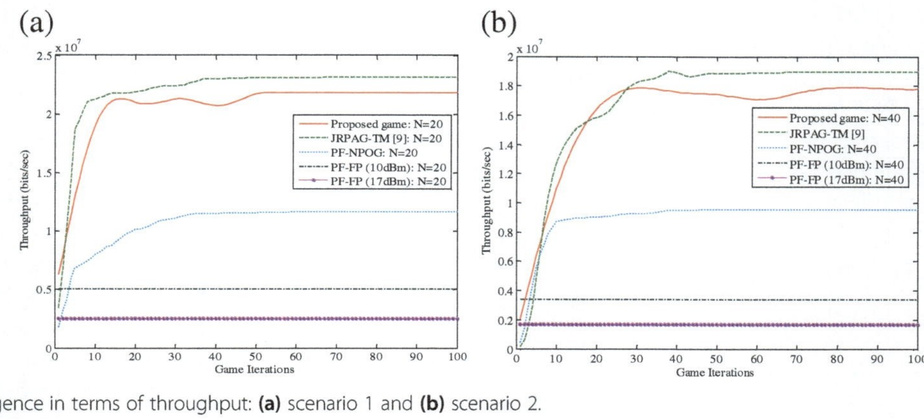

Figure 5 Convergence in terms of throughput: **(a)** scenario 1 and **(b)** scenario 2.

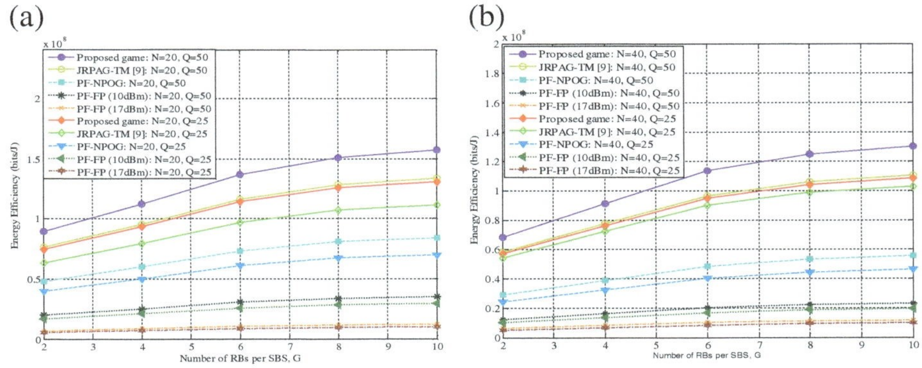

Figure 6 Impact of varying RBs on energy efficiency: **(a)** scenario 1 and **(b)** scenario 2.

Since the concern of the proposed game is to enhance the energy efficiency, a slight loss is observed while comparing with JRPAG-TM. However, the gain achieved in energy efficiency illustrates the supremacy of the proposed game as compared to JRPAG-TM. The throughput performance gap between proposed algorithm and PF-NPOG becomes wider with the increase acquired RBs per SBS. The reason is that in the proposed game, both the RB and power allocations are taken into account in the utility function, whereas in PF-NPOG, random allocation is employed which leads to reduced throughput. As far as the performance of the proposed game and PF-FP is concerned, a significant performance gap of around 490% in scenario 1 and 590% in scenario 2 is observed. The reason is that in PF-FP, both the RB and power levels are exploited without concerning the other players in the networks. Another important observation to be made here is that the reduced power levels in PF-FP (10 dBm) help in achieving high throughput as compared to PF-FP (17 dBm). This is due to the incorporation of the energy-efficient resource management which leads to better performance with reduced power levels.

4.2.3 Impact of varying femtocell cell density

The impact of femtocell density on energy efficiency and throughput is illustrated in Figure 8a,b. A significant performance trend gain about 15% is pertained while comparison of proposed game with JRPAG-TM. The energy efficiency of the proposed game decreases to about 90% with the increasing of femtocells from 10 to 80, whereas a sharp decaying effect is observed for PF-NPOG, i.e., about 250%. This is due to the fact that in the proposed game, both RB and power allocation are catered, whereas in PF-NPOG, only power allocation is exploited in utility function, and RBs are allocated randomly. On the other hand, a significant performance gap in terms of energy efficiency between the proposed game and PF-FP is also elaborated in Figure 8a. There is about 340% improvement with PF-FP (10 dBm) and 690% with PF-FP (17 dBm). In addition, this gap widens with the increase of femtocells.

The comparison in terms of throughput of the proposed game with traditional methods is illustrated in Figure 8b. The loss in the throughput of the proposed game as compared to JRPAG-TM is small of about 5%. The reason being that in JRPAG-TM, the concern is to

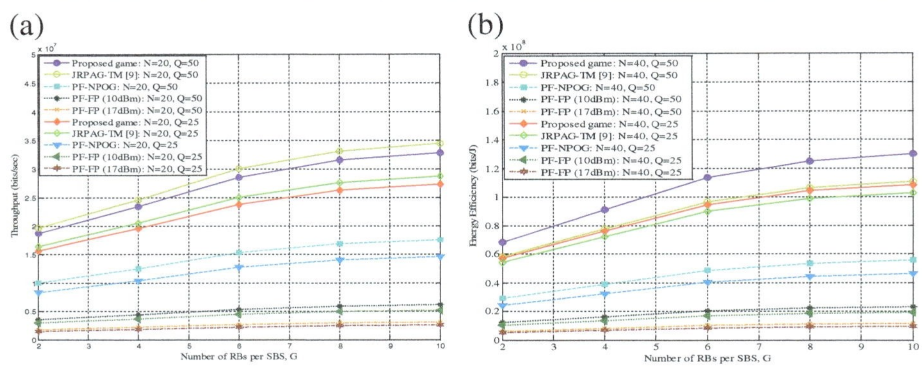

Figure 7 Impact of varying RBs on throughput: **(a)** scenario 1 and **(b)** scenario 2.

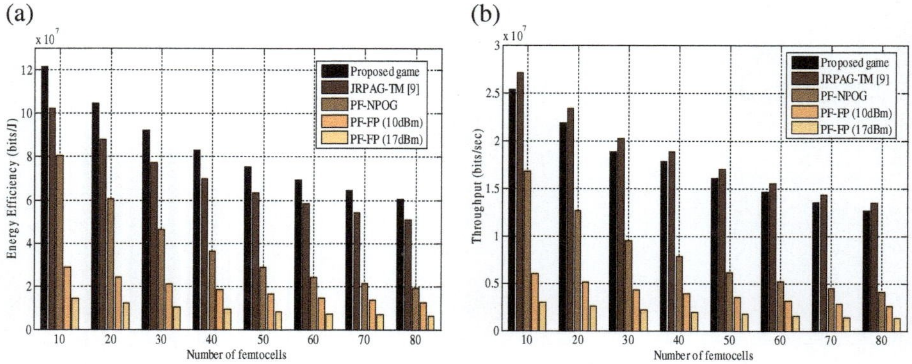

Figure 8 Impact of femtocell density on **(a)** energy efficiency and **(b)** throughput.

maximize the throughput without considering energy efficiency. The throughput of the proposed game decreases gradually to about 90% while a sharp decaying effect is observed for PF-NPOG. However, an important investigation to be made here is that an PF-FP (10 dBm) has high throughput as compared to PF-FP (17 dBM). This is by virtue of the incorporation of energy efficiency in the utility that contributed to high throughput with reduced power values.

4.2.4 Fairness index

Figure 9 illustrates the fairness index comparison of the proposed game with traditional methods. Fairness is computed using Jain's formula [27] as given below,

$$f(n_1, n_2, \ldots, n_N) = \frac{\left(\sum_{i=1}^{N} n_i\right)^2}{N \sum_{i=1}^{N} n_i^2} \tag{18}$$

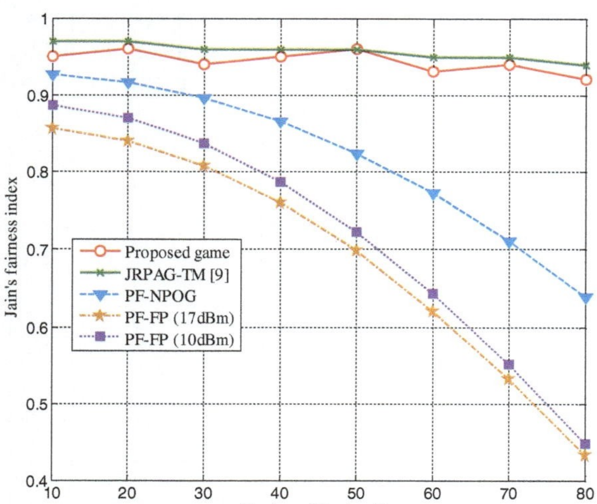

Figure 9 Jain's fairness index.

where $0 \leq f(n_1, n_2, \ldots, n_N) \leq 1$, and this value corresponds to the satisfaction in terms of achieved average throughput. It can be seen that the fairness of the proposed game remains at a high level with the increase of femtocells. The fluctuations in the fairness curve of the propose game are due to the fact that the solution of the proposed game is suboptimal. And due to this, the satisfaction in terms of throughput of SBSs varies somehow which is shown in Figure 9. In other words, all the players involved in the game interact in a manner to obtain a stable Nash equilibrium point. The Nash equilibrium point is classified as a point at which none of the players can deviate profitably, and the requirements of all the players are satisfied in directly. However, the comparison with traditional methods reveals that superior performance of the proposed game exhibits in terms of the fairness index. As far as the comparison with the JRPAG-TM, there is a nominal increase in the fairness index values. However, the high energy efficiency gain illustrates the supremacy of the proposed non-cooperative game. On the other hand, the fairness value of the PF-NPOG drops to about 62% with the increase of femtocells from 10 to 80, and this is by virtue of the fact that RBs are allocated randomly in PF-NPOG. As far as the PF-FP with two power levels are concerned, their fitness values deteriorate progressively to about 43% because of the random and fixed power allocations. On the other hand, it can be seen that RA-FB (10 dBm) performs better as compared to RA-FB (17 dBm). The reason is that the exploitation of energy efficiency criterion leads to the increased energy efficiency and throughput for reduced power levels.

5 Conclusions

In this study, an energy-efficient downlink resource management for two-tier femtocell networks is investigated. Specifically, the joint RB and power allocation are modeled here as a non-cooperative game in which the SBSs are the players and RB and power allocations are the action profiles. Most of the literature focuses only on the throughput enhancement of the two-tier system, while the energy efficiency aspect is largely ignored. To this end, the utility function of the proposed game is designed in a manner that is aligned with energy efficiency of the femtocell networks. The optimization problem is modeled as the maximization of the utility function (energy efficiency) subject to the network constraints C1 to C6. These network constraints not only satisfy the minimum throughput but also include the thresholds for interference components. Under the assumption of the availability of the link gains, the proposed game converges to pure and unique Nash equilibrium. Concerning the complexity of the joint allocation task, the problem is decomposed into two subproblems: an RB allocation and a PSO-based power allocation. The comparison of the proposed game with traditional methods illustrates the superior performance. In addition, the comparison with joint allocation task potential game, considering the throughput maximization, illustrates that significant performance in terms of energy efficiency with slight loss in throughput is achieved. The analysis is carried out in regard to energy efficiency and throughput.

Competing interests
The authors declare that they have no competing interests.

Acknowledgements
This research was supported by the Ministry of Science, ICT and Future Planning (MSIP), Korea, under the Convergence Information Technology Research Center (Convergence-ITRC) support program (NIPA-2014-H0401-14-1006) supervised by the National IT Industry Promotion Agency (NIPA) and the Basic Science Research Program through the National Research Foundation (NRF) of Korea funded by the Ministry of Education, Science and Technology (2012009449).

References
1. V Chandrasekhar, JG Andrews, A Gatherer, Femtocell networks: a survey. IEEE Commun Mag **46**, 59–67 (2008)
2. FP7-ICT-2009-4, FREEDOM, Femtocell-based netwoRk Enhancement by intErference management and coorDination of infOrmation for seaMless connectivity, see official website: http://www.ict-freedom.eu/
3. N Saquib, E Hossain, BL Long, IK Dong, Interference management in OFDMA femtocell networks: issues and approaches. IEEE Wirel Commun **19**, 86–95 (2012)
4. D Kinsely, T Yoshizawa, F Favichia, Standardization of femtocells in 3GPP. IEEE Commu Mag **47**, 68–75 (2009)
5. Y Li, H Celebi, M Daneshmand, C Wang, W Zhao, Energy-efficient femtocells networks: challenges and opportunities. IEEE Wire Commun **20**, 99–105 (2013)
6. A Shahid, S Aslam, HS Kim, K-G Lee, Distributed joint resource and power allocation in self-organized femtocell networks: a potential game approach. J Netw Comput Appl **46**, 280–292 (2014). 7
7. V Chanrasekhar, JG Andrews, T Muharemovic, Z Shen, A Gatherer, Power control in two-tier femtocells networks. IEEE Trans Wirel Commun **8**, 4316–4328 (2009)
8. A Shahid, S Aslam, KG Lee, Decentralized heuristic approach towards resource allocation in femtocells. Entropy **15**, 2524–2547 (2013)
9. A Shahid, S Aslam, HS Kim, KG Lee, A docitive Q-learning for joint resource allocation and power control in self-organized femtocells networks. Trans. Emerging Tel. Tech. (2014) doi:10.1002/ett.2800
10. DT Ngo, S Khakurel, T Le-Ngoc, Joint sub channel assignment and power allocation for OFDMA femtocell networks. IEEE Trans Wire Commun **13**, 342–355 (2014)
11. JR Gállego, M Canales, J Ortín, Distributed resource allocation in cognitive radio networks with a game learning approach to improve aggregate capacity. Ad Hoc Netw **10**, 1076–1089 (2012)
12. T Nadkar, V Thumar, GPS Tej, SN Merchant, UB Desai, Distributed power allocation for secondary users in a cognitive radio scenario. IEEE Trans Wire Commun **11**, 1576–1586 (2012)
13. P Xu, X Fang, M Chen, Y Xu, A Stackelberg game-based spectrum allocation scheme in macro/femtocell hierarchical networks. Computer Commun **36**, 1552–1558 (2013)
14. S Buzzi, G Colavolpe, D Saturnino, A Zappone, Potential game for energy-efficient power control and subcarrier allocation in uplink multicell OFDMA systems. IEEE Sel Topics Sig Proc **6**, 1923–4553 (2011)
15. GW Miao, N Himayat, GY Li, S Talwar, Distributed interference-aware energy-efficient power optimization. IEEE Trans Wire Commun **10**, 1323–1333 (2011)
16. GW Miao, N Himayat, GY Li, D Bormann, in *IEEE International Conference on Communications, 2008. ICC'08.*Energy-efficient design in wireless OFDMA (Beijing, 19-23 May 2008), pp. 3307-3312
17. GW Miao, N Himayat, GY Li, Energy-efficient link adaptation in frequency-selective channels. IEEE Trans Commun **58**, 545–554 (2010)
18. EA Jorswieck, H Boche, S Naik, Energy-aware utility regions: multiple access pareto boundary. IEEE Trans Wire Commun **9**, 2216–2226 (2010)
19. Z Hasan, G Bansal, E Hossain, V Bhargava, Energy efficient power allocation in OFDMA-based cognitive radio systems: a risk-return model. IEEE Trans Wire Commun **8**, 6078–6088 (2009)
20. K Akkarajitsakul, E Hossain, D Niyato, D Kim, Game theoretic approaches for multiple access network in wireless networks. IEEE Commun Surv Tutorials **13**, 372–395 (2011)
21. D Fudenberg, J Tirole, *Game theory* (MIT Press, Cambridge, Massachussets, 1993)
22. J Zhang, T Lv, H Gao, Y Lu, in *IEEE Wireless Communications and Networking Conference (WCNC), 2012.* Joint uplink power and subchannel allocation in cognitive radio network (Paris, 1-4 April 2012), pp. 306–311
23. Z Chong-xian, L Chun-guo, Y Lu-xi, Dynamic resource allocation algorithm for multi-cell OFDMA systems based on noncooperative game theory. J Elect Infor Tech **8**, 1935–1940 (2009)
24. H Izakian, A Abraham, V Snášel, Metaheuristic based scheduling meta-tasks in distributed heterogeneous computing systems. Sensors **9**, 5339–5350 (2009)
25. J Mitola, *Cognitive radio: an integrated architecture for software defined redio.* PhD Thesis (Royal Institute of Technology (KTH), Sweden, 2000)
26. 26. 3GPP TR 36.814, V.9.0.0 "Technical Specification Group Radio Access Network; Evolved Universal Terrestrial Radio Access (E-UTRA): Further advancements for E-UTRA physical layer aspects (Release 9)", March 2010
27. R Jain, D Chiu, W Hawae, A quantitative measure of fairness and discrimination of resource allocation in shared computer system, in *DEC research report*, 1984, p. TR-301

13

Kalman filter with a linear state model for PDR+WLAN positioning and its application to assisting a particle filter

Matti Raitoharju*, Henri Nurminen and Robert Piché

Abstract

Indoor positioning based on wireless local area network (WLAN) signals is often enhanced using pedestrian dead reckoning (PDR) based on an inertial measurement unit. The state evolution model in PDR is usually nonlinear. We present a new linear state evolution model for PDR. In simulated-data and real-data tests of tightly coupled WLAN-PDR positioning, the positioning accuracy with this linear model is better than with the traditional models when the initial heading is not known, which is a common situation. The proposed method is computationally light and is also suitable for smoothing. Furthermore, we present modifications to WLAN positioning based on Gaussian coverage areas and show how a Kalman filter using the proposed model can be used for integrity monitoring and (re)initialization of a particle filter.

Keywords: Pedestrian dead reckoning; Indoor positioning; Wireless LAN; Computational modeling

1 Introduction

Wireless local area network (WLAN) access points (APs) are numerous and ubiquitous in most indoor environments. Although WLAN is meant for data transfer, the WLAN signals can also be used for user localization. Because the WLAN APs are not meant for positioning, they do not usually send information on their own location to clients. WLAN positioning therefore makes use of a 'radio map' that describes certain features of the WLAN signals at given locations. A radio map is created and updated using data collected on-site called fingerprints (FPs). A FP is a report that contains at least the receiver location and the IDs and received signal strength (RSS) values of APs within reception range. A radio map is constructed off-line from the collected FPs. The accuracy of positioning based on WLAN signals depends on the model and environment. In small scale (a few buildings), it is possible to achieve positioning results of the order of a couple of meters [1]. For large-scale positioning (a city or larger), when the size of the database is a limiting factor, the positioning accuracy can be of the order of tens of meters [2].

Pedestrian dead reckoning (PDR) uses an inertial measurement unit (IMU) to detect when a user takes footsteps and how the direction changes between footsteps. The IMU has three-axis accelerometers and gyroscopes. The user heading change is computed by projecting the gyroscope measurements to the horizontal plane, which is estimated from the accelerometer [3]. It is also possible to use a compass for heading determination [4].

The footstep length may also be estimated from the IMU data. If the sensor is mounted on the foot, it is possible to detect when the foot is motionless and then integrate the footstep length from the IMU data [5]. If the IMU is handheld, this method cannot be used, but the footstep length can be inferred from the IMU data also by other methods, see for example [6,7]. A PDR system can greatly improve the positioning locally as the position estimate may be updated at every footstep, but because the errors accumulate over time, PDR is often combined with other sensors that can, at least occasionally, provide absolute position information.

In this paper, we investigate models for fusing heading change information from IMU with WLAN measurements. We propose a linear state model for the state

*Correspondence: matti.raitoharju@tut.fi
Department of Automation Science and Engineering, Tampere University of Technology, P.O.Box 692, FI-33101, Tampere, Finland

evolution, whereas in the literature, the state model used with PDR system is usually nonlinear [7-10] or the heading information is used as nonlinear measurements [4]. In general, there is no closed form optimal algorithm for nonlinear estimation. In this paper, we use the Kalman filter (KF), the extended Kalman filter (EKF), and the unscented Kalman filter (UKF), which are computationally light algorithms, for evaluation of the models.

Particle filters (PFs) are algorithms that can be used for nonlinear estimation. We show how the proposed model with a linear KF can be used for initialization and integrity monitoring of a PF.

Section 2 contains the filtering and smoothing algorithms that are used to estimate the user's kinematic state. In Section 3, we present the WLAN model that is used for positioning. The evaluated PDR models are presented in Section 4. In Section 5, we discuss the possibilities of using the proposed model in a particle filter, and in Section 6, we evaluate the performance of different models with real and simulated data. Section 7 concludes the paper.

This paper is an extended version of a paper [11] presented at the Conference on Design and Architectures for Signal and Image Processing (DASIP 2013). In general, the notations in this paper are changed to be more consistent. We added UKF to Section 2. Section 3 has a new subsection containing modifications to the WLAN coverage area positioning. The first model presented in Section 4 has cross terms in the covariance matrix of the state transition matrix to make it mathematically sound. Section 5 is new. In Section 6, tests are new or done with more test tracks and the UKF is also used.

2 Filtering algorithms

The KF is an algorithm for estimating the state of the system given a time series of measurements in the case of linear state evolution and measurement models. If the measurement and state transition errors are independent, white, and Gaussian processes, the KF computes the posterior mean that is the optimal estimator in minimum mean square error sense (i.e., the KF is the exact Bayesian filter) [12].

The state model is as follows:

$$x_t = F_t x_{t-1} + \epsilon_t^Q, \tag{1}$$

where x_t is the state at the tth time instant, F_t is the state transition matrix, and ϵ_t^Q is a zero mean Gaussian noise term with covariance matrix Q_t. The KF prediction is computed from the previous posterior $\mathrm{N}(\bar{x}_{t-1|t-1}, P_{t-1|t-1})$ via the update formulas:

$$\bar{x}_{t|t-1} = F_t \bar{x}_{t-1|t-1} \tag{2}$$

$$P_{t|t-1} = F_t P_{t-1|t-1} F_t^T + Q_t, \tag{3}$$

that give the new prior $\mathrm{N}(\bar{x}_{t-1|t}, P_{t-1|t})$.

The prior is updated using the linear measurement model:

$$y_t = H_t x_t + \varepsilon_t^R, \tag{4}$$

where matrix H_t defines the relationship between the state and the measurement vector y_t, and ϵ_t^R is a zero mean Gaussian noise term with covariance matrix R_t. In the update part of the KF, the prior is updated via:

$$\begin{aligned} S_t &= H_t P_{t|t-1} H_t^T + R_t && (5)\\ K_t &= P_{t|t-1} H_t^T S_t^{-1} && (6)\\ \bar{x}_{t|t} &= \bar{x}_{t|t-1} + K_t \left(\bar{y}_t - H_t \bar{x}_{t|t-1}\right) && (7)\\ P_{t|t} &= (I - K_t H_t) P_{t|t-1}, && (8) \end{aligned}$$

where $\bar{y}_t$ is the realized measurement value, S_t is the innovation covariance matrix, K_t is the Kalman gain, $\bar{x}_{t|t}$ is the posterior mean, and $P_{t|t}$ the posterior covariance matrix. If there are no measurements, the latest prior is also the new posterior, and if there are several independent measurements in one time instance, the updates can be applied consecutively.

In more general situations, the state transition model is nonlinear, of the form:

$$x_t = f_t\left(x_{t-1}, \epsilon_t^Q\right), \tag{9}$$

where f_t is a nonlinear state transition function and ε_t^Q does not necessarily have zero mean. In the EKF [13], the mean propagation formula (Equation 2) has to be replaced with:

$$\bar{x}_{t|t-1} = f_t\left(\bar{x}_{t-1|t-1}, \bar{\varepsilon}_t^Q\right), \tag{10}$$

where $\bar{\varepsilon}_t^Q$ is the mean of the noise term. The covariance propagation formula (Equation 3) changes to:

$$P_{t|t-1} = \hat{F}_t P_{t-1|t-1} \hat{F}_t^T + \hat{L}_t Q_t \hat{L}_t^T, \tag{11}$$

where

$$\hat{F}_t = \left.\frac{\partial f_t\left(x, \varepsilon^Q\right)}{\partial x}\right|_{\bar{x}_{t-1|t-1}, \bar{\varepsilon}_t^Q} \tag{12}$$

and

$$\hat{L}_t = \left.\frac{\partial f_t\left(x, \varepsilon^Q\right)}{\partial \varepsilon^Q}\right|_{\bar{x}_{t-1|t-1}, \bar{\varepsilon}_t^Q}. \tag{13}$$

Similar approximations can also be done for the measurement update equations, but in this paper, we use only linear measurements.

The EKF requires analytical computation of the partial derivatives and approximates the measurement function in a point. One commonly used algorithm for computing the update that does not need analytical differentiation and approximates the state transition function in larger area is the UKF [14]. In the UKF, a set of sigma points is chosen so that they have the same mean and covariance as the original distribution. The sigma points are propagated through nonlinear functions, and the estimate is updated using the transformed sigma points. The transformation is called unscented transformation. The UKF can be used also for measurement update, but because we use only linear measurements, we present only the state propagation of the UKF.

When computing the state propagation, the UKF uses augmented state:

$$z = \begin{bmatrix} x_{t-1} \\ \varepsilon_t^Q \end{bmatrix} \sim \mathrm{N}(\bar{z}, P^z) = \mathrm{N}\left(\begin{bmatrix} \bar{x}_{t-1|t-1} \\ \bar{\varepsilon}_t^Q \end{bmatrix}, \begin{bmatrix} P_{t-1|t-1} & 0 \\ 0 & Q_t \end{bmatrix}\right). \tag{14}$$

A popular choice of sigma points is as follows:

$$\begin{aligned} \chi_0 &= \bar{z} \\ \chi_i &= \bar{z} + \sqrt{d + \xi_{\mathrm{UKF}}}\sqrt{P^z}_{[:,i]}, \qquad 1 \le i \le d \\ \chi_i &= \bar{z} - \sqrt{d + \xi_{\mathrm{UKF}}}\sqrt{P^z}_{[:,i-d]}, \quad d+1 \le i \le 2d, \end{aligned} \tag{15}$$

where d is the dimension of the augmented state, ξ_{UKF} is an algorithm parameter, and $\sqrt{P}$ is a matrix square root for which:

$$\sqrt{P}\sqrt{P}^T = P. \tag{16}$$

This matrix square root can be computed, for example, with Cholesky decomposition. Parameter ξ_{UKF} is defined as:

$$\xi_{\mathrm{UKF}} = \alpha_{\mathrm{UKF}}^2 (d + \kappa_{\mathrm{UKF}}) - d, \tag{17}$$

where parameters α_{UKF} and κ_{UKF} define how much the sigma points are spread.

The prior mean and covariance computed using sigma points are as follows:

$$\bar{x}_{t|t-1} = \sum_{i=0}^{2d} w_i^s f_t(\chi_i) \tag{18}$$

$$P_{t|t-1} = \sum_{i=0}^{2d} w_i^c \left[f_t(\chi_i) - \bar{x}_{t|t-1}\right]\left[f_t(\chi_i) - \bar{x}_{t|t-1}\right]^T, \tag{19}$$

where the sigma points are computed from the posterior of the previous time step using Equation 15 and the weights are as follows:

$$w_0^s = \frac{\xi_{\mathrm{UKF}}}{d + \xi_{\mathrm{UKF}}} \tag{20}$$

$$\Omega w_0^c = \frac{\xi_{\mathrm{UKF}}}{d + \xi_{\mathrm{UKF}}} + (1 - \alpha_{\mathrm{UKF}}^2 + \beta_{\mathrm{UKF}}) \tag{21}$$

$$w_i^s = w_i^c = \frac{1}{2d + 2\xi_{\mathrm{UKF}}} \quad , i > 0. \tag{22}$$

The parameter β_{UKF} is related to the distribution of the state. In the case of Gaussian distribution, $\beta_{\mathrm{UKF}} = 2$ is optimal [15].

The Rauch-Tung-Striebel smoother [16] can be used to enhance the state estimates when measurements of 'future' time instants are available, for example, when plotting the track over a given time interval. The recursive smoothing equations are as follows:

$$C_t = P_{t|t} F_t^T P_{t+1|t}^{-1} \tag{23}$$

$$\bar{x}_{t|m} = \bar{x}_{t|t} + C_t(\bar{x}_{t+1|m} - \bar{x}_{t+1|t}) \tag{24}$$

$$P_{t|m} = P_{t|t} + C_t(P_{t+1|m} - P_{t+1|t})C_t^T \tag{25}$$

for $t = m-1, m-2, \ldots, 1$, where m is the final time index.

3 WLAN positioning

3.1 Coverage area positioning

In its simplest form, probabilistic coverage area (CA) positioning is a method for radio map construction in which the reception area of each AP is modeled as a bivariate normal distribution. Because the radio map does not contain any raw RSS data, computational, memory, and communication complexity are much lower compared to those of conventional FP positioning methods [17]. The algorithm is explained in [18] and here we only briefly outline it.

The coverage area estimate is as follows:

$$\mu_{i,n_i} = \frac{\sum_{j=1}^{n_i} z_{i,j}}{n_i} \tag{26}$$

$$\Sigma_{i,n_i} = \frac{\sum_{j=1}^{n_i} z_{i,j} z_{i,j}^T + B - n_i \mu_{i,n_i} \mu_{i,n_i}^T}{n_i + 1}, \tag{27}$$

where μ_{i,n_i} and Σ_{i,n_i} are the mean and covariance of the CA estimate of ith AP based on n_i FPs, $z_{i,j}$ is the location of the jth FP where AP i was received and B is the prior covariance. When a new measurement is received, these parameters can be updated by:

$$\mu_{i,n_i+1} = \frac{n_i \mu_{i,n_i} + z_{i,n_i+1}}{n_i + 1} \tag{28}$$

$$\Sigma_{i,n_i+1} = \frac{(n_i+1)\left(\Sigma_{i,n_i} - \mu_{i,n_i+1}\mu_{n_i+1}^T\right) + z_{i,n_i+1} z_{i,n_i+1}^T + n_i \mu_{i,n_i} \mu_{i,n_i}^T}{n_i + 2}. \tag{29}$$

In positioning, the CA model from every received AP could be applied in separate KF updates. To reduce computational cost, it is equivalent to first compute the static position estimate from these CA models using:

$$\hat{\Sigma} = \left(\sum_{i \in \mathrm{FP}} \Sigma_i^{-1}\right)^{-1} \tag{30}$$

$$\hat{\mu} = \hat{\Sigma} \sum_{i \in \mathrm{FP}} \Sigma_i^{-1} \mu_i, \tag{31}$$

where $i \in$ FP is used to indicate AP:s received in the current FP and the second index n_i is not shown because in positioning, the latest CA estimate is always used.

When using CA models in KF update (Equations 5 to 8), the linear and Gaussian measurement model is as follows:

$$\hat{\mu} = H_t x_t + \varepsilon_t^R = \begin{bmatrix} I & 0 \end{bmatrix} x_t + \varepsilon_t^R, \tag{32}$$

where $\varepsilon_t^R \sim \mathrm{N}(0, \hat{\Sigma})$. Here, it is assumed that the position variables are the first components of the state.

In [19], it was shown that using separate CA models for strong RSS values improves the positioning accuracy. That is, in addition to CA models generated using all FPs, one generates CA models using only FPs with strong RSS values. In the positioning stage, the models are chosen according to signal strengths. In [17], the two-level CA positioning was compared with other WLAN positioning methods. Of the compared methods, there was no significantly faster or more accurate method than the two-level CA positioning.

3.2 Modifications to CA positioning

In this section, we present three different heuristics to mitigate some shortcomings of CA positioning. In order to preserve the assumption of independent measurements, the proposed methods do not use the prior information provided by the Kalman filter. The heuristics are applied to measurements in the order they are presented.

3.2.1 *Minimum size*

When there are many FPs, the effect of the prior B used in Equation 27 diminishes. As a result, if many FPs are collected in a small area, the covariance matrix becomes too small. In our experience, this is a problem especially with CAs that model weak RSS in multilevel CA models. In [11], the prior B for weak CAs was set to a very high value $1{,}000^2\mathrm{m}^2 I$ to compensate. Here, we propose to impose a lower limit on the variance of the Gaussian CA model.

The variances along the main axes can be computed using eigendecomposition:

$$V \Lambda V^T = \Sigma_i. \tag{33}$$

Diagonal matrix Λ consists of the variances corresponding to the main axes (columns of V) on its diagonal. We propose to impose a minimum variance $\sigma^2_{\min}$ for main axes by thresholding:

$$\tilde{\Sigma}_i = V \max\left(\Lambda, \sigma^2_{\min} I\right) V^T. \tag{34}$$

3.2.2 *Outlier removal*

Because WLAN APs are freely deployable, the APs may be moved. If the database is not updated continually, the CA models of moved APs provide false information. For this kind of situations, we use outlier removal that is based on squared Mahalanobis distance to detect outliers.

In the outlier removal process, the mean estimate (Equation 31) and each CA model are compared separately. If the mean is outside the 95% ellipse, we discard the CA model. This is done by checking if:

$$d_i = (\mu_i - \hat{\mu})^T \Sigma_i^{-1} (\mu_i - \hat{\mu}) > \chi_2^{-2}(0.95) = 5.9915, \tag{35}$$

where χ_2^{-2} is the inverse cumulative density function of the chi-squared distribution with 2 degrees of freedom.

We propose that the CAs are removed one by one, removing the CA with largest d_i until all are close enough to the mean. After removing a CA, the estimate (Equations 30 to 31) can be updated by:

$$\hat{\Sigma}_+^{-1} = \hat{\Sigma}^{-1} - \Sigma_j^{-1} \tag{36}$$

$$\hat{\mu}_+ = \hat{\Sigma}_+ \left(\hat{\Sigma}^{-1} \hat{\mu} - \Sigma_j^{-1} \mu_j \right), \tag{37}$$

where j is the ID of the least probable CA model and variables with subscript $+$ are the updated variables. If the reduction leads to a case where there are only two CAs left and they are distant to each other according to Equation 35, the WLAN measurements are not to be used at all.

There are examples in the statistics literature where the Mahalanobis distance does not work very well [20], but we have found it satisfactory in the tests reported here. For more advanced treatment, the effect of outliers can be mitigated by doing the positioning using distributions with heavier tails like Student's t-distribution. This could be done by adapting the algorithm presented in [21] to CA positioning.

3.2.3 *MIMO compensation*

Multiple-input multiple-output (MIMO) WLAN devices host several APs on one physical device. Such APs have naturally very similar CAs. If these CAs are used as independent measurements, the estimate covariance becomes too optimistic and the estimate mean gets biased towards the MIMO CA mean. The same problem arises also with non-MIMO APs that have similar coverage areas. This problem could be handled during the construction of the radio map by taking the correlations of received APs into account. This would however make the size of the radio map grow because there would be a need to store the cross correlation terms between every AP pair received simultaneously.

We propose the following heuristic method for compensating the overlapping MIMO CAs in estimation. The method is inspired by [22], where dissimilarity of two components of a Gaussian mixture is measured by:

$$B(i,j) = \frac{1}{2} \left((w_i + w_j) \log \det \Sigma_{i,j} - w_i \log \det \Sigma_i - w_j \log \det \Sigma_j \right), \tag{38}$$

where $\Sigma_{i,j} = \frac{w_i\Sigma_i + w_j\Sigma_j}{w_i + w_j} + (\mu_i - \mu_j)(\mu_i - \mu_j)^T$ and w is a component weight.

We use a slightly modified exponential form of Equation 38 given by:

$$W_{i,j} = \frac{\det\left[\frac{\Sigma_i+\Sigma_j}{2} + (\mu_i - \mu_j)(\mu_i - \mu_j)^T\right]}{\sqrt{\det \Sigma_i \det \Sigma_j}} \tag{39}$$

as the measure of dissimilarity between CAs. This measure is used to scale covariance of CAs such that if there is any number of identical CAs that are not similar with other CAs, they have the same effect as if there was only one CA, and that if two identical APs are further than $1 \cdot \sigma$ from each other, they are considered independent. These properties are achieved by using MIMO-compensated CA variances Σ^{MIMO} in the position estimation (Equations 30 and 31), given by:

$$\Sigma_i^{\text{MIMO}} = \left[\sum_{j=1}^{n} \max(2 - W_{i,j}, 0)\right] \Sigma_i. \tag{40}$$

Proofs that this has the properties stated above are given in the Appendix.

In Figure 1, the outlier detection followed by MIMO compensation is applied in a case of five APs received. One CA is an outlier, and there is a cluster of three similar CAs. The figure shows one-sigma ellipses of CAs and different user location estimates.

4 Pedestrian dead reckoning

We consider positioning systems in which PDR measurements are used in estimation in the state transition models and the WLAN measurements are then used in the update to fuse the measurements together in a Kalman filter. In most of the literature, the state model used in PDR is nonlinear [7-10]. For comparison purposes, we consider two traditional models. In the first one, the state contains the user location r and the direction of the movement θ:

$$x_t = \begin{bmatrix} r_{1,t} \\ r_{2,t} \\ \theta_t \end{bmatrix} = \begin{bmatrix} r_{1,t-1} + s_t \cos\theta_{t-1} \\ r_{2,t-1} + s_t \sin\theta_{t-1} \\ \theta_{t-1} + \Delta\theta_t \end{bmatrix}, \tag{41}$$

where $\Delta\theta_t$ is the change of heading obtained from gyroscopes and s_t is the footstep length estimated from accelerometer data. These are modeled as Gaussian random variables with means $\overline{\Delta\theta_t}$ and $\overline{s_t}$, which are the measured values, and variances $\sigma^2_{\Delta\theta}$ and σ^2_s. In EKF, the predicted mean is as follows:

$$\overline{x}_{t|t-1} = \begin{bmatrix} \overline{r}_{1,t|t-1} \\ \overline{r}_{2,t|t-1} \\ \overline{\theta}_{t|t-1} \end{bmatrix} = \begin{bmatrix} \overline{r}_{1,t-1|t-1} + \overline{s}_t \cos\overline{\theta}_{t-1|t-1} \\ \overline{r}_{2,t-1|t-1} + \overline{s}_t \sin\overline{\theta}_{t-1|t-1} \\ \overline{\theta}_{t-1|t-1} + \overline{\Delta\theta}_t \end{bmatrix}, \tag{42}$$

the linearized state transition matrix is as follows:

$$\hat{F}_t = \begin{bmatrix} 1 & 0 & -\overline{s}_t \sin\overline{\theta}_{t-1|t-1} \\ 0 & 1 & \overline{s}_t \cos\overline{\theta}_{t-1|t-1} \\ 0 & 0 & 1 \end{bmatrix}, \tag{43}$$

and the linearized state transition noise covariance is as follows:

$$\hat{L}Q\hat{L}^T = \begin{bmatrix} \sigma_s^2 \cos^2\overline{\theta}_{t-1|t-1} & \sigma_s^2 \sin\overline{\theta}_{t-1|t-1} \cos\overline{\theta}_{t-1|t-1} & 0 \\ \sigma_s^2 \sin\overline{\theta}_{t-1|t-1} \cos\overline{\theta}_{t-1|t-1} & \sigma_s^2 \sin^2\overline{\theta}_{t-1|t-1} & 0 \\ 0 & 0 & \sigma^2_{\Delta\theta} \end{bmatrix}, \tag{44}$$

where σ^2 is the variance of the variable in the subscript.

In the UKF prediction, the mean of the augmented state is as follows:

$$\overline{z} = \begin{bmatrix} \overline{x}_{t-1|t-1} \\ \overline{\Delta\theta}_t \\ \overline{s}_t \end{bmatrix} \tag{45}$$

and the covariance is as follows:

$$Q_t^z = \begin{bmatrix} P_{t-1|t-1} & 0 & 0 \\ 0 & \sigma^2_{\Delta\theta} & 0 \\ 0 & 0 & \sigma_s^2 \end{bmatrix}. \tag{46}$$

In our second traditional model, the footstep length is also an estimated variable:

$$x_t = \begin{bmatrix} r_{1,t} \\ r_{2,t} \\ \theta_t \\ s_t \end{bmatrix} = \begin{bmatrix} r_{1,t-1} + s_{t-1} \cos\theta_{t-1} \\ r_{2,t-1} + s_{t-1} \sin\theta_{t-1} \\ \theta_{t-1} + \Delta\theta_t \\ s_{t-1} + \Delta s \end{bmatrix}, \tag{47}$$

where Δs is the estimated footstep length change, which is modeled as a zero mean Gaussian noise term with variance $\sigma^2_{\Delta s}$.

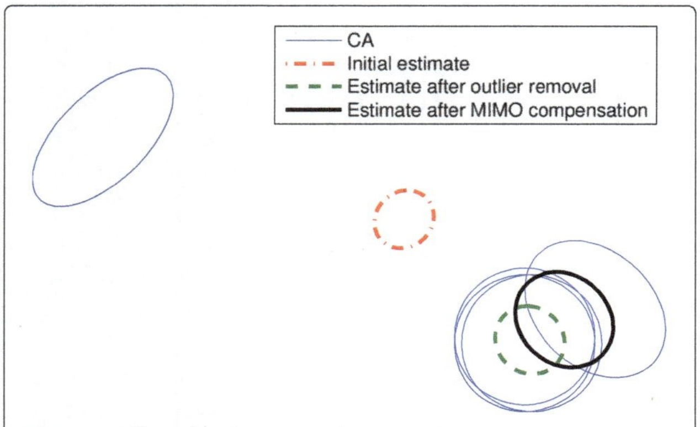

Figure 1 Effect of outlier removal and MIMO compensation on position estimates. The initial estimate (red dash-dot) is heavily influenced by the outlier. After outlier removal, the estimate (green dash) is biased and shrunk by the trio of similar CAs. After MIMO compensation, the estimate (black solid) has reasonable location and scale.

For this model, the EKF prior mean is as follows:

$$\overline{x}_{t|t-1}=\begin{bmatrix}\overline{r}_{1,t|t-1}\\ \overline{r}_{2,t|t-1}\\ \overline{\theta}_{t|t-1}\\ \overline{s}_{t|t-1}\end{bmatrix}=\begin{bmatrix}\overline{r}_{1,t-1|t-1}+\overline{s}_{t|t-1}\cos\overline{\theta}_{t-1|t-1}\\ \overline{r}_{2,t-1|t-1}+\overline{s}_{t-1|t-1}\sin\overline{\theta}_{t-1|t-1}\\ \overline{\theta}_{t-1|t-1}+\overline{\Delta\theta}_t\\ \overline{s}_{t-1|t-1}\end{bmatrix}, \tag{48}$$

the linearized state model is as follows:

$$\hat{F}_t=\begin{bmatrix}1 & 0 & -\overline{s}_{t-1|t-1}\sin\overline{\theta}_{t-1|t-1} & \cos\overline{\theta}_{t-1|t-1}\\ 0 & 1 & \overline{s}_{t-1|t-1}\cos\overline{\theta}_{t-1|t-1} & \sin\overline{\theta}_{t-1|t-1}\\ 0 & 0 & 1 & 0\\ 0 & 0 & 0 & 1\end{bmatrix} \tag{49}$$

and because the state transition noises are additive, we have:

$$\hat{L}_tQ_t\hat{L}_t^T=\begin{bmatrix}0 & 0 & 0 & 0\\ 0 & 0 & 0 & 0\\ 0 & 0 & \sigma^2_{\Delta\theta} & 0\\ 0 & 0 & 0 & \sigma^2_{\Delta s}\end{bmatrix}. \tag{50}$$

In the UKF prediction, the mean of the augmented state is as follows:

$$\bar{z}=\begin{bmatrix}\overline{x}_{t-1|t-1}\\ \overline{\Delta\theta_t}\end{bmatrix} \tag{51}$$

and the covariance is as follows:

$$Q^z_t=\begin{bmatrix}P_{t-1|t-1} & 0\\ 0 & \sigma^2_{\Delta\theta}\end{bmatrix}. \tag{52}$$

Figure 2 shows position estimates after taking one footstep from a known position using this model. The figure shows how the EKF linearization does not take the 'curvature' of the true model into account and how UKF takes it into account by increasing the covariance.

We propose the linear state model:

$$\begin{aligned}x_t=\begin{bmatrix}r_{1,t}\\ r_{2,t}\\ v_{1,t}\\ v_{2,t}\end{bmatrix}&=\begin{bmatrix}1 & 0 & 1 & 0\\ 0 & 1 & 0 & 1\\ 0 & 0 & \cos\overline{\Delta\theta_t} & -\sin\overline{\Delta\theta_t}\\ 0 & 0 & \sin\overline{\Delta\theta_t} & \cos\overline{\Delta\theta_t}\end{bmatrix}\begin{bmatrix}r_{1,t-1}\\ r_{2,t-1}\\ v_{1,t-1}\\ v_{2,t-1}\end{bmatrix}+\begin{bmatrix}0\\ 0\\ \Delta v_{1,t}\\ \Delta v_{2,t}\end{bmatrix}\\ &=F_tx_{t-1}+\varepsilon^Q_t,\end{aligned} \tag{53}$$

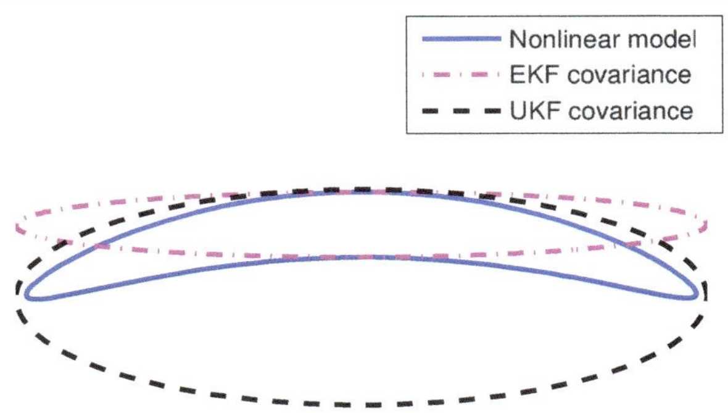

Figure 2 Footstep propagation with nonlinear model and the EKF and UKF approximations.

where v is the footstep vector estimate and Δv is the associated state transition noise. For this model, the predicted mean and covariance are computed using Equations 2 and 3. To make state transition function linear and noise additive and Gaussian, Δv is modeled as a Gaussian that is independent of state and measurement value $\overline{\Delta\theta}_t$. The process noise covariance matrix is as follows:

$$Q_t=\begin{bmatrix}0 & 0 & 0 & 0\\ 0 & 0 & 0 & 0\\ 0 & 0 & \sigma^2_{\Delta v} & 0\\ 0 & 0 & 0 & \sigma^2_{\Delta v}\end{bmatrix}. \tag{54}$$

To compare the amount of state transition noise in traditional model (Equation 47) and linear model (Equation 53), we investigate the amount of noise propagated from step variables to position variables in one step. For simplicity, we use here independent step variables. In linear model, the covariance of the state variables increases every step by:

$$\begin{bmatrix}\sigma^2_v & 0\\ 0 & \sigma^2_v\end{bmatrix} \tag{55}$$

and in the EKF, the position covariance is increased by:

$$\begin{bmatrix}\sigma^2_\theta\bar{s}^2\sin^2\overline{\theta}_{t-1|t-1}+\sigma^2_s\cos^2\overline{\theta}_{t-1|t-1} & (\sigma^2_s-\bar{s}^2\sigma^2_\theta)\sin\overline{\theta}_{t-1|t-1}\cos\overline{\theta}_{t-1|t-1}\\ (\sigma^2_s-\bar{s}^2\sigma^2_\theta)\sin\overline{\theta}_{t-1|t-1}\cos\overline{\theta}_{t-1|t-1} & \sigma^2_\theta\bar{s}^2\cos^2\overline{\theta}_{t-1|t-1}+\sigma^2_s\sin^2\overline{\theta}_{t-1|t-1}\end{bmatrix}. \tag{56}$$

From this, we can see that when:

$$\sigma^2_\theta s^2=\sigma^2_s=\sigma^2_v \tag{57}$$

the amount of propagated error is the same for both models. Because the noise of footstep variables is modeled additive and independent of the state, the increase of position variable variance caused by variance of one footstep is same using the different models when:

$$\sigma^2_{\Delta\theta}s^2=\sigma^2_{\Delta s}=\sigma^2_{\Delta v}. \tag{58}$$

Compared to the traditional models, the proposed model has the benefit that the Kalman filter can be used to compute the optimal estimate for this model when the measurements are linear and Gaussian. In this paper, we use only linear WLAN location measurements presented in the previous section and the information from PDR is used only in the state transition models. A drawback of this model is that it does not have different variances for the heading and footstep length. Constructing a noise model that has different variances for heading change and footstep length is possible by computing θ and s from step vector and adding noise similar to Equation 56 to the step vector. This would, however, make the noise nonlinearly dependent on the state estimate and linearizations would be required. In Section 6.1, we test the performance of linear model in a situation, where $\sigma^2_\theta s^2\neq\sigma^2_s$ and show that the linear model can perform well also in that situation.

Another drawback is that the measured footstep length cannot be used in the linear model. If it was used as a measurement, the model would not be linear. The first presented nonlinear state model uses all the information, and the second presented nonlinear model uses the same amount of information as the proposed linear model. We will test also this in Section 6 and show that even though the proposed method cannot use all the available information, it can outperform the nonlinear model with footstep length when the initial heading is unknown.

5 Integration with a particle filter

Besides footstep length, other information that cannot be used with a linear KF includes the use of floorplan information or nonlinear measurement models in estimation. This information can be such that it cannot be used well in EKF, e.g., floorplan. PFs can be used in these situations [23-25]. Particle filters estimate the probability distribution using a set of weighted point masses called particles. The number of particles required for estimation is case dependent. The larger the uncertainties in the proposal distribution compared to measurement variance, the more particles are needed. Particle filters are usually computationally significantly more expensive than KFs, and so, running a KF alongside a PF does not significantly increase the computational complexity. In [25], the WLAN+PDR KF presented in [11] was compared with a PF that used wall information and a nonlinear state model. The PF with wall information reduced the mean error of KF from 4.8 to 2.0 m, while PF without wall information had slightly worse accuracy than KF. In the following, we present algorithms in a general form so they can be applied to many known PF variants [26].

5.1 Number of particles in initialization

In this section, we consider the situation where the prior covariance is so large that the PF would need an excessively large number of particles to adequately sample the region. We propose to use the KF until the estimate uncertainty is small enough for initialization of a PF that has a given number of particles.

We assume that the PF used is initialized by sampling from the prior distribution. Because the number of particles required for good estimation depends on many factors, there cannot be a simple algorithm for determining the number of particles. We assume that it is known how many particles n_1 is enough with a prior covariance P_1 and then compute the number of particles n_2 that is required with covariance matrix P_2.

In general, a four-dimensional vector x is inside an ellipsoid that contains fraction p of the probability mass of distribution $\mathrm{N}(\bar{x}_1, P_1)$ when:

$$(x-\bar{x}_1)^T P_1^{-1}(x-\bar{x}_1) \le \chi_4^{-2}(p), \tag{59}$$

where χ_4^{-2} is the inverse cumulative density function of the chi-squared distribution with 4 degrees of freedom. The volume of this four-dimensional ellipsoid is as follows:

$$V_1 = \frac{\pi^2}{2}\left[\chi_4^{-2}(p)\right]^2 \sqrt{\det P_1} \tag{60}$$

and the expected number of particles inside it is pn_1, so the expected particle density is $\frac{pn_1}{V_1}$.

We propose to choose the number of particles n_2 in a such way that ellipsoid:

$$(x-\bar{x}_2)^T P_2^{-1}(x-\bar{x}_2) \le \chi_4^{-2}(p) \tag{61}$$

has the same expected particle density as ellipsoid of Equation 59. Thus, we get rule:

$$\frac{pn_2}{V_2} = \frac{pn_1}{V_1} \Leftrightarrow \tag{62}$$

$$n_2 = n_1 \frac{\frac{\pi^2}{2}\left[\chi_4^{-2}(p)\right]^2 \sqrt{\det P_2}}{\frac{\pi^2}{2}\left[\chi_4^{-2}(p)\right]^2 \sqrt{\det P_1}} = n_1 \frac{\sqrt{\det P_2}}{\sqrt{\det P_1}}. \tag{63}$$

5.2 Integrity monitoring

There is a risk that the PF estimate diverges far from the true location. Divergence may be caused by too few particles, error in the floorplan, etc. The linear estimate can be used to detect whether the PF has diverged using the algorithm proposed in [25], as follows.

We assume that the particle filter operates in a state space that contains at least the same information as the linear KF. If the particle filter state has different parametrization, e.g., footstep length and heading instead of step vector, the particle states are converted to the same state space that the linear model uses. First, compute standardized deviances d_i of all positively weighted particles:

$$d_i = (c_i-\bar{x})^T P^{-1}(c_i-\bar{x}), \tag{64}$$

where c_i is the location of ith particle and P and $\bar{x}$ are the parameters of the KF estimate. Then, check if the smallest deviance is below a set threshold. If it is not, then particles are reinitialized. Because also the KF estimate may diverge, the resampling can be done so that some of the original particles are retained and the rest are sampled from the KF estimate.

6 Performance evaluation

6.1 Synthetic data

In simulations, we tested how the linear model performs against the traditional models. If not otherwise specified, the true track is generated by sampling the initial state from prior and then sampling subsequent true states using the model given in Equation 47. In tests with simulated data, the WLAN measurements are independent and the improvements proposed in Section 3.2 are not needed.

Each simulated test track consists of 50 footsteps, and at every time step, there is a 10% probability of receiving a location measurement with variance $10^2\text{m}^2 I$.

First, we test the effect of the initial heading variance on the positioning accuracy of the traditional models and comparing it with estimate obtained with linear model that does not use initial heading information. The traditional models require an initial mean for the heading variable, while the proposed linear model can be initialized with a zero mean step vector. It would also be possible to initialize the linear model with accurate heading and footstep length information, but when the initial heading variance is larger, it is not a straightforward task to select initial covariance for the linear model that is similar to the one used with nonlinear model, but that avoids the problems with linearizations. In the first simulation, the effect of the heading error of the initial mean is tested. The standard deviation of the initial heading is varied from 0° to 180°. The standard deviation of footstep length is set to $\sigma_{\Delta s} = 0.01\text{m}$, and the standard deviation of heading change is $\sigma_{\Delta\theta} = \frac{0.01}{0.7}$ rad. The initial footstep length is set to 0.7m for the true state and for the traditional model. The linear model is initialized to have a zero footstep vector with initial standard deviation $\sigma_{v_0} = 1\text{m}$. If the footstep length does not change much during the track, the linear model, where $\sigma_{\Delta v} = 0.01\text{m}$, has the same amount of propagated error in position as the linearized traditional model. All models are initialized to have correct location mean and the initial variance for location dimensions is 10^2m^2. The standard deviation for initial footstep length is set to $\sigma_{s_0} = 0.2$ m. For the nonlinear model with given footstep length, we use accurate measurements and $\sigma_s = 0$. The footstep length variable is not random in this situation, and in the UKF implementation, it is not used in the augmented state.

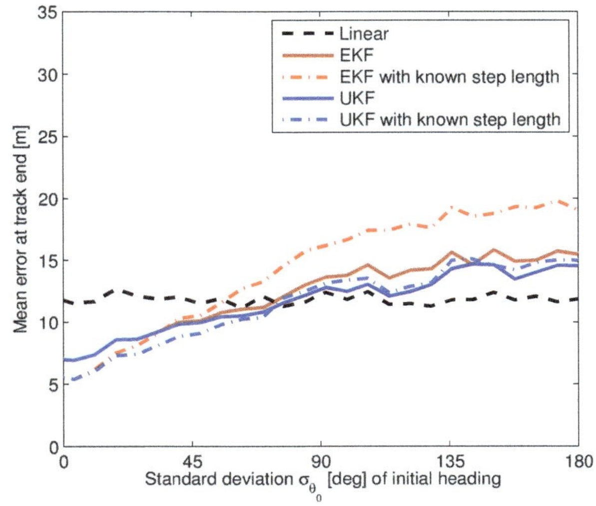

Figure 3 Mean of errors at track end as the function of initial standard deviation of heading.

Figure 3 shows the mean of the final position error of 500 simulated tracks for different standard deviations of the initial heading. The traditional models outperform the linear model when the standard deviation is small. This is expected because the traditional models use prior information of the footstep length and initial heading, unlike the linear model. The models with known footstep lengths are more accurate when the standard deviation is small. When the initial heading variance becomes larger, the linear model outperforms the traditional models. This is caused by the increase of the linearization errors in the traditional models.

In the second simulated test, the standard deviation of the initial heading is fixed to 90°, but the initial angle error of the traditional models is varied. Otherwise, the simulation parameters are the same as in the first test setup. Figure 4 shows how the traditional model is advantageous only if the heading variance is small; small realized error is not enough. This means that to gain the benefit of the known initial heading, the initial heading variance should also be set correctly. Here, the models that use known footstep length have worse accuracy than the models that estimate the footstep length.

Compared to simulation results presented in Figure 2 of [11], the main differences in this test are as follows: in this test, the footstep length variance is smaller, and thus, the traditional model does not end up into situation with negative footstep length that often. In [11], there was error in simulation variance: for location variables, it was 10m^2 instead of 10^2m^2. When using smaller initial variance, the accuracy of the linear model is on the same level as the accuracy of the traditional models.

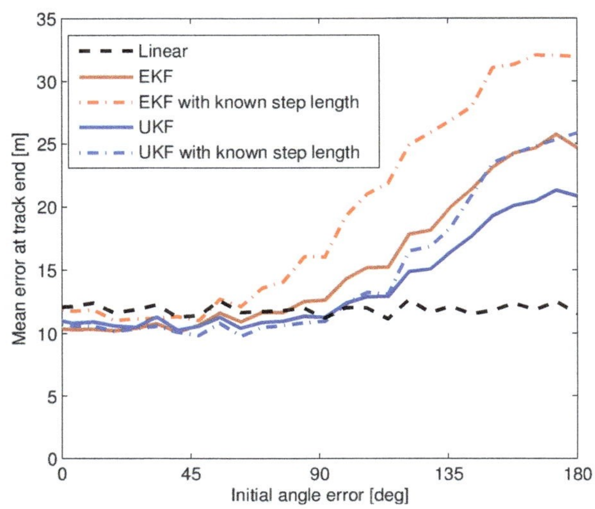

Figure 4 Mean of errors at track end as the function of initial heading error.

Next, we investigate how the different methods perform when the heading error (i.e., gyroscope accuracy) is changed. The traditional models use the correct $\sigma_{\Delta\theta}$ and $\sigma_{\Delta s} = 0.01$ m, and the linear model is tested with $\sigma_{\Delta v} = \sigma_s$ and $\sigma_{\Delta v} = 0.7\sigma_{\Delta\theta}$. In this test, we use initial state noise parameters $\sigma_{r_0} = 0.2$ m, $\sigma_\theta = \frac{0.2}{0.7}$ rad, $\sigma_{s_0} = 0.2$ m, and $\sigma_v = 0.2$ m.

Figure 5 shows the results for this test. The linear model and traditional models that do not use the footstep length have similar accuracy when the heading change error is small, while the models that use footstep length are more accurate. When the noise level increases the linear model that uses the footstep length's variance as the $\sigma_{\Delta v}$ becomes less accurate than the other methods, but the linear model that uses step vector variance computed from the heading change variance with expected footstep length (0.7 m) according to Equation 58 is either on the same level or better than traditional models without known footstep length. Figure 3 of [11] shows a similar test's results. In [11], the initial variance is large compared to this article's test.

Figure 6 shows how the $1 \cdot \sigma$ ellipse of footstep noise is propagated to state noise in one footstep with different models, when the standard deviation for heading change noise is 20°. The linear model that uses variance of the footstep length has a small covariance compared to others, which causes high position errors. On other hand, the linear model with noise level computed from heading noise using Equation 58 has the largest noise covariance. It seems that this does not cause as much problems as having a too small error estimate. Similarly, when the footstep length is known, the EKF linearization does not take the curvature of the true distribution of the nonlinear model into account and has bad accuracy, while the UKF has a better variance approximation, which results in better accuracy.

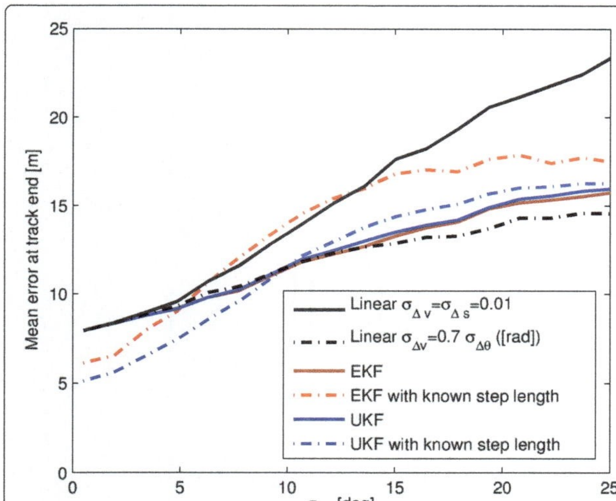

Figure 5 Mean of errors at track end as the function of heading change error.

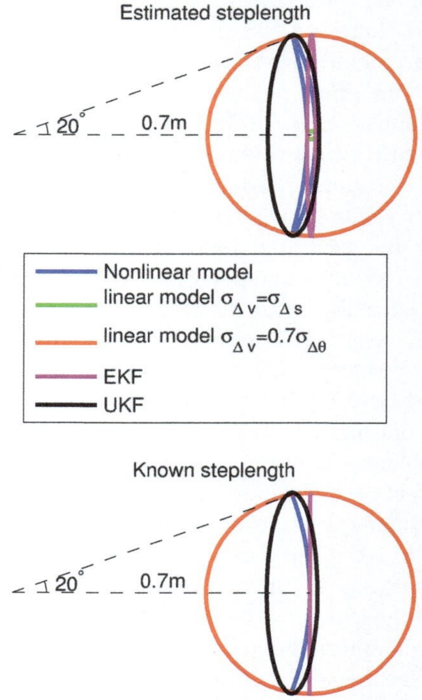

Figure 6 Different estimates after one step when $\sigma\Delta = 2°$.

6.2 True data

In the true-data test, the algorithm is tested in two floors of a building in the Tampere University of Technology. The radio map of both floors contains approximately 300 WLAN APs and 5,500 FPs. The measurements were collected using a XSENS MTi IMU and Acer Iconia tablet. Both devices were carried in hand while doing measurements. Because XSENS was handheld, the errors in heading are caused, in addition to IMU noise, by the small changes in hand orientation compared to the movement direction. We used four test tracks. WLAN scans were done on average every 3.2 s.

For the WLAN positioning, we use the two-level coverage area method presented in Section 3 with all modifications from Section 3.2, if not otherwise stated. We generate two coverage areas: a weak CA that is constructed using all FPs and a strong CA that is constructed with only FPs that have signal strength ≥ -60 dBm. The prior for weak CA is $B = 100^2\,\text{m}^2 I$ and for strong CA $B = 20^2\,\text{m}^2 I$. Minimum variances for main axes were set to $40^2\,\text{m}^2$ for weak CAs and to $5^2\,\text{m}^2$ for strong CAs. Most of the weak CAs had smaller variances than this limit, and they were extended using Equation 34. According to [19],

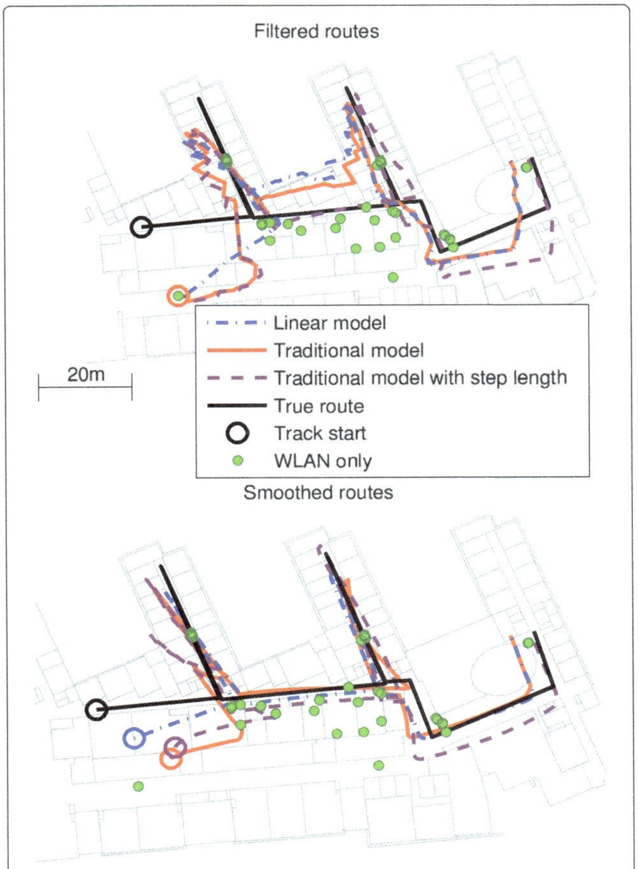

Figure 7 Estimated routes computed with Kalman filter and linear model and with UKF and traditional models.

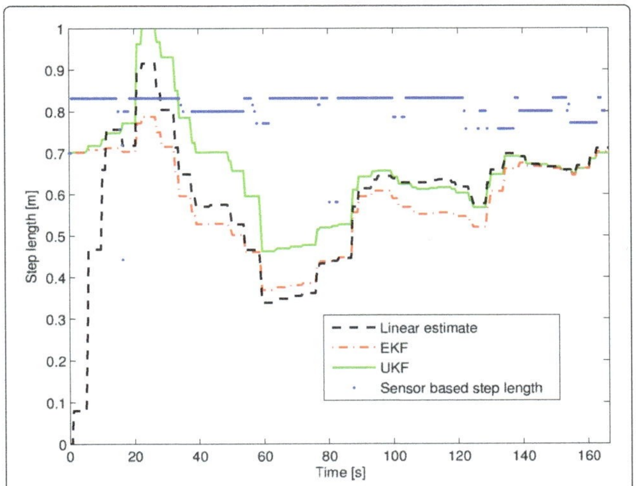

Figure 8 Footstep length estimates.

Table 1 Mean errors [m] of different methods and different initial heading errors

		Filtered		Smoothed	
Linear		7.0		4.1	
Static		9.7			
Linear $\Delta t = 15$ s		9.3		5.2	
Static $\Delta t = 15$ s		12.7			
Traditional models	θ_0	EKF	UKF	EKF	UKF
Estimated footstep length	0°	7.9	6.9	4.6	5.5
Sensor footstep length	0°	7.3	6.3	3.6	4.7
Estimated footstep length	45°	7.4	6.7	4.2	5.3
Sensor footstep length	45°	7.1	6.4	3.9	5.0
Estimated footstep length	90°	10.5	9.8	7.7	7.9
Sensor footstep length	90°	14.7	11.1	10.7	11.0
Estimated footstep length	135°	16.4	20.0	12.8	18.4
Sensor footstep length	135°	17.7	14.7	16.0	16.1
Estimated footstep length	180°	10.1	10.4	7.1	9.2
Sensor footstep length	180°	44.7	37.4	42.8	40.7

the two-level normal CA models have around 10% larger errors than traditional location fingerprinting, but require only storage of ten parameters for each AP.

Figure 7 shows filtered and smoothed routes for a test track. The initial heading for traditional models is 90° error with variance of $(90°)^2$, and the initial footstep length estimate is 0.7 m with variance $(0.2\text{ m})^2$. The traditional method is tested also with footstep lengths estimated from the sensor data with $\sigma_s = 0.1$ m. For the linear model, the initial footstep vector is $\mathbf{0}$ with variance $1\text{ m}^2 I$. The initial position variance is large, i.e., the prior is almost uninformative, which causes all models to have the first estimate at the location of the first WLAN estimate, which happens to be inside a wrong corridor.

In filtered routes, the traditional models start almost equally, and in the end, the traditional model that estimates the footstep length is close to the proposed linear method. The smoothed routes have some difference near the beginning of the route. The traditional models are less accurate than the linear model. This is caused by the wrong linearization of the F matrix in the beginning of the track. The smoothing was done using these F matrices in Equation 23; better results might be obtained with

Table 2 Effect of WLAN positioning modifications on filtering

	Error [m]	50%	95%
All on	7.0	29%	79%
MIMO off	6.9	13%	42%
Outlier off	7.2	28%	76%
Minsize off	7.5	10%	32%
All off	7.9	6%	14%

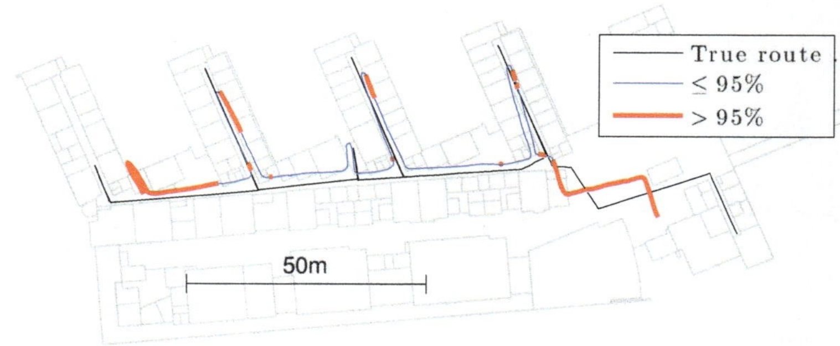

Figure 9 Likely and unlikely parts of a smoothed track.

more complex methods such as the unscented Rauch-Tung-Striebel smoother [27].

Some error statistics are given in Table 1. In the table, θ_0 is the initial error on the heading. 'Static' denotes the WLAN-only position estimates without filtering. When there are only footstep measurements without WLAN measurements, the 'Static' uses the last position computed from WLAN measurements as the position estimate. In the end of the table, there are results for linear model and static estimates when the WLAN measurement interval is increased so that there is always at least 15 s between consecutive measurements. A long WLAN scan interval may be used to conserve energy.

All the filtering methods result in similar accuracy with 0° and 45° initial heading errors. When the initial heading is 90 or more degrees off, neither of the traditional methods improves "Static" when the route is filtered. The UKF is more accurate than the EKF in filtering in almost every test situation, but in smoothing, the results are opposite. This indicates that the UKF linearizations of this model with uncertain prior do not work well with the Rauch-Tung-Striebel smoother. The longer WLAN scan intervals make the static estimate 3.0 m worse, filtered estimate 2.0 m worse, and the smoothed estimate 1.1 m worse.

Footstep length approximations given by different models are shown in Figure 8. The footstep lengths are computed in the case shown in Figure 7 ($\theta_0 = 90°$). Here, we see that the footstep length estimated with the traditional model (s) and the proposed linear model ($||v||$) are rather different in the beginning of the track, but become similar towards the end of the track. The estimated footstep lengths are in general shorter than lengths computed from the sensor data.

In Table 2, the effect of CA positioning improvements presented in Subsection 3.2 is shown. The 50% column shows the 50% consistency value of the estimate, i.e., how often the position estimate mean and true position are closer to each other than 50% of the probability mass. The 95% column is defined analogously. Consistencies are computed taking only position variables into account. The closer the values are to 50% and 95%, the better. Good consistency is important for the integration with the PF because the initialization and integrity checks use the uncertainty of the KF estimate.

The position estimate with and without MIMO compensation are almost the same, but the 50% and 95% columns are much worse without the MIMO compensation. The outlier detection improves all results slightly, but not much. This is probably because the positioning data had only a few CAs that were considered outliers. The use of minimum sizes for coverage areas improved both the positioning accuracy and consistency.

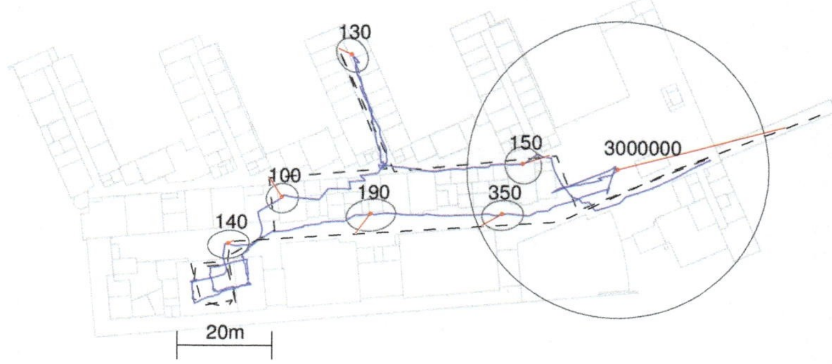

Figure 10 50% ellipses and relative numbers of particles required for particle filter initialization at different points of track.

Even though the estimate with all WLAN improvements has the best consistency values, the error estimates are still too small. This is probably because the radio map is collected inside the building and the CA positioning can be interpreted as computation of a weighted mean. Because there are no APs that have CA center outside the building, the estimates near the edges of the building get biased towards the inside of the building.

Figure 9 contains a smoothed track showing this effect. Position estimates that have the position error within 95% error ellipse are shown with a narrow blue line, and the estimates where error is outside 95% error ellipse are shown with a thicker red line. From the figure, it is evident that most of the red lines are close to edges of the building.

In Figure 10, 50% ellipses are plotted on different parts of a track. The ellipse labels show the relative number of particles required for PF initialization (Equation 63). In the figure, most residuals (red lines) are slightly out from the 50% ellipses as expected from the consistency data in Table 2, but none of them is completely wrong. In the beginning of the track, the required number of particles is $3 \cdot 10^6$ even though the 50% ellipse covers less than half of the building. The next plotted ellipse on the track has 350 particles and has similar size to the remaining ellipses, which all have fewer than 200 particles. This is because the footstep vector part of state has still more uncertainty than in the rest of the track.

7 Conclusions

We proposed in this paper a novel linear model for PDR in indoor personal positioning and compared it to models that are common in the literature. The evaluation shows that although the model is simpler than the traditional methods, it performs well and is especially suited for situations where the initial heading and position are not known. The model assumes that uncertainty of state transition noise is equal in every direction, but our simulations show that the model is not very sensitive to different footstep length and heading variances. Because the proposed model is linear, it can be smoothed optimally with the Rauch-Tung-Striebel smoother.

In this paper, we used linear Gaussian coverage area models to do positioning with WLAN and proposed improvements for the positioning method. The results show that the proposed improvements reduced positioning errors from 7.9 to 7.0 m and significantly improved the estimate's consistency.

In addition to stand-alone position estimation, the proposed model with a KF can be used to initialize and monitor the integrity of PF. Because the linear KF is a computationally light algorithm compared to a PF, running it together with a PF does not significantly increase computational burden.

Appendix

Proofs concerning MIMO compensation

Proof that two identical CAs with $1 \cdot \sigma$ distance are considered independent

Let L be a matrix square root of covariance matrix, i.e., $LL^T = P$. Let means μ_i and μ_j be $1 \cdot \sigma$ from each other, i.e.,

$$(\mu_i - \mu_j)^T P^{-1} (\mu_i - \mu_j) = (\mu_i - \mu_j)^T L^{-T} L^{-1} (\mu_i - \mu_j) = 1. \tag{65}$$

Let unitary matrix U have $L^{-1}(\mu_i - \mu_j)$ as its first column. Now, $Ue_1 = L^{-1}(\mu_i - \mu_j)$, where e_1 is the first column of an identity matrix.

Using these the nominator of Equation 39 becomes:

$$\det\left(P + (\mu_i - \mu_j)(\mu_i - \mu_j)^T\right) = \det\left(LL^T + LUe_1 e_1^T U^T L^T\right) \tag{66}$$

$$= \det\left(L\left(I + Ue_1 e_1^T U^T\right) L^T\right) = \det\left(LU\left(I + e_1 e_1^T\right) U^T L^T\right) \tag{67}$$

$$= \det(L)\det(U)\det\left(I + e_1 e_1^T\right)\det\left(U^T\right)\det\left(L^T\right) = 2\det(P). \tag{68}$$

For such cases, $W_{i,j} = 2$ and the MIMO correction is zero.

Proof that a set of identical CAs act as one

For identical CAs, the MIMO correction is 1. If there are n identical CAs, each covariance Σ is multiplied by n. The resulting compensated covariance matrix of n measurements is as follows:

$$\hat{\Sigma} = \left(\sum_{i=1}^{n} (n\Sigma)^{-1}\right)^{-1} \tag{69}$$

$$= \left((\Sigma)^{-1}\right)^{-1} \tag{70}$$

$$= \Sigma, \tag{71}$$

which is the same as the covariance of one CA.

Competing interests

The authors declare that they have no competing interests.

Acknowledgements

This work was supported by Tampere Doctoral Programme in Information Science and Engineering, TUT's Graduate School, Nokia, Nokia Foundation, and Jenny and Antti Wihuri Foundation. The funding sources were not involved in preparation of this article. We thank Jussi Collin and Jussi Parviainen for codes that estimate the footsteps and headings from the sensor data.

References

1. C Rizos, AG Dempster, B Li, J Salter, in *AusWireless '06*. Indoor positioning techniques based on wireless LAN (Sydney, Australia, 2006)
2. Y-C Cheng, Y Chawathe, A LaMarca, J Krumm, in *Proceedings of the 3rd International Conference on Mobile Systems, Applications, and Services*. MobiSys '05. Accuracy characterization for metropolitan-scale Wi-Fi

localization (ACM New York, NY USA, 2005), pp. 233–245. doi:10.1145/1067170.1067195
3. J Collin, O Mezentsev, G Lachapelle, in *Proceedings of the ION GPS/GNSS 2003 Conference*. Indoor positioning system using accelerometry and high accuracy heading sensors (Portland, OR USA, 2003), pp. 9–12
4. R Jirawimut, P Ptasinski, V Garaj, F Cecelja, W Balachandran, A method for dead reckoning parameter correction in pedestrian navigation system. IEEE Trans. Instrum. Meas. **52**(1), 209–215 (2003). doi:10.1109/TIM.2002.807986
5. S Beauregard, in *4th Workshop on Positioning, Navigation and Communication. (WPNC '07)*. Omnidirectional pedestrian navigation for first responders (Hannover, Germany, 2007), pp. 33–36. doi:10.1109/WPNC.2007.353609
6. SH Shin, CG Park, JW Kim, HS Hong, JM Lee, in *Sensors Applications, Symposium. SAS '07. IEEE*. Adaptive step length estimation algorithm using low-cost MEMS inertial sensors (San Diego, CA USA, 2007), pp. 1–5. doi:10.1109/SAS.2007.374406
7. H Leppäkoski, J Collin, J Takala, in *IEEE International Conference on Acoustics, Speech and Signal Processing (ICASSP)*. Pedestrian navigation based on inertial sensors, indoor map, and WLAN signals (Kyoto, Japan, 2012), pp. 1569–1572. doi:10.1109/ICASSP.2012.6288192
8. F Evennou, F Marx, Advanced integration of WiFi and inertial navigation systems for indoor mobile positioning. EURASIP J. Appl. Signal Process. **2006**, 164–164 (2006). doi:10.1155/ASP/2006/86706
9. W Chen, R Chen, Y Chen, H Kuusniemi, J Wang, in *Position Location and Navigation Symposium (PLANS), IEEE/ION*. An effective pedestrian dead reckoning algorithm using a unified heading error model (Indian Wells, CA USA, 2010), pp. 340–347. doi:10.1109/PLANS.2010.5507300
10. F Zampella, M Khider, P Robertson, A Jimenez, in *Position Location and Navigation Symposium (PLANS) IEEE/ION*. Unscented kalman filter and magnetic angular rate update (MARU) for an improved pedestrian dead-reckoning (Myrtle Beach, SC USA, 2012), pp. 129–139. doi:10.1109/PLANS.2012.6236874
11. M Raitoharju, H Nurminen, Piché R, in *Conference on Design and Architectures for Signal and Image Processing (DASIP)*. A linear state model for PDR+WLAN positioning (Cagliari, Italy, 2013)
12. Y Ho, R Lee, A Bayesian approach to problems in stochastic estimation and control. IEEE Trans. Automatic Control. **9**(4), 333–339 (1964). doi:10.1109/TAC.1964.1105763
13. A Gelb, *Applied Optimal Estimation*. (MIT Press, Cambridge, MA USA, 1974)
14. SJ Julier, JK Uhlmann, in *Signal Processing, Sensor, Fusion, and Target Recognition, VI*. A new extension of the Kalman filter to nonlinear systems, vol. 3068 (Orlando, FL USA, 1997), pp. 182–193. doi:10.1117/12.280797
15. EA Wan, R Van der Merwe, in *Proceedings of the Adaptive Systems for Signal Processing, Communications, and Control Symposium. AS-SPCC*. The unscented Kalman filter for nonlinear estimation (Lake Louise, Canada, 2000), pp. 153–158. doi:10.1109/ASSPCC.2000.882463
16. HE Rauch, C Striebel, F Tung, Maximum likelihood estimates of linear dynamic systems. AIAA J. **3**(8), 1445–1450 (1965)
17. P Müller, M Raitoharju, R Piché, in *17th International Conference on Information Fusion (FUSION)*. A field test of parametric WLAN-fingerprint-positioning methods (Salamanca, Spain, 2014), pp. 1–8
18. L Koski, Piché R, V Kaseva, S Ali-Lötty, M Hännikäinen, in *7th Workshop on Positioning Navigation and Communication (WPNC)*. Positioning with coverage area estimates generated from location fingerprints (Dresden, Germany, 2010), pp. 99–106. doi:10.1109/WPNC.2010.5653409
19. M Raitoharju, M Dashti, S Ali-Löytty, Piché R, in *International Conference on Indoor Positioning and Indoor Navigation (IPIN2012)*. Positioning with multilevel coverage area models (Sydney, Australia, 2012)
20. PJ Rousseeuw, AM Leroy, *Robust Regression and Outlier Detection*. (John Wiley & Sons, Inc, New York, NY USA, 1987)
21. J Ala-Luhtala, R Piché, Gaussian scale mixture models for robust linear multivariate regression with missing data. Commun. Stat. Simul. Comput (2014). doi:10.1080/03610918.2013.875565
22. AR Runnalls, Kullback-Leibler approach to Gaussian mixture reduction. IEEE Trans. Aerospace Electron. Syst. **43**(3), 989–999 (2007). doi:10.1109/TAES.2007.4383588
23. F Evennou, F Marx, E Novakov, in *Wireless Communications and Networking Conference, 2005 IEEE*. Map-aided indoor mobile positioning system using particle filter, vol. 4 (New Orleans, LA USA, 2005), pp. 2490–24944. doi:10.1109/wcnc.2005.1424905
24. Widyawan, M Klepal, S Beauregard. A backtracking particle filter for fusing building plans with PDR displacement estimates (Hannover, Germany, 2008), pp. 207–212. doi:10.1109/wpnc.2008.4510376
25. H Nurminen, A Ristimäki, S Ali-Löytty, R Piché, in *International Conference on, Indoor Positioning and Indoor Navigation (IPIN2013)*. Particle filter and smoother for indoor localization (Montbéliard-Belfort, France, 2013), pp. 137–146
26. B Ristic, S Arulampalam, N Gordon, *Beyond the Kalman Filter: Particle Filters for Tracking Applications*. (Artech House, Boston, MA USA, 2004)
27. S Särkkä, Unscented Rauch–Tung–Striebel smoother. IEEE Trans. Automatic Control. **53**(3), 845–849 (2008). doi:10.1109/TAC.2008.919531

14

Generation bidding game with potentially false attestation of flexible demand

Yuquan Shan[1], Jayaram Raghuram[2], George Kesidis[1,2*], David J Miller[2], Anna Scaglione[3], Jeff Rowe[4] and Karl Levitt[4]

Abstract

With the onset of large numbers of energy-flexible appliances, in particular plug-in electric and hybrid-electric vehicles, a significant portion of electricity demand will be somewhat flexible and accordingly may be responsive to changes in electricity prices. In the future, this increased degree of demand flexibility (and the onset of only short-term predictable intermittent renewable supply) will considerably exceed present level of uncertainty in day-ahead prediction of assumed inelastic demand. For such a responsive demand idealized, we consider a deregulated wholesale day-ahead electricity marketplace wherein bids by generators (or energy traders) are determined through a Nash equilibrium via a common clearing price (i.e., no location marginality). This model assumes the independent system operator (ISO) helps the generators to understand how to change their bids to improve their net revenue based on a model of demand-response. The model of demand-response (equivalently, demand-side bidding day ahead) is based on information from load-serving entities regarding their price-flexible demand. We numerically explore how collusion between generators and loads can manipulate this market. The objective is to learn how to deter such collusion, e.g., how to set penalties for significant differences between stated and actual demand, resulting in higher energy prices that benefit certain generators.

Keywords: Smart grids; Demand response; False demand attestation

Introduction

Game-theoretic approaches to the study of electricity markets, particularly under deregulation, have been explored for decades [1-3]. Recently, problems associated with variations of the optimum power-flow (OPF) problem [4,5] have been considered by several authors for price-elastic demand, e.g., [6,7]. Indeed, demand elasticity for electricity is motivated by the onset of potentially enormous load from flexible appliances, particularly plug-in electric and hybrid-electric vehicles, see, e.g., [8,9] and the references therein, where an electric vehicle represents electricity demand comparable to the rest of the household combined.

In typical models of 'two-settlement' (wholesale day-ahead and real-time) electricity markets commonly used in practice, day-ahead settlements largely account for the real-time supply. In real time, relatively minor adjustments are expected to be made to meet actual current demand. However, significant power supply may need to be quickly secured if there are failures in supply or if real-time (actual) demand is unexpectedly different than that predicted day-ahead, e.g., due to weather conditions. On the other hand, excess supply may have been secured if elastic demand is deferred in real time, or (often intermittent and confidently predictable only in the short-term) renewable supply is employed. So, in the future, actual real-time demand may be additionally difficult to predict day-ahead.

To account for significant flexible demand in the day-ahead market [10,11], we assume that the load-serving entities (or demand aggregators or just consumers) will inform the independent system operator (ISO) regarding their flexible demand - a kind of demand-side bidding. The ISO is assumed to provide sufficient information so

*Correspondence: gik2@psu.edu
[1]Department of Computer Science and Engineering, The Pennsylvania State University, University Park, State College, PA 16802, USA
[2]Department of Electrical Engineering, The Pennsylvania State University, University Park, State College, PA 16802, USA
Full list of author information is available at the end of the article

that the generators (suppliers) can modify their prices to improve upon their net utility, i.e., the ISO determines generation allocation sensitivities to bid prices of the OPF with demand response. In our setting, the ISO solves an OPF which minimizes the common clearing price of supply to the loads. The game is a 'discriminatory' sealed-bid auction in that the generators earn at the price they bid but in quantity determined by the ISO [1,2]. So to simplify matters herein, we do not consider strategic bidding by the generators wherein they may infer demand and/or the bidding strategies of their competitors via a probabilistic model, nor multipart bidding to account for start-up/ramp-up costs, secure contracts involving minimum and maximum supply per generator, and the like[a], nor peak-power consumption penalties [12,13]. Also, we assume each generator has a continuously differentiable and convex cost of supply [3], quadratic in particular [14]. Our model is related to noncooperative Cournot games of electricity markets reversing the direction of day-ahead markets to understand how demand responds to market clearing prices in the long term [15] - here we attempt to understand demand-response on the wholesale (generation-level) market.

How an energy trader (or load-serving entity colluding with one or more generators) can manipulate day-ahead electricity prices by placing uneconomic demand-side bids is studied in [16,17], not considering flexible demand. In this paper, we are interested in how the wholesale day-ahead market can be manipulated by colluding generators and load-serving entities, through misrepresentation of flexible demand by the latter.

This paper is organized as follows. We first specify our model of wholesale electricity market under demand response as a noncooperative game - an argument for the existence of its symmetric Nash equilibrium is given in the Appendix. We then use this model to numerically consider the effects of demand misrepresentation for the example of the benchmark IEEE 14-bus power system. The paper concludes with a summary and discussion of how such computations can inform penalties of significant discrepancies between attested and actual demands, causing the benefit of certain generators or energy traders.

Problem setup

Consider an ISO (single commodity retailer) with generators (suppliers) and loads (consumer demands). As in [6], supposing that each generator $g \in G$ sets its own price p_g\$ per kWh (or other energy commodity units). We model aggregate load to be linear in response to clearing price, P[b]

$$D(P) = (D_{\max} - D_{\min})\left(1 - \frac{P}{P_{\max}}\right) + D_{\min}, \qquad (1)$$

where $D_{\min}$ represents inflexible demand. Note that this is the simplest model of aggregate-demand flexibility and is commonly used in the literature. Although other more sophisticated model may be developed to fit cases in reality, such as plug-in electric and hybrid-electric vehicles, capturing flexibility of real-world aggregate loads is not a contribution of this paper.

Suppose that generators suffer quadratic cost of supply [14][c], so that the net utility/revenue[d] of the gth generator is

$$u_g\left(\underline{p}\right) = p_g S_g\left(\underline{p}\right) - c_g\left(S_g\left(\underline{p}\right)\right) = p_g S_g\left(\underline{p}\right) - a_g S_g^2\left(\underline{p}\right), \qquad (2)$$

i.e., different generators having different a_g parameters.

For noncooperative generator duopoly (two-players), discriminatory, single-part game with no distribution losses or constraints, and assuming that $p_1 \neq p_2$ near the interior Nash equilibria, we can find surprisingly complex plurality of Nash equilibria in closed form [18].

We assume that supply allocations are the result of the optimization of a supply network by a linear program. In electricity markets, the retailer (ISO) is sometimes also the distribution system.

Consider power flow in a power system with the generator set G, load set L, bus set B and branch set R.

- Let $G_b \subset G$ be the set of generators on bus b, each having generated power S_g, price per unit supply p_g, and minimum and maximum supply $S_g^{(min)}$ and $S_g^{(max)}$, respectively.
- Let $L_b \subset B$ be the set of loads on bus b, each having a demand D_l that depends on the clearing price P.
- For each bus $b \in B$, θ_b is its voltage angle.
- Finally, let $r_{i,j}$ be the branch connecting bus b_i and bus b_j, with $x_{i,j}$, the reactance of the branch, $P_{i,j}$, the power 'flowing' from b_j to b_i (if we neglect power loss on the transmission line, we get $P_{i,j} = -P_{j,i}$), and $c_{i,j}$, the maximum tolerable power on the branch.

Optimal power flow problem formulation

Assuming fixed generation prices $\underline{p}$, and associated clearing price P, the total load (consumer demand) is given by $D(P)$. The individual loads are assumed to be some fixed proportion of the total demand, i.e., $D_l(P) = \alpha_l D(P)$, $\forall l \in L$, where $\alpha_l > 0$ and $\sum_{l \in L} \alpha_l = 1$. The ISO solves a constrained optimization problem (linear programming in this case) in order to find $\underline{S} = [S_1, \ldots, S_{|G|}]$, the optimum power generating assignment for the generators, and $\underline{\theta} = [\theta_1, \ldots, \theta_{|B|}]$, the voltage angles on the buses, which minimize the cost charged by the generators. The constrained optimization problem is thus given by:

$$\min_{\underline{S},\underline{\theta}} \sum_{g\in G} p_g S_g$$

such that:

$$\begin{aligned}
&P_{i,j} = \frac{1}{x_{i,j}}(\theta_j - \theta_i) \quad \text{(DC approximation[4])}\\
&\sum_{g\in G_b} S_g - \sum_{l\in L_b} D_l = \sum_{k\in B} P_{k,b},\ \forall b \in B \quad \text{(power balance)}\\
&S_g^{(\min)} \le S_g \le S_g^{(\max)},\ \forall g \in G \quad \text{(generator limits)}\\
&-c_{i,j} \le P_{i,j} \le c_{i,j},\ \forall i,j \in B \quad \text{(branch limits)}
\end{aligned} \tag{3}$$

Here, neglecting the power loss in generation and transmission, we have

$$\sum_{g\in G} S_g = \sum_{l\in L} D_l.$$

Note that when all the generators have the same prices, there exist infinitely many optimal solutions in the above mathematical program. In this case, without violating other constraints, we equally divide allocation to each generator. We will pay particular attention to this situation in the following.

In the above formulation, we used fixed upper and lower bounds on the supply allocations S_g. In addition, the quadratic penalty term in the utility function (2) serves to model cost of supply allocation [3,14]. For a positive a_g and fixed generator prices $\underline{p}$, suppose generator g wants to ensure that its utility function is never smaller than some positive value $u_g^{(\min)}$, then this imposes lower and upper bounds on its supply allocation, given by

$$\frac{1}{2a_g}\left(p_g - \sqrt{p_g^2 - 4a_g u_g^{(\min)}}\right) \le S_g \le \frac{1}{2a_g}\left(p_g + \sqrt{p_g^2 - 4a_g u_g^{(\min)}}\right),$$

provided the generator price satisfies the condition $p_g \ge 2\sqrt{a_g u_g^{(\min)}}$. If $u_g^{\min} = 0$, then we observe that as a_g is made larger, i.e., as the cost of supply allocation increases, the maximum supply allocation (or capacity) decreases, and vice versa.

In practice for power-transmission circuits, thermal losses may determine edge (transmission line) capacities and costs, the latter typically in a power-flow dependent fashion, e.g., 'I^2R' losses (Section 3.1 of [4]). In order to focus on the bidding behavior among the generators, we neglect the power loss on the transmission lines; hence, the cost in power transmission is also neglected, as in the DC approximation.

Setup of generators' iterative game on a platform of demand response

We neither assume that each generator's cost of production is known to the ISO (i.e., the a_g terms), nor, equivalently, that the ISO chooses its allocations to the generators based on this (as in [6]). Besides, the cost function (hence the payoff function) of any generator is opaque to all its competitors, concerning this, our game can be regarded as an incomplete information game. In the following, denote as $\underline{S}(\underline{D}(P), \underline{p})$ the solution of the above optimal power flow allocation problem to determine supply allocations for *fixed* demands (which are based on the clearing price P) $\underline{D}(P) = \{D_l(P) \mid l \in L\}$ and fixed generator prices $\underline{p} = \{p_g \mid g \in G\}$. We propose the following iterative generator game wherein, for fixed generator prices, the clearing price and the consumer demands are adjusted iteratively until they converge to a fixed point. Then each generator $g \in G$ adjusts its price p_g, given the current price for all other generators $\underline{p}_{-g}$, such that its utility function $u_g\left(p_g, \underline{p}_{-g}\right)$ is increased. Given initial prices set by the generators $\underline{p}$, the iterative generator game proceeds as follows:

1. The ISO sets an *initial* mean price of supply (clearing price charged to all consumers), P, say just as the mean of the initial generator prices, p_g, $\forall g \in G$.
2. Determine the price-dependent loads $\underline{D}(P)$, where $D_l(P) = \alpha_l D(P)$, $\forall l \in L$.
3. ISO solves the optimal power flow allocation problem $\underline{S}(\underline{D}(P)), \underline{p})$ given fixed demands $\underline{D}$ and generation/supply costs $\underline{p}$.
4. ISO computes a new mean (clearing) price of supply, $P = \frac{\sum_{g\in G} S_g p_g}{\sum_{g\in G} S_g}$.
5. If the change in clearing price P is significant (larger than some threshold), then go back to Step 2; else continue to Step 6.
6. For the current set of generator prices, consistent supply allocations, loads, and clearing price have been found. Now each generator sets a new price of supply such that there is an increase in its utility function using one of the following two approaches: (i) *Best-response play action*: Each generator g sets a new price of supply based on (an estimate of)[e]

$$\arg\max_{p_g} p_g S_g\left(p_g; \underline{p}_{-g}\right) - c_g\left(S_g\left(p_g; \underline{p}_{-g}\right)\right), \tag{4}$$

where $c_g(x)$ is the cost of supply (assumed $= a_g x^2$ above).

(ii) *Better-response play action*: Each generator g obtains approximate left and right partial derivatives of its utility function with respect to its price p_g, i.e.,

$$\Delta u_g^+ = \frac{u_g\left(p_g + \epsilon, \underline{p}_{-g}\right) - u_g\left(p_g, \underline{p}_{-g}\right)}{\epsilon}$$
$$\Delta u_g^- = \frac{u_g\left(p_g, \underline{p}_{-g}\right) - u_g\left(p_g - \epsilon, \underline{p}_{-g}\right)}{\epsilon},$$

where $\epsilon \searrow 0$[f]. If the left and right derivatives have different signs (a non-differentiable point), then there are two possibilities. If $\Delta u_g^- > 0$ and $\Delta u_g^+ < 0$, the current price p_g is a local maximum and there is no need to change p_g. If $\Delta u_g^- < 0$ and $\Delta u_g^+ > 0$, the current price is a local minimum. In this case, we increase p_g by a small value ζ if $|\Delta u_g^+| > |\Delta u_g^-|$; otherwise, we decrease p_g by ζ. In case the derivatives have the same sign (may still be a non-differentiable point), we increase p_g by ζ if both derivatives are positive and decrease p_g by ζ if both derivatives are negative. The step ζ should increase the price by a small value such that there is an increase in the value of the utility function. It should not make large changes to the price like the best-response play action[g].

7. Exit if there is no change in the generator prices (i.e., if an equilibrium set of prices is obtained); Else go back to Step 1.

We implicitly assume in Step 6 that the ISO helps to compute estimates of supply sensitivities to bid for each generator, $\partial S_g / \partial p_g$. This can be done numerically through difference quotients. The result is a Jacobi iteration representing a 'better response' game [19,20]. Alternatively, the ISO could iteratively compute the 'best response' price for each generator, but this would require knowledge of each generator's net utility.

Numerical study - benchmark IEEE 14-bus power system

We study the iterative generator game described above with the benchmark IEEE 14-bus power system that has 3 generators and 11 loads as shown in Figure 1 [21]. We consider the scenario where there is no cost/loss in delivering power, and the branch capacities are set to fixed values. Here we only focus on three-player best-response game; discussion of a better-response, two-player game for the benchmark IEEE 9-bus system, with numerical results, is given in [22].

For this 14-bus power system, each of the three generators can produce maximum power $\underline{S}_g^{(\max)} = 150$ MW, and minimum power of 0 MW (again, we do not consider ramp-up/down issues herein). The constants in $D(P)$, our model for the total load, were chosen as the possible maximum/minimum power provided by the generators, that is, $D_{\max} = 450$ MW, $D_{\min} = 0$ MW. The total load $D(P)$ is assumed to be proportionally divided among the individual loads, e.g., $\alpha_l = \frac{1}{|L|}$, $\forall l \in L$. The maximum clearing

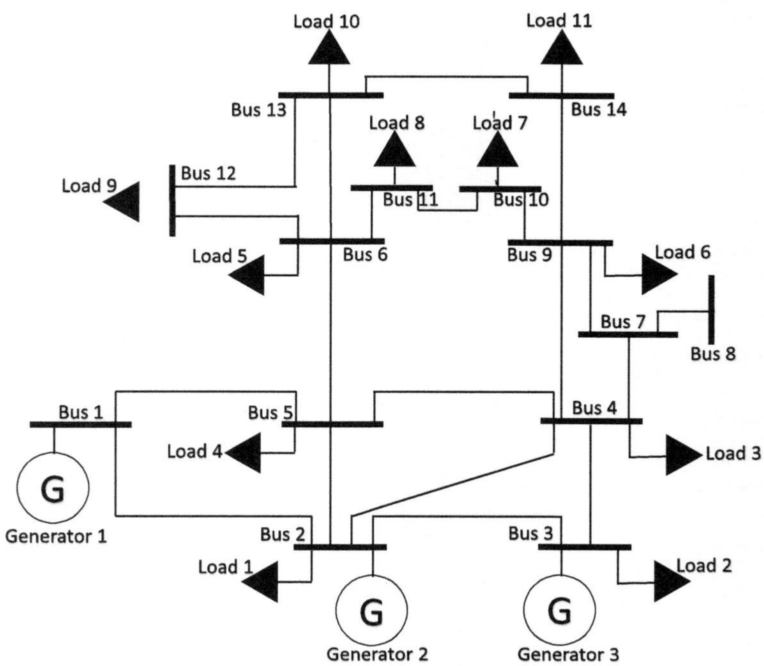

Figure 1 **IEEE 14-bus power system having an asymmetric topology with 3 generators and 11 loads.**

price is set to $P_{\max} = 5$; for clearing prices $P > P_{\max}$, the flexible demand is 0. Other data for this power system can be found in [23]. The constants (a_g) in the utility function (2) of generators are set to 0.02, 0.025, and 0.03, respectively.

We had all three generators participating in the competition, using randomly chosen starting price-points (play-action bids). Most starting points resulted in convergence to the symmetric Nash equilibrium point $\underline{p}^* = [3.53, 3.53, 3.53]$ as shown in Figure 2. Additionally, we numerically found other symmetric Nash equilibrium points forming a line segment, as shown in Figure 3. In the Appendix, we prove the existence of a plurality of symmetric Nash equilibrium points and the lack of asymmetric ones, under certain conditions for our test systems. In [22], we describe a numerical study with similar results for the benchmark IEEE 9-bus power system (having a symmetric topology unlike the 14-bus system).

Effects of false demand attestations

We also numerically explored how false demand attestations may impact the revenues of generators. It is possible that demand aggregators and generators may collude, or noneconomic demand-side bids may be placed by energy brokers, to cause some generator(s) to receive more revenue while possibly reducing the revenue of others.

We first explain that by considering the convex quadratic cost model of (2) how modifying power allocations to generators may improve their revenues. In particular, we can demonstrate in some situations, *decreasing* power consumption could be beneficial for some generators, especially for those with relatively higher cost parameter a_g. Assume all the generators bid the same Nash equilibrium price, p, and equally share the aggregate demand $D(p)$. Clearly, the clearing price is also p and each generator has allocation

$$S_g = \frac{1}{|G|}\left[(D_{\max} - D_{\min})\left(1 - \frac{p}{P_{\max}}\right) + D_{\min}\right].$$

According to (2), the revenue of generator g will be maximized when its allocation is

$$S_g' = \frac{p}{2a_g}.$$

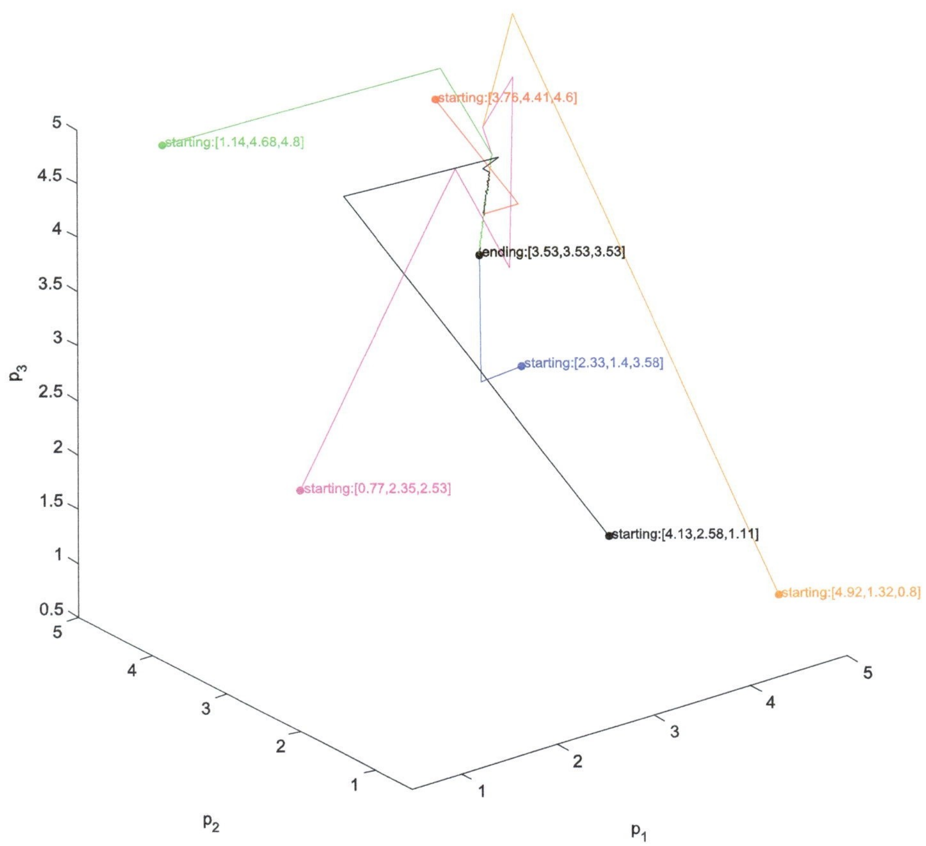

Figure 2 Sample generator-price trajectories which converge to [3.53, 3.53, 3.53]. Trajectories converge to this maximal Nash equilibrium point from most initial price points.

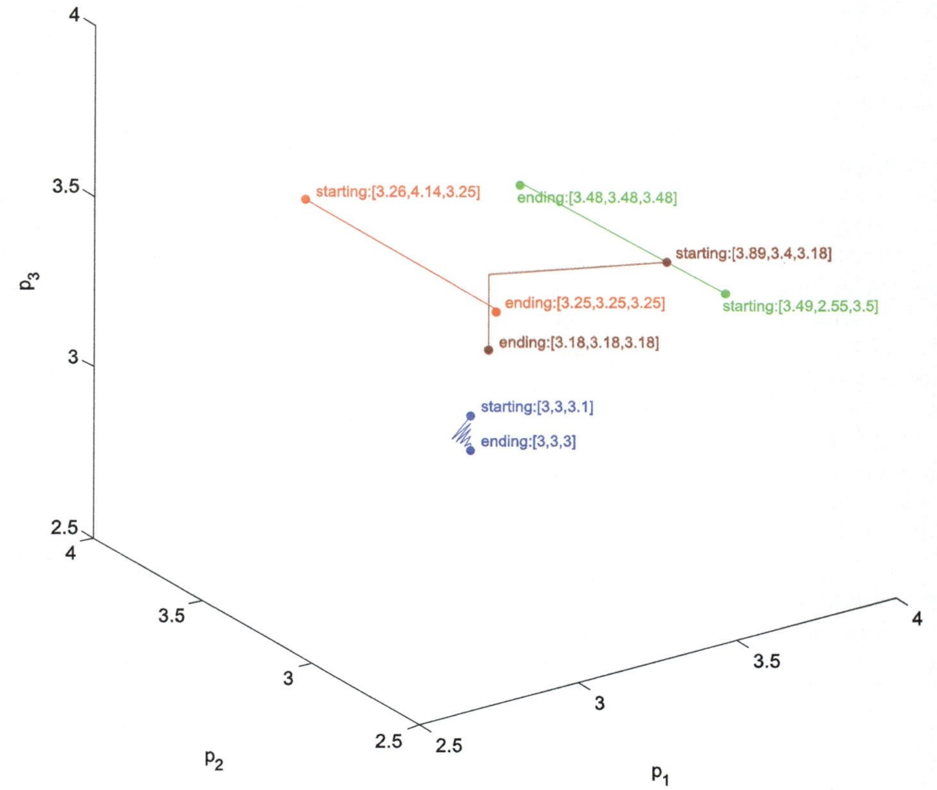

Figure 3 Sample trajectories which converge to non-maximal, symmetric Nash equilibrium points.

Thus, when the current allocation of generator g, S_g, is larger than its optimal allocation, S'_g, *less* allocation brings more revenue. For the same reason, if $S_g < S'_g$ then less allocation S_g leads to financial loss. That is, the following inequality at symmetric Nash equilibrium p will determine whether increased or reduced allocation will be beneficial to a given generator g:

$$S_g > S'_g \iff \frac{1}{|G|}$$
$$\left[(D_{\max} - D_{\min})\left(1 - \frac{p}{P_{\max}}\right) + D_{\min}\right] > \frac{p}{2a_g}$$

Example with numerical study

In the benchmark IEEE 14-bus power system described above, we thereby predicted that decreasing allocation will bring generators 1, 2, and 3 more profit when the current equilibrium price is lower than 2.72, 3, and 3.21, respectively. We verified these thresholds numerically as follows. Denote the factor by which load j has modified its demand as

$$\Gamma_j = \frac{D_{j,false}}{D_{j,true}}.$$

Assume load 1 reports a false demand which is greater than its actual demand, $\Gamma_1 > 1$, while the other loads are true, i.e., $\Gamma_i = 1, \forall i \neq 1$. False demands were used to determine prices while true demands were used when computing the generator net utilities via the power circuit.

The generation net utilities at Nash equilibrium are depicted in Figure 4 as a function of $\Gamma_1 \geq 1$. The upper-left plot depicts the case with (maximal) Nash equilibrium prices $[3.53, 3.53, 3.53]$, in which no generator benefits when $\Gamma_1 > 1$. The upper-right plot depicts the case with equilibrium prices $[3.1, 3.1, 3.1]$, in which only generator 1 will benefit. Lower-left plot depicts the case with equilibrium price $[2.8, 2.8, 2.8]$, in which only generators 2 and 3 will benefit. Finally, the lower-right depicts the case with equilibrium prices $[2.6, 2.6, 2.6]$, in which all the generators will benefit. The effects of false demand attestation in IEEE 9-bus power system is also studied, similar results are observed.

Discussion: when to penalize

Recall the previous discussion of complications in estimating demand day-ahead due to the presence of flexible demand, renewables, and storage devices. This additionally complicates detection of *false* attestations of

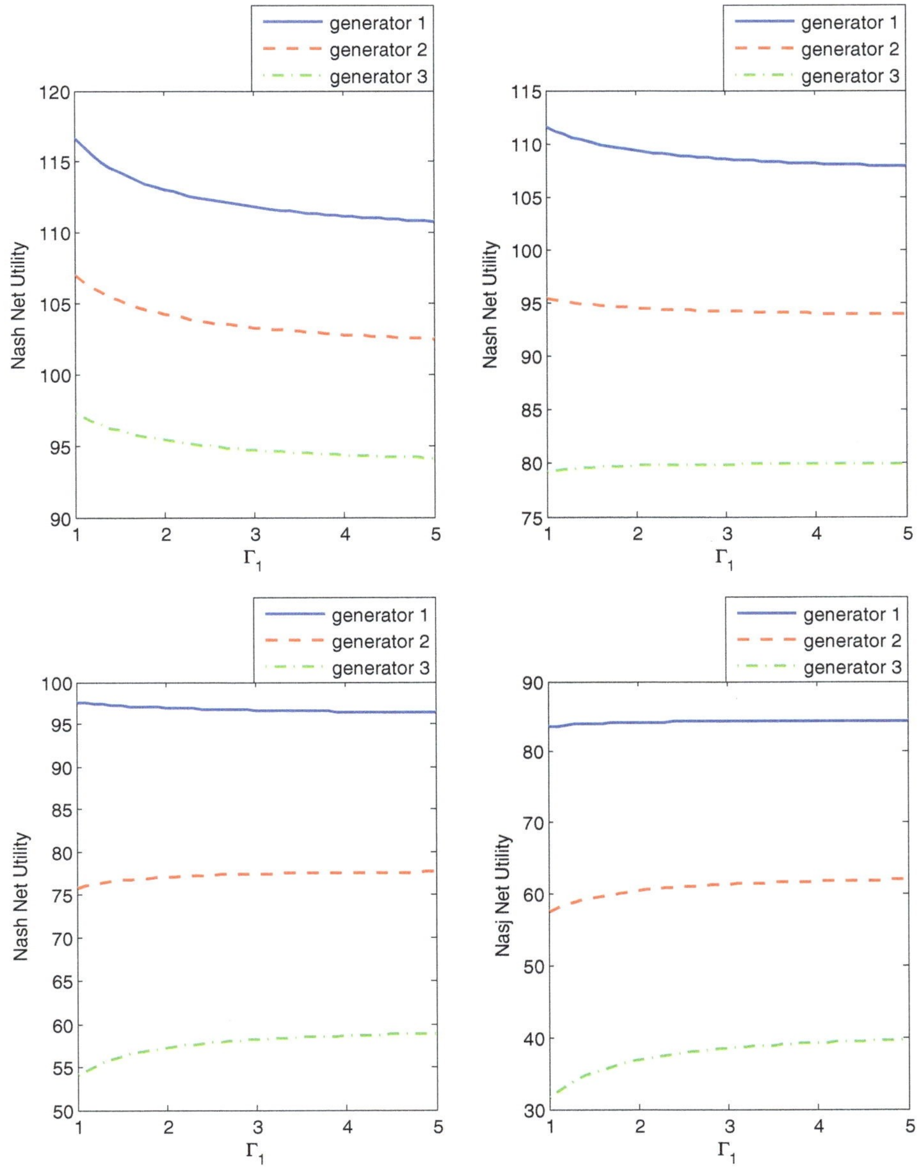

Figure 4 Nash-equilibrium net utilities vs Γ_1 plots under different equilibrium prices $\underline{p}$, clockwise from upper left: $\underline{p} = [3.53, 3.53, 3.53]$, $[3.1, 3.1, 3.1]$, $[2.6, 2.6, 2.6]$, $[2.8, 2.8, 2.8]$.

demand. So, an ISO may need to detect persistent demand-attestation 'biases' over time and let that trigger associated demand-side penalties or reservation costs. General incomplete-information game-theoretic frameworks include Bayesian games and hypergames, e.g., [24]. In both VCG and PSP auctions (e.g., [25,26]), issues of truthfulness in the disclosed bids are considered, i.e., reflecting actual demand response (by marginal valuation). More prosaic approaches simply interpolate and extrapolate from presumed honest bids (by (amount, price)) to obtain a complete estimate of other players' demand response. These frameworks are applicable to iterated (sequential) adversarial (noncooperative) games with or without leaders. Generally, estimates are greatly simplified under the assumption that player strategies are time (play-action iteration) invariant.

The magnitude of such penalties would be informed by calculations such as those described above, i.e., by determining which generators benefit and by how much. By penalizing in this way, the motivation for deliberately false demand-attestations are disincentivized.

Conclusions

In this paper, we studied a day-ahead wholesale electricity market under demand response as a noncooperative game under certain assumptions, in particular no location marginality. We argued for the existence of symmetric Nash equilibria and numerically studied the benchmark IEEE 14-bus power system. We then considered the effects of false demand attestation and again used the 14-bus system to show how generators may (or may not) benefit by false attestation of demands. Such calculations may be used to penalize demand attestations that show persistent bias to the benefit of certain generators.

Endnotes

[a]For example, in [5], an affine single-part bid and associated make-whole "uplift" payments that are part of a joint integer-programming unit commitment and continuous-linear OPF (economic dispatch) problem was considered. In practice, the generation unit-commitment decisions are typically made first, e.g. simply based on mean cost of supply over the capacity range of the generator including its ramp-up costs.

[b]Herein, just the mean price of supply - again, we assume no location marginality.

[c]We assume quadratic cost for tractability in the duopoly studied in [18]. An alternative cost structure could be asymptotic to a maximum, e.g., $c(0)/(s - s_{\max})$ where $c(0)$ is the cost of keeping the generator online even if zero supply is being delivered. In this paper, we do not consider ramp up/ramp-down constraints for generators.

[d]If the net consumer utility is collectively $V(D) - PD$, then for the linear demand-response to price of (1), the utility is quadratic, concave and increasing, $V(D) = (P_{\max}/2)(D_{\max}^2 - (D_{\max} - D)^2)/(D_{\max} - D_{\min})$ for $D_{\min} \leq D \leq D_{\max}$.

[e]Since an explicit closed-form solution is difficult to obtain, here we simply search for objective-optimizing prices by first defining an evenly partitioned price-set from 0 to 5.

[f]We chose a value of $\epsilon = 10^{-6}$.

[g]We chose ζ as follows. Starting with a small trial value of $\zeta = 0.005\,p_g$, if $u_g\left(p_g + \zeta, \underline{p}_{-g}\right) > u_g\left(p_g, \underline{p}_{-g}\right)$ we accept the value of ζ; Else ζ is decreased by a factor of 2 iteratively until $u_g\left(p_g + \zeta, \underline{p}_{-g}\right) > u_g\left(p_g, \underline{p}_{-g}\right)$.

Appendix: symmetry of Nash equilibrium points

In this Appendix, we simply and directly derive conditions on the parameters of our power-system model (in DC approximation) for existence and symmetric Nash equilibria and nonexistence of asymmetric ones. The claims can be directly extended to non-quadratic cost functions for generation at the expense of closed-form expressions for the conditions of the claims in terms of the model parameters.

Proof of existence of symmetric Nash equilibrium points

For existence of a symmetric Nash equilibrium, we need show that $A \cap B \neq \emptyset$, where A is the set of prices, p, in symmetric price vectors $\underline{p} = [p, p, \ldots, p]$ from which decreasing any one price component (of a given generator) is not profitable (for that generator), and similarly B is the set of prices in price vectors from which increasing a price component is not profitable.

Decreasing a price component is not profitable, A

Given a potential symmetric equilibrium price-vector $\underline{p}$ with equal components p, for generator i if $p \geq \hat{p}_i := D^{-1}\left(S_i^{\max}\right)$, i.e., the total demand $D(p) \leq S_i^{(\max)}$, then if generator i decreases its price $p_i < p$, it will be assigned all of D until $D > S_i^{(\max)}$. Since the net revenue of generator i is assumed $u_i(p) = D(p)p - a_i D(p)^2$, i.e., quadratic in p, then without loss of generality we can take $D_{\min} = 0$ so that generator i will receive its maximum revenue $u_i^{(\max)}$, when it bids price

$$\tilde{p}_i = \frac{P_{\max}^2 - 2a_i D_{\max} P_{\max}}{2(P_{\max} + a_i D_{\max})}. \tag{5}$$

However, $\tilde{p}_i$ is feasible only if $\hat{p}_i \leq \tilde{p}_i \leq p$. Therefore, for generators i with $\hat{p}_i \leq p$, the best response p_i^* by decreasing (case A) price from p is

$$u_{A,i}^*(p) = \begin{cases} \hat{p}_i S_i^{(\max)} - a_i\left(S_i^{(\max)}\right)^2 & \text{if } \tilde{p}_i < \hat{p}_i \left(p_i^* = \hat{p}_i\right) \\ pD(p) - a_i D(p)^2 & \text{if } \tilde{p}_i > p \left(p_i^* = p\right) \\ u_i^{(\max)} & \text{Otherwise} \left(p_i^* = \tilde{p}_i\right) \end{cases} \tag{6}$$

Considering equally divided demand at $\underline{p}$ among the $|G|$ generators, define $A_{i,1}$ as follows,

$$A_{i,1} = \left\{ p : p\frac{D(p)}{|G|} - a_i\left(\frac{D(p)}{|G|}\right)^2 \geq u_{A,i}^*(p),\ \hat{p}_i \leq p \right\}.$$

On the other hand, if $p < \hat{p}_i$, if generator i decreases its price from p, its allocation will always be its maximum power, $S_i^{(\max)}$. Therefore, in this case, the optimal revenue in best response for generator i is

$$u'_{A,i}(p) = pS_i^{(\max)} - a_i\left(S_i^{(\max)}\right)^2. \tag{7}$$

Define

$$A_{i,2} = \left\{ p : p\frac{D(p)}{|G|} - a_i\left(\frac{D(p)}{|G|}\right)^2 \geq u'_{A,i}(p),\ \hat{p}_i \geq p \right\},$$

$A_i = A_{i,1} \cup A_{i,2}$, and $A = A_1 \cap A_2 \cap \ldots \cap A_{|G|}$. So, no generators will decrease their prices in the next round when they all bid $\underline{p} \in A$ in the current round.

Increasing price is not profitable, B

Let $S_{-i}^{(\max)}$ denote the sum of maximum capacities of all the generators except for generator i. Given a potential equilibrium price $\underline{p}$, if $p \le \bar{p}_i := D^{-1}\left(S_{-i}^{(\max)}\right)$, then when generator i increase its price, its allocation will not immediately drop to zero. Assume generator i increases its price to p^* from p, and thereby its allocation drops to S_i. Again, w.l.o.g., we assume $D_{\min} = 0$ and have $p' = \left(pS_{-i}^{(\max)} + p_i S_i\right) / \left(S_{-i}^{(\max)} + S_i\right)$ and $S_{-i}^{(\max)} + S_i = D_{\max}\left(1 - \frac{p'}{P_{\max}}\right)$. By simplifying we get

$$p_i S_i = \left(S_{-i}^{(\max)} + S_i\right)\left(1 - \frac{S_{-i}^{(\max)} + S_i}{D_{\max}}\right) P_{\max} - pS_{-i}^{(\max)}. \tag{8}$$

Combining (8) and (2), the revenue of generator i is quadratic in S_i and maximum at a point $\bar{S}_i$ that is independent of p:

$$\bar{S}_i = \frac{D_{\max} - 2S_{-i}^{(\max)}}{2\left(P_{\max} + a_i D_{\max}\right)} P_{\max},$$

with maximum value $u_i^{(\max)\prime}$ that's linear in p.

Let S_i° be the allocation of generator i, when it bids maximum price, $P_{\max}$. At potential symmetric equilibrium point $\underline{p}$, if generator i satisfies $p \le \bar{p}_i$, its optimal revenue in best response can be expressed as

$$u'_{B,i}(p) = \begin{cases} \left(D(p) - S_{-i}^{(\max)}\right)p - \left(D(p) - \left(S_{-i}^{(\max)}\right)^2\right. & \text{if } \bar{S}_i \ge D(p) - S_{-i}^{(\max)} \\ u_i^{(\max)\prime} & \text{if } S_i^{\check{r}} < \bar{S}_i < D(p) - S_{-i}^{(\max)} \\ S_i^{\check{r}} P_{\max} - a_i \left(S_i^{\check{r}}\right)^2 & \text{if } \bar{S}_i \le S_i^{\check{r}} \end{cases}$$

Note that the three conditions can be expressed in terms of best-response price p_i^* (recall S_i° is also known function of p).

Considering equal allocations at symmetric price point $\underline{p}$ as above, define

$$B_{i,1} = \left\{p : p\frac{D(p)}{|G|} - a_i\left(\frac{D(p)}{|G|}\right)^2 \ge u'_{B,i}(p),\ \hat{p}_i \ge p\right\}.$$

When $p > \bar{p}_i := D^{-1}\left(S_{-i}^{(\max)}\right)$, once generator i increases its price, its allocation will immediately drop to zero, and so will its net revenue. So, define

$$B_{i,2} = \left\{p : p\frac{D(p)}{|G|} - a_i\left(\frac{D(p)}{|G|}\right)^2 \ge 0,\ \bar{p}_i < p\right\}.$$

As above, also define $B_i = B_{i,1} \cup B_{i,2}$ and $B = B_1 \cap B_2 \cap \ldots \cap B_{|G|}$. We have therefore proven the following claim.

Claim: If generator parameters $\left\{a_i, S_i^{(\max)}\right\}_{i \in G}$ and aggregate demand parameter $D_{\max}, P_{\max}$ are such that $A \cap B \ne \emptyset$, then there exists price-symmetric Nash equilibria $\underline{p} = [p, p, \ldots, p]$ $(p \in A \cap B)$ with allocation symmetry $\bar{S}_g = D(p)/|G|$ for all $g \in G$.

As an example of a set of parameters where such Nash equilibria exist, we refer to the benchmark IEEE 14-bus power system in DC approximation described above.

Proof of nonexistence of asymmetric Nash equilibrium points

We set $D_{\min} = 0$ and the minimum power allocation of each generator to zero both w.l.o.g. Instead of simply proving nonexistence of asymmetric Nash equilibrium points for the IEEE 14-bus power system under DC approximation, we herein give the conditions for the nonexistence of asymmetric Nash equilibrium points in more general form, which is useful in analyzing the systems with more than three generators.

Case of a unique generator having the lowest price

Recall the clearing price is $P = \sum_i p_i S_i / \sum_g S_g$.

Claim: If there is an asymmetric equilibrium point with the unique lowest price, say p_i, then the total demand $D(P) \le S_i^{(\max)}$.

Proof. For a proof by contradiction, assume a Nash equilibrium pricing point where $D(P) > S_i^{(\max)}$ and p_i is the lowest price. Generator i can increase its price by a small amount while maintaining lowest-price status and the condition $D(P) > S_i^{(\max)}$. Doing so, the allocation S_i of generator i will not be affected after the increase of price but the revenue of generator i will increase. Thus, this hypothetical equilibrium pricing point is not Nash, a contradiction. □

According to the claim above, in such equilibrium, only the generator (i) who bids the only lowest price (p_i) receives nonzero allocation. As a result, the price it bids is the clearing price, $P = p_i$. Thus, $S_i = D(p_i)$ and so we can write the revenue of generator i having the lowest price as follows,

$$\begin{aligned} u_i(p_i) = &-\left(\frac{D_{\max}}{P_{\max}} + \frac{a_i D_{\max}^2}{P_{\max}^2}\right) p_i^2 \\ &+ \left(D_{\max} + \frac{2a_i D_{\max}^2}{P_{\max}}\right) p_i - a_i D_{\max}^2. \end{aligned} \tag{9}$$

Based on (9), generator i can unilaterally change its price p_i to increase its revenue until

$$\begin{aligned} p_i = \hat{p}_i &= \frac{P_{\max} + 2a_i D_{\max}}{2P_{\max} + 2a_i D_{\max}} P_{\max} \\ &= \arg\max_{p_i} U_i(p_i) < P_{\max} \end{aligned}$$

or approaches (is 'infinitely close' to) the second lowest price among generators (see the next case).

On the other hand, given $p_i = \hat{p}_i$, the other generators currently receive no allocation and have no revenue and will consider lowering their prices p_j for $j \neq i$ in the next round to $p_j \leq \hat{p}_i$ in the next round. Doing so, the result will be generators i and j will fairly share the allocation, i.e.,

$$S_i = S_j = \frac{1}{2} D_{\max} \left(1 - \frac{\hat{p}_i}{P_{\max}}\right) > 0.$$

By (2), at price $\hat{p}_i$, generator j will have positive revenue if and only if

$$0 < S_j < \frac{\hat{p}_i}{a_j}. \tag{10}$$

So, generator j has *no motive* to reduce its price p_j to $\hat{p}_i$ if

$$\frac{1}{2} D_{\max} \left(1 - \frac{\hat{p}_i}{P_{\max}}\right) \geq \frac{\hat{p}_i}{a_j} \iff \left(\frac{a_j}{2} - 2a_i\right) \geq \frac{P_{\max}}{D_{\max}} \tag{11}$$

Since condition (11) does not hold for all $i \neq j$ (because the right-hand side is always negative) for our IEEE 14-bus power system under DC approximation, when $\underline{a} = [.02, .025, .03]$), asymmetric Nash equilibria with unique lowest prices do not exist.

Proof of the nonexistence of asymmetric Nash equilibrium point with > 1 generators at the lowest price

Let generators $i \in M \subset G$ have the lowest price $\hat{p}$.

Claim: If there is an asymmetric equilibrium point with $|G| > |M| \geq 2$, the total demand $D(P) \leq |M| \min_{i \in M} \left\{S_i^{(\max)}\right\}$ where the clearing price $P = p_i$ for all $i \in M$.

Proof. Similar to the proof for the first claim, if $D(P) > |M| \min_{i \in M} \left\{S_i^{(\max)}\right\}$ and $S_i^{(\max)} = \min_{j \in M} \left\{S_j^{(\max)}\right\}$, then i will increase its utility by increasing its price a small amount without reducing its allocation, which makes this price point not Nash. □

Again, this claim means that $S_k = 0$ and generator k has zero revenue for all $k \in M$. As with (10), assuming equal allocations among all generators $i \in$M, all generators $i \in M$ are 'satisfied' with nonnegative allocations if

$$\frac{1}{|M|} D_{\max} \left(1 - \frac{\hat{p}}{P_{\max}}\right) \leq \min_{i \in M} \frac{\hat{p}}{a_i}. \tag{12}$$

To give any generator $k \notin M$ no motive to reduce their price to $\hat{p}$, we require

$$\frac{1}{|M|+1} D_{\max} \left(1 - \frac{\hat{p}}{P_{\max}}\right) \geq \frac{\hat{p}}{a_k}. \tag{13}$$

Combining (12) and (13), an asymmetric Nash equilibrium, with more than one generator with minimal price, exists if

$$\min_{k \notin M} a_k \geq \frac{|M|+1}{|M|} \max_{i \in M} a_i. \tag{14}$$

Note that (14) is necessary but not sufficient for existence of asymmetric Nash equilibrium.

Again, for our IEEE 14-bus power system under DC approximation, if utility cost parameters $[a_1, a_2, a_3] = [.02, .025, .03]$ and $|M| = 2$, then (14) is not satisfied; hence, no such asymmetric Nash equilibrium exists.

Competing interests
The authors declare that they have no competing interests.

Authors' contributions
YS and GK established the problem formulation, performed false attestation analysis and carried out numerical experiment and analysis. The other authors critiqued the paper; and YS and GK modified the paper accordingly. All authors read and approved the final manuscript.

Acknowledgements
This research is supported by grant NSF CCF-1228717.

Author details
[1]Department of Computer Science and Engineering, The Pennsylvania State University, University Park, State College, PA 16802, USA. [2]Department of Electrical Engineering, The Pennsylvania State University, University Park, State College, PA 16802, USA. [3]Electrical Engineering, Arizona State University, Tempe, AZ 85287, USA. [4]Department of Computer Science, University of California, Davis, CA 95616, USA.

References
1. AK David, F Wen, Strategic bidding in competitive electricity markets: a literature survey. Power Engineering Society Summer Meeting. **4**, 2168–2173 (2000)
2. N Fabra, N-H von der Fehr, D Harbord, Designing electricity auctions. RAND Jounral of Economics. **31**, 23–46 (Spring 2014)
3. EJ Anderson, H Xu, Supply function equilibrium in electricity spot markets with contracts and price caps. Journal of Optimization Theory and Applications. **124**(2), 257–283 (Dec. 2005)
4. AJ Wood, BF Wollenberg, *Power generation operation and control*, 2nd edn. (John Wiley & Sons, Inc., 605 Third Avenue, New York, NY, 1996), pp. 108–109
5. G Wang, SV Shanbhag, S Meyn, in *Proc. IEEE CDC*. On Nash equilibria of duopolistic power markets subject to make whole uplift (IEEE, Maui, Dec. 2012). doi:10.1109/CDC.2012.6425983
6. M Roozbehani, M Dahleh, S Mitter, in *Proc. IEEE SmartGridComm*. Dynamic pricing and stabilization of supply and demand in modern electric power grids (IEEE, Gaithersburg, MD, 4-6 Oct. 2010)
7. AH Mohsenian-Rad, A Leon-Garcia, Distributed internet-based load altering attacks against smart power grids. IEEE Trans. on Smart Grid. IEEE. **2**(4), 667-674 (2011)
8. M Alizadeh, A Scaglione, A Goldsmith, G Kesidis, in *Proc. IEEE CDC*. Scalable management of large populations of flexible appliances (IEEE, Los Angeles, 2014). Longer version to appear in IEEE Trans. Power Systems. (TPWRS)
9. H Lu, G Pang, G Kesidis, in *Proc. IEEE SmardGrid Comm*. Automated scheduling of deferrable PEV/PHEV load by power-profile unevenness (IEEE, Vancouver, Oct. 2013)
10. C Chen, S Kishore, Z Wang, M Alizadeh, A Scaglione, in *Proc. IEEE Power and Energy Society (PES) General Meeting*. A Cournot game analysis on market effects of queuing energy request as demand response (IEEE, San Diego, CA, July 2012)

11. MG Vaya, G Anderson, in *Proc. Int'l Conf. on European Energy Market (EEM)*. Optimal bidding strategy of a plug-in electric vehicle aggregator in day-ahead electricity markets (IEEE, Stockholm, May 2013)
12. Duke utility bill tariff (2012). http://www.considerthecarolinas.com/pdfs/scscheduleopt.pdf
13. Fort Collins Coincident Peak (2013). http://www.fcgov.com/utilities/business/rates/electric/coincident-peak
14. D Bienstock, M Chertkov, S Harnett, Chance-constrained optimal power flow: risk-aware network control under uncertainty. SIAM Review. **56**(3), 461–495 (2014)
15. B Bobbs, Linear complementarity models of Nash-Cournot Competition in bilateral and POOLCO Power Market. Utilities Policy. **5**(3/4), 194–202 (1996)
16. SD Ledgerwood, PR Pfeifenberger, Using virtual bids to manipulate the value of financial transmission rights. The Electricity Journal. **26**(9), 9–25 (2013)
17. C LoPrete, WW Hogan, *Manipulation of day-ahead electricity prices through virtual bidding. IAEE Energy Forum*. (IAEE, Cleveland, OH, 2014), pp. 31–32
18. G Kesidis, C Griffin, DJ Miller, A marketplace game with neither distribution costs nor distribution-capacity constraints. PSU CSE Dept. Technical Report. **CSE-13-013** (Nov. 2013)
19. Y Jin, G Kesidis, Dynamics of usage-priced communication networks: the case of a single bottleneck resource. IEEE/ACM Trans. Networking. IEEE Press, Piscataway, NJ, USA. **13**(5), 1041–1053 (2005)
20. JS Shamma, G Arslan, Dynamic fictitious play, dynamic gradient play, and distributed convergence to Nash equilibria. IEEE Trans. Auto. Contr. **50**(3), 312–327 (2005)
21. RD Zimmerman, CE Murillo-Sánchez, RJ Thomas, MATPOWER: steady-state operations, planning and analysis tools for power systems research and education. IEEE Trans. Power Syst. **26**(1), 12–19 (2011). See also http://www.pserc.cornell.edu/matpower
22. Y Shan, J Raghuram, G Kesidis, C Griffin, K Levitt, DJ Miller, J Rowe, A Scaglione, Generation bidding game with flexible demand. Arxiv.org Technical Report. **1408.6689** (Aug. 2014). Available at http://arxiv.org/abs/1408.6689
23. IEEE 14-Bus Power-Flow Test Case - System Data. http://www.ee.washington.edu/research/pstca/pf14/pg_tca14bus.htm
24. Y Sasaki, K Kijima, Hypergames and Bayesian games: a theoretical comparison of the models of games with incomplete information. J Syst Sci Complex. **25**, 720–735 (2012)
25. S Bhattacharya, K Kar, JH Chow, A Gupta, in *Proc. Int'l Conf. on Network Games, Control and Optimization (NeTGCoop)*. Progressive second price auctions with elastic supply for pev charging in the smart grid (Trento, Italy, Oct. 2014)
26. X Zhou, H Zheng, in *Proc. IEEE INFOCOM*. Trust: a general framework for truthful double spectrum auctions, (2009), pp. 999–1007

15

Class-specific Gaussian-multinomial latent Dirichlet allocation for image annotation

Zhiming Qian*, Ping Zhong and Runsheng Wang

Abstract

Image annotation has been a challenging problem due to the well-known semantic gap between two heterogeneous information modalities, i.e., the visual modality referring to low-level visual features and the semantic modality referring to high-level human concepts. To bridge the semantic gap, we present an extension of latent Dirichlet allocation (LDA), denoted as class-specific Gaussian-multinomial latent Dirichlet allocation (csGM-LDA), in an effort to simulate the human's visual perception system. An analysis of previous supervised LDA models shows that the topics discovered by generative LDA models are driven by general image regularities rather than the semantic regularities for image annotation. To address this, csGM-LDA is introduced by using class supervision at the level of visual features for multimodal topic modeling. The csGM-LDA model combines the labeling strength of topic supervision with the flexibility of topic discovery, and the modeling problem can be effectively solved by a variational expectation-maximization (EM) algorithm. Moreover, as natural images usually generate an enormous size of high-dimensional data in annotation applications, an efficient descriptor based on Laplacian regularized uncorrelated tensor representation is proposed for explicitly exploiting the manifold structures in the high-order image space. Experimental results on two standard annotation datasets have shown the effectiveness of the proposed method by comparing with several state-of-the-art annotation methods.

Keywords: Image annotation; Latent Dirichlet allocation; Variational EM; Uncorrelated tensor representation; Laplacian regularization

1 Introduction

Automatic image annotation is a challenging work of tasks related to understanding what we see in a visual scene due to the well-known semantic gap [1]. Given an input image, the goal of image annotation is to assign meaningful tags to the image aiming at summarizing its visual contents. Such methods are becoming more and more important given the growing collections of both private and publicly available images. However, challenges for these methods often lie in three aspects: the inter-tag similarity problem that different tags may have similar visual contents, the tag locality problem that most tags are only related to their corresponding semantic regions, and the intra-tag diversity problem that the relevant regions for each tag at different images can be different.

The inter-tag similarity problem reveals the fact that the visual similarity does not always guarantee the semantic similarity, which in general is conflicting with the inherent assumption of many image annotation methods, e.g., some relevant methods [2,3] that perform tag propagations according to their visual similarities. To cope with this problem, it is emergent to develop more discriminative visual features that can be used to separate various visual contents for different tags. However, traditional vector representations in the form of bag-of-features or bag-of-words, such as the visual descriptor that quantizes SIFT local features [3] and the colored pattern appearance model (CPAM) [4], are usually incompetent for the intention. The reason is that these features usually ignore the high-order characteristics of natural images and might lead to the curse of dimensionality problem when requiring a relatively discriminative representation for describing the complex visual world. In practice, an image is intrinsically a two-dimensional or high-order tensor. To

*Correspondence: qianzhiming@nudt.edu.cn
College of Electronic Science and Engineering, National University of Defense Technology, Deya Road, 410073 Changsha, China

fairly evaluate the high-order characteristics of image contents, tensor representations [5,6], which can explicitly describe the multiple interrelated restrictions, might allow us to avoid the problem of curse of dimensionality.

To tackle the tag locality problem, one may employ local image features instead of holistic image features to describe the visual contents of a certain tag. The work in [7] considered each image as a bag of multiple segmented regions and predicted the tag of each region by a multiclass bag classifier. This method, however, heavily depends on the segmentation performance, which is very sensitive to the image noise. Recently, implicit image representations attract much attention on describing local regions. To reveal the tag locality, Bao et al. [8] introduced hidden concepts for decomposing holistic image representation into tag representations. Mesnil et al. [9] learned implicit representations for both the objects and their parts. Although these representations cannot explicitly describe the regions of a certain tag, they implicitly capture the tag's local visual contents by learning from large amount of annotated images. Thus, implicit image representation is nontrivial for tackling the tag locality problem in large-scale datasets.

Considering the problem of intra-tag diversity, a straightforward way is to set up the class-specific techniques [10,11] by treating annotation tags as class labels and learning the visual contents within each class. Although capable of identifying sets of visual contents discriminative for the classes of interest, these straightforward methods do not explicitly model the interclass and intraclass structures of visual distributions due to its lack of hierarchical content groupings. To facilitate the discovery of these structures, various hierarchical generative methods have been recently ported from the text to the vision literature. Among these methods, topic models, such as latent Dirichlet allocation (LDA) [12] and probabilistic latent semantic analysis (pLSA) [13], that consider probabilistic latent variable models for hierarchical learning have caused extensive interest. However, an analysis of previous supervised topic models [14] shows that the topics discovered by these models are driven by general image regularities rather than the semantic regularities for image annotation. For example, it has been noted in [14] that given a collection of movie reviews, LDA might discover topics as movie properties, such as genres, which are not central to the annotation task. Therefore, incorporating a class label variable into a generative model might tackle the intra-tag diversity problem well. Such extensions have been successfully applied into the classification task, such as class LDA (cLDA) [14], supervised LDA (sLDA) [15], class-specific-simplex LDA (css-LDA) [16], and so on.

In this paper, we develop a new extension of LDA coupled with Laplacian regularized uncorrelated tensor representation for learning semantics in the image data. Since tensor representation can well capture the high-order statistics and structures from the training data, the proposed representation method achieves an efficient compressed image representation by imposing noncorrelation constraints and Laplacian regularization in tensor factorization. Based on this representation, a three-level hierarchical probabilistic model, denoted as class-specific Gaussian-multinomial latent Dirichlet allocation (csGM-LDA), is developed by using class supervision at the level of visual features. In csGM-LDA, latent variables or topics are served as middle-level concepts for building the correspondences between visual features and annotation tags.

The core contributions of this paper are listed as follows:

- A novel hierarchical probabilistic model, namely csGM-LDA, is presented by combining the labeling strength of topic supervision with the flexibility of topic discovery, and can be effectively modeled by applying a variational EM algorithm.
- An effective image representation method, namely, Laplacian regularized uncorrelated tensor representation, is developed to explicitly consider the manifold structures in the high-order image space.
- By learning with csGM-LDA, a unified framework is introduced to infer the hierarchies of multiple modalities and predict tags for a new image. Benefiting from the exploration of hierarchical probabilistic inferences, the unified framework can be effectively conducted.

The rest of this paper is organized as follows. We first discuss the related work in Section 2. Then, we present Laplacian regularized uncorrelated tensor representation in Section 3. After that, the proposed model is described in Section 4. Moreover, quantitative experiments validating strong improvements by the proposed method are presented in Section 5. Finally, Section 6 draws the conclusion.

2 Related work

In this section, we outline research contributions which are most related to our work. We first review techniques for tensor-based image representation. Then, topic models are further discussed.

2.1 Tensor-based image representation

It is believed that the specialized structures of a visual object are intrinsically in the form of second or even higher order tensor [5]. To retain these high-order characteristics, tensors or multidimensional arrays become a natural choice for the visual representation. In practice, exact image representation as a full tensor is often redundant and impossible when coping with mass of images.

However, approximative image representation using tensor subspace learning techniques in many cases can be helpful for describing various visual objects. In this paper, we discuss two main kinds of tensor subspace learning (TSL) algorithms: supervised and unsupervised TSL.

Supervised TSL algorithms use concept-driven dimensionality reduction to achieve discriminant tensor subspaces by considering the subsequent classification or recognition tasks. This line of algorithms requires that either manual class labels or object priors in the training set can be applicable to a particular image classification [5,6] or object representation [17,18]. However, as image annotation system needs to handle a large number of classes and most classes may require many training samples due to significant intraclass shape and appearance variations, it is important that the learning does not involve any human interaction. This makes unsupervised TSL algorithms more appealing. Unsupervised TSL algorithms are actively explored for data-driven dimensionality reduction that uses low rank tensors to approximate the exact represented tensors. The extensions of principal component analysis (PCA) and singular value decomposition (SVD) are most familiar methods for the research on this line. By maximizing the variance measure, two-dimensional PCA (2DPCA) [19] represented an image by projecting it to principal components along the vertical direction of the image data. Then, generalized PCA (GPCA) [20] employed bilinear subspace analysis for dimensionality reduction with matrices. Later, the multilinear PCA (MPCA) [21] and uncorrelated MPCA (UMPCA) [22] were proposed for dimensionality reduction with tensors of any order. By minimizing the reconstruction error, the generalized low-rank approximation of matrices (GLRAM) [23] took into account the spatial correlation of image pixels within a localized neighborhood and applied bilinear transforms to the input image matrices. For higher-order tensors, the work in [24] used the high-order SVD (HOSVD) to decompose an ensemble of images into basis images that capture the different underlying factors of variations. Furthermore, concurrent subspaces analysis (CSA) [25] was presented as a generalization of GLRAM for higher-order tensors. Recently, multiple tensor rank-R decomposition (MTRD) [26] was proposed for approximating a higher-order tensor with a series of rank-R tensor approximations.

In this paper, we propose an unsupervised method with Laplacian regularized uncorrelated tensor representation to explicitly consider manifold structures in the high-order image space. That is, data points that are close in the intrinsic geometry of the image space shall thus be close to each other under the factorized tensor basis. By combining unsupervised TSL and Laplacian regularization, we can achieve a more discriminative descriptor which is much important for accurate semantic learning.

2.2 Topic models for image annotation

Topic models annotate images as the samples from a specific mixture of topics, where each topic is a distribution over image observations. Three alternatives of pLSA-based models that provided in [13] were presented by using asymmetric learning for semantic indexing of large image collections. Then, a Gaussian-multinomial pLSA (GM-pLSA) model [27] was presented to learn multimodal correlations from the image data by applying continuous feature vectors. Furthermore, the work in [28] extended pLSA to a higher-order formalism, so as to become applicable for more than two observable variables. However, pLSA-based models are incomplete in that they provide no probabilistic restriction on how to generate the training data. In these models, each image is represented as a list of the mixing proportions for topics, and there is no probabilistic inference for generating these numbers of topics. This leads to two problems: first, the number of modeling parameters grows linearly with the size of the training set, which leads to serious problems with overfitting; second, it is not clear how to assign probability to an image outside of the training set. To overcome these problems, it is much effective to endow the topic model with Dirichlet priors over topic parameters as they are conjugate to the multinomial distribution of the associated tags. The correspondence LDA (Corr-LDA) [29] was first presented for modeling the joint distribution of images and tags. To capture more general forms of association and allow the number of topics in the two data modalities to be different, topic regression multimodal latent Dirichlet allocation (tr-mmLDA) [30] was proposed by introducing a regression module to correlate the two sets of topics. Taking advantage of limited tagged training images and rich untagged images, the work in [31] proposed a regularized semi-supervised latent Dirichlet allocation (r-SSLDA) for learning visual concept classifiers in a semi-supervised way. However, several supervised methods [14-16] show that the topics discovered by LDA models are driven by general image regularities rather than the semantic regularities for image annotation. To address this, we propose a new three-level hierarchical probabilistic model by incorporating supervision into the extended LDA model, making the annotation applications be much effective than previous LDA models.

3 The proposed representation method

In this section, we first give the notations that are necessary in defining the multiway problem. Then, a tensor-based method is proposed for visual representation.

3.1 Notations and definitions

We follow the notation conventions in tensor algebra [32]. Vectors are usually denoted by lowercase letters, e.g., x; matrices by uppercase letters, e.g., X; and tensors by calligraphic letters, e.g., $\mathcal{X}$. Their elements are denoted with indices in parentheses. Table 1 lists the key notations.

The inner product of two tensors of the same size $\mathcal{A}, \mathcal{B} \in \mathbb{R}^{I_1 \times I_2 \times \cdots \times I_N}$ is defined as $\langle \mathcal{A}, \mathcal{B} \rangle = \sum_{i_1} \sum_{i_2} \cdots \sum_{i_N} \mathcal{A}(i_1, i_2, \cdots i_N) \cdot \mathcal{B}(i_1, i_2, \cdots i_N)$. Thus, the Frobenius norm of $\mathcal{A}$ can be denoted by $\|\mathcal{A}\|_F = \sqrt{\langle \mathcal{A}, \mathcal{A} \rangle}$. The n-mode matricization of $\mathcal{A} \in \mathbb{R}^{I_1 \times I_2 \times \cdots \times I_N}$, denoted as $A_{(n)} \in \mathbb{R}^{I_n \times (I_1 \times \cdots \times I_{n-1} \times I_{n+1} \times \cdots \times I_N)}$, are obtained from $\mathcal{A}$ by varying the index i_n while keeping all the other indices fixed. The n-mode product of a tensor $\mathcal{A}$ by a matrix $U \in \mathbb{R}^{J_n \times I_n}$, denoted by $\mathcal{A} \times_n U$, is a tensor with entries:

$$(\mathcal{A} \times_n U)(i_1, \cdots, i_{n-1}, j_n, i_{n+1}, \cdots, i_N) = \sum_{i_n} \mathcal{A}(i_1, \cdots, i_N) \cdot U(j_n, i_n) \tag{1}$$

3.2 Laplacian regularized uncorrelated tensor representation

For image representation, we first represent a two-dimensional image I with an exact tensor representation, $\mathcal{X} \in \mathbb{R}^{I_1 \times I_2 \times I_3}$, where $\{I_1, I_2\}$ denotes the size of the image and I_3 denotes the depth of image feature maps. In this representation, we consider the edge energies and the flow directions as supplementary to pixel-wise color information. At this point, we employ the Gaussian derivative (GD) and the difference of offset Gaussians (DoG), as defined in [33], for defining these features. The filter banks by convolving GD and DoG functions can be defined as:

$$\begin{cases} E_\sigma(I(x,y)) = \sqrt{|I(x,y) * \mathrm{GD}_{\sigma,0}(x,y)|^2 + |I(x,y) * \mathrm{GD}_{\sigma,\pi/2}(x,y)|^2} \\ F_\sigma(I(x,y)) = \sqrt{|I(x,y) * \mathrm{DoG}_{\sigma,0}(x,y)|^2 + |I(x,y) * \mathrm{DoG}_{\sigma,\pi/2}(x,y)|^2} \end{cases} \tag{2}$$

Table 1 List of key notations

Symbol	Description
$\mathcal{X}_n, \tilde{\mathcal{X}}_n, \mathcal{G}_n$	The representations of an original tensor, centered tensor, and its core tensor
$U^{(k)}, \tilde{U}_{(-k)}$	The k-mode transformed matrix and the Kronecker products except the matrix
$\tilde{W}, \tilde{D}, \tilde{L}$	The weight matrix of tensorial features, its diagonal matrix, and its Laplacian matrix
g_n, y_n	Vectorizations of the core tensor and the transformed tensor
$\alpha, \beta, \mu_c, \sigma_c$	Parameters of the Dirichlet distribution for latent topics, multinomial distribution for tags, mean and variance of Gaussian distribution for visual features
w_m, v_d	Symbol of the tag and the visual feature
y_m, z_d	Latent topics for the tag and the visual feature
W, V	Collections of tags and visual features
Y, Z	Collections of latent topics for tags and visual features

where (x, y) denotes the coordinates of the image and σ is a scale parameter. Then, the image can be represented by:

$$\mathcal{X} = [Y\ C_b\ C_r\ E_{\sigma/2}(Y)\ E_\sigma(Y)\ E_{2\sigma}(Y)\ F_{\sigma/2}(Y)\ F_\sigma(Y)\ F_{2\sigma}(Y)] \tag{3}$$

where Y, C_b, C_r are the three color channels obtained by transforming the original RGB image.

Let $\{\mathcal{X}_n | \mathcal{X}_n \in \mathbb{R}^{I_1 \times I_2 \times I_3}, n = 1, \cdots, N\}$ be a set of represented tensors from an image dataset. To learn uncorrelated features without loss of generality, training samples are subtracted to be zero-mean so that the constraint of uncorrelated features is the same as orthogonal projected features. Usually, the image space is always of high dimensionality. However, image contents are typically embedded in a lower dimensional tensor subspace, in analogy to the dimensional reduction problem that considers both feature selection (i.e., give a more informative description to pixels) and spatial correlation. Thus, the task is to find a tensor subspace that captures most of the characteristics in the input space, i.e., define multilinear transformations $\{U^{(k)} \in \mathbb{R}^{I_k \times J_k}, k = 1, 2, 3 | \left(U^{(k)}\right)^T U^{(k)} = I\}$ that rewrite the original tensor as:

$$\tilde{\mathcal{X}}_n \approx \mathcal{G}_n \times_1 U^{(1)} \times_2 U^{(2)} \times_3 U^{(3)} \tag{4}$$

where $\tilde{\mathcal{X}}_n = \mathcal{X}_n - \bar{\mathcal{X}}$, $\bar{\mathcal{X}} = (1/N) \sum_{n=1}^N \mathcal{X}_n$ and $\mathcal{G}_n \in \mathbb{R}^{J_1 \times J_2 \times J_3}$ is a core tensor. To capture the core tensor, we determine the objective function with the following minimization problem:

$$\left\{U^{(k)} \in \mathbb{R}^{I_k \times J_k}, \mathcal{G}_n | \left(U^{(k)}\right)^T U^{(k)} = I\right\} = \arg \min_{U^{(k)}, \mathcal{G}_n} \Psi + \lambda \mathcal{M}_{\mathcal{G}} \tag{5}$$

where $\Psi = \sum_{n=1}^N \left\| \tilde{\mathcal{X}}_n - \mathcal{G}_n \times_1 U^{(1)} \times_2 U^{(2)} \times_3 U^{(3)} \right\|_F^2$, and $\mathcal{M}_G = G^T \tilde{L} G = \sum_{ij} \tilde{W}(i,j) \|g_i - g_j\|_2^2$. Here, $\left\{g_n \in \mathbb{R}^D, D = \prod_{k=1}^3 J_k\right\}$ is the vectorization of $\mathcal{G}_n$, and G is a matrix with the column of g_n. Define the diagonal matrix $\tilde{D}$ whose entries are column sums of the weight matrix $\tilde{W}$, and $\tilde{L} = \tilde{D} - \tilde{W}$ is a Laplacian matrix. The nearest neighbor graph is used to construct the weight matrix by finding the nearest neighbors for each image data. We use $NN(\mathcal{X}_n)$ to denote the set of K_{NN} nearest neighbors of $\mathcal{X}_n$. The weight matrix can be simply defined as:

$$W(i,j) = \begin{cases} 1 & \text{if } \mathcal{X}_i \in NN(\mathcal{X}_j) \text{ or } \mathcal{X}_j \in NN(\mathcal{X}_i) \\ 0 & \text{otherwise} \end{cases} \tag{6}$$

To solve the problem defined in Equation 5, an alternating iteration scheme is applied. Given all the projection matrices $\{\mathcal{U}^{(k)} \in \mathbb{R}^{I_k \times J_k}, k = 1,2,3\}$, we can obtain that:

$$\{\mathcal{G}_n, n = 1, \cdots, N\} = \arg\min_{\mathcal{G}_n} \Psi_{\mathcal{Y}} + \lambda \mathcal{M}_{\mathcal{G}} \quad (7)$$

where $\Psi_{\mathcal{Y}} = \sum_{n=1}^{N} \|y_n - g_n\|_F^2$, $y_n = vec\left(\tilde{\mathcal{X}}_n \times_1 \left(\mathcal{U}^{(1)}\right)^T \times_2 \left(\mathcal{U}^{(2)}\right)^T \times_3 \left(\mathcal{U}^{(3)}\right)^T\right)$. The above function defines a quadratic programming problem and can be solved linearly by using the Newton-Raphson method [12] with the following iteration:

$$g_n^{(t+1)} = g_n^{(t)} - (I + \lambda \tilde{L})^{-1} \left(g_n^{(t)} - y_n + \lambda \sum_j \tilde{L}(n,j) g_j^{(t)} \right) \quad (8)$$

Given the core tensor $\mathcal{G}_n$, we can rewrite Equation 5 as:

$$\{\mathcal{U}^{(k)} \in \mathbb{R}^{I_k \times J_k}, k = 1,2,3 | \left(\mathcal{U}^{(k)}\right)^T \mathcal{U}^{(k)} = I\} = \arg\min_{\mathcal{U}^{(k)}} \Psi \quad (9)$$

Since the projection to a high-order tensor subspace consists of several projections to the corresponding vector subspaces, the optimization can be iteratively solved by finding the k-mode projection that maximizes the scatter in the k-mode vector subspace. To optimize $\mathcal{U}^{(k)}$, we first define two scatter matrices:

$$\begin{cases} \Phi_G^{(k)} = \sum_{n=1}^{N} G_{n(k)} \tilde{\mathcal{U}}_{(-k)}^T \tilde{\mathcal{U}}_{(-k)} G_{n(k)}^T \\ \Phi_X^{(k)} = \sum_{n=1}^{N} \tilde{X}_{n(k)} \tilde{\mathcal{U}}_{(-k)} G_{n(k)}^T \end{cases} \quad (10)$$

where $\tilde{\mathcal{U}}_{(-k)} = \mathcal{U}^{(1)} \otimes \cdots \otimes \mathcal{U}^{(k-1)} \otimes \mathcal{U}^{(k+1)} \otimes \cdots \otimes \mathcal{U}^{(3)}$ and '$\otimes$' denotes the *Kronecker product*. Then, the solution to Equation 9 can be achieved by:

$$\mathcal{U}^{(k)} = \left(\Phi_G^{(k)}\right)^{-1} \Phi_X^{(k)} \quad (11)$$

The pseudo code for the proposed representation method is described in Algorithm 1. For this representation, a full solution is referring to the formalism in Equation 5. However, the alternating solution for this problem is quadratic with respect to the number of the image dataset, which is much expensive for image representation when dealing with a large dataset. In real applications, we perform the above representation method with a much smaller size by first using graph shift [34] for image clustering and then learning the representation for each group. Noticeably, the image data of one group should subtract the projections of previous multilinear transformations to preserve the orthogonality.

Algorithm 1: Laplacian regularized uncorrelated tensor representation
Input: A group of image samples $\{\mathcal{X}_n | \mathcal{X}_n \in \mathbb{R}^{I_1 \times I_2 \times I_3}, n = 1, \cdots, N\}$
Output: Multilinear transformations $\{\mathcal{U}^{(k)} \in \mathbb{R}^{I_k \times J_k}, k = 1,2,3\}$ and core tensors $\{\mathcal{G}_n, n = 1, \cdots, N\}$

Step 1: Center the input image samples as $\{\tilde{\mathcal{X}}_n | \tilde{\mathcal{X}}_n = \mathcal{X}_n - \bar{\mathcal{X}}, n = 1, \cdots, N\}$.
Step 2: Calculate the eigen-decomposition of $S^{(k)} = \sum_{n=1}^{N} \tilde{X}_{n(k)} \tilde{X}_{n(k)}^T$, and set $\mathcal{U}^{(k)}$ to consist of the least number of the eigenvectors corresponding to the largest J_k eigenvalues.
Step 3: Iteratively compute $\{\mathcal{G}_n, n = 1, \cdots, N\}$ according to Equation 8.
Step 4: Update $\{\mathcal{U}^{(k)} \in \mathbb{R}^{I_k \times J_k}, k = 1,2,3\}$ according to Equation 11.
Step 5: Alternately perform steps 3 and 4 until convergence.

4 The proposed annotation method

In this section, we first describe the proposed method for image annotation. Then, we turn our attention to parameter estimation for the modeling problem. Finally, a unified framework is presented to infer the hierarchies of multiple modalities and predict tags for a new image.

4.1 Class-specific Gaussian-multinomial latent Dirichlet allocation

The proposed model, csGM-LDA, is a supervised probabilistic model for learning multiple relationships between images and tags. The basic idea is that the visual and semantic modalities of images are represented as random mixtures over multiple latent topics, where each topic is characterized by a distribution over each modality. In Figure 1, the generative process of csGM-LDA for an image-tag pair with M associated tags and D visual features is given as follows:

1. Draw an image topic proportion $\theta \sim P_\Pi(\theta; \alpha)$
2. For each associated tag $w_m \in \mathcal{W} = \{1, \cdots, C\}, m = 1, \cdots, M$

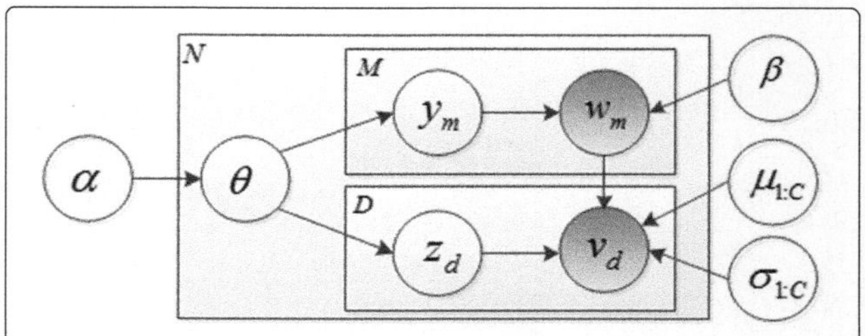

Figure 1 Illustration of csGM-LDA.

(a) Draw a topic assignment, $y_m \sim P_{Y|\Pi}(y_m|\theta), y_m \in \mathcal{T}_y = \{1, \cdots, K\}$
(b) Draw a tag, $w_m \sim P_{W|Y}(w_m|y_m; \beta), w_m \in \mathcal{W} = \{1, \cdots, C\}$

3. For each visual word $v_d, d = 1, \cdots, D$

(a) Draw a topic assignment, $z_d \sim P_{Z|\Pi}(z_d|\theta), z_d \in \mathcal{T}_z = \{1, \cdots, K\}$
(b) Draw a visual description, $v_d \sim P_{V|Z,W}(v_d|z_d, w_m; \mu_{w_m}, \sigma_{w_m})$

Similar to the earlier LDA extensions, $P_{\Pi}(\cdot)$ is a Dirichlet distribution on the topic simplex $\theta \in \mathbb{R}^K$with the parameter $\alpha \in \mathbb{R}^K$, $P_{Y|\Pi}(\cdot)$ and $P_{Z|\Pi}(\cdot)$ are two multinomial distributions over the topic simplex θ, $P_{W|Y}(\cdot)$ is a categorical distribution over a topic y_m with the parameter $\beta \in \mathbb{R}^{K\times C}$ where $\beta(k,c) = p(w_m = c|y_m = k)$, and $P_{V|Z,W}(\cdot)$ is a Gaussian distribution over a topic z_d with the class-dependent parameters $\{\mu_{w_m}(z_d, d), \sigma_{w_m}(z_d, d)|\mu_{w_m}, \sigma_{w_m} \in \mathbb{R}^{K\times D}\}$. In this way, semantic topics generate tags or classes for images with each defining a prior distribution in visual topic space. Taking Bayesian rules, the joint distribution of $\{\theta, y_m, w_m, z_d, v_d\}$ for each image is given by:

$$\begin{aligned} &P_{\Pi,Y,W,Z,V}(\theta, y_m, w_m, z_d, v_d|\alpha, \beta, \mu_{w_m}, \sigma_{w_m}) = \\ &P_{\Pi}(\theta|\alpha)P_{Y|\Pi}(y_m|\theta)P_{W|Y}(w_m|y_m, \beta)P_{Z|\Pi}(z_d|\theta) \\ &P_{V|Z,W}(v_d|z_d, \mu_{w_m}, \sigma_{w_m}) \end{aligned} \tag{12}$$

The modeling problem is then to maximize the log likelihood of the following marginal function:

$$\begin{aligned} P_{W,V}(W, V|\alpha, \beta, \mu_{1:C}, \sigma_{1:C}) = &\int_{\theta_n} \prod_{n=1}^{N} \prod_{m=1}^{M} \prod_{d=1}^{D} \sum_{y_{nm}} \sum_{z_{nd}} \\ &P_{\Pi,Y,W,Z,V}(\theta_n, y_{nm}, w_{nm}, z_{nd}, v_{nd}|\alpha, \beta, \mu_{w_{nm}}, \sigma_{w_{nm}})d\theta_n \end{aligned} \tag{13}$$

In csGM-LDA, the parameters $\{\alpha, \beta, \mu_{1:C}, \sigma_{1:C}\}$ are dataset-level parameters, assumed to be sampled once in the process of generating a set of images. The variables θ_n is an image-level variable, sampled once per image. The variables y_{nm} and w_{nm} are tag-level variables, sampled once for each annotated tag. And the variables z_{nd} and v_{nd} are structure-level variables, sampled once for each visual description. Structural models similar to that shown in Figure 1 are often studied in Bayesian statistical modeling, where they are referred to as conditionally independent hierarchical models. Indeed, as we discuss in the following subsection, we adopt the empirical Bayes approach to estimating parameters with a variational EM algorithm.

4.2 Parameter estimation via variational inference

In this section, we describe a convexity-based variational algorithm for parameter estimation. The basic idea of convexity-based variational inference is to make use of Jensen's inequality to obtain an adjustable lower bound on the log likelihood [13]. Essentially, a family of lower bounds is usually indexed by a set of variational parameters. The variational parameters are chosen by an optimization procedure that attempts to find the tightest possible lower bound. In particular, the objective function in Equation 13 is usually intractable due to the couplings between θ_n and $\{\beta, \mu_{1:C}, \sigma_{1:C}\}$ in the summation over latent topics. By dropping these edges and endowing the simplified graphical model with free variational parameters, we obtain a new family of distributions on the latent variables as seen in Figure 2. The variational distribution $q(\theta, Y, Z|\eta, \phi, \zeta)$ can be characterized by:

$$q(\theta, Y, Z|\eta, \phi, \zeta) = \prod_{n=1}^{N} q(\theta_n|\eta_n) \prod_{m=1}^{M} q(y_{nm}|\phi_{nm}) \prod_{d=1}^{D} q(z_{nd}|\zeta_{nd}) \tag{14}$$

where $\{Y, Z\}$ are the collections of latent topics, the Dirichlet parameter η_n, and the multinomial parameters $\{\phi_{nm}, \zeta_{nd}\}$ are the free variational parameters.

Having specified a simplified family of probability distributions, the next step is to set up an optimization problem that determines the values of the variational parameters:

$$\begin{aligned} \{\eta^*, \phi^*, \zeta^*\} = \arg\min_{\{\eta,\phi,\zeta\}} D_{KL}\,(q(\theta, Y, Z|\eta, \phi, \zeta) \\ ||P_{\Pi,Y,Z}(\theta, Y, Z|W, V, \alpha, \beta, \mu_{1:C}, \sigma_{1:C})) \end{aligned} \tag{15}$$

where $D_{KL}(\cdot)$ is the Kullback-Leibler (KL) divergence and $\{W, V\}$ are the corresponding collections of their lowercase variables. To achieve this minimization, we begin with the expression of the true log-likelihood for an image-tag pair $\{W, V\}$ by bounding the log likelihood of $P_{W,V}(W, V|\alpha, \beta, \mu_{1:C}, \sigma_{1:C})$ using the Jensen's inequality:

$$\begin{aligned} &\log P_{W,V}(W, V|\alpha, \beta, \mu_{1:C}, \sigma_{1:C}) \geq \\ &E_q[\log P_{\Pi,Y,W,Z,V}(\theta, Y, W, Z, V|\alpha, \beta, \mu_{1:C}, \sigma_{1:C})] \\ &-E_q[\log q(\theta, Y, Z|\eta, \phi, \zeta)] \end{aligned} \tag{16}$$

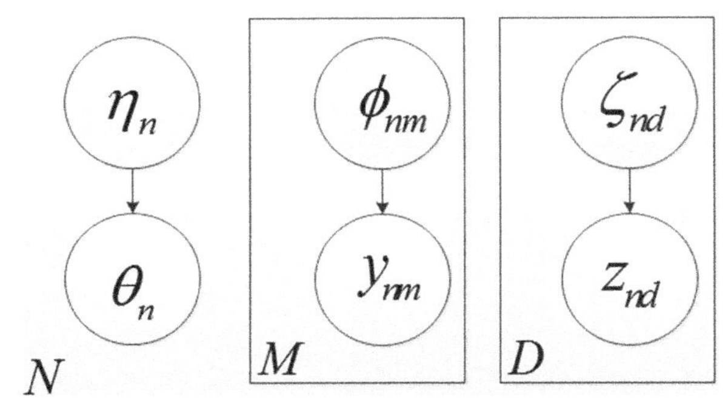

Figure 2 The variational model for approximating csGM-LDA.

It can be easily verified that the difference between the two sides of the above inequation is the KL divergence that provided in Equation 15. Let $\mathcal{L}(\eta,\phi,\zeta;\alpha,\beta,\mu_{1:C},\sigma_{1:C})$ denote the right-hand side of the inequation, we have:

$$\log P_{W,V}(W,V|\alpha,\beta,\mu_{1:C},\sigma_{1:C}) = \mathcal{L}(\eta,\phi,\zeta;\alpha,\beta,\mu_{1:C},\sigma_{1:C}) + D_{KL}\left(q(\theta,Y,Z|\eta,\phi,\zeta)||P_{\Pi,Y,Z}(\theta,Y,Z|W,V,\alpha,\beta,\mu_{1:C},\sigma_{1:C})\right) \quad (17)$$

This shows that maximizing the lower bound $\mathcal{L}(\eta,\phi,\zeta;\alpha,\beta,\mu_{1:C},\sigma_{1:C})$ with respect to $\{\eta,\phi,\zeta\}$ is equivalent to minimizing the above KL divergence. As seen in the Appendix, we obtain pair of updates:

$$\begin{cases} \phi_{nm}(k) \propto \beta(k,w_{nm}) \exp\left(\Psi(\eta_n(k)) - \Psi\left(\sum_{j=1}^{K}\eta_n(j)\right)\right) \\ \zeta_{nd}(k) \propto \left(\prod_{m=1}^{M} P_{V|Z,W}(v_{nd}|k,\mu_{w_{nm}},\sigma_{w_{nm}})\right) \\ \quad \exp\left(\Psi(\eta_n(k)) - \Psi\left(\sum_{j=1}^{K}\eta_n(j)\right)\right) \\ \eta_n(k) = \alpha(k) + \sum_{m=1}^{M}\phi_{nm}(k) + \sum_{d=1}^{D}\zeta_{nd}(k) \end{cases} \quad (18)$$

Based on the above variational inference, parameter estimation with respect to $\{\alpha,\beta,\mu_{1:C},\sigma_{1:C}\}$ yields the following EM algorithm:

1. *E-step:* For each image, find the optimizing values of the variational parameters$\{\eta,\phi,\zeta\}$.
2. *M-step*: Maximize the resulting lower bound $\mathcal{L}(\eta,\phi,\zeta;\alpha,\beta,\mu_{1:C},\sigma_{1:C})$ on the log likelihood of $P_{W,V}(W,V|\alpha,\beta,\mu_{1:C},\sigma_{1:C})$ with respect to the model parameters$\{\alpha,\beta,\mu_{1:C},\sigma_{1:C}\}$.

The update in the M-step corresponds to finding maximum likelihood estimates with expected sufficient statistics for each image under the approximate posterior which is computed by the variational parameters. Thus, we can also maximize the lower bound $\mathcal{L}(\eta,\phi,\zeta;\alpha,\beta,\mu_{1:C},\sigma_{1:C})$ with respect to the parameters $\{\alpha,\beta,\mu_{1:C},\sigma_{1:C}\}$. In Appendix, we show that the update in the M-step for the Dirichlet parameter α can be implemented with an efficient Newton-Raphson method. The parameters $\{\beta,\mu_{1:C},\sigma_{1:C}\}$ can be obtained as:

$$\begin{cases} \beta(k,c) \propto \sum_{n=1}^{N}\sum_{m=1}^{M}\phi_{nm}(k)\delta(c,w_{nm}) \\ \mu_c(k,d) = \frac{\sum_{n=1}^{N}\sum_{m=1}^{M}\zeta_{nd}(k)\delta(c,w_{nm})v_{nd}}{\sum_{n=1}^{N}\sum_{m=1}^{M}\zeta_{nd}(k)\delta(c,w_{nm})} \\ \sigma_c^2(k,d) = \frac{\sum_{n=1}^{N}\sum_{m=1}^{M}\zeta_{nd}(k)\delta(c,w_{nm})(v_{nd}-\mu_c(k,d))^2}{\sum_{n=1}^{N}\sum_{m=1}^{M}\zeta_{nd}(k)\delta(c,w_{nm})} \end{cases} \quad (19)$$

We summarize the parameter estimation algorithm in Algorithm 2. This is a standard EM process, and the lower bound $\mathcal{L}(\eta,\phi,\zeta;\alpha,\beta,\mu_{1:C},\sigma_{1:C})$ is a concave function. Therefore, Algorithm 2 is convergent. From the pseudo code, it is clear that each iteration of the E-step for csGM-LDA requires $\mathcal{O}(NMDK)$ operations. Empirically, we find that the number of iterations in this step is in proportion to the number of tags and the dimensionality of visual features. Parameter estimation in M-step for $\{\beta,\mu_{1:C},\sigma_{1:C}\}$ requires $\mathcal{O}(NCMDK)$ operations. And the number of iterations required for the Newton-Raphson method is linear to the dimensionality of α. Therefore, each EM iteration yields a total number of operations roughly on $\mathcal{O}(NM(C+M+D)DK)$. The number of iterations for the EM algorithm is mainly determined by the number of involved parameters, which is in proportion to $C(1+D)K$. Thus, the complexity for building the proposed model is about $\mathcal{O}\left(NCM(C+M+D)D^2K^2\right)$. When coping with large-scale data (i.e., $N \gg K,C,D$), the complexity of our modeling system is approximately linear to the number of images, which is much effective by comparing with the typical quadratic annotation models (e.g., pLSA [13] that requires $\mathcal{O}\left(N^2KC\right)$ operations, GM-pLSA [27] that yields the number of operations roughly on $\mathcal{O}\left(N^2K^2C\right)$, and so on).

Algorithm 2: Parameter estimation for csGM-LDA
Input: Observations $\{v_{nd},w_{nm},n=1,\cdots,N,m=1,\cdots,M,d=1,\cdots,D\}$
Output: Modeling parameters $\{\alpha,\beta,\mu_{1:C},\sigma_{1:C}\}$

repeat
In E-step:
For each image in the image dataset, initialize $\phi_{nm}(k)=\zeta_{nd}(k)=1/K$ by indexing all *m*, *d*, and *k*;
Initialize $\eta_n(k)=\alpha(k)+M/K$ by indexing all *k*.
for *n* = 1:*N*
repeat
Update $\{\eta,\phi,\zeta\}$ according to Equation 18.
until convergence
In M-step:
Initialize $\alpha(k)=1/K$ for all *k*. Update α with:
$\alpha^{(i+1)}(k)=\alpha^{(i)}(k)-$
$\left(\frac{\partial\mathcal{L}}{\partial\alpha^{(i)}(k)} - \frac{\sum_{j=1}^{K}\frac{\partial\mathcal{L}}{\partial\alpha^{(i)}(j)}\Big/(N\Psi'(\alpha^{(i)}(j)))}{1\Big/\left(N\Psi'\left(\sum_{j=1}^{K}\alpha^{(i)}(j)\right)\right)+\sum_{j=1}^{K}1/(N\Psi'(\alpha^{(i)}(j)))}\right)\Big/(N\Psi'(\alpha^{(i)}(k)))$
Update $\{\beta,\mu_{1:C},\sigma_{1:C}\}$ according to (19).
until convergence

4.3 A unified framework for image annotation

The unified framework is illustrated in Figure 3. Laplacian regularized uncorrelated tensor representation is first performed on the whole dataset. Then, both the semantic and the visual modalities are incorporated into building the proposed model. To annotate a new image, we first get the corresponding core tensor and then predict potential tags with Bayesian inference.

Given an observed visual image, our goal for image annotation is to estimate the posterior distribution of the

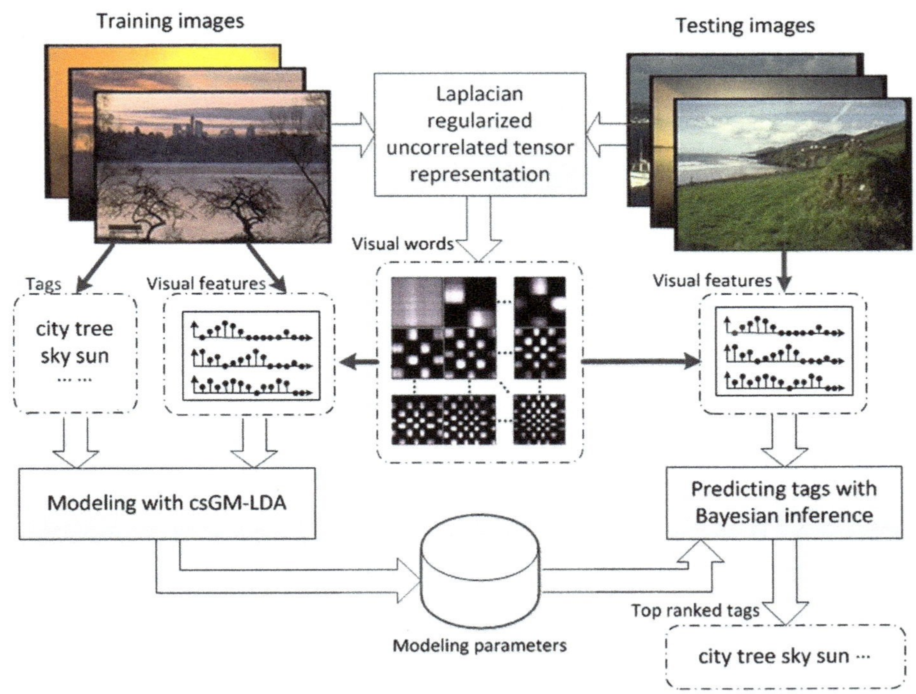

Figure 3 Illustration of the proposed framework for image annotation.

annotated tags. Taking Bayesian rules, we can infer their posterior probability by:

$$P_{W|V}(w|v_{1:D},\alpha,\beta,\mu_w,\sigma_w) = \frac{P_{W,V}(w,v_{1:D}|\alpha,\beta,\mu_w,\sigma_w)}{P_V(v_{1:D}|\alpha,\beta,\mu_w,\sigma_w)}$$
$$= \frac{\int_\theta P_\Pi(\theta|\alpha)\left(\sum_y P_{Y|\Pi}(y|\theta)P_{W|Y}(w|y,\beta)\right)\left(\sum_{z_d} P_{Z|\Pi}(z_d|\theta)P_{V|Z,W}(v_d|z_d,\mu_w(z_d,d),\sigma_w(z_d,d))\right)d\theta}{\int_\theta P_\Pi(\theta|\alpha)\left(\sum_{z_d} P_{Z|\Pi}(z_d|\theta)P_{V|Z,W}(v_d|z_d,\mu_w(z_d,d),\sigma_w(z_d,d))\right)d\theta} \tag{20}$$

For simplicity, we employ Monte Carlo inference [35] to approximate the integral of θ. To get the above probability, we first generate samples $\{\theta^s, s = 1,\cdots,S\}$ from the posterior $\theta^s \sim P_\Pi(\theta|\alpha)$. Then, the tags' probability can be rewritten by:

$$P_{W|V}(w|v_{1:D},\alpha,\beta,\mu_w,\sigma_w) \approx$$
$$\frac{\sum_{s=1}^S\left(\sum_y P_{Y|\Pi}(y|\theta^s)P_{W|Y}(w|y,\beta)\right)\left(\sum_{z_d} P_{Z|\Pi}(z_d|\theta^s)P_{V|Z,W}(v_d|z_d,\mu_w,\sigma_w)\right)}{\sum_{s=1}^S\sum_{z_d} P_{Z|\Pi}(z_d|\theta^s)P_{V|Z,W}(v_d|z_d,\mu_w(z_d,d),\sigma_w(z_d,d))} \tag{21}$$

5 Experiments

To evaluate the performance of our annotation framework, we set up several quantitative experiments. First, we investigate the effects of the setting parameters by conducting cross validation to select the best parameters for our proposed model. Then, we give a comparison of different image representations and validate the effectiveness of our representation method. Finally, we evaluate the proposed method on two benchmark datasets and report the results over state of the art.

5.1 Datasets and representations

We evaluate the proposed framework on two well-known benchmarks: Corel-5K [36] and ESP-Game [37]. The details of the two image datasets are shown in Table 2. The training percentage of each dataset is set as 80%, a validation set occupies 10% of the total images, and the remainder is the test set.

To get a reasonable size that keeps the images from serious deterioration for our representation method, we fix the size of images in Corel-5K and ESP-Game as 128×192 and 225×169, respectively. The vectorization of the core tensor constituting the D-dimensional Laplacian regularized uncorrelated tensorial vector (LGUTV) can be viewed as an image descriptor, with each item corresponding to an uncorrelated elementary multilinear projection. In our experiment, the dimensionality of LGUTV is fixed as 128 for each group in both the two datasets. We further divide the Corel-5K and ESP-Game into five and ten groups by graph shift [34], resulting in a 640-dimensional vector and a 1,280-dimensional vector for these two datasets, respectively. In addition, we

Table 2 Statistics of two image datasets

Dataset	Number of tags	Number of images	Tags per image
Corel-5K	5,000	374	3.4
ESP-Game	20,000	268	4.7

compare the proposed representation method with several common representation methods, i.e., the quantified color histograms with 16 bins in each color channel for RGB, LAB, HSV representations (C-HIST), the quantified SIFT features both extracted densely on a multiscale grid (D-SIFT) or for Harris-Laplacian interest points (H-SIFT) [3], local binary patterns (LBP) [38], and CPAM [4]. To get a proper evaluation of these image descriptors, we set their dimensions equal to that of LGUTV.

5.2 Evaluation criteria and baselines

The performance of image annotation is evaluated by comparing the automatically generated tags for the test set with the human-produced ground truth. In this paper, we give the following two measures for annotation evaluation. Firstly, F_1 score is measured by computing precision (*Prec*) and recall (*Rec*) for fixed annotation length with the five most relevant tags.

$$F_1 = \frac{2 \times Prec \times Rec}{Prec + Rec} \tag{22}$$

Note that each image is forced to be annotated with five tags, although the image might have fewer or more tags in the ground truth. Therefore, even if a model predicts all ground truth tags with a significantly higher probability than other tags, we will not measure perfect precision and recall. Thus, we also measure the precision at different levels of the recall for assessing the general annotation performance. The mean average precision (*mAP*) [13] over tags are found by computing for each tag the average of precisions measured after each relevant image is retrieved.

$$mAP = \frac{1}{N_q} \sum_{q=1}^{N_q} \sum_{i \in rel(q)} \mathrm{Prec}(i) / |rel(q)| \tag{23}$$

where Prec(i) is the precision of the correctly retrieved images at rank i in the ranking results of a query q, $rel(q)$ is the set of relevant images for this query, and N_q is the number of all queries.

For all images in the two standard datasets, our methods are compared with several most relevant and state-of-the-art methods, including TagProp [3], pLSA [13], GM-pLSA [27], GM-LDA [29], Corr-LDA [29], topic regression multimodal latent Dirichlet allocation (tr-mmLDA) [30], and css-LDA [16].

5.3 Investigate the impact of the setting parameters

We first investigate the neighborhood size K_{NN} for the graph construction and the tradeoff parameter λ for evaluating the effectiveness of Laplacian regularization. These two parameters reflect different facets of data construction; we joint discuss the sensitivity of these two parameters. As seen in Figure 4, we measure the F_1 scores for different parameters by setting the number of latent topics equal to the number of tags. Then, we choose to set the two parameters as the most promising for the two datasets in the following experiments, which are $K_{NN} = 10, \lambda = 0.1$ for Corel-5K and $K_{NN} = 15, \lambda = 0.1$ for ESP-Game, respectively. Besides, we observe that Laplacian regularization can achieve a more effective representation for better semantic learning by comparing the case that $\lambda \neq 0$ with the case that $\lambda = 0$.

Regarding topic models, they require the number of latent topics (i.e.,K) to be estimated, as this hyper parameter defines the capacity of the model. We analyze the parameter with a cross validation scheme. As seen in Figure 5, the improvement of annotation performance grows much slowly when the number of latent topics arrives at 100 for both the two datasets. When this number increases over the total number of tag vocabularies, csGM-LDA might suffer from the overfitting problem.

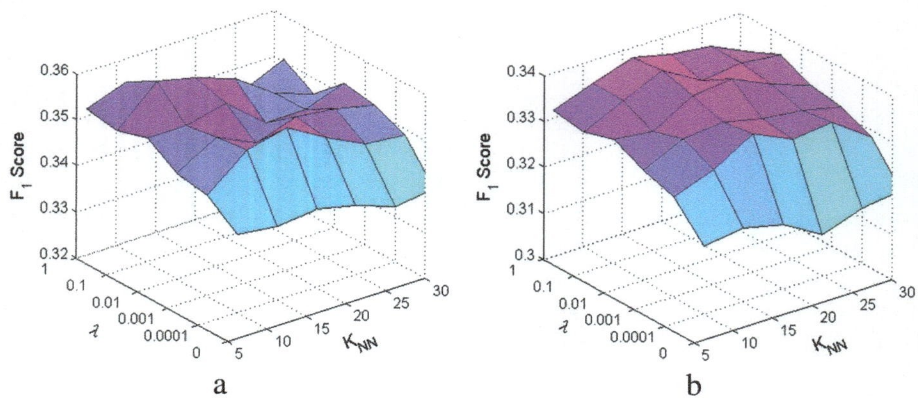

Figure 4 Investigation of the neighborhood size and the tradeoff parameter. F_1 score is measured to evaluate the performance of csGM-LDA by varying the two parameters on **(a)** Corel-5K and **(b)** ESP-Game.

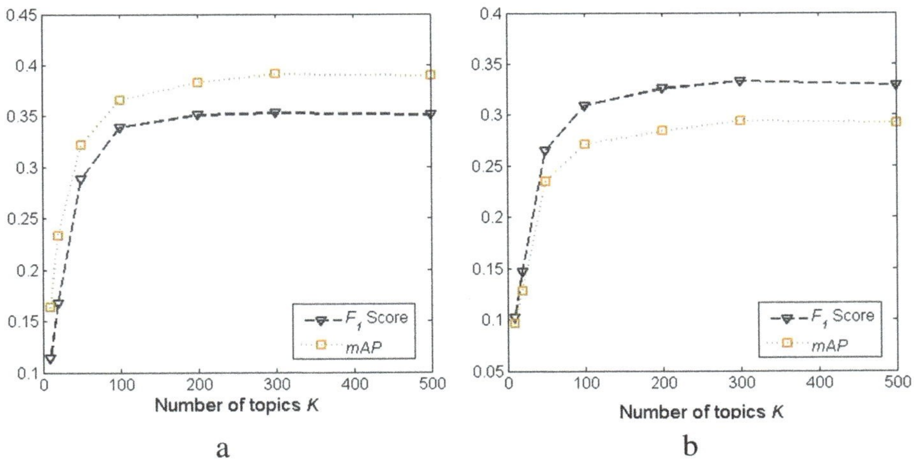

Figure 5 Investigation of the number of latent topics. The performance of csGM-LDA is evaluated by taking different number of latent topics in terms of F_1 score and *mAP* on **(a)** Corel-5K and **(b)** ESP-Game.

For example, the measures of F_1 score and *mAP* in both Figure 5a,b decrease when the number of latent topics is larger than 300. In the following experiments, we set the parameter as $K = 300$, which achieves the best performance for both the two datasets.

5.4 Evaluation of different image representations

In this paper, we argue that traditional vector-based image representations ignore high-order characteristics in image space, and thus, we combine unsupervised TSL and Laplacian regularization for achieving a more discriminative descriptor, i.e., LGUTV. To evaluate its effectiveness, we compare this descriptor with several state-of-the-art image descriptors on the two datasets by measuring F_1 Scores for the results of csGM-LDA. As seen in Figure 6, the performance of modeling csGM-LDA with LGUTV achieves the best performance by comparing with others on both the two datasets, confirming that tensor representation is most likely to provide a discriminative descriptor for recognizing the complex visual scenes.

5.5 Comparison with existing methods

In this section, we perform the annotation tasks on both the Corel-5K and ESP-Game datasets by comparing the proposed method with others. In Table 3, we report the performance by measuring both F_1 score and *mAP* for different methods based on LGUTV. On both two datasets, we observe that class-specific methods (e.g., TagProp, css-LDA, and csGM-LDA) obviously perform better than others. This is consistent with the analysis in the introduction since the generative domains of the complex visual world are much hard to describe. In addition, the measures for ESP-Game are consistently lower than that for Corel-5K because the retrieval tasks on which these measures are now computed are more challenging: an average of nearly 2,000 images for testing (versus 500 images) are ranked.

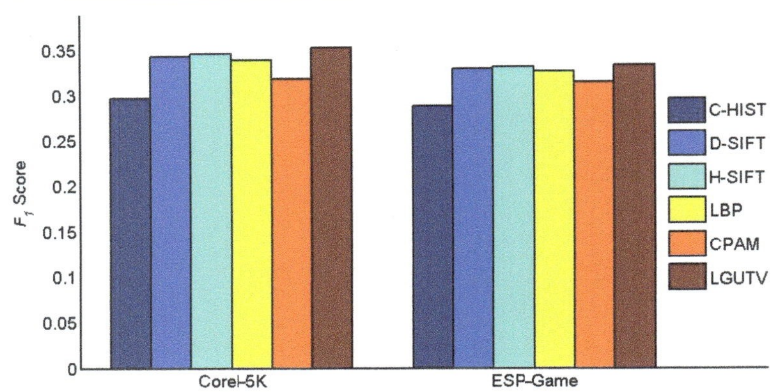

Figure 6 Annotation performance of csGM-LDA with different image representations.

Table 3 Comparison of different methods on both Corel-5K and ESP-Game

	Corel-5K		ESP-game	
Methods	F_1 Score	*mAP*	F_1 score	*mAP*
TagProp	0.342	0.374	0.323	0.281
pLSA	0.193	0.235	0.185	0.156
GM-pLSA	0.254	0.289	0.213	0.179
GM-LDA	0.276	0.303	0.224	0.192
Corr-LDA	0.316	0.354	0.288	0.253
tr-mmLDA	0.323	0.361	0.295	0.259
css-LDA	0.350	0.391	0.325	0.282
csGM-LDA	0.353	0.394	0.334	0.290

The proposed method achieves much more improvement on the ESP-Game dataset by comparing with a bit improvement on the Corel-5K dataset, demonstrating its effectiveness of modeling the large-scale dataset. For all the test sets, our proposed method performs best, and the instantiations of csGM-LDA and css-LDA clearly outperform others. We further compare the time required for training of different LDA models on ESP-Game and Corel-5K on a Intel Core i5-2410M CPU 2.30 GHz processor. Figure 7 gives the results. We observe that csGM-LDA is much faster than css-LDA. The reason is probably that the number of parameters in css-LDA is much larger than that of csGM-LDA. All the above results show the efficiency of our method.

6 Conclusions

In this paper, we propose a novel model, denoted as csGM-LDA, based on Laplacian regularized uncorrelated tensor representation for image annotation. The proposed annotation possesses two characteristics, namely: 1) images are represented by a set of uncorrelated tensorial descriptions and 2) class-specific information is integrated into semantic learning with the extension of the standard LDA model. The entire problem is formulated within the proposed framework, and csGM-LDA is presented to bridge the semantic gap between image contents and annotated tags. The experimental results demonstrate the effectiveness of our proposed method. Following the research on this line, we will further exploit region-based tensorial features for discriminative image representation and discuss the correlation of the class-specific information in a hierarchical LDA formalism.

Appendix

To maximize the lower bound in the E-Step that described in Section 4, we begin by expanding it with the factorizations of $P_{\Pi,Y,W,Z,V}(\theta, Y, W, Z, V|\alpha, \beta, \mu_{1:C}, \sigma_{1:C})$ and $q(\theta, Y, Z|\eta, \phi, \zeta)$:

$$\begin{aligned}\mathcal{L}(\eta,\phi,\zeta;\alpha,\beta,\mu_{1:C},\sigma_{1:C}) &= E_q[\log P_\Pi(\theta|\alpha)] \\ &+E_q[\log P_{Y|\Pi}(Y|\theta)] + E_q[\log P_{W|Y}(W|Y,\beta)] \\ &+E_q[\log P_Z(Z|\theta)] + E_q[\log P_{V|Z,W}(V|Z,W,\mu_{1:C},\sigma_{1:C})] \\ &-E_q[\log q(\theta|\eta)] - E_q[\log q(Y|\phi)] - E_q[\log q(Z|\zeta)]\end{aligned}$$

We unfold each item, and obtain:

$$P_\Pi(\theta|\alpha) = \frac{\Gamma(\sum_k \alpha(k))}{\prod_k \Gamma(\alpha(k))} \prod_k \theta(k)^{\alpha(k)-1},$$

$$P_{Y|\Pi}(y_m|\theta) = \theta(y_m)$$

$$P_{Z|\Pi}(z_d|\theta) = \theta(z_d), P_{W|Y}(w_m|y_m, \beta) = \beta(y_m, w_m)$$

$$P_{V|Z,W}(v_d|z_d, \mu_{w_m}, \sigma_{w_m}) = \frac{1}{\sqrt{2\pi}\sigma_{w_m}(z_d, d)} \times \exp\left(-\frac{(v_d - \mu_{w_m}(z_d, d))^2}{2\sigma^2_{w_m}(z_d, d)}\right)$$

To evaluate $\mathcal{L}(\eta, \phi, \zeta; \alpha, \beta, \mu_{1:C}, \sigma_{1:C})$, we should measure $E_q(\log\theta)$. We find that the sufficient statistic in defining $P_\Pi(\theta|\alpha)$ is $\log\theta(k)$. Since the derivative of the

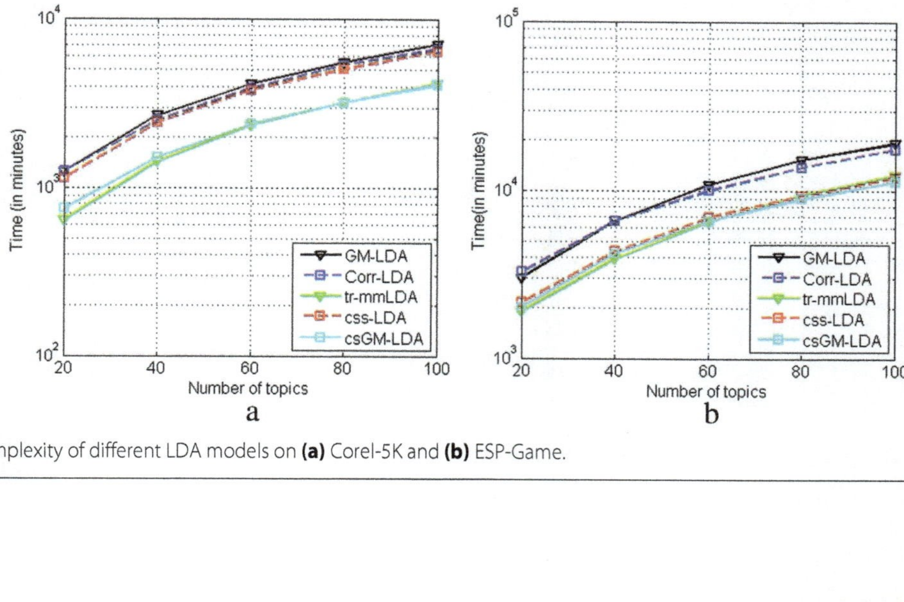

Figure 7 Time complexity of different LDA models on **(a)** Corel-5K and **(b)** ESP-Game.

log normalization factor with respect to the parameter is equal to the expectation of the sufficient statistic, we obtain:

$$E_q[\log\theta(k)] = \Psi(\eta(k)) - \Psi\left(\sum_{k=1}^{K}\eta(k)\right)$$

where $\Psi(\cdot)$ is the first derivative of $\log\Gamma(\cdot)$. Thus, we expand $\mathcal{L}(\eta,\phi,\zeta;\alpha,\beta,\mu_{1:C},\sigma_{1:C})$ as:

$$\begin{aligned}
&\mathcal{L}(\eta,\phi,\zeta;\alpha,\beta,\mu_{1:C},\sigma_{1:C}) = \\
&\textstyle\sum_{n=1}^{N}\Big(\log\Gamma\big(\sum_{k=1}^{K}\alpha(k)\big) - \sum_{k=1}^{K}\log\Gamma(\alpha(k)) \\
&\textstyle+\sum_{k=1}^{K}(\alpha(k)-1)\,E[\log\theta_n(k)|\eta_n]\Big) \\
&\textstyle+\sum_{n=1}^{N}\sum_{m=1}^{M}\sum_{k=1}^{K}\phi_{nm}(k)E[\log\theta_n(k)|\eta_n] \\
&\textstyle+\sum_{n=1}^{N}\sum_{m=1}^{M}\sum_{k=1}^{K}\phi_{nm}(k)\log\beta(k,w_{nm}) \\
&\textstyle+\sum_{n=1}^{N}\sum_{d=1}^{D}\sum_{k=1}^{K}\zeta_{nd}(k)E[\log\theta_n(k)|\eta_n] \\
&\textstyle+\sum_{n=1}^{N}\sum_{m=1}^{M}\sum_{d=1}^{D}\sum_{k=1}^{K}\zeta_{nd}(k)\log P_{V|Z,W}(v_{nd}|k,\mu_{w_{nm}},\sigma_{w_{nm}}) \\
&\textstyle-\sum_{n=1}^{N}\Big(\log\Gamma(\sum_{k=1}^{K}\eta_n(k)) + \sum_{k=1}^{K}\log\Gamma(\eta_n(k)) \\
&\textstyle-\sum_{k=1}^{K}(\eta_n(k)-1)\,E[\log\theta_n(k)|\eta_n]\Big) \\
&\textstyle-\sum_{n=1}^{N}\sum_{m=1}^{M}\sum_{k=1}^{K}\phi_{nm}(k)\log\phi_{nm}(k) \\
&\textstyle-\sum_{n=1}^{N}\sum_{d=1}^{D}\sum_{k=1}^{K}\zeta_{nd}(k)\log\zeta_{nd}(k)
\end{aligned}$$

In the E-Step, we first maximize the above function with respect to $\phi_{nm}(k)$. Observing that this is a constrained maximization since $\sum_{k=1}^{K}\phi_{nm}(k) = 1$, we first take the corresponding derivative:

$$\frac{\partial\mathcal{L}}{\partial\phi_{nm}(k)} = \Psi(\eta_n(k)) - \Psi\left(\sum_{j=1}^{K}\eta_n(j)\right) + \log\beta(k,w_{nm}) - \log\phi_{nm}(k) - 1$$

Adding Lagrange multipliers to $\mathcal{L}$ and setting the derivative of the summation to zero yields the maximizing value of the variational parameter $\phi_{nm}(k)$:

$$\phi_{nm}(k) \propto \beta(k,w_{nm})\exp\left(\Psi(\eta_n(k)) - \Psi\left(\sum_{j=1}^{K}\eta_n(j)\right)\right)$$

Similarly, we can obtain the variational parameter $\zeta_{nd}(k)$ as follows:

$$\zeta_{nd}(k) \propto \left(\prod_{m=1}^{M}P_{V|Z,W}(v_{nd}|k,\mu_{w_{nm}},\sigma_{w_{nm}})\right) \times \exp\left(\Psi(\eta_n(k)) - \Psi\left(\sum_{j=1}^{K}\eta_n(j)\right)\right)$$

Next, we maximize $\mathcal{L}$ with respect to $\eta_n(k)$. Taking the corresponding derivative, we obtain:

$$\begin{aligned}
\tfrac{\partial\mathcal{L}}{\partial\eta_n(k)} &= \Psi'(\eta_n(k))\Big(\alpha(k)-\eta_n(k) \\
&\qquad\textstyle+\sum_{m=1}^{M}\phi_{nm}(k)+\sum_{d=1}^{D}\zeta_{nd}(k)\Big) \\
&\textstyle-\sum_{k=1}^{K}\Psi'\Big(\sum_{j=1}^{K}\eta_n(j)\Big)\Big(\Big(\alpha(k)-\eta_n(k) \\
&\textstyle+\sum_{m=1}^{M_n}\phi_{nm}(k)+\sum_{d=1}^{D}\zeta_{nd}(k)\Big)\Big)
\end{aligned}$$

Setting this equation to zero yields a maximum at:

$$\eta_n(k) = \alpha(k) + \sum_{m=1}^{M}\phi_{nm}(k) + \sum_{d=1}^{D}\zeta_{nd}(k)$$

In the M-Step, we also maximize the resulting lower bound $\mathcal{L}(\eta,\phi,\zeta;\alpha,\beta,\mu_{1:C},\sigma_{1:C})$ with respect to the model parameters $\{\alpha,\beta,\mu_{1:C},\sigma_{1:C}\}$. To estimate the Dirichlet parameter α, we first get the derivatives as:

$$\frac{\partial\mathcal{L}}{\partial\alpha(k)} = \sum_{n=1}^{N}\left(\Psi\left(\sum_{j=1}^{K}\alpha(j)\right) - \Psi(\alpha(k)) + \left(\Psi(\eta_n(k)) - \Psi\left(\sum_{j=1}^{K}\eta_n(j)\right)\right)\right)$$

This derivative depends on $\alpha(j)$, for $j \neq k$, and we therefore must use an iterative method to find the maximal α. In particular, we can invoke the Newton-Raphson method with the update:

$$\alpha^{(i+1)}(k) = \alpha^{(i)}(k) - \left(\frac{\partial\mathcal{L}}{\partial\alpha^{(i)}(k)} - \frac{\sum_{j=1}^{K}\frac{\partial\mathcal{L}}{\partial\alpha^{(i)}(j)}/(N\Psi'(\alpha^{(i)}(j)))}{1/\left(N\Psi'\left(\sum_{j=1}^{K}\alpha^{(i)}(j)\right)\right)+\sum_{j=1}^{K}1/(N\Psi'(\alpha^{(i)}(j)))}\right) \Big/ \left(N\Psi'(\alpha^{(i)}(k))\right)$$

To maximize $\mathcal{L}(\eta,\phi,\zeta;\alpha,\beta,\mu_{1:C},\sigma_{1:C})$ with respect to β, we add Lagrange multipliers and set the derivative of the summation to zero, we can obtain:

$$\beta(k,c) \propto \sum_{n=1}^{N}\sum_{m=1}^{M}\phi_{nm}(k)\delta(c,w_{nm})$$

As to $\{\mu_{1:C},\sigma_{1:C}\}$, we set the corresponding derivatives of $\mathcal{L}(\eta,\phi,\zeta;\alpha,\beta,\mu_{1:C},\sigma_{1:C})$ to zero and can exactly obtain:

$$\mu_c(k,d) = \frac{\sum_{n=1}^{N}\sum_{m=1}^{M}\zeta_{nd}(k)\delta(c,w_{nm})v_{nd}}{\sum_{n=1}^{N}\sum_{m=1}^{M}\zeta_{nd}(k)\delta(c,w_{nm})}$$

$$\sigma_c^2(k,d) = \frac{\sum_{n=1}^{N}\sum_{m=1}^{M}\zeta_{nd}(k)\delta(c,w_{nm})\left(v_{nd}-\mu_c(k,d)\right)^2}{\sum_{n=1}^{N}\sum_{m=1}^{M}\zeta_{nd}(k)\delta(c,w_{nm})}$$

Competing interests

The authors declare that they have no competing interests.

Acknowledgement

This research was conducted with the support of National Natural Science Foundation of China (Grant No. 61271439).

References

1. H Ma, J Zhu, M-T Lyu, I King, Bridging the semantic gap between image contents and tags. IEEE Trans. Multimedia. **12**(5), 462–473 (2010)
2. J Liu, M Li, Q Liu, H Lu, S Ma, Image annotation via graph learning. Pattern Recognit. **42**, 218–228 (2009)
3. M Guillaumin, T Mensink, J Verbeek, C Schmid, in *Proc. IEEE Conf. Comput. Vis. Recognit.* Tagprop: Discriminative metric learning in nearest neighbor models for image auto-annotation, (2009), pp. 309–316

4. N Zhou, W Cheung, G Qiu, X Xue, A hybrid probabilistic model for unified collaborative and content-based image tagging. IEEE Trans. Pattern Anal. Mach. Intell. **33**(7), 1281–1294 (2011)
5. S Yan, D Xu, Q Yang, L Zhang, X Tang, H-J Zhang, in *Proc. IEEE Conf. Comput. Vis. Recognit.* Discriminant analysis with tensor representation, (2005), pp. 526–532
6. Y Liu, Y Liu, S Zhong, K Chan, Tensor distance based multilinear globality preserving embedding: a unified tensor based dimensionality reduction framework for image and video classification. Expert Syst. Appl. **39**, 10500–10511 (2012)
7. Z Zhou, M Zhang, in *Proc. Adv. Neural Inf. Process. Syst.* Multi-instance multi-label learning with application to scene classification, (2006), pp. 1609–1616
8. B-K Bao, T Li, S Yan, Hidden-concept driven multilabel image annotation and label ranking. IEEE Trans. Multimedia. **14**(1), 199–210 (2012)
9. G Mesnil, A Bordes, J Weston, G Chechik, Y Bengio, Learning semantic representations of objects and their parts. Mach Learn. **94**(2), 281–301 (2014)
10. J Li, J Wang, Automatic linguistic indexing of pictures by a statistical modeling approach. IEEE Trans. Pattern Anal. Mach. Intell. **25**(9), 1075–1088 (2003)
11. Q Mao, I-H Tsang, S Gao, Objective-guided image annotation. IEEE Trans. Image Process. **22**(4), 1585–1597 (2013)
12. D Blei, A Ng, M Jordan, Latent dirichlet allocation. J. Mach. Learn. Res. **3**, 993–1022 (2003)
13. F Monay, D Gatica-Perez, Modeling semantic aspects for cross-media image indexing. IEEE Trans. Pattern Anal. Mach. Intell. **29**(10), 1802–1817 (2007)
14. D Blei, J McAuliffe, in *Proc. Adv. Neural Inf. Process. Syst.* Supervised topic models, (2008), pp. 121–128
15. Q Guo, N Li, Y Yang, G Wu, in *IEEE International Conference on Systems, Man, and Cybernetics*. Supervised LDA for image annotation, (2011), pp. 471–476
16. N Rasiwasia, N Vasconcelos, Latent Dirichlet allocation models for image classification. IEEE Trans. Pattern Anal. Mach. Intell. **35**(11), 2665–2679 (2013)
17. S Yan, D Xu, Q Yang, L Zhang, X Tang, H-J Zhang, Multilinear discriminant analysis for face recognition. IEEE Trans. Image Process. **16**(1), 212–220 (2007)
18. D Tao, X Li, X Wu, S Maybank, General tensor discriminant analysis and gabor features for gait recognition. IEEE Trans. Pattern Anal. Mach. Intell. **29**(10), 1700–1715 (2007)
19. J Yang, D Zhang, A Frangi, J Yang, Two-dimensional PCA: a new approach to appearance-based face representation and recognition. IEEE Trans. Pattern Anal. Mach. Intell. **26**(1), 131–137 (2004)
20. J Ye, R Janardan, Q Li, in *Proc. ACM SIGKDD Int'l Conf. Knowledge Discovery and Data Mining*. Gpca: An efficient dimension reduction scheme for image compression and retrieval, (2004), pp. 354–363
21. H Lu, K Plataniotis, A Venetsanopoulos, MPCA: multilinear principal component analysis of tensor objects. IEEE Trans. Neural Netw. **19**(1), 18–39 (2008)
22. H Lu, K Plataniotis, A Venetsanopoulos, Uncorrelated multilinear principal component analysis for unsupervised multilinear subspace learning. IEEE Trans. Neural Netw. **20**(11), 1820–1836 (2009)
23. J Ye, Generalized low rank approximations of matrices. Mach. Learn. **16**(1), 167–191 (2005)
24. M Vasilescu, D Terzopoulos. Eur. Conf. Comput. Vis, (2002), pp. 447–460
25. D Xu, S Yan, L Zhang, S Lin, H-J Zhang, T Huang, Reconstruction and recognition of tensor-based objects with concurrent subspaces analysis. IEEE Trans. Circuits Syst. Video Technol. **18**(1), 36–47 (2008)
26. B Zhou, F Zhang, L Peng, Compact representation for dynamic texture video coding using tensor method. IEEE Trans. Circuits Syst. Video Technol. **23**(2), 280–288 (2013)
27. Z Li, Z Shi, X Liu, Z Shi, Modeling continuous visual features for semantic image annotation and retrieval. Pattern Recognit. Lett. **32**, 516–523 (2011)
28. S Nikolopoulos, S Zafeiriou, I Patras, I Kompatsiaris, High-order pLSA for indexing tagged images. Signal Process. **93**, 2212–2228 (2013)
29. D Blei, M Jordan, in *Proc. 26nd Ann. Int'l ACM SIGIR Conf. Research and Development in Information Retrieval*. Modeling annotated data, (2003), pp. 127–134
30. D Putthividhya, H Attias, S Nagarajan, in *Proc. IEEE Conf. Comput. Vis. Recognit.* Topic regression multi-modal latent Dirichlet allocation for image annotation, (2010), pp. 3408–3415
31. L Zhuang, H Gao, J Luo, Z Lin, Regularized semi-supervised latent Dirichlet allocation for visual concept learning. Neurocomputing. **119**, 26–32 (2013)
32. H Lu, K Plataniotis, A Venetsanopoulos, A survey of multilinear subspace learning for tensor data. Pattern Recognit. **44**, 1540–1551 (2011)
33. W-Y Ma, B Manjunath, Edgeflow: a technique for boundary detection and image segmentation. IEEE Trans. Image Process. **9**(8), 1375–1388 (2000)
34. H Liu, S Yan, in *International Conference on Machine Learning*. Robust graph mode seeking by graph shift, (2010), pp. 671–6783
35. K Murphy, *Machine Learning: A Probabilistic Perspective*. (The MIT Press, 2012)
36. H Müller, S Marchand-Maillet, T Pun, in *Proc. ACM Int'l Conf. Image and Video Retrieval*. The truth about Corel evaluation in image retrieval, (2002), pp. 38–49
37. L Ahn, L Dabbish, in *Proc. ACM SIGCHI Conf. on Human Factors in Computing Systems*. Labeling images with a computer game, (2004), pp. 354–363
38. T Ahonen, A Hadid, M Pietikainen, Face description with local binary patterns: application to face recognition. Mach. Learn. **28**(12), 2037–2041 (2006)

Optimization of a two-way MIMO amplify-and-forward relay network

Ying Zhang[1*], Yinjiang Chen[1], Chuanyi Pan[1], Huapeng Zhao[2] and Ning Kang[1]

Abstract

In this paper, we consider optimization of a two-way multiple-input multiple-output (MIMO) amplify-and-forward relay network which consists of a pair of transceivers and several relay nodes. Multiple antennas are equipped on the transceivers and relays. Multiple access broadcast scheme which finishes communication in two time slots is considered. In the first time slot, signals received by the relays are scaled by several beamforming matrices. In the second time slot, the relays transmit the scaled signals to the two transceivers. Upon receiving these signals, a MIMO equalizer is implemented at each transceiver to recover the desired signal. In this paper, zero forcing equalizers are used. Joint optimization of the beamforming matrices and the equalizers are realized using the following criteria: 1) the total relay transmission power is minimized subject to the minimal output signal-to-noise ratio (SNR) constraint at each transceiver, 2) the minimal output SNR of the two transceivers is maximized subject to total relay transmission power constraint, and 3) the minimal output SNR of the two transceivers is maximized subject to individual relay transmission power constraint. It is shown that the proposed optimization problems can be formulated as the second-order cone programming problems which can be solved efficiently. The validity of the proposed algorithm is verified by computer simulations.

Keywords: Amplify-and-forward; Beamforming; Multiple-input multiple-output; Relaying; Second-order cone programming; Zero-forcing

Introduction

Relaying technique is capable of extending communication range and coverage by providing link to shadowed users via relay nodes, and it received extensive study in recent years. For collaborative relaying technique, the network with a single pair of users and multiple relay nodes equipped with single antenna has been widely investigated [1-7]. Zheng [1] assumed perfect knowledge of channel-state information (CSI) and proposed to optimize the beamforming vector by maximizing the destination signal-to-noise ratio (SNR) subject to total and local relay transmission power constraints. In [2], similar optimization criterion was used in the case that only the second-order statistics of CSI are available. In [3,4], quantized CSI was considered. Quantizer at each relay and beamforming vectors at destination were optimized to minimize the uncoded bit error rate. In [5], the minimizing mean square error (MMSE) criterion was adopted to optimize the beamforming coefficients. The advantage of this algorithm lies in its ability to adaptively allocate transmission power of each relay. In recent, a so-called filter-and-forward distributed beamforming technique was proposed in [6]. Different from previous algorithms, this method solved the problem of relay beamforming in frequency selective environments, where a finite impulse response filter is used at each relay. Apart from the above mentioned one-way scheme, a two-way relaying strategy was proposed in [7]. In a two-way relaying scheme, the relays cooperate with each other to establish the connection between two transceivers. The design of the beamformer should simultaneously satisfy the requirements from the two transceivers. In [8], an optimization strategy was proposed to optimize the performance of a two-way single-input single-output relaying network.

Multiple relay nodes create a virtual multiple-input multiple-output (MIMO) environment at the relay layer.

*Correspondence: zh0045ng@e.ntu.edu.sg
[1]College of Electronic Engineering, University of Electronic Science and Technology of China, Xiyuan Avenue, 610066 Chengdu, China
Full list of author information is available at the end of the article

With multiple antennas equipped at transmitting and receiving nodes, user nodes can also employ advantages of MIMO techniques, such as spatial multiplexing, space-time coding, and beamforming. Attracted by these merits, more and more algorithms are proposed to optimize MIMO relay networks [9-14] and virtual MIMO relay networks [15-18]. Most of these algorithms consider one-way communication scheme, where various optimization criteria are adopted, such as maximizing the destination SNR and minimizing the total system/relay transmission power, MMSE, ZF (zero force), etc. For a single pair of users and multiple relays, a unified algorithm which computes the optimal linear transceivers jointly at the source node and the relay nodes for two-way amplify-and-forward (AF) protocols was proposed [14]. The optimization algorithm was designed based on maximization of sum rate and MMSE. In [16], a two-way scheme was considered for a relay network with multiple users and single relay. The network was optimized using MMSE criterion at the destination subject to power constraint on relays.

In this paper, we consider a two-way MIMO relay network with one pair of transceivers and multiple relays, where the two transceivers are each equipped with M antennas, and every relay is equipped with N antennas. We assume that relays receive the mixture of signals from two transceivers in the first time slot. With perfect CSI, relays scale the received signals and then transmit these signals to the transceivers in the second time slot. Finally, a MIMO equalizer is used at each transceiver to recover the desired signal. To achieve power-efficient communication, beamforming coefficient matrices are optimized based on three criteria which are designed based on the minimal output SNR and relay transmission power. Meanwhile, the MIMO equalizer is optimized by imposing the ZF constraint. It is shown that the proposed optimization problems can be formulated as the second-order cone programming (SOCP) problems, which can be efficiently solved using the cvx toolbox [19]. Contributions of this paper are summarized as follows. First, two-way communication scheme of a MIMO relay network consists of multiple relays with multiple antennas is firstly considered. Second,the proposed optimization problem is formulated as an SOCP problem which can be solved efficiently.

The rest of this paper is organized as follows. The 'Problem formulation' section presents the problem formulation. The 'Mathematical approximation' section develops mathematical preparation for optimizing the considered relay network, and 'Optimization of the proposed relay network' section gives a detailed optimization procedure to derive the optimal beamforming and equalization matrices. In the 'Computer simulations' section, computer simulations are conducted to demonstrate validity of the proposed algorithm. Finally, conclusions and discussions are presented in the 'Conclusions' section.

Notation

Bold lower case is used for vectors, while bold capital letters for matrices. $\mathbf{I}$ denotes the unit matrix, $*$ denotes the complex conjugate operation, T denotes the matrix transposition operation, and H denotes the complex conjugate transposition operation. $\mathrm{tr}(\mathbf{A})$ is the trace of matrix $\mathbf{A}$. $\mathrm{vec}(\mathbf{A})$ stacks the columns of $\mathbf{A}$ into a single column vector. $\otimes$ denotes the Kronecker product, and $(\mathbf{A}_0 \oplus \ldots \oplus \mathbf{A}_N)$ yields a block diagonal matrix with block elements given by $\mathbf{A}_i$.

Problem formulation

Figure 1 depicts a two-way MIMO relay scheme, where $\mathbf{W}_i$ $(i = 1, 2, \ldots, L)$ is the beamforming matrix for the ith relay and $\mathbf{D}_j$ $(j = 1, 2)$ is the equalizer for the jth transceiver. Transceiver 1 and transceiver 2 are both equipped with M antennas. It is supposed that L relays equipped with N antennas are used. Flat fading channels are considered. We assume that no direct link exists between the two transceivers. The channel matrix from transceiver 1 to the ith relay is denoted as $\mathbf{H}_i$ $(i = 1, 2, \ldots, L)$, and the one from the ith relay to transceiver 2 is denoted as $\mathbf{G}_i$, where $\mathbf{H}_i \in \mathbb{C}^{N \times M}$ and $\mathbf{G}_i \in \mathbb{C}^{M \times N}$ consist of independent complex Gaussian variables. It is assumed that these channels are reciprocal, i.e., the channel matrix from the ith relay to transceiver 1 is $\mathbf{H}_i^T$, and the one from transceiver 2 to the ith relay is $\mathbf{G}_i^T$. In the first time slot, each transceiver sends messages to the L relays. With the knowledge of $\mathbf{H}_i$ and $\mathbf{G}_i$ (which can be obtained via training), $\mathbf{W}_i$ is computed. In the second time slot, the relays scale the received signals according to $\mathbf{W}_i$ and then transmit these signals to the two transceivers. After receiving signals, a MIMO equalizer denoted as $\mathbf{D}_j$, $j = 1, 2$ is used at each transceiver. The aim of this paper is to optimize the performance of the relay network by designing $\mathbf{W}_i$ and $\mathbf{D}_j$.

The mixture of signals received by the ith relay can be expressed as

$$\mathbf{r}_i = \mathbf{H}_i \mathbf{s}_1 + \mathbf{G}_i^T \mathbf{s}_2 + \mathbf{v}_i, \tag{1}$$

where $\mathbf{s}_1$ and $\mathbf{s}_2$ are transmitted signals with covariance matrices $P_1\mathbf{I}$ and $P_2\mathbf{I}$, respectively, and they are independent from each other. $\mathbf{v}_i$ denotes the additive Gaussian noise (AGN) with covariance matrix $\mathbf{R}_{vi}$ at the ith relay. In this paper, it is assumed that AGN is white, and its covariance matrix is identical for all relays, i.e., $\mathbf{R}_{vi} = \sigma_v^2\mathbf{I}$, $\forall i = 1, \ldots, L$.

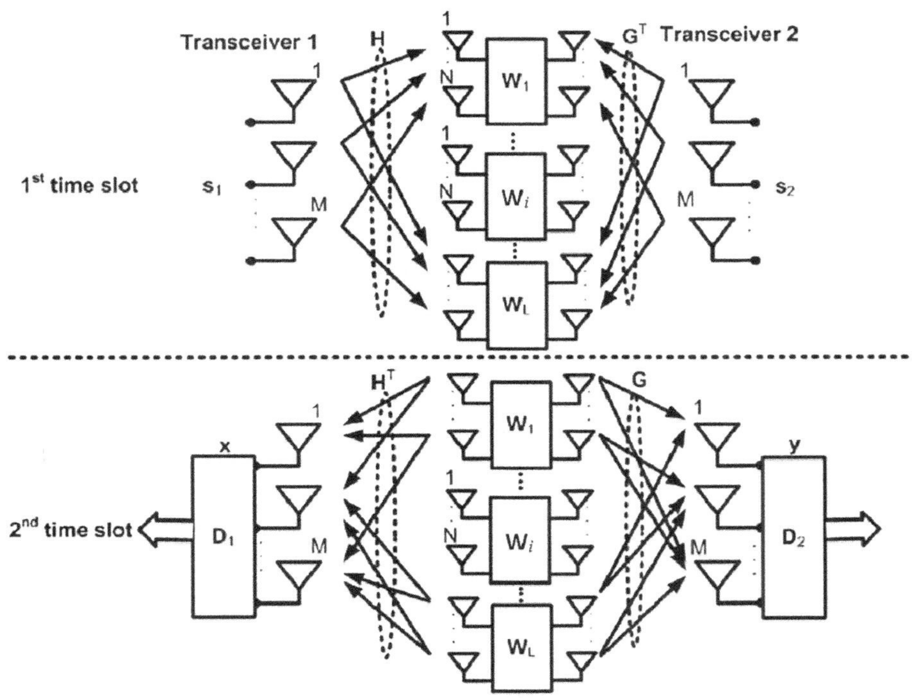

Figure 1 A two-way MIMO relay scheme.

In the second time slot, transceiver 1 and transceiver 2 each receives

$$\mathbf{x} = \sum_{i=1}^{L} \mathbf{H}_i^T \mathbf{W}_i \mathbf{r}_i + \mathbf{v}_x, \tag{2a}$$

$$= \sum_{i=1}^{L} \mathbf{H}_i^T \mathbf{W}_i \mathbf{H}_i \mathbf{s}_1 + \sum_{i=1}^{L} \mathbf{H}_i^T \mathbf{W}_i \mathbf{G}_i^T \mathbf{s}_2 + \sum_{i=1}^{L} \mathbf{H}_i^T \mathbf{W}_i \mathbf{v}_i + \mathbf{v}_x,$$

$$\mathbf{y} = \sum_{i=1}^{L} \mathbf{G}_i \mathbf{W}_i \mathbf{r}_i + \mathbf{v}_y \tag{2b}$$

$$= \sum_{i=1}^{L} \mathbf{G}_i \mathbf{W}_i \mathbf{H}_i \mathbf{s}_1 + \sum_{i=1}^{L} \mathbf{G}_i \mathbf{W}_i \mathbf{G}_i^T \mathbf{s}_2 + \sum_{i=1}^{L} \mathbf{G}_i \mathbf{W}_i \mathbf{v}_i + \mathbf{v}_y,$$

where $\mathbf{v}_x$ and $\mathbf{v}_y$ denote AGN at transceiver 1 and transceiver 2, and their covariance matrices are assumed to be $\sigma_x^2\mathbf{I}$ and $\sigma_y^2\mathbf{I}$, respectively. $\mathbf{s}_1$ and $\mathbf{s}_2$ are known by transceiver 1 and transceiver 2, respectively. If $\mathbf{G}_i$, $\mathbf{H}_i$, and $\mathbf{W}_i$ are available to transceivers, terms containing $\mathbf{s}_1$ and $\mathbf{s}_2$ can be subtracted from Equations 2a and 2b, respectively. Equations 2a and 2b are then rewritten as

$$\bar{\mathbf{x}} = \sum_{i=1}^{L} \mathbf{H}_i^T \mathbf{W}_i \mathbf{G}_i^T \mathbf{s}_2 + \sum_{i=1}^{L} \mathbf{H}_i^T \mathbf{W}_i \mathbf{v}_i + \mathbf{v}_x, \tag{3a}$$

$$\bar{\mathbf{y}} = \sum_{i=1}^{L} \mathbf{G}_i \mathbf{W}_i \mathbf{H}_i \mathbf{s}_1 + \sum_{i=1}^{L} \mathbf{G}_i \mathbf{W}_i \mathbf{v}_i + \mathbf{v}_y. \tag{3b}$$

A MIMO equalizer is used at each transceiver; thereby, the restored signal after equalization is expressed as

$$\widehat{\mathbf{s}}_2 = \mathbf{D}_1 \bar{\mathbf{x}}, \tag{4a}$$

$$\widehat{\mathbf{s}}_1 = \mathbf{D}_2 \bar{\mathbf{y}}. \tag{4b}$$

Based on Equations 1 to 4, the total relay transmission power and terminal SNRs are defined as follows:

1) Total relay transmission power:

$$P_\mathrm{r} = \sum_{i=1}^{L} E\left[\|\mathbf{W}_i \mathbf{r}_i\|_2^2\right] \tag{5}$$

$$= \sum_{i=1}^{L} \left[P_1 \mathrm{tr}\left(\mathbf{W}_i \mathbf{H}_i \mathbf{H}_i^H \mathbf{W}_i^H\right) + P_2 \mathrm{tr}\left(\mathbf{W}_i \mathbf{G}_i^T \mathbf{G}_i^* \mathbf{W}_i^H\right) + \sigma_v^2 \mathrm{tr}\left(\mathbf{W}_i \mathbf{W}_i^H\right)\right]$$

2) Terminal SNR at transceiver 1:

$$\mathrm{SNR}_1 = \frac{E\left[\left\|\mathbf{D}_1\sum_{i=1}^{L}\mathbf{H}_i^T\mathbf{W}_i\mathbf{G}_i^T\mathbf{s}_2\right\|_2^2\right]}{E\left[\left\|\mathbf{D}_1\sum_{i=1}^{L}\mathbf{H}_i^T\mathbf{W}_i\mathbf{v}_i+\mathbf{D}_1\mathbf{v}_x\right\|_2^2\right]} = \frac{P_2\mathrm{tr}\left(\mathbf{D}_1\left(\sum_{i=1}^{L}\mathbf{H}_i^T\mathbf{W}_i\mathbf{G}_i^T\right)\left(\sum_{i=1}^{L}\mathbf{G}_i^*\mathbf{W}_i^H\mathbf{H}_i^*\right)\mathbf{D}_1^H\right)}{\mathrm{tr}\left(\mathbf{D}_1\left(\sigma_v^2\sum_{i=1}^{L}\mathbf{H}_i^T\mathbf{W}_i\sum_{i=1}^{L}\mathbf{W}_i^H\mathbf{H}_i^*+\sigma_x^2\mathbf{I}\right)\mathbf{D}_1^H\right)} \tag{6}$$

3) Terminal SNR at transceiver 2:

$$\mathrm{SNR}_2 = \frac{E\left[\left\|\mathbf{D}_2\sum_{i=1}^{L}\mathbf{G}_i\mathbf{W}_i\mathbf{H}_i\mathbf{s}_1\right\|_2^2\right]}{E\left[\left\|\mathbf{D}_2\sum_{i=1}^{L}\mathbf{G}_i\mathbf{W}_i\mathbf{v}_i+\mathbf{D}_2\mathbf{v}_y\right\|_2^2\right]} = \frac{P_1\mathrm{tr}\left(\mathbf{D}_2\left(\sum_{i=1}^{L}\mathbf{G}_i\mathbf{W}_i\mathbf{H}_i\right)\left(\sum_{i=1}^{L}\mathbf{H}_i^H\mathbf{W}_i^H\mathbf{G}_i^H\right)\mathbf{D}_2^H\right)}{\mathrm{tr}\left(\mathbf{D}_2\left(\sigma_v^2\sum_{i=1}^{L}\mathbf{G}_i\mathbf{W}_i\sum_{i=1}^{L}\mathbf{W}_i^H\mathbf{G}_i^H+\sigma_y^2\mathbf{I}\right)\mathbf{D}_2^H\right)}. \tag{7}$$

In subsequent sections, P_r, SNR_1, and SNR_2 will be used to optimize beamforming matrices and MIMO equalizers.

Mathematical approximation

From (5) to (7), it seems difficult to directly evaluate $\mathbf{W}_i$ and $\mathbf{D}_j$ because they appear in both signal and noise terms. Therefore, before designing beamformers, three lemmas are derived to make $\mathbf{W}_i$ and $\mathbf{D}_j$ solvable.

Lemma 1. *Total relay transmission power can be expressed as a quadratic function of* $\mathbf{w}$:

$$P_r = P_1\mathbf{w}^H\overline{\mathbf{H}}\mathbf{w} + P_2\mathbf{w}^H\mathbf{G}\mathbf{w} + \sigma_v^2\mathbf{w}^H\mathbf{w}, \tag{8}$$

where

$$\mathbf{w}_i = vec(\mathbf{W}_i),\ i = 1,\ldots,L,$$
$$\mathbf{w} = \left(\mathbf{w}_1^T,\ldots,\mathbf{w}_L^T\right)^T,$$
$$\overline{\mathbf{H}}_{ik} = \left(\mathbf{h}_{ik}^T\otimes\mathbf{I}_N\right),$$
$$\overline{\mathbf{H}} = \sum_{k=1}^{M}\overline{\mathbf{H}}_{1k}^H\overline{\mathbf{H}}_{1k}\oplus\ldots\oplus\sum_{k=1}^{M}\overline{\mathbf{H}}_{Lk}^H\overline{\mathbf{H}}_{Lk},$$
$$\mathbf{G}_{ik} = \left(\overline{\mathbf{g}}_{ik}^T\otimes\mathbf{I}_N\right),$$
$$\mathbf{G} = \sum_{k=1}^{M}\mathbf{G}_{1k}^H\mathbf{G}_{1k}\oplus\ldots\oplus\sum_{k=1}^{M}\mathbf{G}_{Lk}^H\mathbf{G}_{Lk},$$

are defined. The notation of $\mathbf{h}_{ik}$ *and* $\overline{\mathbf{g}}_{ik}$ *is given in the proof.*

Proof. From (5), the kth column of $(\mathbf{W}_i\mathbf{H}_i)$ can be expressed as $\left(\mathbf{h}_{ik}^T\otimes\mathbf{I}_N\right)\mathbf{w}_i$, where $\mathbf{h}_{ik}$ denotes the kth column of matrix $\mathbf{H}_i$. Therefore, the trace of $\left(\mathbf{W}_i\mathbf{H}_i\mathbf{H}_i^H\mathbf{W}_i^H\right)$ is given by the Frobenious norm of $(\mathbf{W}_i\mathbf{H}_i)$, which can be computed as

$$\mathrm{tr}\left(\mathbf{W}_i\mathbf{H}_i\mathbf{H}_i^H\mathbf{W}_i^H\right) = \mathbf{w}_i^H\sum_{k=1}^{M}\left(\mathbf{h}_{ik}^T\otimes\mathbf{I}_N\right)^H\left(\mathbf{h}_{ik}^T\otimes\mathbf{I}_N\right)\mathbf{w}_i. \tag{9}$$

Expressing the trace of $\left(\mathbf{W}_i\mathbf{G}_i^T\mathbf{G}_i^*\mathbf{W}_i^H\right)$ in the similar way of (9) and substituting it and (9) into (5) yield

$$P_r = P_1\sum_{i=1}^{L}\mathbf{w}_i^H\sum_{k=1}^{M}\left(\mathbf{h}_{ik}^T\otimes\mathbf{I}_N\right)^H\left(\mathbf{h}_{ik}^T\otimes\mathbf{I}_N\right)\mathbf{w}_i + P_2\sum_{i=1}^{L}\mathbf{w}_i^H\sum_{k=1}^{M}\left(\overline{\mathbf{g}}_{ik}^T\otimes\mathbf{I}_N\right)^H\left(\overline{\mathbf{g}}_{ik}^T\otimes\mathbf{I}_N\right)\mathbf{w}_i + \sigma_v^2\sum_{i=1}^{L}\mathbf{w}_i^H\mathbf{w}_i,$$

where $\overline{\mathbf{g}}_{ik}$ denotes the kth column of matrix $\mathbf{G}_i^T$.

Using the definitions in *Lemma 1*, (8) can be derived. □

SNR constraint is usually used in optimizing a relay network. For our problem, constraints on destination SNRs are expressed as

$$\mathrm{SNR}_1 \geq \gamma_1, \tag{10a}$$

$$\mathrm{SNR}_2 \geq \gamma_2, \tag{10b}$$

where γ_1 and γ_2 are required SNRs at transceiver 1 and transceiver 2, respectively. From (6) and (7), it is seen that (10) is related to $\mathbf{W}_i$ and $\mathbf{D}_j$ in a complicated form. In the rest of this section, two lemmas are derived to transform (10) into a manageable form.

The ZF constraint requires that

$$\mathbf{D}_1\sum_{i=1}^{L}\mathbf{H}_i^T\mathbf{W}_i\mathbf{G}_i^T = \mathbf{I}, \tag{11a}$$

$$\mathbf{D}_2\sum_{i=1}^{L}\mathbf{G}_i\mathbf{W}_i\mathbf{H}_i = \mathbf{I}, \tag{11b}$$

where $\mathbf{D}_1$ and $\mathbf{D}_2$ are defined as the left pseudoinverse of $\sum_{i=1}^{L}\mathbf{H}_i^T\mathbf{W}_i\mathbf{G}_i^T$ and $\sum_{i=1}^{L}\mathbf{G}_i\mathbf{W}_i\mathbf{H}_i$, respectively.

Lemma 2. *From definitions of* $\mathbf{D}_1$ *and* $\mathbf{D}_2$ *given above, if* $\mathbf{D}_1^H\mathbf{D}_1$ *and* $\mathbf{D}_2^H\mathbf{D}_2$ *are diagonal matrices, we have*

$$tr\left(\mathbf{D}_1^H\mathbf{D}_1\right) = \sum_{i=1}^{M} \frac{1}{|\phi_{ii}|^2 \sigma_i}, \tag{12a}$$

$$tr\left(\mathbf{D}_2^H\mathbf{D}_2\right) = \sum_{i=1}^{M} \frac{1}{|\widetilde{\phi}_{ii}|^2 \widetilde{\sigma}_i}, \tag{12b}$$

where the definitions of ϕ_{ii}, $\widetilde{\phi}_{ii}$, σ_i, *and* $\widetilde{\sigma}_i$ *are given in the proof.*

Proof. It is straightforward to show that

$$\mathbf{D}_1\mathbf{D}_1^H = \left(\sum_{i=1}^{L} \mathbf{G}_i^*\mathbf{W}_i^H\mathbf{H}_i^* \sum_{i=1}^{L} \mathbf{H}_i^T\mathbf{W}_i\mathbf{G}_i^T\right)^{-1}. \tag{13}$$

If we define $\underline{\mathbf{H}} = \left[\mathbf{H}_1^T, \ldots, \mathbf{H}_L^T\right]^T$, $\underline{\mathbf{W}} = \mathbf{W}_1 \oplus \ldots \mathbf{W}_L$, and $\underline{\mathbf{G}} = [\mathbf{G}_1, \ldots, \mathbf{G}_L]$, (13) can be expressed as

$$\mathbf{D}_1\mathbf{D}_1^H = \left(\underline{\mathbf{G}}^*\underline{\mathbf{W}}^H\underline{\mathbf{H}}^*\underline{\mathbf{H}}^T\underline{\mathbf{W}}\,\underline{\mathbf{G}}^T\right)^{-1}. \tag{14}$$

Suppose that the eigendecomposition of $\underline{\mathbf{H}}^*\underline{\mathbf{H}}^T$ is given by

$$\begin{aligned}\underline{\mathbf{H}}^*\underline{\mathbf{H}}^T &= \left(\mathbf{U}\ \mathbf{U}_\parallel\right)\begin{pmatrix}\mathbf{\Lambda} & \mathbf{0}\\ \mathbf{0} & \mathbf{0}\end{pmatrix}\begin{pmatrix}\mathbf{U}^H\\ \mathbf{U}_\parallel^H\end{pmatrix} \\ &= \mathbf{U}\mathbf{\Lambda}\mathbf{U}^H.\end{aligned} \tag{15}$$

In (15), the diagonal of $\mathbf{\Lambda}$ consists of M nonzero eigenvalues. The matrix $\mathbf{U}$ consists of all the eigenvectors corresponding to these nonzero eigenvalues. $\mathbf{U}_\parallel$ consists of column vectors which are linearly dependent on columns of $\mathbf{U}$. The dependence of eigenvectors is caused by rank deficiency of $\underline{\mathbf{H}}^*\underline{\mathbf{H}}^T$ whose effective rank is M.

We define $\underline{\underline{\mathbf{W}}} = \underline{\mathbf{W}}\,\underline{\mathbf{G}}^T$, and assume that $\underline{\underline{\mathbf{W}}}$ can be represented by the complete orthogonal basis in the NL-dimensional space, where $\mathbf{U}$ is contained in the complete orthogonal basis, i.e.,

$$\underline{\underline{\mathbf{W}}} = \left(\mathbf{U}\ \mathbf{U}_\perp\right)\begin{pmatrix}\mathbf{\Phi}\\ \mathbf{\Phi}_\perp\end{pmatrix}. \tag{16}$$

In (16), $\mathbf{U} \in \mathbb{C}^{NL\times M}$, $\mathbf{\Phi} \in \mathbb{C}^{M\times M}$, and $\mathbf{U}_\perp$ consist of $N-M$ orthogonal basis of the NL-dimensional space, which can be obtained via Gram-Schmidt procedure based on $\mathbf{U}$.

Substituting (15) and (16) into (14) yields

$$\begin{aligned}\mathbf{D}_1\mathbf{D}_1^H = &\left(\left(\mathbf{\Phi}^H\ \mathbf{\Phi}_\perp^H\right)\begin{pmatrix}\mathbf{U}^H\\ \mathbf{U}_\perp^H\end{pmatrix}\mathbf{U}\mathbf{\Lambda}\mathbf{U}^H\right.\\ &\left.\left(\mathbf{U}\ \mathbf{U}_\perp\right)\begin{pmatrix}\mathbf{\Phi}\\ \mathbf{\Phi}_\perp\end{pmatrix}\right)^{-1} = \left(\mathbf{\Phi}^H\mathbf{\Lambda}\mathbf{\Phi}\right)^{-1}.\end{aligned} \tag{17}$$

From (17), it is seen that

$$\mathrm{tr}|\left(\mathbf{D}_1^H\mathbf{D}_1\right) = \mathrm{tr}\left(\mathbf{D}_1\mathbf{D}_1^H\right) \le \sum_{i=1}^{M} \frac{1}{|\phi_{ii}|^2 \sigma_i}, \tag{18}$$

where ϕ_{ii} and σ_i are the ith diagonal element of $\mathbf{\Phi}$ and $\mathbf{\Lambda}$, respectively.

Similarly, for $\mathbf{D}_2$, we have

$$\mathrm{tr}\left(\mathbf{D}_2^H\mathbf{D}_2\right) \le \sum_{i=1}^{M} \frac{1}{|\widetilde{\phi}_{ii}|^2 \widetilde{\sigma}_i}, \tag{19}$$

where $\widetilde{\sigma}_i$ is the ith nonzero eigenvalue of $\underline{\mathbf{G}}^H\underline{\mathbf{G}}$. $\widetilde{\phi}_{ii}$ is the ith diagonal element of $\widetilde{\mathbf{\Phi}}$, where $\underline{\mathbf{WH}} = \left(\widetilde{\mathbf{U}}\ \widetilde{\mathbf{U}}_\perp\right)\begin{pmatrix}\widetilde{\mathbf{\Phi}}\\ \widetilde{\mathbf{\Phi}}_\perp\end{pmatrix}$, and $\widetilde{\mathbf{U}}$ consists of eigenvectors of $\underline{\mathbf{G}}^H\underline{\mathbf{G}}$ corresponding to its nonzero eigenvalues. □

Lemma 3. *Inequalities (10) can be relaxed as*

$$\frac{\sigma_v^2\mathbf{w}^H\mathbf{H}\mathbf{w} + M\sigma_x^2}{\mathbf{w}^H\mathbf{d}_i\mathbf{d}_i^H\mathbf{w}\sigma_i} \le \frac{1}{\gamma_1},\ \forall i = 1, \ldots, M, \tag{20a}$$

$$\frac{\sigma_v^2\mathbf{w}^H\overline{\mathbf{G}}\mathbf{w} + M\sigma_y^2}{\mathbf{w}^H\widetilde{\mathbf{d}}_i\widetilde{\mathbf{d}}_i^H\mathbf{w}\widetilde{\sigma}_i} \le \frac{1}{\gamma_2},\ \forall i = 1, \ldots, M, \tag{20b}$$

$$\mathbf{w}^H\mathbf{e}_{ij}^* = 0,\ \forall i = 1, \ldots, M, j \ne i, \tag{20c}$$

$$\mathbf{w}^H\widetilde{\mathbf{e}}_{ij}^* = 0,\ \forall i = 1, \ldots, M, j \ne i, \tag{20d}$$

where

$$\begin{aligned}\mathbf{H}_{ik} &= \left(\mathbf{I}_N \otimes \mathbf{h}_{ik}\right)^*,\\ \mathbf{H} &= \sum_{k=1}^{M}\mathbf{H}_{1k}\mathbf{H}_{1k}^H \oplus \ldots \oplus \sum_{k=1}^{M}\mathbf{H}_{Lk}\mathbf{H}_{Lk}^H,\\ \overline{\mathbf{G}}_{ik} &= \left(\mathbf{I}_N \otimes \overline{\mathbf{g}}_{ik}\right)^*,\\ \overline{\mathbf{G}} &= \sum_{k=1}^{M}\overline{\mathbf{G}}_{1k}\overline{\mathbf{G}}_{1k}^H \oplus \ldots \oplus \sum_{k=1}^{M}\overline{\mathbf{G}}_{Lk}\overline{\mathbf{G}}_{Lk}^H,\\ \mathbf{d}_i &= \mathbf{Q}\left(\mathbf{I}\otimes\mathbf{u}_i^*\right)\underline{\mathbf{g}}_i,\\ \widetilde{\mathbf{d}}_i &= \mathbf{Q}\left(\mathbf{I}\otimes\widetilde{\mathbf{u}}_i^*\right)\underline{\mathbf{h}}_i,\\ \mathbf{e}_{ij} &= \mathbf{Q}\left(\mathbf{I}\otimes\mathbf{u}_i^*\right)\underline{\mathbf{g}}_j,\\ \widetilde{\mathbf{e}}_{ij} &= \mathbf{Q}\left(\mathbf{I}\otimes\widetilde{\mathbf{u}}_i^*\right)\underline{\mathbf{h}}_j,\end{aligned}$$

and $\mathbf{u}_i$, $\widetilde{\mathbf{u}}_i$, $\underline{\mathbf{g}}_i$, $\underline{\mathbf{h}}_i$ *and* $\mathbf{Q}$ *are defined in the following proof.*

Proof. With the ZF constraint, (6) and (7) can be simplified as

$$\mathrm{SNR}_1 = \frac{P_2M}{\mathrm{tr}\left(\mathbf{D}_1\left(\sigma_v^2\sum_{i=1}^{L}\mathbf{H}_i^T\mathbf{W}_i\sum_{i=1}^{L}\mathbf{W}_i^H\mathbf{H}_i^* + \sigma_x^2\mathbf{I}\right)\mathbf{D}_1^H\right)}$$

$$\mathrm{SNR}_2 = \frac{P_1M}{\mathrm{tr}\left(\mathbf{D}_2\left(\sigma_v^2\sum_{i=1}^{L}\mathbf{G}_i\mathbf{W}_i\sum_{i=1}^{L}\mathbf{W}_i^H\mathbf{G}_i^H + \sigma_y^2\mathbf{I}\right)\mathbf{D}_2^H\right)}.$$

From the property of tr(.), we may relax the inequality of SNR_1 as

$$\begin{aligned}\mathrm{SNR}_1 &= \frac{P_2 M}{\mathrm{tr}\left(\mathbf{D}_1\mathbf{B}\mathbf{D}_1^H\right)} \\ &= \frac{P_2 M}{\mathrm{tr}\left(\mathbf{B}\mathbf{D}_1^H\mathbf{D}_1\right)} \\ &\geq \frac{P_2 M}{\mathrm{tr}\left(\mathbf{B}\right)\mathrm{tr}\left(\mathbf{D}_1^H\mathbf{D}_1\right)},\end{aligned} \tag{21}$$

where $\mathbf{B} = \left(\sigma_v^2 \sum_{i=1}^{L} \mathbf{H}_i^T \mathbf{W}_i \sum_{i=1}^{L} \mathbf{W}_i^H \mathbf{H}_i^* + \sigma_x^2 \mathbf{I}\right)$.

Substituting (18) into (21) yields

$$\mathrm{SNR}_1 \geq \frac{P_2 M}{\mathrm{tr}(\mathbf{B}) \sum_{i=1}^{M} \frac{1}{|\phi_{ii}|^2 \sigma_i}}. \tag{22}$$

From (16) and the definition of $\overline{\mathbf{W}}$, we have

$$\begin{aligned}\mathbf{\Phi} &= \mathbf{U}^H \overline{\mathbf{W}} = \mathbf{U}^H \underline{\mathbf{W}}\,\underline{\mathbf{G}}^T \\ &= \begin{pmatrix} \mathrm{vec}(\underline{\mathbf{W}})^H \left(\mathbf{I} \otimes \mathbf{u}_1^*\right) \\ \vdots \\ \mathrm{vec}(\underline{\mathbf{W}})^H \left(\mathbf{I} \otimes \mathbf{u}_M^*\right) \end{pmatrix} \underline{\mathbf{G}}^T.\end{aligned} \tag{23}$$

Therefore, the elements of $\mathbf{\Phi}$ can be represented by

$$\phi_{ij} = \mathrm{vec}(\underline{\mathbf{W}})^H \left(\mathbf{I} \otimes \mathbf{u}_i^*\right) \underline{\mathbf{g}}_j, \tag{24}$$

where $\mathbf{u}_i$ denotes the ith column of $\mathbf{U}$ and $\underline{\mathbf{g}}_j^T$ denotes the jth row of $\underline{\mathbf{G}}$.

Similarly, for SNR_2, we have

$$\mathrm{SNR}_2 \geq \frac{P_1 M}{\mathrm{tr}(\widetilde{\mathbf{B}}) \sum_{i=1}^{M} \frac{1}{|\widetilde{\phi}_{ii}|^2 \widetilde{\sigma}_i}}, \tag{25}$$

where $\widetilde{\mathbf{B}} = \left(\sigma_v^2 \sum_{i=1}^{L} \mathbf{G}_i \mathbf{W}_i \sum_{i=1}^{L} \mathbf{W}_i^H \mathbf{G}_i^H + \sigma_y^2 \mathbf{I}\right)$, and $\widetilde{\phi}_{ij}$ can be expressed by

$$\widetilde{\phi}_{ij} = \mathrm{vec}(\underline{\mathbf{W}})^H \left(\mathbf{I} \otimes \widetilde{\mathbf{u}}_i^*\right) \underline{\mathbf{h}}_j, \tag{26}$$

where $\widetilde{\mathbf{u}}_i$ denotes the ith column of $\widetilde{\mathbf{U}}$ and $\underline{\mathbf{h}}_j$ denotes the jth column of $\underline{\mathbf{H}}$. It is assumed that the eigendecomposition of $\underline{\mathbf{G}}^H \underline{\mathbf{G}}$ is $\widetilde{\mathbf{U}}\widetilde{\mathbf{\Lambda}}\widetilde{\mathbf{U}}^H$.

Similar to *Lemma 1*, the trace of $\mathbf{B}$ and $\widetilde{\mathbf{B}}$ can be expressed as

$$\mathrm{tr}(\mathbf{B}) = \sigma_v^2 \mathbf{w}^H \mathbf{H} \mathbf{w} + M\sigma_x^2, \tag{27a}$$

$$\mathrm{tr}(\widetilde{\mathbf{B}}) = \sigma_v^2 \mathbf{w}^H \overline{\mathbf{G}} \mathbf{w} + M\sigma_y^2, \tag{27b}$$

where the definitions of $\mathbf{H}$ and $\overline{\mathbf{G}}$ are given in *Lemma 3*.

Substituting (24) and (27a) into (22) and (26), and (27b) into (25) yields

$$\mathrm{SNR}_1 \geq \frac{P_2 M}{\left(\sigma_v^2 \mathbf{w}^H \mathbf{H} \mathbf{w} + M\sigma_x^2\right) \sum_{i=1}^{M} \frac{1}{\left|\mathrm{vec}(\underline{\mathbf{W}})^H (\mathbf{I} \otimes \mathbf{u}_i^*) \underline{\mathbf{g}}_i\right|^2 \sigma_i}}, \tag{28a}$$

$$\mathrm{SNR}_2 \geq \frac{P_1 M}{\left(\sigma_v^2 \mathbf{w}^H \overline{\mathbf{G}} \mathbf{w} + M\sigma_y^2\right) \sum_{i=1}^{M} \frac{1}{|\mathrm{vec}(\underline{\mathbf{W}})^H (\mathbf{I} \otimes \widetilde{\mathbf{u}}_i^*) \underline{\mathbf{h}}_i|^2 \widetilde{\sigma}_i}}. \tag{28b}$$

From (28), (10) can be relaxed as

$$\frac{P_2 M}{\gamma_1} \geq \sum_{i=1}^{M} \frac{\sigma_v^2 \mathbf{w}^H \mathbf{H} \mathbf{w} + M\sigma_x^2}{\left|\mathrm{vec}(\underline{\mathbf{W}})^H \left(\mathbf{I} \otimes \mathbf{u}_i^*\right) \underline{\mathbf{g}}_i\right|^2 \sigma_i}, \tag{29a}$$

$$\frac{P_1 M}{\gamma_2} \geq \sum_{i=1}^{M} \frac{\sigma_v^2 \mathbf{w}^H \overline{\mathbf{G}} \mathbf{w} + M\sigma_y^2}{\left|\mathrm{vec}(\underline{\mathbf{W}})^H \left(\mathbf{I} \otimes \widetilde{\mathbf{u}}_i^*\right) \underline{\mathbf{h}}_i\right|^2 \widetilde{\sigma}_i}. \tag{29b}$$

If every term on the right side of (29a) and (29b) is smaller than $\frac{P_2}{\gamma_1}$ and $\frac{P_1}{\gamma_2}$, respectively, i.e.,

$$\frac{P_2}{\gamma_1} \geq \frac{\sigma_v^2 \mathbf{w}^H \mathbf{H} \mathbf{w} + M\sigma_x^2}{\left|\mathrm{vec}(\underline{\mathbf{W}})^H \left(\mathbf{I} \otimes \mathbf{u}_i^*\right) \underline{\mathbf{g}}_i\right|^2 \sigma_i}, \ \forall i = 1, \ldots, M, \tag{30a}$$

$$\frac{P_1}{\gamma_2} \geq \frac{\sigma_v^2 \mathbf{w}^H \overline{\mathbf{G}} \mathbf{w} + M\sigma_y^2}{\left|\mathrm{vec}(\underline{\mathbf{W}})^H \left(\mathbf{I} \otimes \widetilde{\mathbf{u}}_i^*\right) \underline{\mathbf{h}}_i\right|^2 \widetilde{\sigma}_i}, \ \forall i = 1, \ldots, M, \tag{30b}$$

(29) can be satisfied.

Because $\underline{\mathbf{W}}$ is block diagonal matrices, there are many zero elements in $\mathrm{vec}(\underline{\mathbf{W}})$, which do not contribute to the calculation of (30). Suppose $\mathbf{Q}$ is chosen such that

$$\mathbf{w} = \mathbf{Q}\mathrm{vec}(\underline{\mathbf{W}}) \tag{31}$$

holds.

To derive (30), we have make assumption that $\mathbf{D}_1^H \mathbf{D}_1$ and $\mathbf{D}_2^H \mathbf{D}_2$ should be diagonal. From (17), we may achieve this by forcing $\mathbf{\Phi}$ to be a diagonal matrix. Therefore, the following equations should be satisfied:

$$\phi_{ij} = \mathrm{vec}(\underline{\mathbf{W}})^H \left(\mathbf{I} \otimes \mathbf{u}_i^*\right) \underline{\mathbf{g}}_j = 0, \ \forall i \neq j, \tag{32}$$

$$\widetilde{\phi}_{ij} = \mathrm{vec}(\underline{\mathbf{W}})^H \left(\mathbf{I} \otimes \widetilde{\mathbf{u}}_i^*\right) \underline{\mathbf{h}}_j = 0, \ \forall i \neq j. \tag{33}$$

With (30) to (33) and definitions given in Lemma 3, (20) can be derived. □

Optimization of the proposed relay network

In this section, we introduce optimization of the proposed two-way MIMO relay network using the following three criteria.

Minimizing the total relay transmission power subject to individual minimal output SNR constraint and ZF constraint

Using this criterion, the optimization problem is formulated as

$$\min_{\mathbf{w}} P_r, \tag{34a}$$

$$\text{subject to } (\text{SNR}_1)_{\text{lower}} \geq \gamma_1,\ (\text{SNR}_2)_{\text{lower}} \geq \gamma_2. \tag{34b}$$

where $(\text{SNR}_1)_{\text{lower}}$ and $(\text{SNR}_2)_{\text{lower}}$ denote the minimal output SNR at transceiver 1 and transceiver 2, respectively.

Theorem 1. *(34) can be approximated as an SOCP problem given as*

$$\min_{\mathbf{w}} \|\mathbf{U}_1\mathbf{w}\|_2, \tag{35a}$$

subject to

$$\left\|\mathbf{U}_2\mathbf{w}'\right\|_2 \leq \frac{Real\{\mathbf{w}^H\mathbf{d}_i\}}{\sqrt{\gamma_1}},\ \forall i = 1,\ldots,M \tag{35b}$$

$$\left\|\mathbf{U}_3\mathbf{w}'\right\|_2 \leq \frac{Real\{\mathbf{w}^H\widetilde{\mathbf{d}}_i\}}{\sqrt{\gamma_2}},\ \forall i = 1,\ldots,M, \tag{35c}$$

$$Imag\{\mathbf{w}^H\mathbf{d}_i\} = 0,\ \forall i = 1,\ldots,M, \tag{35d}$$

$$Imag\{\mathbf{w}^H\widetilde{\mathbf{d}}_i\} = 0,\ \forall i = 1,\ldots,M, \tag{35e}$$

$$\mathbf{w}^H\mathbf{e}_{ij}^* = 0,\ \forall i = 1,\ldots,M, j \neq i, \tag{35f}$$

$$\mathbf{w}^H\widetilde{\mathbf{e}}_{ij}^* = 0,\ \forall i = 1,\ldots,M, j \neq i, \tag{35g}$$

$$\mathbf{w}' = \begin{pmatrix} \mathbf{w} \\ 1 \end{pmatrix}. \tag{35h}$$

and $\mathbf{d}_i$, $\widetilde{\mathbf{d}}_i$, $\mathbf{e}_{ij}$, *and* $\widetilde{\mathbf{e}}_{ij}$ *are given in Lemma 3.*

Proof. Define $\mathbf{A}_i = \begin{pmatrix} \overline{\mathbf{H}}_{i1}^H \\ \vdots \\ \overline{\mathbf{H}}_{iM}^H \end{pmatrix}$, $\mathbf{B}_i = \begin{pmatrix} \mathbf{G}_{i1}^H \\ \vdots \\ \mathbf{G}_{iM}^H \end{pmatrix}$ and $\mathbf{U}_1 = \begin{pmatrix} \sqrt{P_1}(\mathbf{A}_1 \oplus \ldots \oplus \mathbf{A}_L) \\ \sqrt{P_2}(\mathbf{B}_1 \oplus \ldots \oplus \mathbf{B}_L) \\ \sigma_v \mathbf{I}_{NL} \end{pmatrix}$. From *Lemma 1*, P_r can represented as $\|\mathbf{U}_1\mathbf{w}\|_2^2$. If we define $\mathbf{C}_i = \begin{pmatrix} \mathbf{H}_{i1}^H \\ \vdots \\ \mathbf{H}_{iM}^H \end{pmatrix}$ and $\mathbf{U}_2 = \begin{pmatrix} \sigma_v(\mathbf{C}_1 \oplus \ldots \mathbf{C}_L) & \mathbf{0} \\ \mathbf{0}^T & \sqrt{M}\sigma_x \end{pmatrix}$, the nominator of the left side of (20a) can be represented as $\left\|\mathbf{U}_2\mathbf{w}'\right\|_2^2$. Similarly, we define $\mathbf{E}_i = \begin{pmatrix} \overline{\mathbf{G}}_{i1}^H \\ \vdots \\ \overline{\mathbf{G}}_{iM}^H \end{pmatrix}$ and $\mathbf{U}_3 = \begin{pmatrix} \sigma_v(\mathbf{E}_1 \oplus \ldots \mathbf{E}_L) & \mathbf{0} \\ \mathbf{0}^T & \sqrt{M}\sigma_y \end{pmatrix}$, the nominator of the left side of (20b) can be represented as $\left\|\mathbf{U}_3\mathbf{w}'\right\|_2^2$. Using these definitions, (20a) and (20b) can be expressed as

$$\left\|\mathbf{U}_2\mathbf{w}'\right\|_2^2 \leq \frac{\mathbf{w}^H\mathbf{d}_i\mathbf{d}_i^H\mathbf{w}}{\gamma_1},\ \forall i = 1,\ldots,M, \tag{36a}$$

$$\left\|\mathbf{U}_3\mathbf{w}'\right\|_2^2 \leq \frac{\mathbf{w}^H\widetilde{\mathbf{d}}_i\widetilde{\mathbf{d}}_i^H\mathbf{w}}{\gamma_2},\ \forall i = 1,\ldots,M. \tag{36b}$$

From the fact that Real$\{x\} \leq |\, x\,|$, (36) can be relaxed by

$$\left\|\mathbf{U}_2\mathbf{w}'\right\|_2 \leq \frac{\text{Real}\{\mathbf{w}^H\mathbf{d}_i\}}{\sqrt{\gamma_1}},\ \forall i = 1,\ldots,M, \tag{37a}$$

$$\left\|\mathbf{U}_3\mathbf{w}'\right\|_2^2 \leq \frac{\text{Real}\{\mathbf{w}^H\widetilde{\mathbf{d}}_i\}}{\sqrt{\gamma_2}},\ \forall i = 1,\ldots,M, \tag{37b}$$

$$\text{Imag}\{\mathbf{w}^H\mathbf{d}_i\} = 0,\ \forall i = 1,\ldots,M, \tag{37c}$$

$$\text{Imag}\{\mathbf{w}^H\widetilde{\mathbf{d}}_i\} = 0,\ \forall i = 1,\ldots,M. \tag{37d}$$

Then, with *Lemma 3*, (35) can be derived. □

Maximizing the minimal output SNR of transceivers subject to total relay transmission power constraint and ZF constraint

Assuming that the minimum SNR required by the two transceivers is t, the optimization problem can be formulated as

$$\max_{\mathbf{w}} t, \tag{38a}$$

$$\text{subject to } (\text{SNR}_1)_{\text{lower}} \geq t,\ (\text{SNR}_2)_{\text{lower}} \geq t,\ P_r \leq P. \tag{38b}$$

where P denotes the maximal total relay transmission power.

Theorem 2. *(38) can be approximated as an SOCP problem:*

$$\max_{\mathbf{w}} t, \tag{39a}$$

$$\textit{subject to } \|\mathbf{U}_1\mathbf{w}\|_2 \leq \sqrt{P}, \tag{39b}$$

$$\left\|\mathbf{U}_2\mathbf{w}'\right\|_2 \leq \frac{Real\{\mathbf{w}^H\mathbf{d}_i\}}{\sqrt{t}},\ \forall i = 1,\ldots,M, \tag{39c}$$

$$\left\|\mathbf{U}_3\mathbf{w}'\right\|_2 \leq \frac{Real\{\mathbf{w}^H\widetilde{\mathbf{d}}_i\}}{\sqrt{t}},\ \forall i = 1,\ldots,M, \tag{39d}$$

$$Imag\{\mathbf{w}^H\mathbf{d}_i\} = 0,\ \forall i = 1,\ldots,M, \tag{39e}$$

$$Imag\{\mathbf{w}^H\widetilde{\mathbf{d}}_i\} = 0,\ \forall i = 1,\ldots,M, \tag{39f}$$

$$\mathbf{w}^H\mathbf{e}_{ij}^* = 0,\ \forall i = 1,\ldots,M, j \neq i, \tag{39g}$$

$$\mathbf{w}^H\widetilde{\mathbf{e}}_{ij}^* = 0,\ \forall i = 1,\ldots,M, j \neq i, \tag{39h}$$

$$\mathbf{w}' = \begin{pmatrix} \mathbf{w} \\ 1 \end{pmatrix}. \tag{39i}$$

Proof. It can be easily obtained from *Lemmas 1* to *3* and *Theorem 1*. □

Because (39) is quasi-convex, for any given value of t, it becomes the following SOCP problem:

$$\text{find } \mathbf{w}, \tag{40a}$$

$$\text{subject to (39b) to (39i).} \tag{40b}$$

The bisection search procedure can be applied to solve (40).

Maximizing the minimal output SNR of transceivers subject to individual relay transmission power constraint and ZF constraint

The optimization problem is given as

$$\max_{\mathbf{w}} t, \tag{41a}$$

$$\text{subject to } (\text{SNR}_1)_{\text{lower}} \geq t,\ (\text{SNR}_2)_{\text{lower}} \geq t, \tag{41b}$$

$$P_{ri} \leq P_i,\ i = 1, .., L, \tag{41c}$$

where P_i denotes the maximal transmission power of the ith relay, and $P_{ri} = P_1 \text{tr}\left(\mathbf{W}_i \mathbf{H}_i \mathbf{H}_i^H \mathbf{W}_i^H\right) + P_2 \text{tr}\left(\mathbf{W}_i \mathbf{G}_i^T \mathbf{G}_i^* \mathbf{W}_i^H\right) + \sigma_v^2 \text{tr}\left(\mathbf{W}_i \mathbf{W}_i^H\right)$.

Theorem 3. *(41) can be approximated as an SOCP problem:*

$$\max_{\mathbf{w}} t, \tag{42a}$$

$$\textit{subject to } \left\|\mathbf{U}_1^i \mathbf{w}\right\|_2 \leq \sqrt{P_i},\ i = 1, \ldots, L, \tag{42b}$$

$$\left\|\mathbf{U}_2 \mathbf{w}'\right\|_2 \leq \frac{Real\{\mathbf{w}^H \mathbf{d}_i\}}{\sqrt{t}},\ \forall i = 1, \ldots, M, \tag{42c}$$

$$\left\|\mathbf{U}_3 \mathbf{w}'\right\|_2 \leq \frac{Real\{\mathbf{w}^H \widetilde{\mathbf{d}}_i\}}{\sqrt{t}},\ \forall i = 1, \ldots, M, \tag{42d}$$

$$Imag\{\mathbf{w}^H \mathbf{d}_i\} = 0,\ \forall i = 1, \ldots, M, \tag{42e}$$

$$Imag\{\mathbf{w}^H \widetilde{\mathbf{d}}_i\} = 0,\ \forall i = 1, \ldots, M, \tag{42f}$$

$$\mathbf{w}^H \mathbf{e}_{ij}^* = 0,\ \forall i = 1, \ldots, M, j \neq i, \tag{42g}$$

$$\mathbf{w}^H \widetilde{\mathbf{e}}_{ij}^* = 0,\ \forall i = 1, \ldots, M, j \neq i, \tag{42h}$$

$$\mathbf{w}' = \begin{pmatrix} \mathbf{w} \\ 1 \end{pmatrix}, \tag{42i}$$

where $\mathbf{U}_1^i$ *is defined as* $\mathbf{U}_{ri} = \begin{pmatrix} \sqrt{P_1}\mathbf{A}_i \\ \sqrt{P_2}\mathbf{B}_i \\ \sigma_v \mathbf{I}_N \end{pmatrix}$.

Proof. It can be easily obtained from *Lemmas 1* to *3* and *Theorem 1*. □

For any given value of t, (42) reduces to the following SOCP probelm:

$$\text{find } \mathbf{w}, \tag{43a}$$

$$\text{subject to (42b) to (42i).} \tag{43b}$$

Similar to the solution of (40), (43) is solved by the bisection search procedure.

Computer simulations

In order to verify the validity of the proposed algorithm, we devise the following simulation scenario. The number of antennas of transceiver 1, transceiver 2, and relays is assumed to be $M = N = 3$, and the number of relays is $L = 10$. The communication channel coefficients are modeled by complex Gaussian variables with zero mean and variance σ_h^2 and σ_g^2. The two transceivers transmit independent data streams from different antennas with $P_1 = P_2 = 0$ dB. AGN on each antenna is assumed to be complex Gaussian variable with zero mean and unit variance, i.e., $\sigma_x^2 = \sigma_y^2 = \sigma_v^2 = 0$ dB. Sources are generated from a QPSK constellation. The values of SNR are computed from 100 independent trials for each plot. Furthermore, the power consumption to increase the minimal output SNR_2 for 2 dB becomes smaller as the value of γ_1 increases, which means the derived minimal output SNR approaches the output SNR as the value of SNR increases. Therefore, less additional power consumption is needed to increase the same amount of output SNR. This phenomenon can also be demonstrated by Figure 2.

Minimizing the total relay transmission power subject to individual minimal output SNR constraint and ZF constraint

We assume that $\sigma_h^2 = \sigma_g^2 = 0$ dB. Figure 3 depicts the total relay transmission power P_r against the value of γ_1. It is observed that the required transmission power increases as the value of γ_1 increases. Also, for a given γ_1, the total relay transmission power increases with the increase of γ_2.

Figure 2 shows the cumulative distribution function (CDF) of the output SNR at transceiver 2 with different values of γ_2. In Figure 2, the value of γ_2 is assumed to vary from −6 to 6 dB with 2 dB stepsize. From the figure, we see that for a given γ_2, the output SNR at transceiver 2 does not change significantly with the variation of γ_1, and the output SNR_2 is about 2 to 3 dB higher than the value of γ_2 with 90% probability. This is reasonable since the proposed optimization problem uses the minimal output SNR instead of the real output SNR. It can be seen that the difference between the real output SNR_2 and γ_2 decreases as the value of γ_2 increases, which is in accordance with the phenomenon observed in Figure 3.

Figure 4 plots the CDF of output SNR_1 with different values of γ_2. It is found that the average output SNR_1

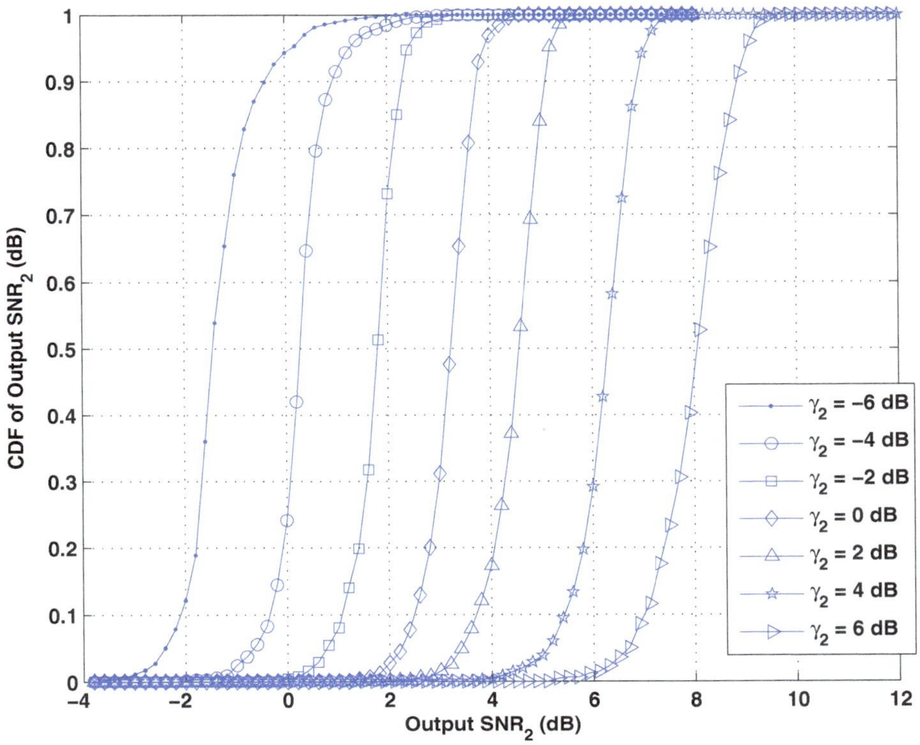

Figure 2 CDF of the output SNR at transceiver 2 with different values of γ_2.

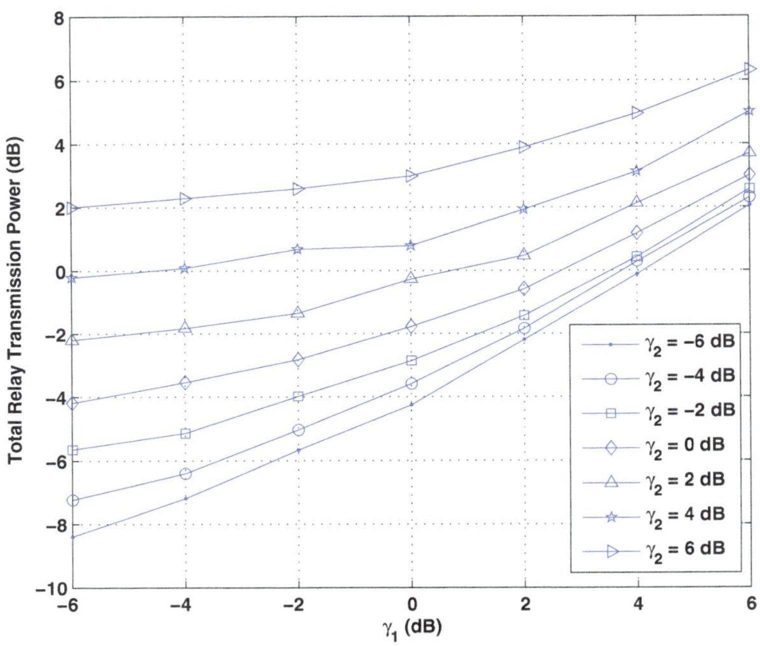

Figure 3 Total relay transmission power P_r versus the value of γ_1.

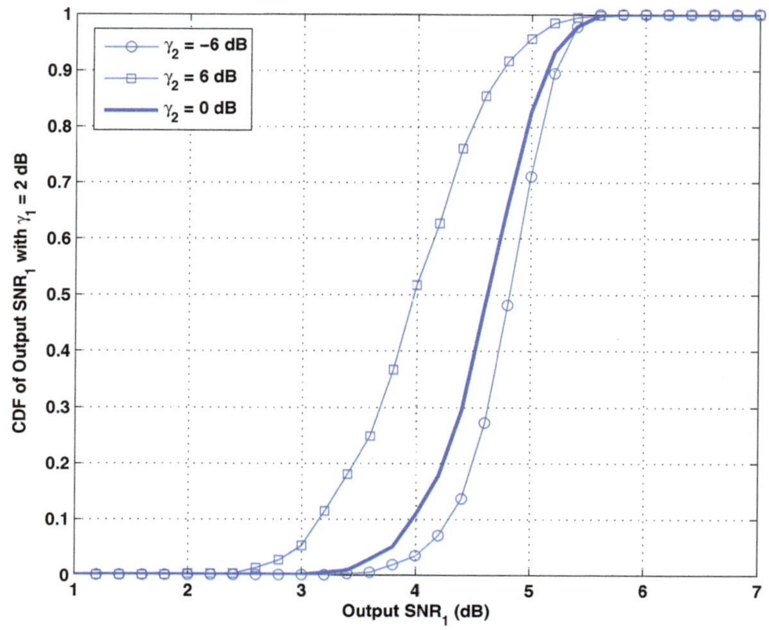

Figure 4 CDF of the output SNR at transceiver 1 with different values of γ_2.

with $\gamma_2 = 6$ dB is less than that with $\gamma_2 = -6$ dB. It is known that P_r is allocated to the relays such that the two transceivers can simultaneously meet the required SNR. It can be concluded from Figure 4 that P_r allocation tends to emphasize maximizing the output SNR which has higher requirement under the condition that the lower SNR requirement can be satisfied. Therefore, SNR_1 can achieve a higher average value when $\gamma_2 = -6$ dB than when $\gamma_2 = 6$ dB.

Maximizing the minimal output SNR of transceivers subject to total relay transmission power constraint and ZF constraint

Figure 5 depicts the output SNR at transceiver 1 with the value of σ_h^2 changing from 0 to 10 dB. Total relay transmission powers of 0 and 5 dB are considered. It is found that for a given σ_h^2, the output SNR_1 increases with the increase of σ_g^2, while for a given σ_g^2, SNR_1 does not keep increasing with the increase of σ_h^2. This is because as the quality of channels between transceiver 1 and the relays improves, i.e., σ_h^2 increases, the desired transmission power at transceiver 2 to guarantee its output SNR increases [2]. Due to limitation of total relay transmission power, the output SNR at transceiver 1 can not increase consistently. When the quality of channels between transceiver 2 and the relays improves, the output SNR_1 increases with the increase of σ_h^2.

Figure 6 shows the same plot for output SNR_2. It is observed that the output SNR_2 increases with the increase of σ_h^2, it while does not increase with the increase of σ_g^2 especially when σ_g^2 is high and σ_h^2 is relatively low. The reason is the same as that for SNR_1 versus σ_h^2. Also, as noticed from Figures 5 and 6, the output SNR_1 and SNR_2 increases with the increase of the total relay transmission power.

Maximizing the minimal output SNR of transceivers subject to individual relay transmission power constraint and ZF constraint

In this simulation, we assume that the total relay transmission power is uniformly allocated to the relays. Figures 7 and 8 show the output SNR versus the value of σ_h^2 with individual relay powers of -10 and 0 dB. It is noted that these plots are similar to those with total relay transmission power constraint. With the increase of individual relay transmission power, output SNRs at transceiver 1 and transceiver 2 increase. Compared with Figures 5 and 6, it is found that the output SNR_1 and SNR_2 are slightly lower with individual power constraint than those with total power constraint. This is because individual power constraint is more restrictive than the total power constraint.

Conclusions

In this paper, we focus on the optimization of a two-way MIMO relay network. The proposed optimization criteria yield three SOCP problems which can be solved

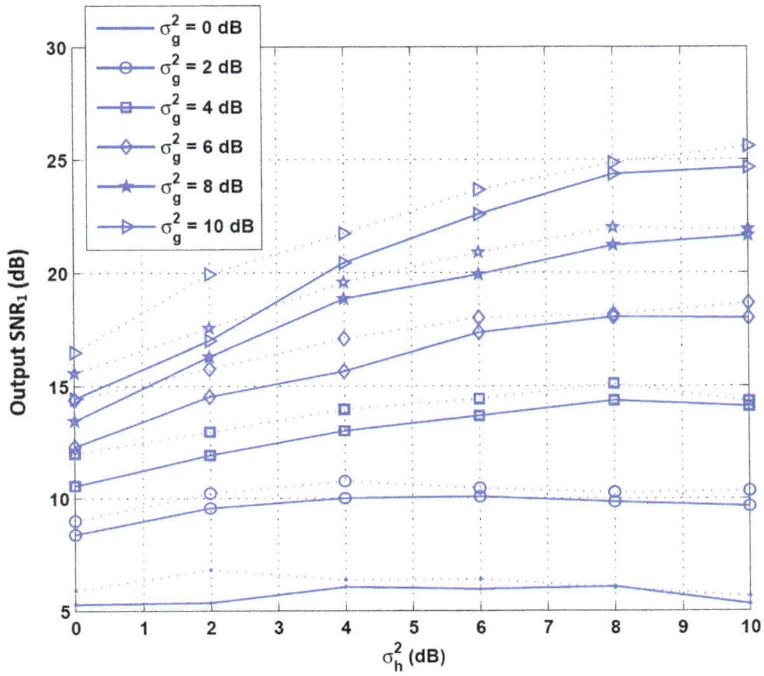

Figure 5 Output SNR_1 versus the value of σ_h^2. Solid line: with total relay transmission power of 0 dB, dash line: with total relay transmission power of 5 dB.

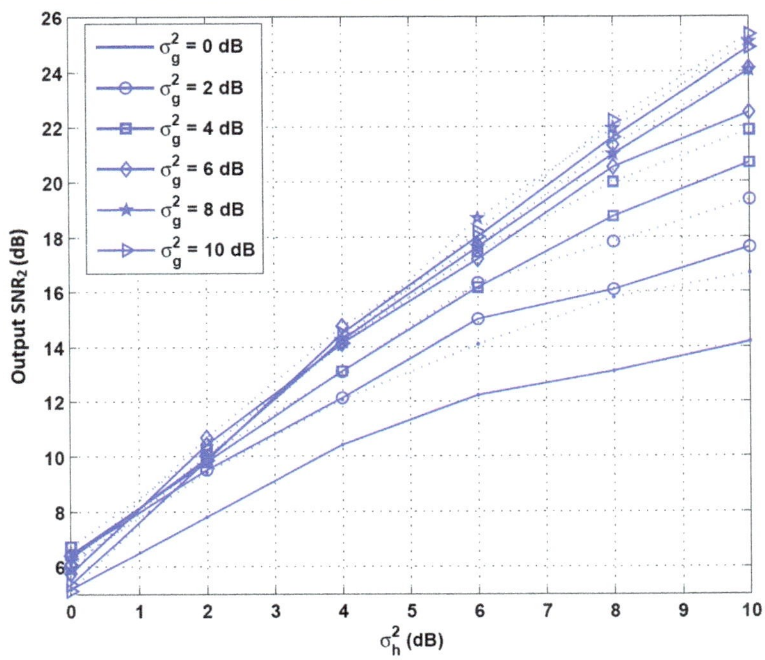

Figure 6 Output SNR_2 versus the value of σ_h^2. Solid line: with total relay transmission power of 0 dB, dash line: with total relay transmission power of 5 dB).

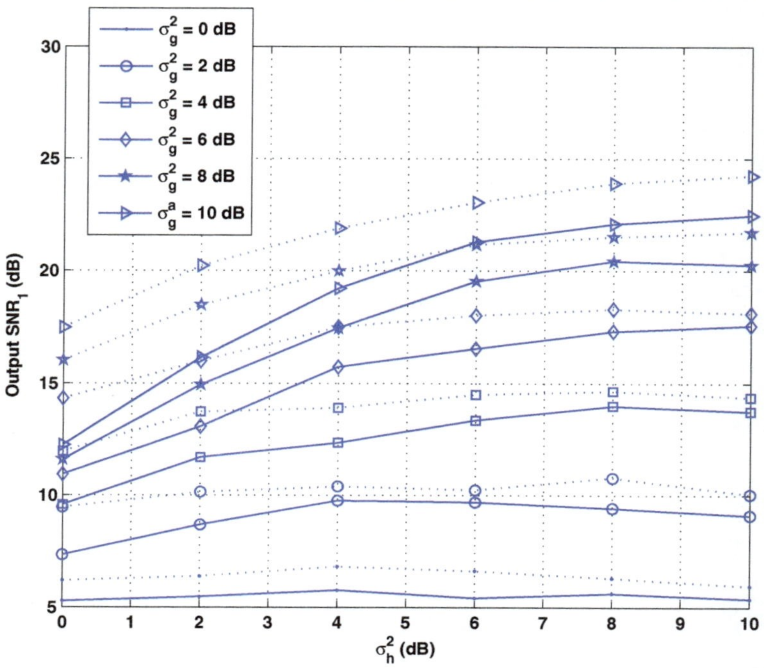

Figure 7 Output SNR$_1$ versus the value of σ_h^2. Solid line: with individual relay transmission power of −10 dB, dash line: with individual relay transmission power of 0 dB.

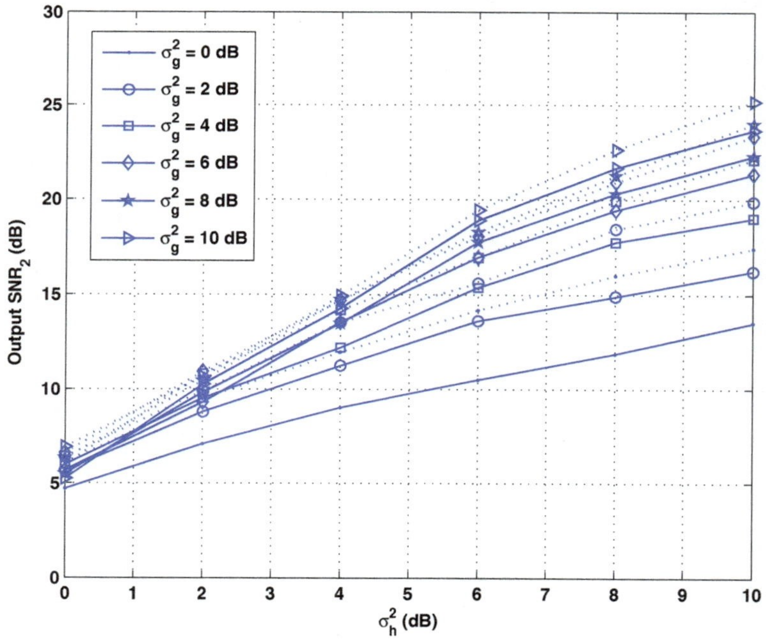

Figure 8 Output SNR$_2$ versus the value of σ_h^2. Solid line: with individual relay transmission power of −10 dB, dash line: with individual relay transmission power of 0 dB.

efficiently. Computer simulation demonstrates validity of the proposed algorithm. Furthermore, it is straightforward to see that the proposed algorithm can be implemented distributively as long as $\mathbf{U}$ and $\widetilde{\mathbf{U}}$ are broadcasted to all the relays. With $\mathbf{w}$ replaced by $\mathbf{w}_i$, $\mathbf{U}_1$ replaced by $\mathbf{U}_{ri}$, $\mathbf{U}_2$ replaced by $\begin{pmatrix} \sigma_v \mathbf{C}_i & \mathbf{0} \\ \mathbf{0}^T & \sigma_x \end{pmatrix}$, and $\mathbf{U}_3$ replaced by $\begin{pmatrix} \sigma_v \mathbf{D}_i & \mathbf{0} \\ \mathbf{0}^T & \sigma_y \end{pmatrix}$, (35), (39), and (42) can be solved at each relay. The performance of distributed implementation will be analyzed in our future work.

Competing interests
The authors declare that they have no competing interests.

Acknowledgements
The author wishes to acknowledge the financial support of the National Science Foundation of China through Grant No. 61101094 and No. 61201275.

Author details
[1]College of Electronic Engineering, University of Electronic Science and Technology of China, Xiyuan Avenue, 610066 Chengdu, China. [2]Department of Electronics and Photonics, Institute of High Performance Computing, Fusionopolis Way, Singapore 138632, Singapore.

References
1. G Zheng, Collaborative-relay beamforming with perfect CSI: optimum and distributed implementation. IEEE Signal Process Lett. **16**(4), 257 (2009)
2. V Havary-Nassab, S Shahbazpanahi, A Grami, Z-Q Luo, Distributed beamforming for relay networks based on second-order statistics of the channel information. IEEE Trans. Signal Process. **56**(9), 4306 (2008)
3. MM Abdallah, HC Pagadopoulos, Beamforming algorithms for implementation relaying in wireless sensor networks. IEEE Trans. Signal Process. **56**(10), 4772 (2008)
4. E Koyuncu, Y Jing, H Jafarkhani, Distributed beamforming in wireless relay networks with quantized feedback. IEEE J. Selected Areas Commun. **26**(8), 1429 (2008)
5. N Khajehnouri, AH Sayed, Distributed MMSE relay strategies for wireless sensor networks. IEEE Trans. Signal Process. **55**(7), 3336 (2007)
6. H Chen, AB Gershman, S Shahbazpanahi, Filter-and-forward distributed beamforming in relay networks with frequency selective fading. IEEE Trans. Signal Process. **58**(3), 1251 (2010)
7. V Havary-Nassab, S Shahbazpanahi, A Grami, Optimal distributed beamforming for two-way relay networks. IEEE Trans. Signal Process. **58**(3), 1238 (2010)
8. H Chen, Beamforming Optimization for Two-Way Relay Channel (2014). p. 28–31
9. Y Rong, Optimal joint source and relay beamforming for MIMO relays with direct link. IEEE Commun. Lett. **14**(5), 390 (2010)
10. B Khoshnevis, W Yu, R Adve, Grassmannian beamforming for MIMO amplify-and-forward relaying. IEEE J. Selected Areas Commun. **26**(8), 1397 (2008)
11. W Guan, H Luo, Joint MMSE transceiver design in non-regenerative MIMO relay systems. IEEE Commun. Lett. J. **12**(7), 517 (2008)
12. Y Rong, F Gao, Optimal beamforming for non-regenerative MIMO relays with direct link. IEEE Commun. Lett. J. **13**(12), 927 (2009)
13. AS Behbahani, R Merched, AM Eltawil, Optimization of a MIMO relay network. IEEE Trans. Signal Process. **56**(10), 5063 (2008)
14. K-J Lee, H Sung, E Park, I Lee, Joint optimization for one and two-way MIMO AF multiple-relay systems. IEEE Trans. Wireless Commun. **9**(12), 3671 (2010)
15. A El-Keyi, B Champagne, Adaptive linearly constrained minimum variance beamforming for multiuser cooperative relaying using the Kalman filter. IEEE Trans. Wireless Commun. **9**(2), 641 (2010)
16. J Joung, AH Sayed, Multiuser two-way amplify-and-forward relay processing and power control methods for beamforming systems. IEEE Trans. Signal Process. **58**(3), 1833 (2010)
17. R Zhang, C Choy Chai, Y-C Liang, Joint beamforming and power control for multiantenna relay broadcase channel with QoS constraints. IEEE Trans. Signal Process. **57**(2), 726 (2009)
18. O Oyman, AJ Paulray, Design and analysis of linear distributed MIMO relaying algorihms. IEEE Proc. Commun. **153**(4), 565 (2006)
19. M Grant, S Boyd, cvx user's guide for cvx version 1.21 (build 790). http://cvxr.com/cvx/. Accessed 16 July 2013

Permissions

All chapters in this book were first published in EURASIP-JASP, by Springer; hereby published with permission under the Creative Commons Attribution License or equivalent. Every chapter published in this book has been scrutinized by our experts. Their significance has been extensively debated. The topics covered herein carry significant findings which will fuel the growth of the discipline. They may even be implemented as practical applications or may be referred to as a beginning point for another development.

The contributors of this book come from diverse backgrounds, making this book a truly international effort. This book will bring forth new frontiers with its revolutionizing research information and detailed analysis of the nascent developments around the world.

We would like to thank all the contributing authors for lending their expertise to make the book truly unique. They have played a crucial role in the development of this book. Without their invaluable contributions this book wouldn't have been possible. They have made vital efforts to compile up to date information on the varied aspects of this subject to make this book a valuable addition to the collection of many professionals and students.

This book was conceptualized with the vision of imparting up-to-date information and advanced data in this field. To ensure the same, a matchless editorial board was set up. Every individual on the board went through rigorous rounds of assessment to prove their worth. After which they invested a large part of their time researching and compiling the most relevant data for our readers.

The editorial board has been involved in producing this book since its inception. They have spent rigorous hours researching and exploring the diverse topics which have resulted in the successful publishing of this book. They have passed on their knowledge of decades through this book. To expedite this challenging task, the publisher supported the team at every step. A small team of assistant editors was also appointed to further simplify the editing procedure and attain best results for the readers.

Apart from the editorial board, the designing team has also invested a significant amount of their time in understanding the subject and creating the most relevant covers. They scrutinized every image to scout for the most suitable representation of the subject and create an appropriate cover for the book.

The publishing team has been an ardent support to the editorial, designing and production team. Their endless efforts to recruit the best for this project, has resulted in the accomplishment of this book. They are a veteran in the field of academics and their pool of knowledge is as vast as their experience in printing. Their expertise and guidance has proved useful at every step. Their uncompromising quality standards have made this book an exceptional effort. Their encouragement from time to time has been an inspiration for everyone.

The publisher and the editorial board hope that this book will prove to be a valuable piece of knowledge for researchers, students, practitioners and scholars across the globe.

List of Contributors

Turkey N Alotaiby
KACST, Riyadh 11442, Kingdom of Saudi Arabia

Saleh A Alshebeili
Electrical Engineering Department, KACST-TIC in Radio Frequency and Photonics for the e-Society (RFTONICS), King Saud University, Riyadh 11362, Saudi Arabia

Tariq Alshawi
Department of Electrical Engineering, King Saud University, Riyadh 11362, Saudi Arabia

Ishtiaq Ahmad
Institute for Telecommunications Research, University of South Australia, South Australia 5095, Australia

Fathi E Abd El-Samie
KACST-TIC in Radio Frequency and Photonics for the e-Society (RFTONICS), King Saud University, Riyadh 11362, Saudi Arabia
Faculty of Electronic Engineering, Menoufia University, Menouf, 32952, Egypt

Soheil Khorram and Hossein Sameti
Department of Computer Engineering, Sharif University of Technology, Azadi Avenue, Tehran 14588, Iran

Simon King
The Centre for Speech Technology Research, University of Edinburgh, Informatics Forum, 10 Crichton Street, Edinburgh EH8 9AB, UK

Khaled M Mohamed and Russell C Hardie
University of Dayton, Department of Electrical and Computer Engineering, 300 College Park, Dayton, 45469-0226 OH, USA

Sungeun Lee and Xiaoli Ma
School of Electrical and Computer Engineering, Georgia Institute of Technology, 777 Atlantic Drive NW, Atlanta, GA 30332-0250, USA

Luigi Feola, Roberto Langella and Alfredo Testa
Department of Industrial and Information Engineering, Second University of Naples, DIII-SUN, Via Roma, 29, 81031 Aversa, Caserta, Italy

Rosi Asirvatham and Seshasayanan Ramachandran
Department of Electronics and Communication Engineering, College of Engineering, Guindy, Chennai 600025, India

Zhirui Hu
Beijing Key Laboratory of Network System Architecture and Convergence, Beijing University of Posts and Telecommunications, Beijing 100876, China
School of Communication Engineering, Hangzhou Dianzi University, Hangzhou 310018, China

Chunyan Feng, Tiankui Zhang and Qin Niu
Beijing Key Laboratory of Network System Architecture and Convergence, Beijing University of Posts and Telecommunications, Beijing 100876, China

Yue Chen
School of Electronic Engineering and Computer Science, Queen Mary University of London, London E1 4NS, UK

Krzysztof Kołek and Krzysztof Piątek
AGH University of Science and Technology, al. A.Mickiewicza 30, Kraków 30-059, Poland

Mohammed M Olama, Stephen M Killough, Teja Kuruganti and Stephen F Smith
Computational Sciences and Engineering Division, Oak Ridge National Laboratory, P.O. Box 2008, MS 6085, Oak Ridge, TN 37831, USA

Xiao Ma
Department of Electrical Engineering and Computer Science, University of Tennessee, 1520 Middle Drive, Knoxville, TN 37996, USA

Seddik M Djouadi
Department of Electrical Engineering and Computer Science, University of Tennessee, 1520 Middle Drive, Knoxville, TN 37996, USA
Department of Electrical Engineering and Computer Science, Masdar Institute of Science and Technology, Masdar City, Abu Dhabi, UAE

Mehmet Tahir Ozden
Piri Reis University, Tuzla, 34940 Istanbul, Turkey

Chunxiao Jiang
Department of Electronic Engineering, Tsinghua University, Beijing 100084, China

Yan Chen and K J Ray Liu
Department of Electrical and Computer Engineering, University of Maryland, College Park, MD 20742, USA

Adnan Shahid, Saleem Aslam, Hyung Seok Kim and Kyung-Geun Lee
Department of Information and Communication Engineering, Sejong University, 98 Gunja-dong, Gwangjin-gu, Seoul 143-747, Republic of Korea

Matti Raitoharju, Henri Nurminen and Robert Piché
Department of Automation Science and Engineering, Tampere University of Technology, P.O.Box 692, FI-33101, Tampere, Finland

Yuquan Shan
Department of Computer Science and Engineering, The Pennsylvania State University, University Park, State College, PA 16802, USA

Jayaram Raghuram and David J Miller
Department of Electrical Engineering, The Pennsylvania State University, University Park, State College, PA 16802,USA

George Kesidis
Department of Computer Science and Engineering, The Pennsylvania State University, University Park, State College, PA 16802, USA
Department of Electrical Engineering, The Pennsylvania State University, University Park, State College, PA 16802, USA

Anna Scaglione
Electrical Engineering, Arizona State University, Tempe, AZ 85287, USA

Jeff Rowe and Karl Levitt
Department of Computer Science, University of California, Davis, CA 95616, USA

Zhiming Qian, Ping Zhong and Runsheng Wang
College of Electronic Science and Engineering, National University of Defense Technology, Deya Road, 410073 Changsha, China

Ying Zhang, Yinjiang Chen, Chuanyi Pan and Ning Kang
College of Electronic Engineering, University of Electronic Science and Technology of China, Xiyuan Avenue, 610066 Chengdu, China

Huapeng Zhao
Department of Electronics and Photonics, Institute of High Performance Computing, Fusionopolis Way, Singapore 138632, Singapore